Epidemic Modeling

Epidemic Modeling

Principles, Algorithms, and Applications

Teddy Lazebnik
Department of Computing
Jönköping University
Jönköping, Sweden

ELSEVIER
ACADEMIC PRESS
An imprint of Elsevier

Academic Press is an imprint of Elsevier
125 London Wall, London EC2Y 5AS, United Kingdom
50 Hampshire Street, 5th Floor, Cambridge, MA 02139, United States

ISBN: 978-0-443-55525-1

For information on all Academic Press publications
visit our website at https://www.elsevier.com/books-and-journals

Publisher: Megan Ball
Acquisitions Editor: Elizabeth A. Brown
Editorial Project Manager: Priyanshi Peelwan
Production Project Manager: Fahmida Sultana
Cover Designer: Mark Rogers

Typeset by VTeX

Contents

Preface

The catastrophic global impact of recent epidemics and pandemics, from HIV to COVID-19, has underscored the need for sophisticated, adaptable, and practical modeling tools. *Epidemic Modeling: Principles, Algorithms, and Applications* is designed as a direct response to this challenge. This book's goal is to provide a comprehensive, interdisciplinary toolkit that enables you to move beyond abstract theory and apply computational power to design data-driven, transparent, and robust policy decisions for current and future epidemics.

While many existing resources focus on the pure mathematics of epidemiological theory and other focus entirely on the applied end of known models' utilization, this book integrates three crucial components side-by-side: core principles (understanding disease dynamics), modern algorithms (the computational methods needed to run, scale, and fit models), and real-world applications (how to translate model output into policy). This book aims to teach you not just to design and solve a model, but how also to extend, adapt, and critically assess a model to address a novel public health crisis.

Thus, this book is tailored for an interdisciplinary audience working at the intersection of epidemiology, public health, computer science, and mathematics. Epidemiologists and public health researchers will learn modern computational methods, including agent-based simulation and time-series data-driven models, to enhance your existing mechanistic models. Applied mathematicians will discover how classic mathematical techniques, such as SIR models, are adapted and extended to manage real-world complexities like heterogeneous populations and spatiotemporal spread. Data scientists and artificial intelligence engineers will find guidance to applying advanced algorithms, such as optimal control and deep learning, to biological systems, bridging the gap between theoretical AI and computational public health. Most importantly, students, I hope this book will be the first in a series you read, giving both an overview and an anchor point for your future epidemiological modeling efforts and education.

Across this book, it is assumed that readers have a basic background in linear algebra, calculus, ordinary differential equations, stochastic processes, programming (Python), and computer simulations. The text is structured to build your understanding from the ground up, so while basic definitions are provided, some classical theorems and coding practices are used with only a reference. Technically, the book is organized into three parts to guide you through the modeling journey.

Foundations (Part I): Covers core epidemiological concepts, from the biology of pandemics to the mathematics of classic compartmental models (SIR). This sets the conceptual stage.

Algorithms and Inference (Part II): This is the computational heart of the book. It details simulation methods, parameter estimation techniques (including modern Bayesian approaches), uncertainty quantification, and the crucial step of integrating real-world data (surveillance, mobility, genomic) into your models.

Applications and Practice (Part III): Presents case studies and policy-focused chapters, culminating in a practical guide for the epidemiological modeler, which synthesizes the entire workflow – from framing a policy question to reporting and communicating robust, actionable results to decision-makers.

I believe the next generation of effective epidemic response will be a partnership between epidemiology and computation. This book is your manual for that partnership.

Acknowledgments

First and foremost, my deepest gratitude goes to my beloved wife. Through every stage of this journey, protecting the time I needed, giving me the quiet and peace of mind to keep going, and, perhaps most of all, being the steady presence who listened as I poured out every unfiltered thought and half-formed idea along the way. This book carries your fingerprints in more ways than I can count. I am also profoundly grateful to my closest academic collaborators (and I am lucky enough to say, my friends), who stood beside me both personally and professionally throughout this work: Labib Shami, Ariel Rosenfeld, Svetlana Bunimovich-Medrazitsky, Alexander Liberzon, and Einav Peretz-Andersson. Your insight, generosity, and companionship made the hard parts lighter and the good parts richer. My sincere thanks to Stephan Beck and Avner Friedman, whose guidance, often through personal example more than words, has shown me what scientific excellence should look like. I have learned from the standards you set and the integrity with which you uphold them. To my research students, past and present: thank you for walking part of this path with me. You trusted me with your time, your talent, and a piece of your future, and I do not take that lightly. Your curiosity and persistence are what move this field forward – one careful step, one question, one epidemiological model at a time. Finally, I thank my family, who from my earliest years pushed me to seek knowledge, work hard, and aim for excellence. A special thank you to Svetlana Hardak-Nissan, Eddie Lazebnik, Mark Lazebnik, and Galina Fudim. Your support and belief in me have been a quiet force behind everything I have been able to build.

CHAPTER

The history of pandemics and epidemiological modeling

1

The term "pandemic" was first documented in 1666 to describe a disease that spreads continuously throughout a country [1]. During the seventeenth and eighteenth centuries, the terms epidemic and pandemic were often used interchangeably in both medical and social discourse. Over time, however, epidemiological terminology has become more precise, giving rise to a clearer conceptual framework that distinguishes between endemic, outbreak, epidemic, and pandemic based on disease frequency and geographical distribution [2]. An endemic disease refers to a condition that maintains a relatively constant presence within a defined geographic area, which may range from a small community to an entire continent. An outbreak denotes an unexpected increase in disease incidence or the appearance of cases in a previously unaffected location. An epidemic is characterized by disease spread beyond the expected baseline within a region, while a pandemic represents an epidemic that extends across multiple continents [3].

Multiple factors have contributed to the historical occurrence and increasing likelihood of pandemics. For instance, climate change significantly influences the transmission dynamics of zoonotic diseases (infectious diseases that spread between animals and humans) by altering animals' habitats and ecological conditions [4]. Similarly, expanding land use driven by population growth reshapes the spatial distribution of animals and increases human exposure [5]. Intensified animal–human interactions further elevate the risk of zoonotic spillover events [6]. In parallel, antimicrobial resistance poses a critical threat by reducing the effectiveness of existing treatments and amplifying the potential impact of emerging infections [7]. The deliberate misuse of microorganisms as biological weapons also represents a non-negligible pandemic risk [8]. Finally, shortages of healthcare workers and inadequately prepared health systems undermine the capacity to contain widespread outbreaks [9]. These factors and concerns are modern in nature, rooted in recent technological and socio-economic shifts; however, they are only a natural extension of the historical processes taking place over three millennia ago, where the first pandemics are recorded. Below, we present pandemic over history and the evolution of pandemics modeling as a tool to understand and countermeasure their negative social effects.

1.1 Pandemics over history

Historically, pandemics have imposed profound health burdens and posed existential threats to human populations. Rapidly spreading infections have overwhelmed healthcare systems, restricted access to essential services, and increased mortality from both communicable and non-communicable diseases. Beyond health outcomes, pandemics have exerted substantial social, economic, and political impacts [10–12]. Roughly, one can divide the history of pandemics into seven main periods, different from each

Epidemic Modeling. https://doi.org/10.1016/B978-0-44-355525-1.00007-2

other in a combination of for properties: (1) how people are connected, (2) how people live day to day, (3) what societies believe causes disease, and (4) what tools and institutions exist to interrupt the pandemic spread. When these background conditions shift, the same basic problem (i.e., contagion) produces very different kinds of pandemics and responses.

First, the early agrarian and imperial era, from the first cities to late antiquity. Once humans began farming, keeping livestock, and living in dense settlements, diseases that require a steady supply of susceptible hosts became easier to maintain. Empires and long-distance trade linked far-apart populations, so outbreaks could move across regions even when travel was slow by modern standards. Written sources from this era are uneven, and most societies lacked a biological explanation for contagion, so responses tended to be social, such as flight, exclusion, or rituals, rather than targeted medical control. Late-antique pandemics, including major plague waves, show how shocks can be amplified when ecology, trade, and politics line up [13,14].

Much later, and for a long period, the medieval plague era and the rise of quarantine – roughly 6th–18th centuries, peaking after the 14th-century Black Death – marked a time when the Black Death and other plague outbreaks shaped public life, altering culture, trade dynamics, and even governmental agendas and objectives. A distinctive feature of this era was that governments, especially in cities, began to build tools that look like early public health: health boards, reporting systems, travel restrictions, and dedicated isolation spaces. "Quarantine" became a recognizable strategy, and it carried a built-in tension that persists today: protecting the public often meant limiting movement and targeting outsiders, which could fuel stigma and conflict [15–17].

In particular, during the same period, but mainly outside Europe, the oceanic expansion and "virgin soil" epidemics take place between the 15th and 18th centuries. When sustained transoceanic travel connected the Americas (North and South), Europe, Africa, and Asia, pandemics were increasingly shaped by colonialism and forced migration [18]. A key property of this era was immunological asymmetry in which populations that had not been exposed to "Old World" infections experienced catastrophic "virgin soil" epidemics, turning disease into a major driver of conquest, labor systems, and demographic collapse [19]. At the same time, pathogens (microorganisms that cause disease in humans, animals, and plants) and epidemiological vectors (living organisms that transmit pathogens) circulated through Atlantic trade networks, showing that pandemic history cannot be separated from the history of empire and global exchange [20].

Right after, the "industrial, urban, and sanitary" era took place during the 19th century. During this time, rapid urban growth, crowded housing, and unsafe water created ideal conditions for explosive outbreaks, especially of waterborne diseases, while population's spatial densities were on the rise due to large migration from villages to cities. Cholera is a central example for a large-scale pandemic during this period because it repeatedly exposed the weak points of industrial cities and forced arguments about causation (bad air vs. specific contagion) to become arguments about infrastructure and governance [21]. A hallmark of this era is that public health increasingly meant building such as sewers, clean water systems, waste removal, and new forms of municipal administration. Data and mapping also became more important, as governments tried to measure risk and justify interventions towards the second half of the century [22].

Partially parallel to the industrial, urban, and sanitary era, a novel biological theory was proposed by Louis Pasteur, suggesting germs (microorganisms) cause disease and fermentation, overturning the theory of spontaneous generation (a [false] theory holding that living organisms can develop from non-living matter, such as maggots from rotting meat or mice from moldy grain) [23], and sharply pivoting into the germ theory and mass mobilization era during the late 19th and up to the early 20th centuries. With the rise of laboratory microbiology, pandemics entered an era in which identifying a pathogen became plausible, and prevention could be designed around that knowledge. But this was also the era of mass armies, mass transportation, and mass communication – conditions that could accelerate the spread. For instance, the 1918 influenza pandemic shows a core feature of modern respiratory pandemics: even without a contaminated water supply, a fast-moving virus can travel with people and overwhelm health systems quickly, especially during wartime disruption. This period also saw growing expectations that states should manage population health, even when the tools were still limited, a social norm that began to shift pandemic preparation and mitigation [24].

Shortly after, the Postwar biomedical optimism and emerging infections era arrived during the mid–late 20th century [25]. Antibiotics and vaccines encouraged a belief in many countries that infectious disease was a shrinking problem, yet the period also revealed new vulnerabilities. For example, HIV/AIDS highlighted how stigma, inequality, and blame can shape an epidemic's trajectory as strongly as biology does [26]. At the same time, ecologic disruption, intensive farming, and expanding global travel increased opportunities for zoonotic "spillover" and for regional outbreaks to become international threats. A defining property of this era was the return of surprise associated with the first area with outbreaks that experts did not expect, in places and forms that challenged existing health systems [27–29].

Finally, the 21st-century genomic and information era (2000s to the present). Today's pandemics unfold in a world of dense air travel networks, real-time news, and fast laboratory tools. Governments can sequence pathogens, model spread, and publish dashboards quickly [30]. Modern response is also shaped by international rules and institutions, including legal expectations for reporting and coordination across borders [31]. At the moment, there is no better example than COVID-19 that made these tensions visible: the world had unprecedented scientific capacity, yet still struggled with preparedness, equity, and sustained public cooperation, even as the World Health Organization (WHO) publicly characterized the situation as a pandemic in March 2020 [32–34].

Conceptually, what separates these eras is not that microbes suddenly changed their nature (for most of the cases), but that the human systems they move through changed. These changes can be attributed to three main drivers:

- Connectivity – in early agrarian and imperial worlds, long-distance contact existed, but it was slower and more episodic. In the oceanic expansion period, whole continents became connected through sustained transoceanic travel. In the 20th–21st centuries, mass rail, shipping, and then aviation created dense, rapid networks that allow respiratory viruses, in particular, to spread globally within days to weeks. So each era is defined by a different "map" of human mixing.
- Living conditions – operating as the local "fuel" for transmission. Namely, dense settlements are the key conceptual break from the pre-urban world: once people live close together – often near animals – some infections can persist and recur rather than burn out. Later, industrial urbanization adds a new kind of fuel: overcrowded housing and unsafe water, which made diseases like cholera explode and turned public health into an infrastructure problem (clean water, waste removal). In other words, some eras are dominated by crowding and proximity, while others are dominated by sanitation and environment.
- State capacity and knowledge – Before modern science, explanations were often religious, moral, or environmental, and interventions were blunt: flight, exclusion, ad hoc isolation. In the medieval

plague era, a conceptual shift occurred even without germ theory: governments begin to build administrative tools like quarantine, health boards, and surveillance – early public health as governance. With germ theory and laboratory medicine, the focus expanded from managing people to targeting pathogens (identification, vaccines, and later antivirals), opening the door to a completely novel set of mitigation strategies.

As a whole, this conceptual-historical framework provides the lens through which we now turn to specific historical pandemics. The detailed case studies that follow are not presented merely as chronological episodes, but as moments where biology interacted with a particular configuration of connectivity, living conditions, knowledge, and political capacity.

1.1.1 The Athenian plague

The outbreak, conventionally called the "Athenian Plague", struck Athens around 430 BCE (Before the Common Era), early in the Peloponnesian War (this war was fought in the Eastern Mediterranean between the Athenian-led Delian League and the Spartan-led Peloponnesian League from 431 to 404 BC for hegemony over Ancient Greece) [35]. There is no clear historical evidence, and the plague taking place is based on a single eyewitness – Thucydides – who places the epidemic within a wartime city under extreme demographic pressure, where rural inhabitants had crowded behind the long walls for protection, while Athens remained tied to the wider Mediterranean through its port at Piraeus [36]. Thucydides frames his description with an unusually "clinical" purpose, as he reports the sequence of symptoms so that the disease might be recognized should it ever recur, and he notes that he himself contracted the illness and survived. He also emphasizes patterns that matter for epidemiological reasoning, especially the high risk borne by caregivers and physicians and the apparent resistance of survivors to reinfection; observations later scholars have treated as central evidence, but also as evidence filtered through a literary and historiographical agenda [37,38].

In symptom terms, Thucydides describes a disease that begins abruptly and violently, with intense heat and inflammation "in the head", reddened eyes, and severe involvement of the throat and tongue, followed by respiratory distress (sneezing, hoarseness, cough) and then systemic collapse. He stresses unquenchable thirst, profound restlessness and insomnia, visible skin manifestations, and, in many cases, severe diarrhea later in the course. Among survivors, he reports lasting sequelae such as loss of extremities, blindness, and memory impairment. Modern clinicians and classicists repeatedly warn that every one of these features depends on how Greek medical vocabulary is translated and contextualized; nonetheless, the clustering of respiratory, cutaneous, and gastrointestinal signs is one reason the epidemic has resisted confident retrospective diagnosis [39].

Thucydides' narrative and later syntheses describe multiple waves over several years, with mortality so heavy that it strained Athens' military and civic capacity. Because ancient populations are hard to reconstruct, modern estimates of total deaths remain uncertain; what is secure is the convergence of war, crowding, and disruption of ordinary life as enabling conditions. Epidemiologically oriented work has therefore focused less on "naming the pathogen" and more on constraints implied by the outbreak's apparent speed, seasonality, and persistence in a dense, immunologically naïve population [40].

Interestingly, Thucydides also treats the plague as a social and political event, depicting the breakdown of customary funerary practices under the weight of mass death, a collapse of confidence in traditional religious explanations, and a turn toward short-term self-preservation that undermined law and norms. These themes have made the plague episode a foundational text for later writing about epidemic psychology, but they also pose a methodological challenge, since Thucydides is not only reporting what happened; he is crafting an explanation of how war and disease together can unravel a polis (city-state). Modern historians who focus on institutions and power relations still see the epidemic as politically consequential, even while cautioning against reducing Athens' eventual defeat to a single biological shock [41].

1.1.2 The Black Death

The epidemic known today as the "Black Death" was the most lethal disease event recorded in late-medieval Europe and the Mediterranean world [42]. It is best understood not as a single, isolated outbreak, but as the opening shock of the *Second Plague Pandemic*, a long era of recurrent plague waves that continued in many regions into the early modern period [43]. Most reconstructions place the pandemic's westward movement within the connected trading and shipping systems of the fourteenth century. By the mid-1340s, severe mortality was reported across parts of Eurasia; by 1347–1348, the disease had entered the Mediterranean maritime network and spread rapidly along major ports and commercial corridors [44]. Italian port cities figure prominently in narrative chronologies because they were hubs where shipborne movement, dense urban residence, and constant market exchange made sustained transmission easier. Notably, the Black Death was not only a European story, as major outbreaks occurred across North Africa and the Middle East as well, indicating that the relevant unit of analysis is the broader Afro-Eurasian disease ecology and the trade routes that linked it.

Contemporary observers described a terrifying illness with sudden onset, high fever, and painful swellings (commonly referred to as *buboes*) in the groin or armpit, often followed by delirium and death within a few days. Some accounts also refer to coughing of blood and rapid death, which later medical writers associated with pneumonic plague; however, historians emphasize that medieval descriptions are shaped by genre, theology, and the limits of premodern medical language [45]. For this reason, careful scholarship treats symptom reports as evidence that must be interpreted alongside patterns of spread, seasonality, and mortality rather than as a simple checklist that yields a single, certain diagnosis [45].

A major shift in Black Death scholarship came with biomolecular work that recovered ancient *Yersinia pestis* DNA from fourteenth-century human remains, strongly supporting plague as the causative agent [46]. This evidence helped settle earlier debates in which some scholars questioned whether the Black Death could have been plague because of its apparent speed and scale. Even with strong evidence for Yersinia pestis, important questions remain about how transmission worked in different places. The classic model emphasizes rodent reservoirs and flea vectors, but historians and interdisciplinary researchers continue to debate the relative contribution of rat-flea cycles, human ectoparasites, direct pneumonic transmission in certain clusters, and local ecological conditions [45]. Conceptually, the key point is that plague is not a single social mechanism; rather, the same pathogen can spread differently depending on housing, sanitation, animal ecologies, and the structure of everyday contact [47].

In parallel, estimating mortality is difficult, but there is broad agreement that the demographic shock was extraordinary. Regional outcomes varied sharply, yet many syntheses place overall mortality in Europe in the range of roughly one-third to one-half of the population, with some localities experiencing substantially higher losses [15]. To this end, mortality was not only a short-term catastrophe; it changed the baseline conditions of late-medieval society for generations. Population loss altered land use, settlement patterns, and the balance between labor and resources, contributing to long transitions that historians now analyze in combination with climate variability, agrarian structures, and political institutions [47].

The social consequences were equally complex. At the household level, the Black Death produced an acute crisis of care, where sick people needed nursing, but close contact increased risk, and normal funerary practice could collapse under the pressure of mass death. At the community level, fear and uncertainty often generated moral interpretation and scapegoating. Violence against minorities, especially anti-Jewish persecution in parts of Europe, forms one of the darkest and best-documented social responses to the pandemic, demonstrating how epidemic threat can be redirected into existing prejudices and political anxieties [45]. At the same time, many communities created practical responses that were not purely irrational, including ad hoc isolation, regulation of burial, and attempts to manage movement and markets.

Economically, the sudden scarcity of labor reshaped bargaining power and intensified conflict over wages, rents, and obligations. In many regions, workers could demand higher pay and better terms, while elites and states attempted to restrict mobility or cap wages; the outcomes differed by local legal regimes and the strength of landlords and cities [48]. These changes unfolded over decades and were repeatedly disturbed by later plague outbreaks, which is why historians stress that the Black Death cannot be fully analyzed without attention to its long aftermath of recurrent epidemics [43].

1.1.3 The cholera pandemics

Cholera is a severe diarrheal disease caused by toxigenic strains of *Vibrio cholerae*, and it can kill quickly through dehydration if untreated [49,50]. While cholera had long existed as an endemic disease in parts of South Asia, historians typically date the beginning of the "cholera pandemic era" to the early nineteenth century, when repeated waves spread far beyond the Bay of Bengal and became a truly global problem [51]. This is why cholera is often treated as a defining pandemic disease of the industrial and imperial world – it was shaped not only by microbial biology, but also by expanding trade, colonial armies, steamship travel, and fast-growing cities. Most accounts locate the *homeland* of pandemic cholera in the Ganges delta region, Bengal, where the organism could persist in a complex ecology of rivers, estuaries, and human settlement. What changed in the nineteenth century was not simply the presence of cholera, but the scale of connectivity that linked endemic zones to distant populations. Military movement, labor migration, pilgrimage routes, and maritime commerce repeatedly carried infection outward, turning what had been regionally familiar into a recurring international crisis. In this sense, the cholera pandemics are a clear example of how pandemics emerge from the interaction of local ecology with global networks [52,53].

The first cholera pandemic (often dated from 1817 into the 1820s) spread from South Asia across parts of Southeast Asia, the Middle East, and beyond, producing a new awareness among European and imperial authorities that fast-moving epidemic disease could accompany global exchange [22]. Later waves, commonly counted as the second through sixth pandemics across the nineteenth and early twentieth centuries, repeatedly entered Europe and North America and reshaped public debate about whether cholera was an imported contagion or a locally generated environmental problem. These debates mattered because they implied different interventions: if cholera was imported, quarantine and border controls looked rational; if it was generated by local filth or bad air, then sanitation and urban reform became the priority.

The second pandemic, beginning in 1829 and reaching many European and American cities in the early 1830s, became a turning point for public health politics [54]. In the United States, for instance, cholera epidemics exposed tensions between commercial interests and public safety, between local autonomy and centralized authority, and between moral explanations and structural explanations [54]. Across Europe, cholera also generated social conflict and distrust, especially when interventions were coercive or when the poor were portrayed as the *source* of disease rather than as those most exposed to hazardous living conditions.

A distinctive feature of the cholera pandemics is how directly they tied epidemic control to urban infrastructure. Because cholera is usually transmitted through contaminated water and food, epidemics tended to reveal the invisible systems that made nineteenth-century cities function: water supply, sewage disposal, drainage, and housing density [22]. The repeated appearance of cholera therefore helped transform public health from a set of temporary emergency actions into a long-term engineering and governance project, in which prevention increasingly meant building and maintaining safe urban environments.

By the late nineteenth century, bacteriology reshaped cholera control. Laboratory approaches helped identify *Vibrio cholerae* and supported the idea that prevention should focus on interrupting fecal–oral transmission – especially through safe water, sanitation, and (later) targeted vaccination strategies [49]. Yet the cholera story also shows that "germ theory" did not replace sanitation; rather, it gave sanitation a clearer mechanism and helped justify infrastructure investment as a scientifically grounded form of prevention. In other words, cholera's modern history is not a simple narrative of scientific progress but a layered transition in which engineering, administration, and laboratory science reinforced each other unevenly across different societies.

The seventh cholera pandemic, associated with the El Tor lineage and often dated from 1961 to the present, demonstrates that cholera is still a global threat when water and sanitation systems fail. Modern outbreaks are strongly associated with structural vulnerability – rapid urbanization, conflict and displacement, and fragile public services – rather than with "mystery" or unavoidable fate. At the same time, twentieth-century clinical advances (especially effective rehydration therapy) drastically reduced cholera mortality when treatment is accessible, shifting cholera from a disease that was often rapidly fatal to one that is highly survivable under adequate care [50].

1.1.4 1918 Spanish influenza

The influenza pandemic of 1918–19 was a global catastrophe that coincided with the final year of the First World War and its immediate aftermath. Despite the fact that it is widely referred to as the "Spanish influenza", this name is historically misleading: Spain's prominent association with the outbreak largely reflects wartime information politics rather than a proven geographic origin. As historians and medical scholars emphasize, belligerent states often limited or shaped press reporting during wartime, whereas neutral Spain reported more openly, helping to create the impression that Spain was uniquely affected [55]. The pandemic is therefore best approached not as a "Spanish" event but as a world event amplified by the extraordinary mobilization, crowding, and movement of people characteristic of total war [56].

A central characteristic of this pandemic is its wave pattern, which was both temporally denser than previous pandemics and also parallel in different geographical locations. Specifically, many countries experienced an initial wave in early 1918 that was comparatively mild, followed by a far more lethal wave in late 1918, and then additional outbreaks into 1919. The second wave is often treated as the decisive phase because it combined high transmissibility with strikingly elevated fatality in many settings, associated with the emergence of a new strain. These waves unfolded in a world connected by ships, railways, and troop movements, so the pandemic spread at a speed that was historically unusual even when compared to earlier epidemic crises. Moreover, the pandemic's mortality pattern also differed from that of typical seasonal influenza. Instead of producing deaths mainly among infants and older adults, 1918 influenza produced unusually high mortality among young and middle-aged adults, creating a "W-shaped" age-mortality curve. The reasons for this pattern remain debated, but it is commonly linked to a complex mix of immunological history, the intensity of viral infection, and the role of secondary bacterial complications in an era before antibiotics [57].

Johnson and Mueller's influential historical-demographic reassessment highlighted both the plausibility of approximately 50 million deaths globally and the uncertainty that surrounds any single point estimate, given uneven data quality and undercounting [58]. Notably, the key point is not only the scale of death but also its uneven distribution: crowded urban neighborhoods, military camps, and institutions with close-contact living conditions often experienced especially intense transmission and high case-fatality rates. Toward this objective, a particularly important line of scholarship argues that many deaths in 1918 were driven by secondary bacterial pneumonia. In the pre-antibiotic era, influenza infection often set the stage for bacterial invasion of the lungs, and autopsy series from 1918 are consistent with bacterial pneumonia being a major proximate cause of fatal outcomes. Thus, the implication for pandemic history is that severity is partly a function of therapeutic capacity, not only of the pathogen's intrinsic danger.

The 1918 pandemic is also a foundational case for evaluating non-pharmaceutical interventions. Because vaccines and antivirals were not available, cities relied on layered measures such as school closures, bans on mass gatherings, isolation of cases, and public messaging. Comparative historical studies of U.S. cities suggest that earlier, sustained, and combined non-pharmaceutical interventions were associated with lower peak mortality and, in some analyses, reduced cumulative mortality, although results varied by timing, adherence, and local conditions. These findings are significant not because they imply a simple policy formula, but because they show that epidemic outcomes can be shaped by administrative choices and public cooperation even when biomedical tools are limited or unutilized [59].

1.1.5 HIV/AIDS

HIV/AIDS is a pandemic with an unusual temporal shape since it is driven by an infectious agent that can spread efficiently through bodily fluids, yet it often unfolds over years, not days [60]. This long course (i.e., a prolonged asymptomatic period followed by progressive immune damage) means that HIV reshaped public health in a different way than acute respiratory pandemics. Instead of short emergency waves, HIV demanded sustained systems for testing, long-term treatment, and social support, while also provoking intense debate about sexuality, drug use, and stigma [61]. Untreated infection typically causes gradual immune deterioration, eventually permitting opportunistic infections and malignancies that define AIDS clinically. This mechanism explains why HIV prevention is tightly linked to the routes of transmission-sexual exposure, blood-to-blood exposure (including needle sharing and unsafe medical injections), and vertical transmission from mother to child; and why effective control requires both behavioral interventions and durable biomedical care [61,62].

The HIV/AIDS epidemic was first recognized clinically in 1981, when physicians reported clusters of rare opportunistic infections and profound immune dysfunction in previously healthy patients. A landmark early clinical description documented *Pneumocystis* pneumonia with other infections and striking depletion of helper T cells, helping to establish that a new acquired immunodeficiency syndrome was occurring and that it was likely transmissible [63]. In the early years, however, public discussion often framed AIDS through risk groups, which encouraged blame and misunderstanding. Subsequent scholarship has emphasized that such framing shaped policy, delayed protective responses for some populations, and intensified discrimination [28,64]. Notably, research on origins and early spread has reframed HIV/AIDS as a zoonotic emergence shaped by twentieth-century social change. Comparative phylogenetic work indicates that pandemic HIV-1 group M resulted from cross-species transmission of a simian immunodeficiency virus in central Africa, followed by sustained human-to-human transmission [62].

Social response has been central to HIV/AIDS history from the beginning. Stigma and discrimination did not merely accompany the disease; they often increased vulnerability by discouraging testing, driving infection underground, and justifying exclusion or coercive policies. A widely used conceptual framework argues that HIV-related stigma is not only an individual attitude but also a social process connected to power, inequality, and the policing of marginalized identities [64].

As the pandemic globalized, its epidemiology diversified rather than converging into a single pattern. In some settings, generalized heterosexual epidemics became dominant; in others, transmission was concentrated in specific populations, including men who have sex with men, people who inject drugs, sex workers, and incarcerated communities. Because HIV transmission risk is influenced by both biology (viral load, co-infections) and social structure (poverty, gender inequality, criminalization, access to prevention), the same virus produced very different national and regional trajectories [61,64]. To this end, the introduction of effective combination antiretroviral therapy in the mid-1990s transformed HIV in many settings from a frequently fatal condition into a manageable chronic illness. Clinical cohort evidence from the United States documented steep declines in opportunistic infections and mortality as more intensive regimens, especially those including protease inhibitors, became widespread [65]. Yet the benefits of antiretroviral therapy arrived unevenly across the world, making HIV/AIDS a lasting example of how pharmaceutical innovation can widen health disparities when access is unequal.

1.1.6 2002 SARS outbreak

The 2002–03 SARS outbreak is considered as the first major, rapidly globalized epidemic of the twenty-first century. SARS mattered not only because of its severity, but because it exposed how quickly a novel respiratory pathogen could move through air travel and healthcare systems, while also demonstrating that aggressive public health measures could contain a new disease under certain biological and social conditions [66–68].

Most reconstructions trace the earliest recognized cases to southern China in late 2002, where clusters of severe atypical pneumonia appeared before the syndrome was clearly identified as a distinct entity [66]. What turned a regional outbreak into a global emergency was the interaction between urban density, hospital-based amplification, and international mobility. A key early dissemination event occurred when an infected traveler stayed in a Hong Kong hotel in February 2003 and infected multiple other guests, who then carried the virus to several countries, seeding outbreaks in healthcare settings and communities [69,70], indicating the role of tourism in pandemic spread.

Clinically, SARS typically presented with fever and systemic symptoms, followed by respiratory complaints and radiographic pneumonia; severe cases could progress to hypoxemia and respiratory failure [71,72]. From a population perspective, SARS combined moderate-to-high transmissibility with a case-fatality risk that rose steeply with age and comorbidities, making it both a clinical crisis and a public health emergency. The global total burden in 2003 was far smaller than later pandemics, but the outbreak's seriousness came from its severity, its concentration in hospitals, and its potential to expand if control failed [66,68].

A defining feature of SARS was the central role of hospitals. A substantial fraction of early cases occurred among healthcare workers or were linked to healthcare exposure, reflecting the virus's ability to spread through close contact during clinical care, especially when infection control practices were not yet optimized for a novel airborne/respiratory threat [69,71]. This hospital amplification was not uniform: transmission was highly overdispersed, meaning that many infected individuals transmitted to few or no others, while a small number of cases generated large *superspreading* clusters [66,67].

These historical records are later used by healthcare policy makers to considers indoor settings design as a central mechanism of pandemic spread. For instance, Hong Kong's Amoy Gardens outbreak became an emblematic example of how built environments can reshape respiratory epidemic dynamics. Careful investigation supported the plausibility that unusual environmental conditions (including plumbing and airflow patterns) contributed to transmission beyond routine close-contact spread, underscoring that droplet transmission and airborne spread are not always cleanly separable in real buildings. This episode pushed SARS analysis beyond person-to-person contact alone and into questions of architecture, ventilation, and infrastructure vulnerability [73].

Transmission modeling and field epidemiology converged on several properties that made SARS unusually "controllable" compared with many other respiratory viruses. Incubation periods were typically several days, and (critically) transmission appeared to occur largely after symptom onset in most cases, rather than being dominated by presymptomatic spread [67,69]. This meant that rapid identification of symptomatic cases, immediate isolation, and systematic contact tracing could substantially reduce onward transmission, especially when paired with strict infection control in healthcare settings [67,70]. As such, policy responses leaned heavily on nonpharmaceutical interventions. Jurisdictions used case isolation, quarantine of exposed contacts, closure or restriction of high-risk institutional spaces, enhanced hospital infection control, and (in some instances) travel-related measures [68]. These interventions were socially costly and not free of controversy, but SARS is frequently cited as a case where such measures plausibly played a decisive role, partly because of the virus's symptom-linked infectiousness and the concentration of transmission in identifiable settings.

1.1.7 2009 H1N1 influenza pandemic

The 2009 H1N1 influenza pandemic was the first influenza pandemic officially recognized in the twenty-first century, caused by a novel influenza A(H1N1) virus, now commonly labeled H1N1pdm09 [74]. It was widely called "swine flu", a name that reflected its genetic ancestry in swine influenza lineages rather than any transmission from eating pork. The pandemic was detected in spring 2009 through surveillance that linked unusual human influenza cases in North America with outbreaks in

Mexico [75]. Virological and evolutionary analyses quickly showed that the new virus had a distinctive reassortant genome, with gene segments derived from multiple swine influenza lineages rather than from a single recent animal-to-human jump [76,77].

One of the most distinctive epidemiological features of 2009 H1N1 was its age pattern. Illness and hospitalization were concentrated in children, adolescents, and younger adults, while many older adults were relatively less affected than expected for influenza [78]. Immunological studies supported the idea that a portion of older adults had some preexisting cross-reactive antibodies, plausibly because of exposure decades earlier to antigenically related H1N1 viruses [79].

Conceptually, the event is important because it sat between two familiar frames: it spread globally with modern speed, yet its average severity was generally closer to seasonal influenza than to worst-case historical pandemics, forcing public health agencies to make high-stakes decisions under deep uncertainty [74,80]. Early in the outbreak, the crucial unknowns were transmissibility and severity. Rapid analyses using outbreak data and early international spread suggested sustained human-to-human transmission, but also emphasized how unstable early severity estimates can be when detection is biased toward the sickest patients [80]. Simply put, the pandemic began as a real-time lesson in inference, decision makers had to act before they could reliably know whether the virus would resemble a mild pandemic or a catastrophic one. Moreover, differences in the pandemic spread in different age groups illustrate a key principle of pandemics where the impact of a new virus or strain depends not only on novelty in virological terms, but also on the population's layered immune history with respect to this pathogen [74].

Hence, pandemic spread control relied on layered strategies rather than a single decisive technology. Antivirals were used for treatment and (in some settings) prophylaxis, nonpharmaceutical interventions were applied variably depending on local circumstances, and vaccines were developed and deployed during the pandemic rather than before it, presenting one of the first multi-mitigation and data-driven strategy in history [81]. Retrospective analyses of the response treated 2009 H1N1 as a stress test of pandemic governance revealed real strengths, such as rapid identification and global communication of a new virus, while also persistent weaknesses, like unequal access to countermeasures, difficulties in communicating changing risk assessments, and the challenge of scaling vaccination quickly enough to shape the early waves [74].

In historical perspective, the 2009 H1N1 pandemic is best understood as a moderate-severity but high-uncertainty event whose main legacy was institutional rather than purely demographic. It refined ideas about how to estimate severity in real time, clarified the importance of immunity landscapes in shaping age risk, and demonstrated that global mortality can be heavily underestimated if one relies on laboratory confirmation alone [80,82]. It also left a biological imprint, as the H1N1pdm09 lineage persisted and became part of the seasonal influenza ecosystem, reminding us that the endpoint of a pandemic is often not disappearance but incorporation into ongoing endemic circulation.

1.1.8 2013 Ebola outbreak

The Ebola virus disease epidemic that began in late 2013 in Guinea and then spread across West Africa is widely treated as a turning point in modern outbreak history. Unlike earlier Ebola events that were typically smaller and more geographically contained, this epidemic became a prolonged, multi-country emergency that reached major cities and required a sustained international response. Retrospective accounts emphasize transmission likely began in a rural area in late 2013, but laboratory confirmation and full recognition came months later in 2014, during which time chains of transmission were able to expand and cross borders [83].

Early clinical-epidemiologic investigations linked confirmed cases to a presumed first fatality in December 2013 and identified the causative agent as Zaire ebolavirus in Guinea [84]. Follow-up ecological investigation and reconstruction of early events focused attention on the likely spillover setting and on candidate reservoir species, especially bats, while being careful about what could be concluded definitively from limited early evidence [85]. Interestingly, this epidemic's magnitude was not driven primarily by new Ebola biology, but rather by social shifts such as population mobility, porous borders, and limited surge capacity in health systems [86,87].

1.1.9 COVID-19

The coronavirus disease 2019 (COVID-19) pandemic, caused by the novel severe acute respiratory syndrome coronavirus 2 (SARS-CoV-2), represented a systemic shock to global health, economics, and society unmatched in recent history. First identified in Wuhan (China), in late 2019, the virus was characterized by efficient human-to-human transmission, including significant spread from presymptomatic and asymptomatic individuals, which rendered containment strategies based solely on symptomatic screening ineffective [88]. By the time the WHO declared a pandemic in March 2020, the virus had already established cryptic transmission chains globally, exposing the fragility of just-in-time global supply chains and the limitations of pandemic preparedness plans that were largely designed for influenza [33,89]. COVID-19 defied simple categorization as a respiratory viral illness. While severe cases were marked by acute respiratory distress syndrome (ARDS), the virus also behaved as a multi-system vascular disease, causing endothelial dysfunction, hypercoagulability, and cytokine storms that led to multi-organ failure [90].

Epidemiologically, SARS-CoV-2 demonstrated a capacity for rapid evolutionary adaptation. While early lineages had a relatively mid-level of spread, subsequent variants of concern, such as the Delta and Omicron strains, evolved increased transmissibility and immune escape phenotypes, deemed more aggressive as a result [89]. A pivotal conceptual shift during the pandemic was the recognition of airborne transmission via aerosols as a primary driver of spread, challenging early public health guidance that focused heavily on large droplets and fomites [91]. This evolution in understanding necessitated structural interventions, such as improved ventilation and high-quality masking, rather than merely behavioral adjustments or deep cleaning.

The pandemic's toll was unevenly distributed, acting as a magnifier of existing societal inequities. In the United States and United Kingdom, systematic reviews found that Black, Hispanic, and Asian populations experienced disproportionately higher rates of infection and mortality compared to White populations, disparities driven principally by structural determinants such as occupation, housing density, and access to healthcare rather than intrinsic biological susceptibility. For example, data showed that Black individuals in the UK were at significantly higher risk of ICU admission and death, highlighting how pandemic stressors interact with long-standing health stratifications [92]. Mortality assessment for COVID-19, like previous pandemics, required statistical reconstruction beyond reported counts. While official dashboards recorded approximately 6 to 7 million deaths by 2023, excess mortality studies suggest the true global toll was likely between 14.8 million and 18.2 million deaths during the first two years alone [93]. This gap was particularly pronounced in lower- and middle-income countries where diagnostic capacity was limited, reinforcing the utility of all-cause mortality monitoring as a less biased surveillance tool [94].

The scientific response was marked by the unprecedented speed of vaccine development, particularly the validation of mRNA platforms (Pfizer-BioNTech, Moderna), which demonstrated high efficacy against severe disease less than a year after the viral sequence was published [95]. However, this technological success was juxtaposed with severe failures in global equity, as high-income nations monopolized early supply [96]. Ultimately, COVID-19 transitioned from a global emergency to an endemic pathogen, leaving a legacy of advanced surveillance technologies but also unresolved questions regarding the long-term burden of post-acute disease and the resilience of health systems against future zoonotic spillovers [89].

1.2 The evolution of pandemics modeling

Long before laboratory diagnostics, governments and cities produced administrative records of births and deaths, and those records invited comparative reasoning: which causes rose or fell, which seasons were worse, which neighborhoods suffered more. A pivotal moment occurred in 1854 when John Snow (not the fictional character from the popular TV serial "Game of Thrones" [97]) traced a cholera outbreak in London to the contaminated Broad Street water pump, establishing a data-driven, spatial approach to disease investigation that predated and ultimately helped usher in the germ theory of disease [98]. Snow's investigation is often remembered for the dramatic clarity of its conclusion, but its deeper historical importance for modeling lies in the way it intertwined data, mechanism, and action. He did not merely compile case counts. He used their arrangement in space, their timing, and the social practices that linked households to water sources to argue that cholera propagation was not random, nor simply an atmospheric misfortune, but a patterned process that could be interrupted. That logic – reducing a frightening, complex event into an interpretable pattern and then using that pattern to motivate an intervention – captures a core aspiration of epidemic modeling that has persisted ever since [99].

Related innovations in demographic accounting were equally consequential. Edmond Halley's life table, built from municipal data in Breslau, exemplified how population-level records could be transformed into expectations about survival and mortality [100]. This kind of baseline thinking – what mortality would look like in the absence of crisis – later became essential to interpreting pandemic impact through concepts such as "excess deaths", even when the direct attribution of cause was uncertain. In other words, early demographic tools did not model epidemics in the contemporary sense, but they provided a quantitative scaffolding that made epidemic events legible against the background of ordinary population turnover [101].

Epidemic modeling became more explicitly connected to policy when prevention itself became a subject of structured comparison. Smallpox, with its dramatic mortality and long social history, generated one of the earliest influential examples of using a formal argument to evaluate a public health intervention with Daniel Bernoulli's analysis of inoculation (variolation) [102]. Bernoulli asked not only whether inoculation worked for an individual, but how widespread use might shift population survival over time. Even without adopting his assumptions, the episode illustrates a lasting theme – epidemics are not only natural phenomena to be described, but also moving targets reshaped by human behavior and public health choices – choices that can be reasoned about systematically [103].

The nineteenth century provided repeated shocks (led by the cholera pandemic) that accelerated both the administrative capacity to measure disease and the practical demand to explain it [104]. Alongside Snow, William Farr's work stands out as a bridge between routine vital statistics and interpretive epidemiology [105]. Farr and his contemporaries helped normalize the practice of assembling large-scale mortality and morbidity data, visualizing epidemic trajectories, and debating competing explanatory frameworks using population evidence. Even when interpretations were contested, the method was transformative since epidemics could be treated as population processes with discernible structure, rather than as collections of unrelated tragedies [106]. In this period, "modeling" often meant developing disciplined ways to compare quanta – comparing districts, occupational groups, water supplies, elevations, or time periods. The key innovation was not mathematical sophistication but an evolving evidentiary standard. This new line of thought argued that hypotheses about transmission should leave footprints in the data. Snow's spatial comparison of households and water sources is exemplary precisely because it translated an invisible process into observable implications. This is one of the enduring lessons for pandemic modeling – when direct measurement of transmission is impossible, models are often judged by the plausibility and consistency of the patterns they imply.

Slightly later in the historical records, as bacteriology and laboratory identification of pathogens strengthened the concept of specific causes, a parallel development sought more general principles of epidemic behavior [107]. Early twentieth-century public health thinkers increasingly treated epidemics as dynamic phenomena shaped by contact between people (and, in some cases, vectors), by environmental context, and by changing susceptibility in populations. William H. Hamer's Milroy lectures are representative of this transition – they framed epidemic waves as outcomes of population interactions rather than as mere reflections of weather, chance, or moral failing [108]. Such work helped crystallize a style of reasoning in which the epidemic curve itself became evidence of underlying processes. In vector-borne disease, Ross's work on malaria combined entomology, public health practice, and population reasoning to argue that transmission depended not simply on the presence of disease, but on ecological and behavioral linkages that could, in principle, be disrupted. Here again, the historical significance lies less in technical apparatus than in a conceptual shift, as control strategies could be evaluated through their hypothesized effect on the chain of transmission [109].

In parallel, the basic compartmental models to describe the transmission of communicable diseases are contained in a sequence of three papers by W.O. Kermack and A.G. McKendrick in 1927, 1932, and 1933 [110–112]. The first of these papers described epidemic models in a formally mathematical model, starting the revolution of formal and mechanistic explanation of pandemic spread. Conceptually, the central idea across these three works is that epidemic trajectories can be driven by the changing composition of a population (particularly the depletion of those still vulnerable) so that the epidemic itself transforms the conditions that allow it to spread. This conceptualization became a cornerstone for later pandemic reasoning, influencing how public health officials interpreted why waves crest and decline, and why timing matters for interventions. At roughly the same time, modeling also evolved as a pedagogical and institutional practice. The Reed–Frost epidemic theory [113], which calculates new infections based on the probability of contact between susceptible and infectious individuals over successive generations, developed and taught at Johns Hopkins, is significant not only for its simplifying assumptions but for how it helped train epidemiologists to think analytically about transmission in populations [114]. This was not merely a technical development but rather it contributed to shaping epidemiology's self-understanding as a discipline that could move from descriptive accounts toward controlled reasoning with explicit assumptions [114,115]. In this sense, "modeling" became part of epidemiological education and professional identity, not just a set of calculations, forever shifting the way nations think, understand, and react to pandemic.

Another major historical pivot took place in the second half of the twentieth century with the institutionalization of surveillance. As epidemiological models become better and more accurate, a clear challenge emerged in the form of response time. These models where as good as the data they fitted on. Thus, the ability to model epidemics depends on timely knowledge of what is happening, where, and to whom; conversely, surveillance systems often develop in response to the recognition that epidemic dynamics demand systematic monitoring. Alexander Langmuir's articulation of communicable disease surveillance made explicit that ongoing data collection, consolidation, analysis, and dissemination should exist to *trigger public health action*, not merely to archive statistics. This emphasis on action-oriented information flows influenced how epidemic intelligence was framed in practice: outbreaks were events to be detected early, tracked, and managed, requiring data pipelines that could support rapid interpretation [116]. Indeed, historical reviews of U.S. public health surveillance describe how these infrastructures expanded and diversified, shaped by changing public health priorities and by evolving relationships between local, national, and international authorities. Arguably, for pandemic modeling, the importance is twofold. First, surveillance determines what can be known in real time – whether models must rely on mortality, clinical syndromes, laboratory confirmation, or special studies. Second, the very existence of surveillance changes how societies respond, since models and policy debates can proceed with a shared empirical frame, even if interpretations differ [117].

By the late twentieth century, epidemic modeling increasingly served as a bridge between scientific understanding and planning. At the time, the HIV/AIDS pandemic demanded long-horizon thinking, attention to heterogeneity in risk and behavior, and explicit evaluation of prevention strategies due to its unique contagiousdynamics compared to previous pandemics. In this broader context, Anderson and May's synthesis became emblematic of a mature field that treated infectious disease dynamics as intelligible population processes that could be linked to data and public health decisions. The historical value of such syntheses was not only technical consolidation but also the establishment of a shared language for discussing transmission, control, and uncertainty across disciplines and institutions [118].

Following these works, in the twenty-first century, three forces reshaped pandemic modeling: rapid global connectivity, expanding data sources, and rising expectations for real-time guidance. The 2003 SARS outbreak exemplified the new tempo [119–121]. Models were developed under pressure, using incomplete and evolving case data, to interpret transmission patterns, and the likely impact of interventions. This period made visible a recurring challenge – models must more often than not inform decisions before the most informative data exist. Therefore, these models must be built around uncertainty, partial observation, and shifting case definitions, outlining the need for a new generation of stochastic models and optimization under uncertainty methods [70,122,123]. Similarly, planning for influenza pandemics further pushed modeling toward explicit scenario analysis of interventions. Work on mitigation strategies highlighted how choices such as household measures, social distancing, and vaccine deployment could be compared in terms of plausible consequences, even when the future pathogen was unknown [124]. Here, the modeling process functioned as a structured way to reason about preparedness, and powered by recent developments in more advanced simulation methods [124].

These new capabilities are driven by the dramatic changes in the data landscape, where digital traces and novel mobility measurements offered new proxies for contact and movement, creating opportunities to link epidemic dynamics to behavioral patterns at scale [125]. These developments did not eliminate older data sources; rather, they layered new forms of evidence on top of classical surveillance, reviving enduring debates about representativeness, bias, and ethical governance of sensitive information [125].

The COVID-19 pandemic intensified all these trends with an unprecedented volume of data, rapid dissemination of results, and high-stakes decision-making under deep uncertainty [126,127]. Models were used to compare the likely effects of non-pharmaceutical interventions across countries, often relying on mortality and other routinely reported indicators to infer changing transmission conditions [128]. COVID-19, probably more than any other pandemic in history, highlighted how pandemic modeling is inseparable from communication and governance: model outputs travel through media, politics, and public trust, and their influence depends not only on methodological choices but on transparency about assumptions and limits [128].

Taken jointly, the evolution of epidemiological modeling over history share three basic truths. First, epidemic modeling has repeatedly advanced when new forms of data made previously hidden structure visible. For example, spatial clustering in cholera [99], time series in vital statistics [106], routine surveillance streams [116,117], and phylogenetic relationships in the genomic era [129,130]. Second, models have often been most influential when they made a practical claim about leverage, such as which link in the transmission process might be weakened, and with what plausible consequences and not focused on prediction alone. Third, modeling has always operated in tension between simplification and realism. Its purpose has never been to reproduce the world in full detail, but to construct interpretable representations that can be confronted with evidence and used to guide action. In this sense, the evolution of pandemic modeling is less a story of a single method steadily improving than a story of an expanding toolkit of ways to connect data to decisions.

1.3 Pandemic modeling approaches

Based on the recent developments in pandemic modeling, pandemic models can be understood as a family of strategies for translating limited information about infections, contacts, and immunity into disciplined statements about what is happening, what may happen next, and what might happen under alternative actions [131]. Starting from the bottom line – Table 1.1 summarizes the different modeling approaches commonly used nowadays. These models are not competing *schools* so much as different answers to the same design question – which features of transmission, heterogeneity, and uncertainty must be represented to serve the model's intended role. For example, a model built for explanation must make mechanisms legible, while a model built for prediction is not necessarily measured by the same metric and rather must perform reliably under the data-generating process in force. These roles impose different standards of adequacy and motivate different modeling choices [132].

The most common starting point, and often the most useful baseline, is the deterministic compartmental model. It represents the population as grouped states and treats change as the smooth evolution of average (or mean) rates. This class earns its prominence because it makes the logic of epidemic growth and control transparent, as transmission is not merely a count of cases, but a process that recruits new infections through contacts between infectious and susceptible people, and that process is self-limiting as susceptibility declines. From this representation one can develop an

Table 1.1 Summary of common epidemic modeling strategies in terms of their properties, strengths, limitations, and typical uses.

Modeling approach / Strategy (name)	Key properties	Pros	Cons	Typical uses
Deterministic compartmental	Population grouped into compartments; dynamics follow average rates; often assumes homogeneous mixing or simple stratification	Transparent structure; fast to simulate; supports clear reasoning about thresholds, feedback, and intervention levers; convenient for sensitivity analysis	Can hide heterogeneity and clustering; may misrepresent early outbreak chance effects; may require strong assumptions to link to observed data	Mechanistic explanation; scenario analysis; rapid policy comparison at aggregate level; teaching and conceptual understanding
Stochastic compartmental	Compartments but with random transitions; captures demographic and transmission stochasticity; variability across realizations	Captures fadeout vs. takeoff; supports probability statements (e.g., elimination likelihood); better for small numbers and early outbreak phases	Harder to fit and communicate; more computational cost than deterministic; results can be sensitive to distributional assumptions	Emergence and extinction questions; elimination strategies; uncertainty-aware short-horizon projections when counts are low
Branching process	Transmission represented as random offspring counts per case; often focuses on early phase; can include overdispersion (superspreading)	Minimal, analytically tractable; clarifies thresholds and extinction probabilities; natural for cluster risk and superspreading interpretation	Limited beyond early phase; weak for depletion of susceptibility and long-term dynamics; often abstracts away interventions and time-varying behavior	Early outbreak assessment; importation risk; outbreak size distributions; evaluating superspreading and control via rapid isolation
Metapopulation models	Multiple subpopulations (regions, institutions) linked by movement or contact flows; within-patch dynamics often compartmental	Captures spatial spread and seeding; supports travel and connectivity scenarios; remains relatively interpretable compared to agent-based models	Quality depends on mobility and mixing data; still averages within patches; may miss within-setting clustering (households, events)	Regional planning; importation and travel policies; allocation of resources across areas; understanding spread between connected communities
Network models	Individuals as nodes, contacts as edges; transmission depends on network structure; can be static or dynamic networks	Represents clustering and degree heterogeneity; clarifies role of hubs and community structure; supports targeted interventions (vaccinate hubs, cut bridges)	Network data rarely complete; calibration can be hard; results can be brittle to network misspecification; less transparent than compartments	Targeted control strategies; assessing clustering effects; modeling settings where contact structure dominates (schools, workplaces, households)

continued on next page

Table 1.1 *(continued)*

Modeling approach / Strategy (name)	Key properties	Pros	Cons	Typical uses
Agent-based simulation	Explicit individuals with attributes and rules; rich heterogeneity; can encode behavior, households, institutions, movement, compliance; often stochastic	Operational realism for targeted interventions; captures nonlinearities from local saturation and behavioral feedback; can produce distributional outcomes	High assumption burden; calibration/validation can be difficult; computationally intensive; opacity and reproducibility challenges	Decision support for operational policies; evaluating intervention packages; distributional impact and equity analyses; scenario exploration with implementation constraints
Statistical time-series forecasting	Models incidence or related observables directly (e.g., autoregressive, state-space, ML); may incorporate covariates; limited mechanistic claims	Strong for short-term prediction in stable regimes; can handle reporting artifacts; can be rapidly updated as data arrive	Weak for counterfactuals; vulnerable to regime shifts (new variants, behavior changes, policy changes); interpretability may be limited	Near-term forecasts; nowcasting; anomaly detection; operational planning where short horizon is sufficient
Semi-mechanistic	Combines mechanistic core (e.g., latent infections, time-varying transmission) with flexible statistical components and explicit observation model	Balances interpretability and fit; can incorporate delays and underreporting; supports uncertainty quantification while retaining causal hooks	Still depends on identifiability and priors; may blur mechanism vs. fit; counterfactual validity limited if drivers of transmission are not modeled	Routine situational awareness; estimating time-varying transmission; short- to medium-term projections under stated scenarios
Inference-focused mechanistic models	Mechanistic model coupled with explicit measurement process; parameters estimated from data (Bayesian or likelihood-based) with uncertainty	Supports explanation grounded in data; provides credible intervals; enables principled comparison of hypotheses (e.g., waning vs. behavior change)	Can be sensitive to misspecification; requires careful treatment of observation biases; computational complexity for rich models	Parameter estimation (e.g., susceptibility, duration, reporting rates); testing mechanistic hypotheses; retrospective analysis to inform preparedness

intuition for thresholds and feedback without appealing to technicalities. Below a critical level of transmission potential, chains tend to peter out; above it, growth becomes self-reinforcing until negative feedback from immunity accumulation, behavior change, or intervention pushes the system back toward decline. The same simplicity that makes deterministic compartmental models explanatory also defines their limits. By averaging over individuals and chance events, they can understate the importance of randomness early in an outbreak, and they can conceal the consequences of clustering and heterogeneity that often dominate real epidemics.

To address questions where "whether" matters as much as "how much", stochastic models enter naturally. A stochastic compartmental model keeps the compartmental structure but allows transitions to occur as random events; the consequence is not mere statistical noise, but qualitatively different behavior in small populations or at low incidence. Near the threshold, chance can determine whether an introduction fades out or takes off, and this is exactly the type of question that deterministic averages can answer only misleadingly. A closely related but even more specialized strategy is the branching-process or offspring-distribution model, which focuses on the early phase by describing how many secondary cases arise from each primary case. Such models are valuable because they isolate a critical feature of many pathogens: transmission is often overdispersed, with many cases generating no onward transmission and a few generating many. This lens clarifies why outbreaks can appear quiescent and then surge, and why control measures that prevent rare high-transmission events can be disproportionately effective. The trade-off is scope, where branching-process models are not designed to represent the later-phase dynamics shaped by susceptibility depletion, repeated exposure, and long-term intervention programs.

When the question depends on where transmission occurs and how epidemics move across space or institutions, models must represent structure beyond a single well-mixed population. Metapopulation and spatio-temporal models do this by dividing the population into subpopulations – cities, regions, schools, or facilities – and linking them through movement or contact flows. This strategy sits between simple compartmental aggregation and fully individual simulation. It preserves interpretability while capturing seeding, synchrony, and the possibility that outbreaks in one subpopulation can reignite another. The cost is that within each patch, individuals are still averaged together, so local clustering and setting-specific dynamics may be smoothed out in ways that matter for targeted interventions. Network models push further by representing individuals as nodes and contacts as edges, making clustering, hubs, and community structure explicit determinants of spread. This makes them well suited for analyzing interventions that act on contact structure – targeted vaccination, removal of bridging contacts, or policies aimed at specific settings – but their credibility depends on how well the network corresponds to reality, and such data are often incomplete, context-specific, and change over time.

Agent-based simulations generalize these ideas on both the temporal and spatial axes by making individuals and their interactions the primary objects of representation. In this modeling approach, heterogeneity is not an adjustment term but the organizing principle since individuals differ in behavior, contact patterns, risk, compliance, and mobility. This can be a decisive advantage when decision support requires operational fidelity, such as evaluating policies for schools, workplaces, contact tracing, or testing regimes that depend on logistics and behavior. The same realism, however, carries a predictable burden. Agent-based models require many assumptions, are harder to calibrate, and can become opaque in the sense that causal attribution (what mechanism produced what outcome) can be difficult to establish without careful experimental design within the simulation. They should be used when the decisions truly depend on the details they can represent, and when data and transparency practices can support that complexity.

Alongside mechanistic approaches, statistical forecasting strategies represent a different commitment. Rather than encoding the pathways of transmission explicitly, statistical time-series models learn relationships between recent observations and near-term outcomes, sometimes enriched by covariates such as mobility, weather, or policy indicators. Their strength is practical performance when the system is stable enough that recent patterns remain informative. They can absorb reporting artifacts and can be updated quickly, which is often essential for operational planning. Their weakness becomes apparent when the environment changes in ways not seen in the training window: new variants, shifts in testing and reporting, changes in contact behavior, or the introduction of interventions that alter the causal structure. In such moments, a purely statistical relationship can fail without warning because it has no representation of why the relationship held in the first place. Semi-mechanistic models occupy a middle ground, combining a minimal mechanistic core with an explicit observation model to connect latent dynamics to reported cases, hospitalizations, or deaths. This strategy is common in real-time situational awareness because it retains interpretable levers (changes in transmission) while accommodating measurement delays and under-ascertainment.

Finally, decision-analytic models emphasize that *decision support* is not merely a matter of projecting cases. In many settings, the relevant output is a comparison of options under constraints: how interventions trade off health outcomes, costs, social disruption, and feasibility; how limited resources should be allocated; how sensitive the preferred policy is to uncertain parameters. Decision-analytic frameworks formalize these comparisons by linking epidemic trajectories to utilities, costs, and objectives. They can be coupled to any of the epidemic model strategies described above, but the addition is conceptual rather than computational: it forces explicit articulation of what counts as success and what trade-offs are acceptable. This class highlights a frequent misconception: a model can be excellent for decision support even when it cannot "predict" the exact future, because the decision problem depends on relative outcomes under plausible scenarios rather than on a single-point forecast.

Taken together, these strategies support a practical cookbook for model selection. One begins by naming the role – explanation, prediction, or decision support – and the quantity that matters: mechanism identification, near-term accuracy, elimination probability, distributional impact, or policy comparison under constraints. One then chooses whether to treat randomness as central (favoring stochastic and branching approaches) or secondary (favoring deterministic baselines), and whether the causal mechanism must be explicit (favoring mechanistic and semi-mechanistic designs) or whether stable empirical mapping is sufficient (favoring statistical forecasts). Next comes the decision about structure: whether aggregation is acceptable, whether subpopulation coupling is essential, whether network clustering must be represented, or whether operational policies require an agent-based representation. At each step, the guiding principle is alignment between model properties and the inferential and practical demands of the use-case, as reflected in the table's pros, cons, and typical uses. A well-chosen epidemic model is not the most detailed model available, but the one whose abstractions are appropriate to the question, whose uncertainties are honest, and whose outputs can be interpreted and acted upon without confusing explanation with prediction or prediction with decision.

In addition, Fig. 1.1 provides a compact, qualitative comparison of the main epidemic modeling strategies discussed in this section. Each axis corresponds to a practical criterion that frequently drives model choice: interpretability, data frugality, compute efficiency, counterfactual fidelity, short-term forecasting performance, heterogeneity detail, operational realism, and uncertainty handling. Each modeling approach is represented by a closed polygon whose distance from the center indicates relative strength on that criterion. All axes are oriented so that larger values are preferable; in particular, data frugality indicates that a method can be deployed with fewer or less granular data, and "compute efficiency" indicates that analyses can be run quickly and repeatedly for exploration and sensitivity checks. The visual purpose is not to produce a definitive ranking, but to make trade-offs explicit: strategies that emphasize transparency and speed tend to abstract away heterogeneity and operational detail, whereas strategies that represent individuals or contact structure more explicitly typically demand richer assumptions, more data, and greater computational effort. In this way, the figure complements the table by highlighting how different modeling families occupy different "profiles" of strengths that align with distinct use-cases in explanation, prediction, and decision support.

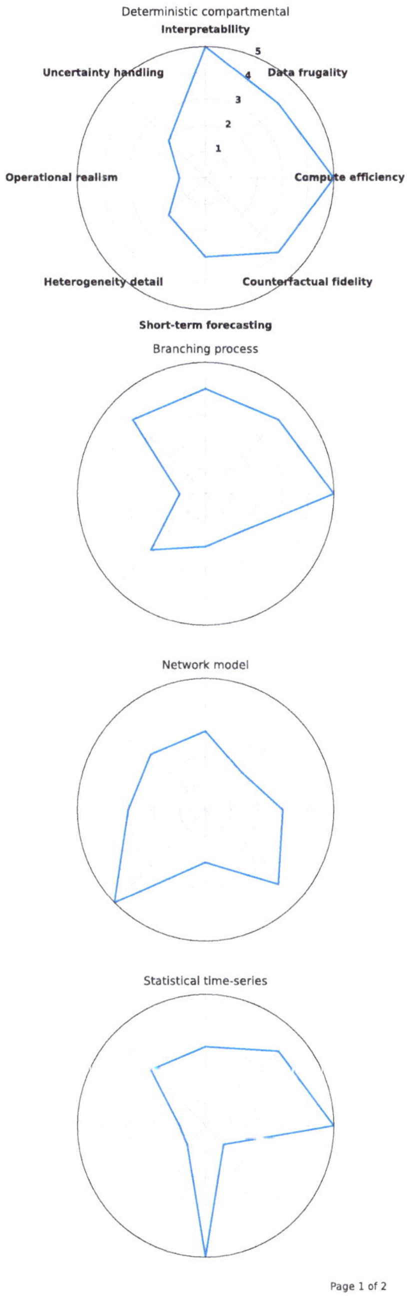

FIGURE 1.1

Qualitative strength profiles of common epidemic modeling strategies across practical criteria used in model selection. Scores are on a 1–5 scale and are oriented so that higher values are preferable on every axis.

CHAPTER

2 The biology, epidemiology, psychological, economic, and governance aspects of pandemics

Historical records and evidence from recent pandemics repeatedly show that pandemics aren't driven by a single mechanism, but by a complex web of feedback loops linking population dynamics, pathogen characteristics, and environmental conditions. Namely, what ultimately becomes visible as rapid growth in cases is the emergent result of (i) the epidemiological architecture of contacts and movement through which transmission opportunities arise, (ii) biological features of a pathogen and its interaction with hosts, (iii) psychological processes that shape protective and risky behaviors, (iv) economic incentives and constraints that govern what individuals and institutions can realistically do, and (v) governance capacities that coordinate collective action, surveillance, and communication. Even when two pathogens share broadly similar clinical presentations, differences across these domains can produce sharply different epidemic trajectories and societal consequences [133–135].

2.1 How infections spread through populations

If biology explains how a pathogen survives, enters the body, and produces infection, epidemiology explains how those individual infections become a population-level outbreak [136]. Epidemiology focuses on patterns of who becomes infected, when, where, and under what circumstances. It connects everyday life to transmission by asking a simple but powerful question: what opportunities does a pathogen have to move from one host to another, and how do those opportunities change as people adapt, immunity accumulates, and institutions respond.

A useful way to think about epidemic spread is as a competition between transmission and interruption. Transmission requires contact between an infectious person and a susceptible person in a way that allows the pathogen to pass. Interruption occurs when those links are broken or weakened, for example, because susceptible people become immune, because infectious people are isolated sooner, or because contact patterns change. The balance between these forces is often summarized by the reproduction number (usually denoted by $R_0(t)$ for some time t), meaning the average number of secondary infections caused by a typical infected individual under specified conditions. When that average is above one, chains of transmission tend to grow; when it falls below one, chains tend to shrink. Importantly, this is not a fixed property of the pathogen alone. It depends on the population context, including how people mix, how quickly infections are detected, and what control measures are in place [137,138].

Epidemic Modeling. https://doi.org/10.1016/B978-0-44-355525-1.00008-4

The structure of contact patterns is central. Real populations do not mix uniformly [139]. People have repeated close contacts within households, clustered contacts at work and school, and occasional contacts in community settings. Contact surveys show strong age patterns, with many contacts occurring among people of similar ages and with distinct differences by setting, such as home, school, and workplaces [140]. Because these mixing patterns vary across societies, researchers often use demographic and contact data to project plausible contact matrices for places where direct surveys are unavailable, which matters because interventions that target specific settings or age groups can have very different effects depending on the underlying social structure [141]. Heterogeneity also appears at the level of individual connectivity where some people have many more contacts than others, and certain environments enable one infectious person to infect many people in a short time (commonly referred to as "super-spreaders)" [142,143]. These clustered and highly uneven opportunities for transmission help explain why outbreaks can accelerate suddenly or appear to spread in bursts rather than smoothly [144].

Epidemics also unfold in space. Even when a pathogen spreads efficiently within a city or town, the broader pattern across regions depends on mobility: commuting, travel between settlements, and longer-range movement through transportation networks. Models that represent populations as interconnected subpopulations show that long-range travel can seed new outbreaks, while short-range commuting can synchronize nearby areas and change the timing of peaks, shaping the overall geography of risk [145]. From a practical perspective, spatial structure means that the same national-level case count can conceal very different local realities, including simultaneous surges in some regions and declines in others [146,147].

Epidemiology grapples with what is observable versus what is really happening. Infections occur before symptoms, testing, diagnosis, and reporting, and each step adds delay and distortion. Reported case counts therefore reflect both transmission and the surveillance system itself, including who seeks care, who gets tested, and how quickly results are processed. To track changing transmissibility during an outbreak, epidemiologists often estimate time-varying reproduction numbers from incidence data, while accounting for delays and the typical spacing between linked infections. Such tools are widely used because they provide an interpretable way to judge whether transmission is accelerating or slowing, even when the underlying data are imperfect [148–150].

2.2 The biology of infectious diseases

Taking a step deeper into the rabbit hole, an infectious disease starts as a biological interaction between an infectious agent and a host [151]. What later appears in public discourse as case counts, policy debates, and pressure on hospitals is rooted in a small set of underlying biological processes that determine whether infection occurs, when an infected person becomes contagious, how severe illness becomes, how long protection lasts after recovery or vaccination, and how readily the pathogen adapts over time [152,153].

Pathogens come in several major forms, including viruses, bacteria, fungi, protozoa, helminths (parasitic worms), and prions. The type matters because it shapes how replication happens and how the immune system detects and controls the invader. Viruses must enter host cells to replicate, whereas many bacteria can grow independently within tissues or in the environment. In classical infectious disease thinking, many pandemic threats are driven by microparasites such as viruses and many bacteria, which tend to generate acute infections and can spread rapidly when transmission opportunities are frequent [118]. This is one reason why timing is so central in pandemic response as a short delay between infection and contagiousness can allow spread to accelerate before illness is even recognized. It is important to separate infection from disease. Infection refers to the presence and replication or persistence of the pathogen in the host. Disease refers to the clinical consequences, such as symptoms and organ dysfunction. A person can be infected without feeling ill (i.e., asymptomatic) and still transmit to others. Several time intervals shape this process. The incubation period is the time between infection and symptom onset in those who develop symptoms. The latent period is the time between infection and becoming infectious. The infectious period is the time window during which transmission can occur. These intervals vary among individuals and are best thought of as distributions rather than fixed values, and systematic evidence shows substantial differences across respiratory viruses [154]. For control, the relative ordering of these intervals is crucial. When infectiousness typically begins after clear symptoms, identifying and isolating symptomatic cases can be highly effective. When substantial infectiousness occurs before symptoms, or among people who never develop symptoms, symptom-based control becomes far less reliable, and outbreaks are harder to contain [155].

Transmission requires that pathogens exit an infected host, survive transit, and reach a susceptible host in a dose sufficient to initiate infection. Biology shapes the dominant routes of transmission, including respiratory spread, direct contact, indirect contact through contaminated objects, fecal-oral spread, blood or sexual transmission, and vector-borne transmission through another species. These routes matter because they imply different vulnerabilities and different interventions. Respiratory transmission is strongly shaped by indoor air, crowding, and ventilation, whereas waterborne transmission is shaped by sanitation and water treatment. Even for the same pathogen, the environment can shift, which route contributes most. A critical biological factor is how long a pathogen remains infectious outside the body, which depends on temperature, humidity, sunlight, and the surface or medium. For instance, studies about *influenza A* highlight that environmental inactivation can influence the relative contribution of airborne, droplet or spray, and contact transmission [156]. During and after COVID-19, public health agencies have also moved toward clearer language that focuses on transmission through the air without forcing sharp categories that do not reflect the continuous behavior of exhaled particles across real-world settings [157].

Infectiousness is rarely uniform across people or across time. Within a single individual, pathogen shedding often rises and falls, producing a characteristic infectiousness profile [158,159]. The timing of peak infectiousness relative to symptoms affects whether testing and isolation can interrupt spread quickly enough. Across individuals, transmission can be highly uneven. Namely, a minority of infections may generate a large fraction of onward spread. This pattern can arise from biological differences (for example, pathogen load dynamics), from behavior (numbers and types of contacts), and from environment (crowded indoor events). A foundational analysis of individual variation shows how such heterogeneity can make outbreaks rarer but more explosive when they do occur, and why averages alone can mislead planning [160]. In practical terms, this means that preventing high-risk contexts can sometimes reduce spread more effectively than measures that assume everyone contributes equally.

Biology also influences illness severity. Virulence refers to the degree of harm a pathogen causes to its host. High severity increases the demand for healthcare and raises mortality, but it can also feed back on transmission by changing behavior and prompting stronger interventions. Evolutionary theory often frames virulence in terms of constraints and trade-offs, in which traits that increase replication and shedding may also increase host damage, while severe host damage can reduce opportunities for transmission [161]. Reviews of virulence evolution emphasize that there is no universal rule that pathogens become milder, because the direction of selection depends on the ecology of transmission and host behavior [162]. For pandemic modeling, the main lesson is that pathogen traits can change, and the surrounding social and environmental context helps determine which variants spread.

Immunity is central to why epidemics slow down and why they can return. The immune response includes rapid, non-specific defenses and slower, more targeted responses that can leave immunological memory. Immunity can reduce the chance of reinfection, reduce infectiousness if reinfection occurs, reduce severity, or combine these effects. Introductory immunology texts stress that immunity is not a single on–off state but a collection of mechanisms that vary by pathogen and by host [163]. Over longer time horizons, waning protection and pathogen evolution that alter immune recognition can both allow renewed waves.

Another important point, many pandemics begin with zoonotic spillover, in which a pathogen crosses from a non-human vertebrate reservoir into humans and then achieves sustained human-to-human transmission [164,165]. Spillover is not a single event but a sequence of barriers involving ecological exposure, pathogen release and survival, within-human susceptibility, and the capacity for onward transmission. A widely used framework describes these pathways and highlights why human activities and environmental change can alter the probability of emergence, even when the pathogen itself has not changed [166]. These biological foundations will be used throughout the chapter to connect mechanistic understanding to observed epidemic patterns and to the practical limits and opportunities for control [167].

2.3 Psychological elements of epidemiology

Infectious-disease spread is shaped not only by biology and contact opportunities, but also by how people perceive risk, interpret information, and decide whether to adopt protective actions [168]. These psychological processes matter epidemiologically because they influence everyday behaviors that create or remove transmission opportunities, such as staying home when ill, reducing close contacts, wearing masks in crowded indoor spaces, improving ventilation, seeking testing, isolating after exposure, and accepting vaccination [169,170]. As people observe rising illness around them or receive new guidance, they may change what they do, and these changes can feed back into the epidemic trajectory itself. This kind of two-way coupling between disease dynamics and behavior helps explain why outbreaks can accelerate, slow, plateau, or surge again even when the pathogen is unchanged [133,135].

Risk perception is a central driver of protective behavior, but it is not a simple reflection of objective danger. When a novel pathogen emerges, most individuals must make decisions under uncertainty, relying on limited personal experience and imperfect information. In such settings, people often use mental shortcuts that are efficient but can systematically mislead [171]. For example, events that are vivid, recent, or heavily covered in the media tend to feel more likely than they are, while slow-moving risks can be underestimated until they become personally salient. These patterns are well documented in the broader psychology of judgment under uncertainty and are directly relevant to epidemics, because they influence when and how strongly communities respond to rising transmission [172]. The practical implication is that two populations facing similar epidemiological conditions can behave very differently depending on what they believe is happening, how confident they feel in recommended measures, and whether they interpret the situation as controllable.

Decisions are also shaped by motivation and perceived ability to act. Even if individuals believe a disease is dangerous, they may not change their behavior if they doubt that protective actions work, if they feel they cannot perform them consistently, or if they anticipate high personal costs. Behavioral science emphasizes that intentions arise from a combination of attitudes toward the action, perceptions of what others expect, and perceived control over the behavior. These ingredients help explain why compliance is often high early in a crisis and then becomes uneven, and why the same policy can produce different outcomes across communities with different constraints and social expectations [173].

Because infection risk is partly collective, social influence is especially powerful during pandemics. People look to others to decide what is normal, acceptable, and worth the effort. If mask use, testing, or staying home when sick becomes a visible social norm, adoption can spread quickly through workplaces, schools, and friendship networks. Conversely, if protective behavior becomes rare or socially contested, individuals may abandon it even when risk remains substantial. Classic work on compliance and conformity explains how norms, cues from authority, and perceived group expectations shape behavior, and these mechanisms operate at scale during health crises. Social identity can further intensify these effects such as in cases when public-health measures become associated with group membership, behavior may track identity and loyalty as much as health beliefs, producing polarized responses that complicate control.

Trust and communication sit at the center of these dynamics. When people trust health authorities and local messengers, they are more likely to accept guidance, tolerate inconvenience, and update beliefs when recommendations change [174,175]. When trust is low, the same information can be discounted, reinterpreted as self-serving, or treated as evidence of hidden motives. Trust is fragile during crises because uncertainty is high and decisions are consequential. Guidance from the World Health Organization on vaccination communication emphasizes that confidence can be strengthened through preparedness, consistent messaging, responsiveness to concerns, and transparent discussion of uncertainty, all of which generalize beyond vaccines to many pandemic decisions [176].

Modern information environments add another layer of influence on the population psychological aspect. Misinformation and low-quality content can spread rapidly, shaping beliefs and behavior in ways that may increase transmission or reduce uptake of effective interventions. Large-scale analyses of online diffusion show that false information can propagate broadly and quickly, not simply because of automated accounts but because people preferentially share novel or emotionally engaging claims [177]. In a pandemic, this can distort perceived risks, amplify distrust, and promote ineffective or harmful practices. From a modeling perspective, information spread and disease spread can become entangled, because beliefs influence behavior and behavior influences transmission.

Sustained crises impose cognitive and emotional burdens, and many people experience declining motivation to maintain protective routines, especially when benefits are less visible than costs. The World Health Organization has described pandemic fatigue as an expected response to prolonged public-health measures and emphasizes that maintaining adherence requires strategies that support motivation, capability, and opportunity, rather than assuming people will comply indefinitely through willpower alone [178]. For epidemiological thinking, the key point is that adherence is dynamic: it rises and falls with perceived threat, trust, lived experience, and feasibility. Epidemic control, therefore, depends not only on what measures exist, but on how human psychology shapes whether, when, and for how long those measures are actually practiced.

2.4 Economic aspects of pandemics

Pandemics are often described as biological events with social consequences, but they are also economic shocks in a precise sense, as they disrupt how people allocate time, money, and attention, and they change what is feasible for households, firms, and governments [179–181]. These choices then feed back into transmission, because most infections occur through ordinary economic and social activity, such as working, shopping, commuting, caregiving, and leisure [182–184]. The result is a coupled system in which disease risk alters behavior and behavior alters disease risk. This coupling is the central idea of economic epidemiology.

At the individual level, economic decisions shape exposure opportunities [185]. A person who can work from home can reduce contacts without losing income, while someone in a job that requires physical presence faces a much harsher trade-off between health risk and earnings. Even within the same city, two households can face very different constraint sets: one may be able to stock up on essentials, isolate in a separate room, and take time off work, while another may live in crowded housing, rely on public transport, and need daily income to cover basic needs. From a modeling perspective, these constraints matter because they determine which protective behaviors are realistically available. When we observe unequal infection burdens across occupations, neighborhoods, or income groups, it is rarely enough to attribute the pattern to individual preferences. Differences in constraint and opportunity are often the deeper mechanism.

A key economic insight is that epidemics create externalities, meaning one person's decisions impose costs or benefits on others that are not fully reflected in private incentives. When an infectious individual goes to work, they may earn wages and keep a business running, but they also increase risk for coworkers and customers. Conversely, when a person chooses to stay home, they lose income or opportunities, but they may reduce the chance of passing the pathogen onward. Because these spillover effects are not fully priced in normal market transactions, private choices need not align with what is best for society as a whole, even when everyone is acting rationally and with good intentions [179,186,187].

This externality logic helps explain why governments intervene in pandemics, but it also clarifies why intervention is difficult [188]. Many protective actions are not only costly, they are also costly in ways that are unevenly distributed. Asking a whole population to reduce contacts is not a single policy lever; it is a set of millions of household-level decisions, each embedded in employment contracts, childcare responsibilities, debt obligations, and access to safety nets. Policies that ignore these economic constraints can fail even if the public understands the health rationale, because knowledge alone does not create feasible options. For example, if workers do not have paid sick leave, symptomatic individuals may continue working not because they doubt the disease, but because they cannot afford to stop. Empirical work during COVID-19 found that expanding emergency paid sick leave in the United States was associated with reduced case growth, consistent with the idea that relaxing financial constraints can change transmission-relevant behavior [189].

A second core economic mechanism is risk-driven voluntary behavior change [190]. During outbreaks, people often reduce travel, avoid crowded venues, and shift consumption patterns even in the absence of formal restrictions. This matters because it means economic contraction is not solely the result of mandated policies; it can also arise endogenously from fear of infection and from uncertainty about what is safe. Studies using high-frequency data from the early COVID-19 period found large declines in consumer activity that were strongly associated with local disease conditions, and that legal shutdown orders explained only a portion of the observed drop in foot traffic [191,192]. In coupled epidemic-economic models, this voluntary response can reduce transmission by lowering contact rates, but it simultaneously deepens a recession by cutting labor supply and demand for in-person services [192].

Pandemics also affect the economy through multiple macroeconomic channels. Illness and caregiving reduce available labor [193]. Disruptions to schooling can reduce parental labor supply in the short run and human capital formation in the long run. Demand falls in contact-intensive sectors such as restaurants, tourism, and entertainment, while demand rises in other sectors such as delivery logistics, remote communication technologies, and some health-related goods. Financial stress can spread as households reduce spending, firms lose revenue, and credit constraints tighten. In this way, a pandemic resembles a combined demand and supply shock, with the balance between them varying across time and place [194,195].

These macroeconomic dynamics are amplified by the structure of production networks. Modern economies are highly interconnected: a reduction in output in one sector can propagate through input-output linkages to many others [196]. A factory closure can ripple through downstream assembly lines; a transport bottleneck can delay intermediate goods; a shortage of a specialized component can halt production even if most inputs remain available [197]. For instance, a study that explicitly models the COVID-19 crisis in a multi-sector framework shows how sectoral disruptions can generate spillovers and how complementarities across sectors can magnify aggregate losses [198]. In this context, distributional consequences are not a side issue; they shape both fairness and epidemic dynamics. Lower-income workers are more likely to hold jobs that cannot be performed remotely, and they often have fewer buffers, such as savings, stable contracts, or access to paid leave. Children from disadvantaged backgrounds may experience larger educational losses when schools close, with long-term consequences for earnings and health. Women often shoulder disproportionate caregiving burdens, affecting labor force participation. Small firms can be more vulnerable to liquidity shocks than large firms with easier credit access. Research summarizing pandemic-era inequalities emphasizes that many of these patterns are structural and can persist after the acute health crisis ends [199]. At the global scale, international organizations documented large disruptions to employment and working hours, and major setbacks to poverty reduction, particularly in countries with limited fiscal capacity and large informal sectors [200].

Economic reasoning is also central to the design of targeted interventions. When risk is not uniform across age groups, occupations, or settings, the socially optimal allocation of restrictions and protections may be highly non-uniform as well. Models that combine multi-group epidemic structure with economic activity have been used to formalize the intuition that reducing contacts in high-risk or high-transmission contexts can yield larger health benefits per unit of economic cost than untargeted lockdowns [201]. The practical lesson is not that targeting is always easy or ethically uncontroversial, but that the trade-offs are shaped by heterogeneity. Ignoring heterogeneity tends to produce blunt policies that are simultaneously more costly and less effective than they could be.

To this end, pandemics force explicit consideration of how societies value health outcomes relative to other outcomes. This is not because lives should be reduced to prices, but because resources are finite and policies have multiple consequences. Public finance and health economics offer frameworks for comparing costs and benefits, including the use of the value of a statistical life in some policy contexts [202]. During COVID-19, some analyses attempted to monetize mortality reductions from distancing policies to compare them with economic losses, highlighting both the potential scale of health benefits and the sensitivity of conclusions to assumptions [203]. Simply put, the main point is that quantitative comparisons require careful attention to what is being measured. Mortality is only one outcome; morbidity, long-term disability, mental health, educational loss, and delayed non-pandemic care can also be substantial. Moreover, the distribution of costs and benefits across groups often matters as much as the aggregate totals [186,187].

2.5 The role of governments in pandemics

Pandemics are collective action problems. Even if every individual understands that reducing contact, improving ventilation, isolating when sick, and getting vaccinated can lower their personal risk, these actions are hard to sustain without coordination, trustworthy information, and practical support. Governments exist, in part, to solve problems that individuals and markets cannot solve well on their own: providing public goods such as surveillance systems, laboratory networks, emergency logistics, and clear rules; managing shared resources such as hospitals and medicines; and protecting people from harms that arise when one person's choices affect many others. In infectious disease, these spillover effects are the norm, rather than the exception. This is why governance is not just a backdrop to epidemic dynamics. It is one of the engines that shapes the epidemic itself.

A useful way to understand governance is to separate what must be in place before a crisis from what must happen during a crisis. Before a pathogen begins spreading widely, countries need foundations: legal authority to act, trained public health staff, laboratories that can test and sequence pathogens, systems that collect and share data, and relationships between agencies that will later be asked to cooperate under pressure. Internationally, countries also have agreed obligations to build and maintain core capacities and to report unusual events that may pose cross-border risks. These obligations matter because delays in detection and reporting can turn local outbreaks into regional crises. In practice, the strength of these foundations varies widely across countries and often within countries, and this heterogeneity helps explain why the same pathogen can produce very different trajectories and outcomes in different places.

These foundations are not purely technical. They also include the legitimacy of institutions. People must believe that the public health system is competent, that it has their interests in mind, and that it will treat them fairly. When legitimacy is low, even well-designed policies can be undermined by non-compliance, avoidance of testing, or resistance to isolation and vaccination. When legitimacy is higher, governments can often achieve the same epidemiological effect with less coercion because voluntary cooperation is stronger. This is one reason why trust is not an abstract political concept in a pandemic; it is a variable that changes transmission.

During a crisis, governments face a sequence of decisions that are tightly coupled to epidemic time. Early on, the main governance task is to reduce uncertainty fast. That requires expanding testing, establishing case definitions and reporting pipelines, integrating clinical reports with laboratory data, and building situational awareness about who is infected, where transmission is occurring, and which settings are driving spread. It also requires making risk information usable. Data that are technically correct but slow, fragmented, or hard to interpret will not lead to timely action. Governments, therefore, have to build mechanisms that transform raw signals into decisions, such as incident management structures, standing advisory committees, and rapid procedures for updating guidance when evidence changes.

Once sustained transmission begins, the policy challenge shifts from detection to interruption. Governments cannot directly stop a virus, but they can change the environment in which transmission happens. They can influence how often people meet, where meetings occur, how risky those encounters are, and how quickly infectious individuals are identified and separated from others. These policy levers include measures such as isolation and quarantine support, contact tracing, testing access, improvements to indoor air, masking policies, limits on crowding, and temporary restrictions on high-risk activities. They also include health system actions such as expanding hospital surge capacity and protecting health workers. Importantly, the epidemiological impact of these measures depends on timing and adherence, not just on their formal announcement. A late or weakly complied-with intervention can have a smaller effect than an earlier, more targeted action that people understand and accept.

Governments also have to decide how to combine multiple interventions over time. A common mistake in public debate is to evaluate one policy at a time, as if epidemics respond to single switches. In reality, policies interact. Improving access to testing increases the value of isolation guidance, because people who cannot confirm infection are less likely to isolate. Financial support increases compliance with quarantine and reduces the incentive to hide symptoms or continue working while infectious. Communication policies affect every other policy because they shape whether people recognize a recommendation as credible, relevant, and feasible. For this reason, the effectiveness of governance is often less about choosing a perfect measure and more about assembling a coherent package that fits the local context and can be executed.

Execution is the hard part. A policy written on paper must become a set of actions performed repeatedly by many actors: health workers, school administrators, employers, local government officials, border agents, and households. Implementation failures commonly arise from capacity constraints and coordination problems. Testing may be authorized but not accessible. Isolation may be recommended but not supported with paid leave or food delivery. Contact tracing may be attempted but overwhelmed when case numbers rise. Hospitals may have beds but lack staff. These gaps matter because pathogens exploit inconsistency. If a small fraction of infectious people cannot isolate, or if high-risk workplaces remain unprotected, transmission continues and can rebound even after strong initial progress. In this context, a further governance responsibility is balancing public health goals with other societal goals while maintaining ethical standards. Pandemic policies can constrain liberty, disrupt education, and impose unequal burdens. Governments must therefore justify why intrusive measures are necessary, ensure that measures are proportionate to the risk, and build safeguards against misuse. They must also protect vulnerable groups who may have fewer resources to comply or who face greater exposure risk, such as people in crowded housing, those with precarious work, and those with limited access to health care. Equity is not only a moral consideration; it affects epidemic outcomes. If high-risk groups cannot protect themselves, transmission remains high and can spill over into the broader population.

Resource governance becomes especially visible during a pandemic. When demand spikes, markets alone often cannot allocate scarce supplies in ways that minimize transmission and mortality. Governments may need to coordinate procurement and distribution of diagnostics, protective equipment, treatments, and vaccines; expand manufacturing; and manage logistics such as cold chains and last-mile delivery. They also play a key role in regulation, including emergency authorization of new diagnostics and vaccines while maintaining safety standards and monitoring adverse events. These functions are frequently invisible when they work well, but they are decisive when they fail. Delays in procurement or distribution can shift the epidemic curve by weeks, and in a growing epidemic, weeks matter.

Pandemics also expose the limits of purely national governance. Pathogens cross borders, and so do the consequences of policy choices. International norms and institutions exist to facilitate reporting, coordination, and mutual assistance, but they depend on cooperation by states. Global governance also matters for access to countermeasures. Vaccines and treatments are often developed and manufactured in a small number of countries, which creates geopolitical and economic pressures when supply is limited. Mechanisms for pooled procurement and equitable allocation attempt to reduce these disparities, but they face political constraints because governments are accountable first to their own populations. This tension between national accountability and global interdependence is a recurring feature of pandemics and helps explain why global solidarity is difficult precisely when it is most needed.

2.6 Interconnected dynamics

The biological, epidemiological, psychological, economic, and governance dimensions of a pandemic are best understood as a single coupled system rather than five, among other, separate topics. Each domain changes the conditions for the others, often through feedback loops, delays, and thresholds. What makes pandemics difficult to anticipate is not only that each component is complicated on its own, but that the components interact in ways that can amplify or dampen change, sometimes rapidly. Modeling is an attempt to represent this coupled system in a simplified form that is still useful for decision-making, but the simplification itself introduces uncertainty and disagreement about what should be included, how it should be represented, and what can reliably be inferred from available data [204,205].

Consider the biological aspect, within-host processes shape when infectiousness begins, how long it lasts, and how much it varies across individuals, while immunity and evolution reshape susceptibility and transmissibility over time. Those biological features become epidemiological only when they meet a contact structure: who meets whom, where, and for how long [140,154]. Contact structures are not static, however, because people respond psychologically to perceived risk and socially to what they see others doing. As cases rise, fear, concern for family, social norms, and trust in information sources can change behavior, reducing or redirecting contact opportunities. Those behavioral shifts then change transmission, which changes case counts, which feeds back into perceptions of risk [133,135]. Economic constraints determine the strength and direction of this behavioral feedback. A worker with secure paid leave can isolate when symptomatic; a worker without it may continue working despite understanding the risk. In this situation, transmission is not merely a biological probability but a consequence of incentives and constraints. In economic-epidemiological terms, risk influences activity, activity influences risk, and neither can be treated as fixed [134,191].

Governance enters this loop as both a coordinator and a signal generator. Policies and public communication can change behavior directly by restricting settings, enabling testing and isolation, or improving indoor air, and indirectly by shaping beliefs about what is happening and what is expected. At the same time, governments respond to the epidemic as it unfolds, using delayed and imperfect indicators such as reported cases, hospitalizations, and deaths. This creates an additional feedback loop: data influence policy, policy influences behavior, behavior influences transmission, and transmission then appears in the data after lags introduced by incubation, testing, reporting, and clinical progression [206,207]. These delays can create a persistent sense that the system is always behind reality. Decisions may be made using information that reflects transmission from weeks earlier, while the true state of infection and behavior has already changed.

The same coupling that makes interventions possible also makes trajectories highly sensitive. In the early phase of spread, small differences in timing can matter disproportionately because exponential growth converts small initial gaps into large differences in later outcomes [208,209]. In addition, transmission is often uneven, with a minority of events or individuals contributing a large share of onward infections. This means that chance and local context can dominate what happens next, especially when case numbers are low: whether an early cluster happens to occur in a crowded indoor setting or in a low-contact environment can alter the apparent growth rate and the perceived urgency of response [160]. From a modeling standpoint, this creates a form of chaos in the everyday sense of the word: outcomes can change sharply due to small differences in context, timing, and behavior, even when the underlying pathogen has not changed.

These interdependencies create several practical modeling challenges. First, many of the most important variables are not directly observed. True infections are only partially measured by testing, and testing behavior changes with policy, access, and risk perception. Contact rates, adherence to isolation, and changes in work or travel are difficult to measure precisely and often require proxies that are imperfect and uneven across groups. Second, different combinations of assumptions can reproduce the same observed case curve, especially when data are sparse or biased. This means that matching historical data does not guarantee that a model has captured the correct mechanism, and it does not guarantee reliable extrapolation under new policies or new variants [204,210]. Third, model results must be communicated with their uncertainty. Uncertainty is not a minor technical detail. It arises from unknown parameters, structural choices about what mechanisms matter, measurement error, reporting delays, and the fact that human behavior and policy are adaptive. If uncertainty is not communicated clearly, decision-makers and the public may treat single numbers as precise predictions rather than conditional scenarios [205,211].

For these reasons, the most defensible role of models in pandemics is often not to predict a single future, but to map plausible futures under explicitly stated assumptions, to test what would have to be true for a policy to succeed, and to identify which uncertainties matter most for decisions. Models remain indispensable because they provide a disciplined way to reason about feedback loops that are otherwise invisible, but their usefulness depends on treating pandemics as coupled systems of biology, contact structure, psychology, economics, and governance, and on being explicit about what has been simplified away [212,213].

CHAPTER

Compartmental temporal models

3

Compartmental temporal models describe infectious disease transmission by partitioning a population into a (small) number of epidemiological states (also known as "compartments") and specifying how individuals flow between these states over time. This approach is deliberately reductive, as it replaces detailed individual-level histories with population-level bookkeeping, aiming to capture *dominant mechanisms* that drive incidence, peaks, and persistence. Two modeling decisions define this family of models: state representation and temporal dynamics. State representation describes which compartments are used (e.g., Susceptible, Infectious, Recovered), and what biological or administrative meaning each carries. In a complementary manner, temporal dynamics describe how flows depend on time and current state sizes, typically through deterministic ordinary differential equations (ODEs), discrete-time difference equations, or stochastic jump processes [214–216].

The simplest and most influential example of a compartmental model is the Susceptible-Infected-Recovered (SIR) framework, which partitions a host population into three states: susceptible individuals who can acquire infection, infectious individuals who can transmit the pathogen, and recovered (sometime also called "removed") individuals who are no longer contributing to transmission. The last class typically includes those who have recovered with immunity, those who have been isolated, and those who have died, depending on the disease context and modeling objective. At any time t, the corresponding compartment sizes are denoted by $S(t)$, $I(t)$, and $R(t)$, respectively. In its classical closed-population formulation, the SIR model rests on several simplifying assumptions that jointly define its behavior. First, individuals are assumed to mix homogeneously (also known as the "well-mixture" assumption), meaning that every individual is equally likely to contact every other individual at any given time. Second, new infections arise through mass-action incidence, so that the rate of transmission is proportional to the product of the numbers of susceptible and infectious individuals. Third, the total population size is assumed to remain constant over the modeled time horizon, excluding births, deaths unrelated to the disease, and migration. Fourth, recovery is assumed to confer permanent immunity, so that individuals do not return to susceptibility. Finally, the duration of infectiousness is assumed to be exponentially distributed, implying a constant per-capita recovery rate. These assumptions should not be interpreted as literal descriptions of reality; rather, they define a transparent baseline against which more detailed and realistic extensions can be assessed [217–219].

Under these assumptions, the dynamics of the SIR model in continuous time are governed by the following system of ordinary differential equations (ODEs):

$$\frac{dS}{dt} = -\beta S I, \tag{3.1}$$

$$\frac{dI}{dt} = \beta S I - \gamma I, \tag{3.2}$$

Epidemic Modeling. https://doi.org/10.1016/B978-0-44-355525-1.00009-6

$$\frac{dR}{dt} = \gamma I, \tag{3.3}$$

where the parameter β is the transmission rate and aggregates both the frequency of contacts and the probability of transmission per contact across all relevant interaction contexts and the parameter γ is the per-capita recovery or removal rate, so that under the exponential assumption, the mean infectious period is given by $1/\gamma$. We denote $N := S + I + R$ to be the total population size, which is consistent at all times.

In Eqs. (3.1)–(3.2) the incidence term βSI corresponds to what is often called the density-dependent form, reflecting the rate or probability an infected individuals infects susceptible individuals. In contrast, some applications adopt frequency-dependent incidence, reflecting the assumption that individuals make a fixed number of potentially infectious contacts per unit time, independent of population size, and formally write $\beta SI/N$. The distinction between these formulations is not merely technical, as it determines the scaling of transmission with population size and the interpretation and units of the transmission parameter, and it should be chosen to reflect the underlying contact process [214]. Fig. 3.1 presents a schematic view of the SIR model. The boxes indicates the states of the model, and the arrows the potential transitions between them.

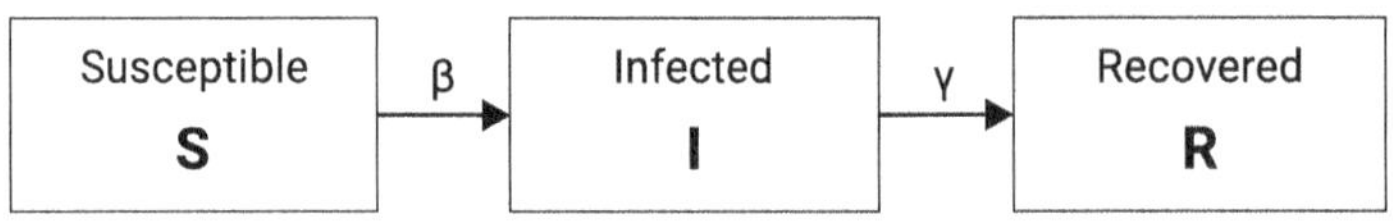

FIGURE 3.1

A schematic view of the SIR model, where susceptible individuals become infected and afterward recovered over time.

To complement the analytical formulation, it is often useful to compute numerical solutions of the SIR system. Below is a minimal Python implementation that evaluates the right-hand side of the SIR equations and integrates them forward in time using a standard ODE solver.

```
import numpy as np
from scipy.integrate import solve_ivp

def sir(t, y, beta, gamma):
    S, I, R = y
    dSdt = -beta * S * I / N
    dIdt =  beta * S * I / N - gamma * I
    dRdt =  gamma * I
    return [dSdt, dIdt, dRdt]

# parameters
beta = 0.3
gamma = 0.1

# initial conditions
S0, I0, R0 = 0.99, 0.01, 0.0
y0 = [S0, I0, R0]
```

```

# time grid
t_span = (0, 160)
t_eval = np.linspace(*t_span, 500)

# solve the system
solution = solve_ivp(
    sir_rhs,
    t_span,
    y0,
    args=(beta, gamma),
    t_eval=t_eval
)
S, I, R = solution.y
```

Listing 3.1: Python implementation of the SIR model solving the system numerically.

This code mirrors Eqs. (3.1)–(3.3) directly. The function `sir` defines the vector field of the SIR system, taking as input the current time t, the state vector (S, I, R), and the parameters β and γ. The call to `solve_ivp` numerically integrates the system from the specified initial condition over a chosen time interval, returning the full time series of the three compartments. When initialized with nonnegative states summing to one, the numerical solution preserves the epidemiological interpretation of the model, producing the characteristic depletion of susceptibles, rise and fall of infection, and monotone accumulation of removed individuals. Fig. 3.2 presents a plot based on the code's output.

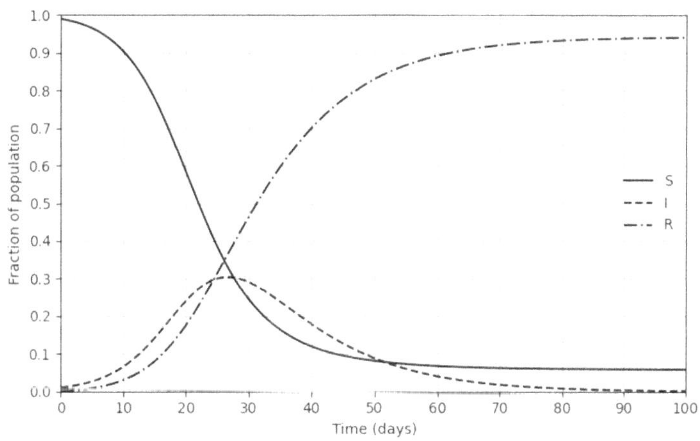

FIGURE 3.2

A numerical solution of the SIR model for $S(0) = 0.99$, $I(0) = 0.01$, $R(0) = 0$ and $\beta = 0.3$, $\gamma = 0.1$.

A central quantity emerging from the SIR model is the basic reproduction number, denoted R_0, which is defined as the expected number of secondary infections caused by a typical infectious individual introduced into an otherwise fully susceptible population [220]. For the closed-population SIR system defined by Eqs. (3.1)–(3.3), and under the approximation that $S \approx N$ early in an outbreak, the basic reproduction number takes the simple form

$$R_0 = \frac{\beta}{\gamma}. \tag{3.4}$$

This quantity determines the epidemic threshold. From Eq. (3.2), the number of infectious individuals initially increases when

$$\frac{S(t)}{N} > \frac{1}{R_0}. \tag{3.5}$$

Thus, sustained epidemic growth is possible only when the susceptible fraction exceeds a critical value given by the inverse of R_0. If this condition is not met, introductions of infection lead only to minor outbreaks that quickly fade out [214]. As the epidemic progresses and susceptible individuals are depleted, the relevant quantity becomes the effective reproduction number,

$$R_t = R_0 \frac{S(t)}{N}. \tag{3.6}$$

This time-dependent reproduction number decreases monotonically as susceptibility declines and crosses below unity at the epidemic peak, marking the transition from increasing to decreasing prevalence [138].

An essential consistency requirement of the SIR model is that the state variables $S(t)$, $I(t)$, and $R(t)$ represent population counts or population fractions and therefore must remain nonnegative at all times. Negative values in any compartment have no epidemiological interpretation, since a negative number of susceptible, infectious, or recovered individuals is meaningless and indicates either an invalid model formulation or a numerical or analytical error. For this reason, any solution of the SIR equations that leaves the nonnegative region should be regarded as epidemiologically inadmissible. Importantly, the continuous-time SIR system preserves nonnegativity provided it is initialized with valid conditions. Suppose that at the initial time $t = 0$, the state satisfies $S(0) \geq 0$, $I(0) \geq 0$, and $R(0) \geq 0$. From Eq. (3.1), the derivative of $S(t)$ is nonpositive whenever $S(t) \geq 0$ and $I(t) \geq 0$, implying that $S(t)$ can decrease but cannot cross below zero. From Eq. (3.3), the derivative of $R(t)$ is nonnegative whenever $I(t) \geq 0$, so the removed compartment is monotone increasing and cannot become negative. Finally, when $I(t) = 0$, Eq. (3.2) reduces to $\frac{dI}{dt} = 0$, showing that the infectious compartment cannot cross from zero into negative values. Together, these observations imply that the nonnegative orthant is a forward-invariant region of the SIR dynamics: any solution starting with epidemiologically meaningful initial conditions remains epidemiologically meaningful for all future times. This invariance property underlies the internal consistency of the SIR model and is a basic requirement for both analytical work and numerical simulation.

Despite its simplicity, the SIR model yields two qualitative results that extend far beyond its original assumptions. First, epidemics typically terminate while a substantial fraction of the population remains susceptible. The decline of infection is driven not by the exhaustion of all susceptibles, but by the reduction of the susceptible pool below the level required to sustain transmission. Second, the total epidemic size is governed by a transcendental relationship rather than by the timing of the epidemic peak. By eliminating time between Eqs. (3.1) and (3.3), one obtains the classic final-size equation. Writing $s(t) = S(t)/N$ and letting s_∞ denote the limiting susceptible fraction as $t \to \infty$, this relationship can be expressed as

$$\log\left(\frac{s_\infty}{s_0}\right) = -R_0\left(1 - s_\infty\right), \tag{3.7}$$

which implicitly determines the total attack rate $1 - s_\infty$ given the basic reproduction number and the initial conditions [110,214]. These properties are often used as consistency checks when developing extensions of the SIR framework; departures from them typically signal either a deliberate modeling choice or a potential mis-specification.

In practical applications, the SIR compartments are rarely observed directly. Surveillance data usually consist of reported cases, hospitalizations, or deaths, which are indirect and often biased proxies for the underlying infection process. Reported incidence reflects only a fraction of true infections and is shaped by testing availability, reporting delays, and clinical case definitions. Consequently, applied SIR models commonly incorporate an explicit observation layer that links latent model states to observed data. Such observation models may include reporting fractions applied to true incidence, delay distributions implemented through convolution kernels, and stochastic noise models, often with negative binomial overdispersion. While these components do not alter the mechanistic flows of the SIR system itself, they are essential for reliable inference and realistic uncertainty quantification [221,222]. We will cover such methods later in this book.

3.1 States extensions

The SIR framework can be viewed as a minimal model capturing the core mechanism of susceptible depletion in a closed population. Most extensions of the model arise by relaxing one or more of its simplifying assumptions, and it is therefore useful to adopt a systematic perspective on how such extensions are constructed. One common class of extensions introduces additional biological stages into the infection process, such as latent or exposed states, presymptomatic phases, or asymptomatic infection, in order to better represent within-host progression and delays between infection and infectiousness. A second class augments the set of possible outcomes or fates of infected individuals, for example, by explicitly modeling hospitalization, severe disease, or death as distinct compartments rather than subsuming them into a generic removed class. Other extensions introduce new time scales into the system, allowing demographic turnover through births and deaths, waning immunity with return to susceptibility, or externally driven temporal variation such as seasonality in transmission. Further generalizations relax the assumption of homogeneous mixing by incorporating population structure, including stratification by age, risk group, or spatial location, often resulting in coupled systems of compartmental equations. Below, we analyze each of these state extensions.

3.1.1 Demography – birth and natural death

When epidemics are considered over longer temporal horizons, demographic turnover becomes an essential component of transmission dynamics. Births continuously replenish the susceptible pool, while natural deaths remove individuals from all epidemiological states. This demographic renewal fundamentally alters the qualitative behavior of the system by allowing infection to persist indefinitely at a constant level, giving rise to endemic equilibria rather than the single-wave burnout characteristic of the closed-population SIR model [214]. In such settings, the epidemic is no longer a transient event but a long-run population process maintained by the balance between transmission, recovery, and demographic turnover. A commonly used formulation incorporates a per-capita birth and natural death rate ν, with all births entering the susceptible compartment. The resulting dynamics are described by

$$\frac{dS}{dt} = \underbrace{\nu N}_{\text{births}} - \beta S I - \underbrace{\nu S}_{\text{natural death}}, \tag{3.8}$$

$$\frac{dI}{dt} = \beta S I - \gamma I - \underbrace{\nu I}_{\text{natural death}}, \tag{3.9}$$

$$\frac{dR}{dt} = \gamma I - \underbrace{\nu R}_{\text{natural death}}, \tag{3.10}$$

where the total population size N remains constant provided that births exactly balance deaths. Compared to the closed SIR system, demographic turnover introduces an additional exit route from every compartment and a continual inflow into susceptibility, thereby preventing permanent depletion of the susceptible pool. Fig. 3.3 presents a schematic view of the SIR with demographics model.

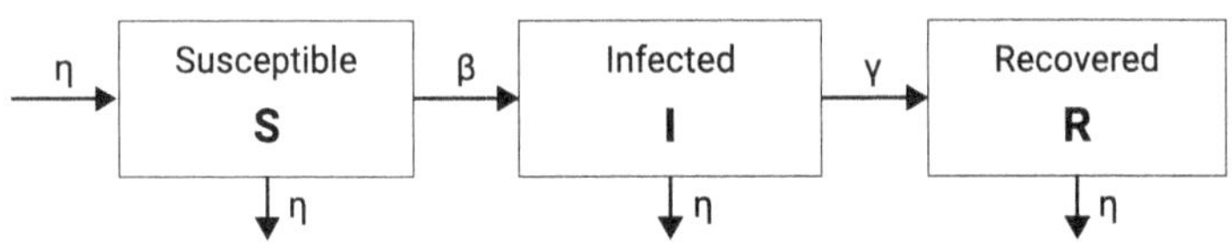

FIGURE 3.3

A schematic view of the SIR model with demographics dynamics.

The presence of natural mortality modifies the effective infectious period. Infectious individuals now leave the infectious compartment either through recovery or through death, so the average duration of infectiousness becomes $1/(\gamma + \nu)$. As a consequence, the basic reproduction number takes the form

$$R_0 = \frac{\beta}{\gamma + \nu}. \tag{3.11}$$

This expression highlights that demography reduces transmission potential by shortening the expected time during which an infected individual can generate secondary cases. The epidemic threshold retains its familiar interpretation: if $R_0 \leq 1$, the disease-free equilibrium is globally attracting, and infection cannot persist, whereas if $R_0 > 1$, a unique endemic equilibrium exists with strictly positive prevalence $I^* > 0$ [214,223]. At endemic equilibrium, the susceptible fraction converges to $S^*/N = 1/R_0$, a result that mirrors the epidemic threshold condition in the closed SIR model, but now holds as a long-run balance rather than a transient turning point. The remaining population is split between the infectious and recovered compartments, with equilibrium prevalence determined by the balance between demographic inflow and removal through recovery and death. This equilibrium interpretation provides an intuitive understanding of endemicity where susceptibility is continually replenished by births at exactly the rate required to sustain ongoing transmission.

The inclusion of demography also introduces a clear separation of time scales. Transmission and recovery typically operate on the scale of days or weeks, whereas births and deaths occur over years or decades. When these scales are well separated, short-term outbreak dynamics may resemble those of the closed SIR model, while long-term behavior converges toward the endemic equilibrium. In contrast, when demographic turnover is rapid relative to infection dynamics, demographic effects shape both transient and equilibrium behavior, blurring the distinction between epidemic waves and endemic persistence.

From a modeling perspective, demographic SIR models form a conceptual bridge to related frameworks. In the limit where recovery confers no lasting immunity, and individuals return to susceptibility, the model approaches an SIS-type structure, while the inclusion of waning immunity yields SIRS dynamics with qualitatively similar endemic behavior. Vaccination can also be interpreted naturally in this setting as reducing the effective inflow into the susceptible compartment, leading to a modified threshold condition and a critical vaccination coverage analogous to the herd immunity threshold derived in the closed-population case.

Although demographic processes are often negligible for acute outbreaks such as influenza pandemics or emerging pathogens, they are indispensable for understanding childhood infections, immunizing diseases, and long-run population-level patterns. Importantly, the inclusion of demographic turnover can change the interpretation of parameters inferred from data. If the observation window is long enough that births and deaths are non-negligible, estimates of β and γ obtained under a closed-population assumption may be biased, as part of the observed removal from infection is attributable to mortality rather than recovery. In practice, demographic parameters such as μ can often be fixed using external demographic data, reducing the inferential burden and improving identifiability of transmission-related parameters [118].

In the SIR with demographics model, endemicity is naturally understood through the geometry of the flow in the (S, I) plane, as presented in Fig. 3.4a, where the nullcline $dI/dt = 0$ consists of the axis $I = 0$ together with the vertical line $S = (\gamma + \mu)/\beta$, while the nullcline $dS/dt = 0$ forms a curve determined by the balance between births and loss of susceptibility through infection and natural death. Notably, when $R_0 = \beta/(\gamma + \mu) > 1$, these nullclines intersect at a unique interior equilibrium (S, I) with $I^* > 0$, and trajectories from a wide range of initial conditions converge to this endemic point rather than returning to the disease-free state. This provides a direct dynamical explanation for why demographic renewal prevents permanent depletion of susceptibles and sustains long-run transmission.

Moreover, the separation between epidemiological and demographic time scales is illustrated in Fig. 3.4b. For small μ (slow turnover), the initial dynamics closely resemble the closed SIR model: $I(t)$ exhibits an outbreak peak followed by a long decline as susceptibles are depleted. However, because births continuously replenish susceptibility, the decline does not proceed to extinction; instead, the system gradually approaches a nonzero endemic level. When μ is larger (faster turnover), demographic renewal operates on a comparable time scale to infection and recovery, so the approach to endemic equilibrium is more rapid, and the distinction between a single epidemic wave and long-run persistence becomes less pronounced. In both regimes, the long-run limit is governed by the endemic balance implied by $R_0 > 1$, but the transient behavior—and hence what is observed over finite study windows—depends strongly on the magnitude of μ.

3.1.2 Reinfection

Reinfection becomes relevant whenever infection does not place individuals into a permanently protected state on the time scale of interest. In many host-pathogen systems, immune protection is

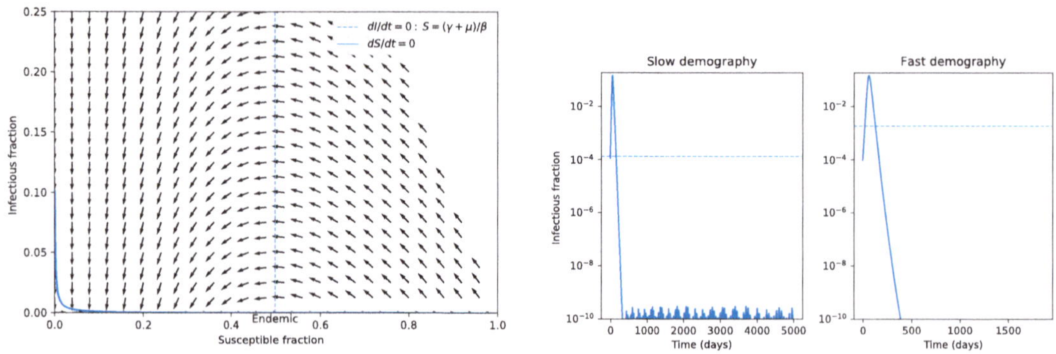

(a) Phase plane and nullclines in the (S, I) plane. (b) Slow vs. fast demographic turnover.

FIGURE 3.4

Geometric and temporal effects of demographic turnover in the SIR with demographics model. Left: trajectories converge to a unique endemic equilibrium when $R_0 > 1$ (with nullclines shown). Right: the same transmission and recovery rates can exhibit closed-SIR-like short-run outbreak dynamics while converging to a long-run endemic level, with the rate of convergence controlled by μ.

incomplete (reducing severity rather than preventing infection), short-lived, strain-specific, or highly heterogeneous across individuals [224,225]. From a modeling standpoint, these realities can be represented in two common ways that form a useful spectrum of assumptions. At one extreme, infection confers no meaningful immunity, and individuals become susceptible again immediately after recovery. On the other hand, infection confers temporary immunity that wanes over time, so reinfection is possible but delayed. These two options correspond naturally to the SIS and SIRS models, respectively, and they often provide a parsimonious alternative to adding demographic turnover when the primary driver of persistence is repeated infection rather than births.

The SIS model is appropriate when prior infection does not produce durable protective immunity against reinfection, or when immunity is so weak that it can be ignored at the modeled horizon. In this setting, "recovered" individuals are best interpreted as returning directly to susceptibility: the population is always partitioned into susceptible and infectious states, and individuals can experience many infection episodes over their lifetime. This structure is commonly used for bacterial colonization or infections where immune memory is limited, and reinfection is frequent. A realistic example is *gonorrhea*, where infection does not reliably induce long-lasting protective immunity, and reinfections are common in the same individuals [226]. Another frequent use is for hospital-acquired colonization when clearance does not imply protection and individuals can be recolonized repeatedly. In such cases, the central question is whether transmission can sustain itself via a steady circulation between susceptibility and infection.

In a closed population with size N and no recovery class, we have $S + I = N$ and the dynamics reduce into one equation:

$$\frac{dI}{dt} = \beta \underbrace{(N - I)}_{\text{susceptible}} I - \gamma I, \tag{3.12}$$

where β is the effective transmission rate and γ is the recovery (or clearance) rate, like in the SIR model. The SIS model always admits a disease-free equilibrium $I^* = 0$. When the basic reproduction number $R_0 = \beta/\gamma$ satisfies $R_0 > 1$, there is also a stable endemic equilibrium that takes the form:

$$I^* = N\left(1 - \frac{1}{R_0}\right), \tag{3.13}$$

indicating persistent circulation without the need for births. Unlike SIR, there is no absorbing recovered compartment and therefore no "final size" concept, since persistence is the default long-run outcome whenever transmission exceeds the threshold [214]. Conceptually, endemicity in SIS arises because infections continuously regenerate susceptibility through recovery, creating a self-sustaining feedback loop. Fig. 3.5 presents a schematic view of the SIS model.

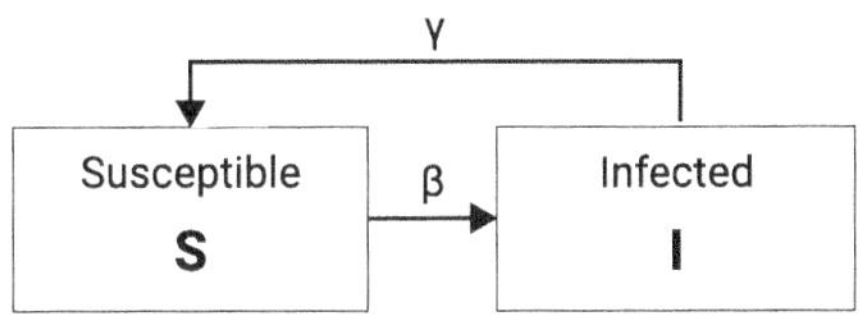

FIGURE 3.5

A schematic view of the SIS model.

However, with a waning-immunity flow $R \to S$, the model becomes

$$\frac{dS}{dt} = -\beta SI + \underbrace{\omega R}_{\text{immune decayed}}, \tag{3.14}$$

$$\frac{dI}{dt} = \beta SI - \gamma I, \tag{3.15}$$

$$\frac{dR}{dt} = \gamma I - \underbrace{\omega R}_{\text{immune decaying}}, \tag{3.16}$$

where $1/\omega$ is the mean duration of immunity in the model. This single additional timescale controls how strongly the system departs from SIR and SIS behaviors. When waning is slow (i.e., small ω value), short term dynamics can resemble an SIR outbreak—incidence rises and falls as susceptibles are depleted—yet the eventual return of immune individuals to susceptibility enables later resurgence. When waning is fast (large ω value), the immune class becomes short-lived, and the model approaches SIS-like behavior in which endemicity is maintained primarily by rapid recycling back into S but with "delay". Fig. 3.6 presents a schematic view of the SIRS model.

To this end, both SIS and SIRS can support endemic equilibria in a closed population, but the mechanisms are distinct. In SIS, the *renewal* of susceptibility is immediate via recovery, so persistence reflects a direct circulation between S and I. On the other hand, in SIRS, susceptibility is replenished more slowly via waning, so the system can exhibit multi-year oscillations or recurrent waves even with constant parameters, depending on the relative sizes of β, γ, and ω. From an epidemiological standpoint, SIS corresponds to reinfections that occur readily after clearance, while SIRS corresponds to reinfections that are possible but typically delayed by a finite period of protection.

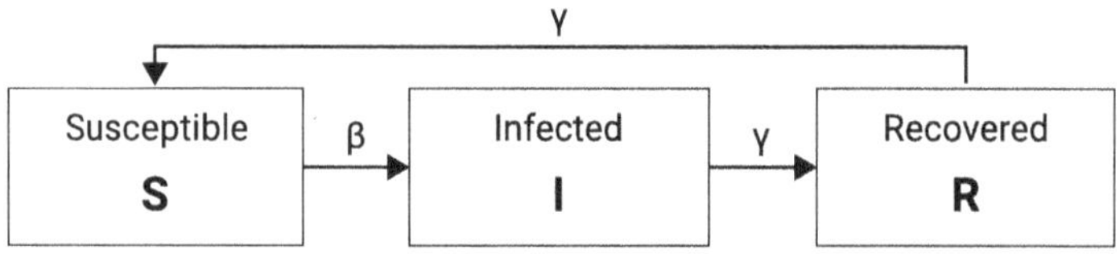

FIGURE 3.6

A schematic view of the SIRS model.

Notably, observed repeated waves do not by themselves uniquely identify reinfection mechanisms. SIS-like persistence can mimic recurrent epidemics only if reinfection is sufficiently rapid or if transmission parameters vary over time through seasonality, behavior change, interventions, or stochastic forcing. Conversely, SIRS can generate recurrent outbreaks even with constant parameters, but similar patterns can also arise from time-varying contact rates, immune escape, or reporting artifacts. In practice, distinguishing the *no immunity* from the *waning immunity* cases often requires information beyond case incidence alone – such as longitudinal serology, reinfection cohort studies, or immunological measurements that constrain the plausible range of $1/\omega$ [221]. Thus, when such evidence is available, it can substantially improve identifiability and prevent the model from attributing resurgence to waning immunity when the true driver is changing transmission conditions.

In order to illustrate the differences between SIS and SIRS, Fig. 3.7 contrasts SIS and SIRS dynamics under two transmission regimes. The *top row* uses a small transmission rate β, while the bottom row uses a large β; the recovery rate γ is held fixed, so changing β primarily changes $R_0 = \beta/\gamma$ and hence the force of infection. In the left column (SIS), where no protective immunity is present, individuals return directly to susceptibility after recovery, allowing infection to recycle continuously and approach a persistent endemic level even when transmission is only modestly above threshold. On the other hand, in the right column (SIRS), where there is long-lasting but temporary immunity, implemented by a low waning rate ω, the dynamics more SIR-like on shorter horizons as prevalence can drop to very low levels after an initial wave and then rebound only gradually as immunity wanes and susceptibles are replenished. The contrast is most pronounced in the aggressive case (bottom row): SIS rapidly settles into sustained circulation at relatively high prevalence, whereas SIRS produces sharper boom–bust behavior with a deeper post-wave trough and slower re-emergence driven by waning immunity.

3.1.3 Latent infection

Many pathogens exhibit a non-negligible delay between the acquisition of infection and the onset of infectiousness. For example, influenza and SARS-CoV-2 often have short latent periods on the order of a few days, where individuals may begin transmitting before clear symptoms appear [227,228]. Measles also illustrates this phenomenon, with infectiousness typically starting several days before the characteristic rash, following a longer incubation period [229]. Other infections can involve much longer or more variable delays, like varicella (chickenpox), which generally becomes infectious about one to two days before rash onset after a roughly 10–21 day incubation period, while tuberculosis may involve infection persisting for months or years before a person becomes infectious (and some never do) [230,231].

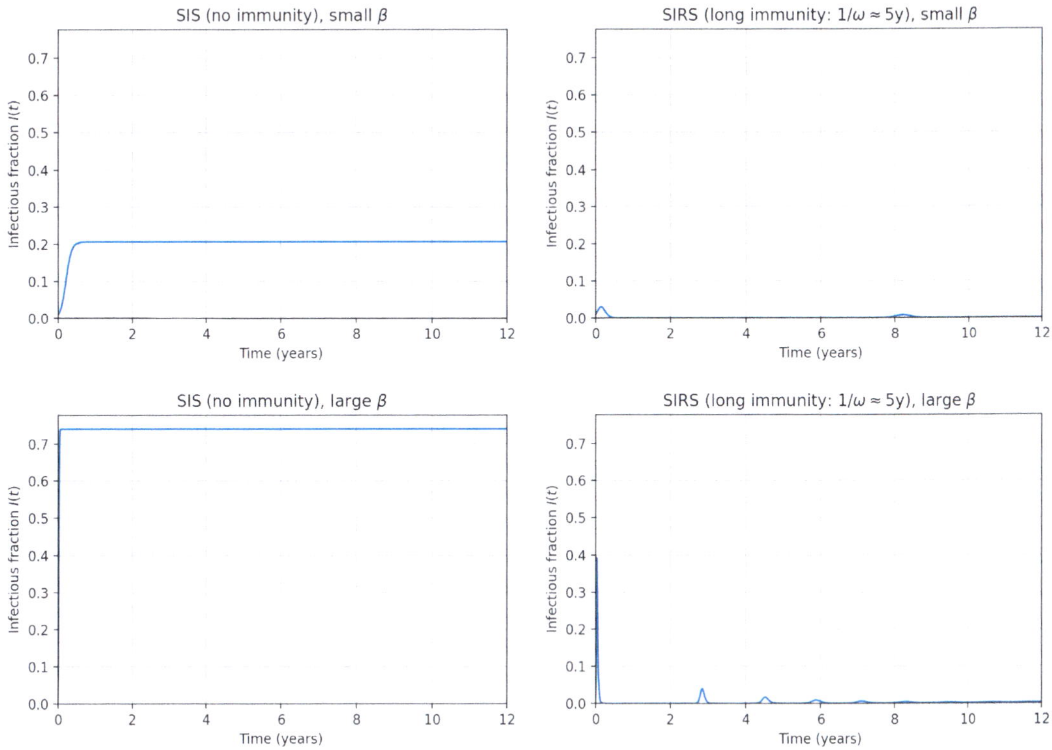

FIGURE 3.7

SIS (no immunity) versus SIRS (long-lasting but temporary immunity) under mild and aggressive transmission. The top row uses a small transmission rate $\beta = 0.18$ (non-aggressive outbreak) and the bottom row uses a large $\beta = 0.55$ (aggressive outbreak), with $\gamma = 0.14$ held fixed. The left column (SIS) returns recovered individuals directly to susceptibility, producing sustained circulation and convergence toward an endemic level. The right column (SIRS) uses slow waning (small $\omega = 1/(365 \cdot 5)$, here corresponding to long immunity), which delays replenishment of susceptibles and yields more SIR-like boom–bust dynamics: prevalence can drop to very low levels after the initial wave and later rebound as immunity wanes.

The standard SIR model effectively assumes that individuals become infectious immediately upon infection, an approximation that can misrepresent early epidemic growth and shift the timing and height of the incidence peak when the latent period is comparable to the infectious period [214]. Introducing a latent state is therefore often essential when modeling such infections, where individuals typically experience a period of incubation/latency before they begin to transmit. For that purpose, one can add a latent state between susceptible and infected individuals, often called "exposed" (E) individuals, which are already infected but not infectious yet. In this SEIR model, newly infected individuals enter the exposed compartment ($E(t)$). They progress to the infectious compartment at a rate δ, so that $1/\delta$ is the mean latent period. A standard formulation is

$$\frac{dS}{dt} = -\beta S I, \tag{3.17}$$

$$\frac{dE}{dt} = \underbrace{\beta SI - \delta E}_{\text{infected and becoming infectious}}, \tag{3.18}$$

$$\frac{dI}{dt} = \underbrace{\delta E}_{\text{becoming infectious}} - \gamma I, \tag{3.19}$$

$$\frac{dR}{dt} = \gamma I. \tag{3.20}$$

Fig. 3.8 presents a schematic view of the SEIR model.

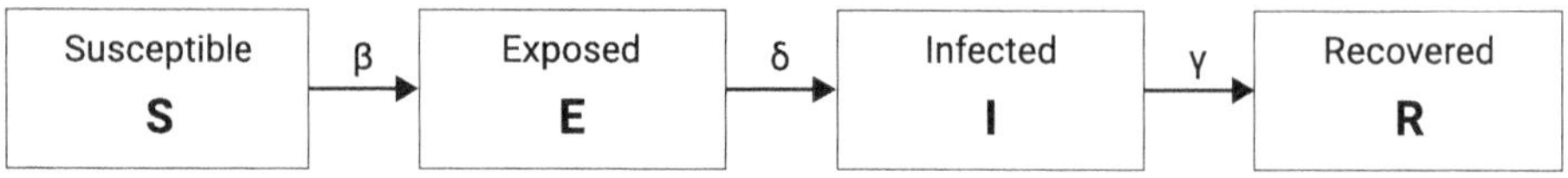

FIGURE 3.8

A schematic view of the SEIR model.

As in SIR, transmission is generated by I at rate $\beta SI/N$, and infectious individuals recover at rate γ. Notably, when infectiousness occurs only in the I compartment, the basic reproduction number in a fully susceptible population remains $R_0 = \frac{\beta}{\gamma}$, because the expected number of secondary infections is still governed by the per-capita infection rate β acting over an average infectious duration $1/\gamma$. However, while R_0 is unchanged by the introduction of E, the early exponential growth rate does depend on δ. Linearization around the disease-free state yields the characteristic equation

$$(r + \delta)(r + \gamma) = \beta\delta, \tag{3.21}$$

demonstrating that the mapping from an observed initial growth rate r to an implied R_0 depends on the latent-period time scale. This reflects a more general principle: reproduction numbers and epidemic growth are linked through the *generation-interval structure* of transmission, not merely through mean infectious duration [232].

From a practical standpoint, the SEIR structure is often crucial when fitting to incidence curves that show a clear delay between infection and onward transmission, or when the observation process is itself delayed (e.g., symptom onset and testing). Nevertheless, if only case reports are observed, the latent compartment can become statistically confounded with reporting delays, because both create similar shifts between true infection events and observed counts. For this reason, careful separation of biological latency (infection $\rightarrow$ infectiousness) from observation latency (infection $\rightarrow$ detection/report) is critical when interpreting fitted δ values [221].

Notably, the SEIR model assumes that individuals leave the exposed class at a constant per-capita rate δ, implying an exponentially distributed latent period, similar to the SIR model. The same "memoryless" assumption appears in the infectious compartment via γ. Empirically, however, latent and infectious periods are often better represented by distributions with less variance than the exponential one. Specifically, incubation and infectiousness durations are frequently more concentrated around their mean.

Thus, from a modeling perspective, a standard extension that preserves a Markovian ODE framework while improving realism is to replace a single compartment (e.g., E) with multiple sequential stages, yielding an Erlang (gamma) distribution for the residence time [233]. This construction can be viewed as a direct refinement of SEIR, as it keeps the same biological interpretation (infection → latency → infectiousness → recovery) but relaxes the overly variable exponential assumption.

Formally, we replace E with k exposed stages $E_1, \ldots, E_k$ with equal progression rates $k\sigma$:

$$\frac{dS}{dt} = -\beta S I, \tag{3.22}$$

$$\frac{dE_1}{dt} = \underbrace{\beta S I - k\sigma E_1}_{\text{initial infection}}, \tag{3.23}$$

$$\frac{dE_j}{dt} = \underbrace{k\sigma(E_{j-1} - E_j), \quad j = 2, \ldots, k,}_{\text{latent phase transition}} \tag{3.24}$$

$$\frac{dI}{dt} = \underbrace{k\sigma E_k}_{\text{latent to infectious}} - \gamma I, \tag{3.25}$$

$$\frac{dR}{dt} = \gamma I. \tag{3.26}$$

This preserves the mean latent period at $1/\sigma$ (since an Erlang distribution with shape k and rate $k\sigma$ has mean $1/\sigma$), while reducing the variance as k increases. Intuitively, a larger k makes the latent period more tightly concentrated around its mean, thereby better matching incubation distributions observed in many diseases. The same idea can be applied to the infectious compartment by splitting I into stages, enabling more realistic infectious-period distributions without changing the overall modeling framework. Importantly, although this formalization is appealing from a representational standpoint, it complicates numerical computation by substantially increasing the number of equations that must be solved [234,235].

These distributional refinements can have measurable dynamical consequences. Even when the mean latent and infectious durations are held fixed, changing the *shape* of the residence-time distribution can alter three properties: first, the timing and sharpness of epidemic peaks; second, the relationship between the growth rate r and R_0; and finally, the tendency to generate or damp oscillations in endemic settings [233]. Specifically, models with more concentrated residence times often produce more coherent wavefronts and can shift the inferred R_0 needed to match an observed initial growth rate. Thus, answering the question "how long people stay latent or infectious" is not only about the mean duration but also the variance (and more broadly, the distributional shape) can matter for transient and long-run dynamics.

From a data-analysis perspective, stage-structured models frequently improve fit to epidemic curves without resorting to ad hoc time variation in β. The tradeoff is increased model dimension (additional state variables) and, if one introduces stage-specific parameters or allows heterogeneity across stages, potentially weaker identifiability from incidence time series alone. In practice, parsimonious choices (e.g., modest k with shared rates) often provide a good balance between biological realism and statistical tractability, especially when supported by external information on incubation or infectiousness distributions.

In order to illustrate the computationally convenient *stage* approximation of a fixed latent delay, Fig. 3.9 compares the infectious prevalence $I(t)$ produced by an SEIR model in which the exposed compartment is split into k sequential stages $(E_1, \ldots, E_k)$ while keeping the mean latent period fixed. As k increases, the implied latent-time distribution becomes more concentrated (approaching a deterministic delay), which shifts and reshapes the epidemic curve relative to the single-stage model.

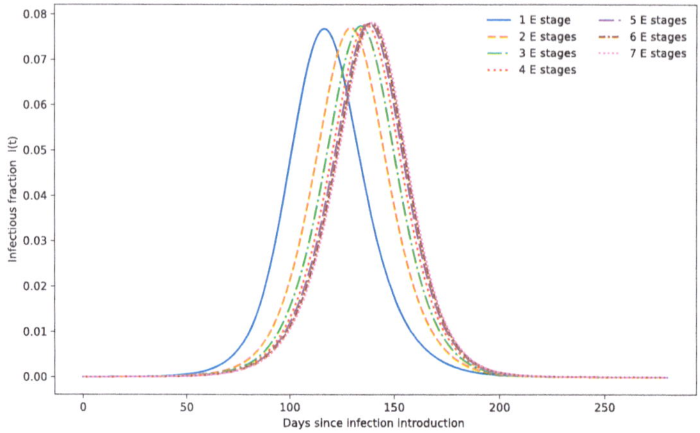

FIGURE 3.9

Infectious prevalence over time, $I(t)$, under an SEIR model with an Erlang (stage-structured) latent period. The exposed state is represented by $k \in \{1, \ldots, 7\}$ sequential compartments $(E_1, \ldots, E_k)$ with identical transition rate $\sigma = k/\bar{\tau}$, so that the mean latent period remains fixed at $\bar{\tau}$ while its variance decreases with k.

3.1.4 Multiple infectious classes

Transmission potential can differ markedly by clinical presentation and by time since infection. For many respiratory pathogens, infectiousness may peak before symptom onset, and individuals who remain asymptomatic can still contribute to spread through continued community contact. Contrariwise, severe cases may reduce contacts due to illness, hospitalization, or isolation, and may be detected more reliably by surveillance. A single infectious compartment I therefore conflates heterogeneous contact patterns, heterogeneous per-contact infectiousness, and heterogeneous detection probabilities, which can bias interpretation and intervention evaluation [150].

Thus, a common modeling pattern is to split infectiousness into subclasses. For example, presymptomatic infectious (I_p), asymptomatic infectious (I_a), and symptomatic infectious (I_s). Each of these stages is associated with stage-specific transmission coefficients, stage durations (progression rates), and potentially different recovery/removal rates. A simple flow is $S \to E \to I_p \to I_s \to R$ or $S \to E \to I_a \to R$, with a branching fraction π_a into the asymptomatic route. The force of infection becomes a weighted sum over infectious subclasses,

$$\lambda(t) = \frac{1}{N}\Big(\beta_p I_p(t) + \beta_a I_a(t) + \beta_s I_s(t)\Big), \qquad \frac{dS}{dt} = -\lambda(t)\, S. \tag{3.27}$$

To represent severity strata, one may further split I_s into mild vs severe (or community vs hospitalized) classes, e.g. I_m and I_h, with reduced community transmission for hospitalized individuals and altered infectious periods:

$$I_p \to I_m \to R, \qquad I_p \to I_h \to R, \qquad \beta_h \ll \beta_m,$$

and optionally an explicit isolation compartment Q capturing diagnosis and quarantine:

$$I_s \xrightarrow{\delta(t)} Q \to R,$$

where $\delta(t)$ can vary over time to encode improving testing and contact tracing.

As a concrete instance consistent with the discussion above, consider an SEIR-type model in which exposed individuals branch into asymptomatic vs. presymptomatic infection, presymptomatic individuals branch into mild vs. severe symptomatic infection, and symptomatic cases may be diagnosed and isolated. Let σ^{-1} denote the mean latent period, κ^{-1} the mean presymptomatic infectious period, π_a the asymptomatic fraction, and π_h the fraction of symptomatic infections that become "severe". Isolation occurs at rates δ_m, δ_h for mild and severe symptomatic classes, respectively. The force of infection aggregates contributions from all infectious subclasses,

$$\lambda(t) = \frac{1}{N}\Big(\beta_p I_p(t) + \beta_a I_a(t) + \beta_m I_m(t) + \beta_h I_h(t) + \beta_q Q(t)\Big), \tag{3.28}$$

where typically $\beta_h \ll \beta_m$ and $\beta_q \ll \beta_m$ reflect reduced community contact in hospital and isolation. Therefore, the full ODE system representation of this system takes the form

$$\frac{dS}{dt} = \underbrace{-\lambda(t)\,S}_{\text{new infections}}, \tag{3.29}$$

$$\frac{dE}{dt} = \underbrace{\lambda(t)\,S}_{\text{enter latent (exposed)}} - \underbrace{\sigma E}_{\text{latent progression}}, \tag{3.30}$$

$$\frac{dI_p}{dt} = \underbrace{(1-\pi_a)\sigma E}_{\text{latent} \to \text{presymptomatic (symptomatic route)}} - \underbrace{\kappa I_p}_{\text{progression to symptoms}}, \tag{3.31}$$

$$\frac{dI_a}{dt} = \underbrace{\pi_a \sigma E}_{\text{latent} \to \text{asymptomatic route}} - \underbrace{\gamma_a I_a}_{\text{recovery (asymptomatic)}}, \tag{3.32}$$

$$\frac{dI_m}{dt} = \underbrace{(1-\pi_h)\kappa I_p}_{\text{presymptomatic} \to \text{mild symptomatic}} - \underbrace{(\gamma_m + \delta_m) I_m}_{\text{recovery + diagnosis/isolation}}, \tag{3.33}$$

$$\frac{dI_h}{dt} = \underbrace{\pi_h \kappa I_p}_{\text{presymptomatic} \to \text{severe symptomatic}} - \underbrace{(\gamma_h + \delta_h) I_h}_{\text{recovery + diagnosis/isolation}}, \tag{3.34}$$

$$\frac{dQ}{dt} = \underbrace{\delta_m I_m + \delta_h I_h}_{\text{symptomatic} \to \text{isolated}} - \underbrace{\gamma_q Q}_{\text{removal from isolation}}, \tag{3.35}$$

$$\frac{dR}{dt} = \underbrace{\gamma_a I_a + \gamma_m I_m + \gamma_h I_h + \gamma_q Q}_{\text{all recoveries/removals}}. \tag{3.36}$$

Fig. 3.10 presents a schematic view of the $SEI_pI_aI_hQR$ model.

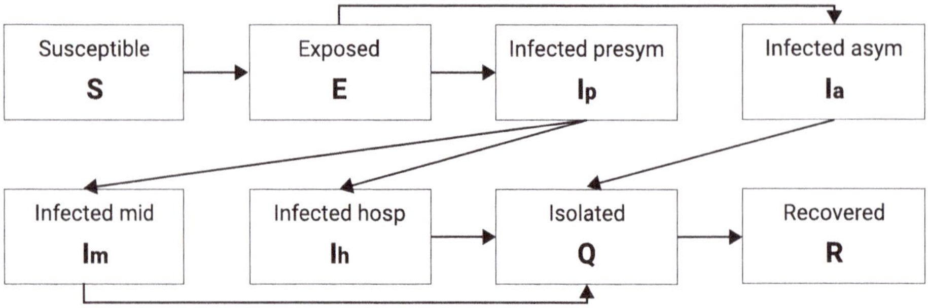

FIGURE 3.10

A schematic view of the $SEI_pI_aI_hQR$ model.

Fig. 3.11 illustrates this $SEI_pI_aI_hQR$ multi-class infectiousness model by showing (left) the time series of each infectious subclass and (right) an S–I_{tot} projection of trajectories, where $I_{\text{tot}}(t) := I_p(t) + I_a(t) + I_m(t) + I_h(t) + Q(t)$. The simulation uses population fractions ($N = 1$) and the following parameter values (per day): $\beta_p = 0.55$, $\beta_a = 0.30$, $\beta_m = 0.65$, $\beta_h = 0.08$, $\beta_q = 0.03$, $\sigma = 1/4$, $\kappa = 1/2$, $\pi_a = 0.30$, $\pi_h = 0.12$, $\gamma_a = 1/6$, $\gamma_m = 1/6$, $\gamma_h = 1/10$, $\gamma_q = 1/7$, $\delta_m = 1/4$, and $\delta_h = 1/2$. The initial condition is $(S, E, I_p, I_a, I_m, I_h, Q, R) = (0.9995, 3 \times 10^{-4}, 2 \times 10^{-4}, 0, 0, 0, 0, 0)$ (with R set to ensure mass conservation). These plots highlight how class structure separates transmission-relevant states (e.g., presymptomatic and asymptomatic infection) from observed/isolated pathways, and how isolation can shift burden into Q without eliminating earlier-stage transmission.

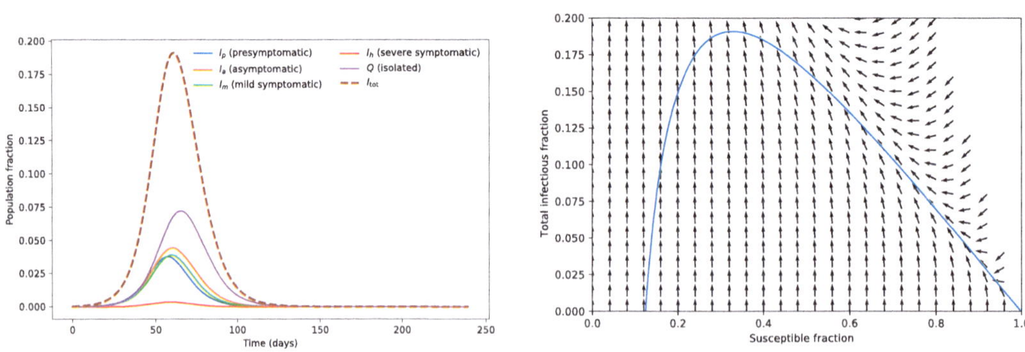

(a) Time series of infectious subclasses and total infectious fraction.

(b) S–I_{tot} projection: trajectory projections + illustrative vector field.

FIGURE 3.11

Illustrative dynamics of the multi-class infectiousness model.

Splitting the infectious compartment changes both interpretation and model behavior. First, it enables attribution of transmission to distinct biological or clinical stages. Under standard assumptions, the reproduction number can be decomposed into additive contributions from each infectious class (e.g.,

$R_0 = R_p + R_a + R_s$), making it possible to quantify how much spread arises from presymptomatic versus symptomatic or asymptomatic infection, and therefore where leverage points for control are likely to be most effective. Second, class structure provides a direct mapping from interventions to mechanisms as symptom-based isolation primarily reduces transmission from symptomatic individuals (e.g., I_s or I_m/I_h), but may leave presymptomatic and asymptomatic transmission largely unaffected unless paired with rapid testing, screening, or contact tracing. In contrast, interventions that reduce per-contact risk (masking, ventilation, behavior change) act broadly on all infectious classes, while targeted programs (e.g., workplace screening) may disproportionately reduce transmission from otherwise undetected asymptomatic infections. Third, splitting clarifies observation mismatch and reporting bias as surveillance data often disproportionately reflect symptomatic (and especially severe) infections, whereas onward transmission may be driven partly by presymptomatic or asymptomatic individuals. Consequently, calibrating a model to reported case counts alone can mis-estimate transmission parameters unless the observation process is modeled explicitly [221]. Fourth, stage-specific infectiousness alters the generation-interval structure. Namely, by distributing infectiousness across presymptomatic, symptomatic, and asymptomatic stages, the model effectively reshapes the infectiousness kernel over time since infection. This changes the relationship between epidemic growth rate and reproduction number, and can lead to different inferred R_t from the same incidence curve when the assumed stage structure differs. Finally, the added structure introduces identifiability challenges and confounding since stage-specific transmissibility and stage durations can trade off, such as larger β_p with shorter presymptomatic duration can resemble smaller β_p with longer duration.

3.1.5 Clinical outcome

Public health decisions are often driven more by severe outcomes, hospitalizations, ICU occupancy, and deaths than by infections per se. Extending transmission models with explicit outcome compartments supports burden planning (e.g., capacity forecasting), improves the connection between latent infections and observed endpoints, and enables calibration to signals that may be more stable than case counts when testing practices vary [150]. Moreover, separating transmission states from clinical progression states helps avoid conflating how fast infection spreads with how long it takes for severe outcomes to manifest.

Methodologically, a minimal extension splits the recovered process into recovery and death. In its simplest form,

$$\frac{dS}{dt} = -\beta S I, \tag{3.37}$$

$$\frac{dI}{dt} = \beta S I - (\gamma + \alpha) I, \tag{3.38}$$

$$\frac{dR}{dt} = \underbrace{\gamma I}_{\text{recoveries}}, \tag{3.39}$$

$$\frac{dD}{dt} = \underbrace{\alpha I}_{\text{deaths}}, \tag{3.40}$$

where γ is the recovery rate and α is the disease-induced mortality rate. This system is attractive for its parsimony, since it preserves the transmission mechanism while introducing a mortality endpoint that can be compared to reported deaths. Fig. 3.12 presents a schematic view of the $SIRD$ model.

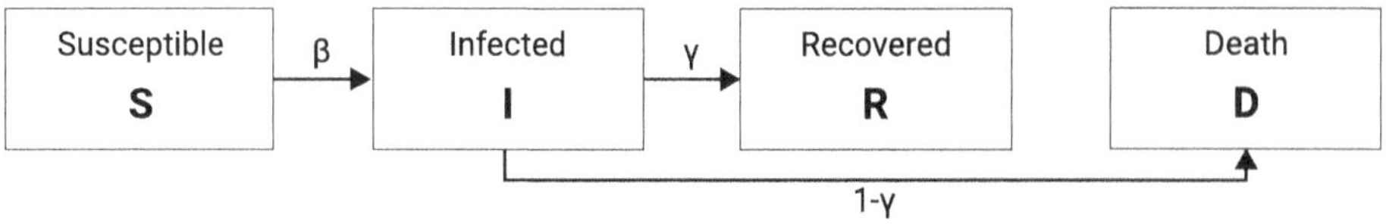

FIGURE 3.12

A schematic view of the $SIRD$ model.

In this setting, an infectious individual exits I either by recovery or by death, implying that the fraction who die is

$$\Pr(\text{death} \mid \text{infectious}) = \frac{\alpha}{\alpha + \gamma},$$

which resembles an infection fatality risk only under strong assumptions: homogeneous risk, no explicit infection-to-death delay beyond the exponential residence time in I, and a direct mapping between infections and deaths without stratification or reporting biases. In practice, death typically occurs after a non-negligible delay following infection, and mortality risk is strongly heterogeneous (e.g., by age, comorbidities, vaccination, and care availability). On the other hand, burden-oriented models often insert explicit clinical progression stages to represent both timing and selection into severe outcomes. A common refinement introduces hospitalization (and optionally ICU) compartments, for example,

$$I \to H \to D, \qquad I \to H \to R,$$

so that transmission is governed largely by I (community infectiousness), while H captures downstream burden and the lag between infection and death [150]. This separation is especially useful when interventions affect not only transmission (changing β) but also clinical outcomes. A simple SIRD with hospitalization (SIHRD) model illustrates the principle:

$$\frac{dS}{dt} = -\beta S I, \tag{3.41}$$

$$\frac{dI}{dt} = \beta S I - (\gamma_I + \eta) I, \tag{3.42}$$

$$\frac{dH}{dt} = \underbrace{\eta I}_{\text{admissions}} - \underbrace{(\gamma_H + \mu) H}_{\text{discharge + death}}, \tag{3.43}$$

$$\frac{dR}{dt} = \gamma_I I + \gamma_H H, \tag{3.44}$$

$$\frac{dD}{dt} = \mu H, \tag{3.45}$$

where η is the rate of progression from infectiousness to hospitalization, γ_I is the non-hospital recovery rate, γ_H is the discharge/recovery rate from hospital, and μ is the hospital mortality rate. Fig. 3.13

presents a schematic view of the $SIHRD$ model. This structure decouples who transmits from who generates burden, and permits realistic lags by placing deaths downstream of transmission.

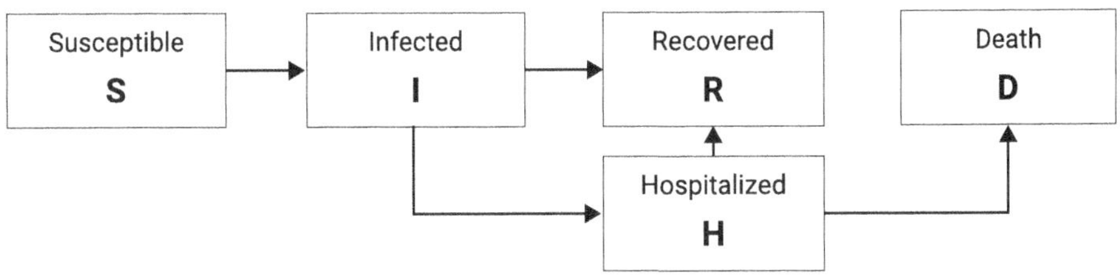

FIGURE 3.13

A schematic view of the $SIHRD$ model.

Principally, outcome compartments can improve identifiability of epidemic timing and magnitude because deaths and hospital admissions may be less sensitive than cases to testing policies, and because multiple endpoints (cases, admissions, ICU, deaths) jointly constrain latent trajectories. However, these benefits depend critically on modeling delay distributions realistically. If the infection-to-hospitalization and hospitalization-to-death delays are misspecified (e.g., forced to be exponential when data imply a more concentrated distribution), the model may misattribute lag structure to changes in transmission (incorrectly shifting $\beta(t)$) rather than to clinical timing. Practically, this motivates either using staged compartments (Erlang chains) for H or for infection-to-death delays, or imposing informative priors from clinical studies on lengths of stay and outcome probabilities when fitting to mortality and hospitalization time series.

To illustrate the difference, Fig. 3.14 compares the baseline SIR model with two burden-oriented extensions, SIRD (explicit mortality) and SIHRD (hospitalization-mediated mortality), while keeping the transmission and recovery parameters identical across all panels. Specifically, we use $\gamma = 1/7$ day^{-1} and $\beta = R_0\gamma$ with $R_0 = 2.5$, hence $\beta \approx 0.357$ day^{-1}. For SIRD, deaths occur directly from I at rate $\alpha = 1/30$ day^{-1} and for SIHRD, infectious individuals are hospitalized at rate $\eta = 1/10$ day^{-1}, hospitalized individuals recover at rate $\gamma_H = 1/10$ day^{-1}, and die at rate $\mu = 1/12$ day^{-1}. The comparison highlights that even when the transmission process is held fixed, introducing explicit outcome states changes both the timing and magnitude of cumulative deaths and shifts transient burden into intermediate compartments (e.g., hospital occupancy $H(t)$), which is essential for planning and for aligning model outputs with observed severe endpoints.

3.1.6 Structured populations

Homogeneous mixing, also known as "well-mixing" is the assumption that every individual is equally likely to contact every other individual. This assumption is rarely realistic, especially for long periods of time or large spatial configurations. Specifically, contact opportunities differ by age, household structure, school/work attendance, occupation, and health-related risk factors, so epidemic trajectories can vary substantially across subpopulations even when the overall epidemic curve appears smooth. Multi-group compartmental models preserve the bookkeeping clarity of SIR while introducing heterogeneity by replicating states per group and coupling them through a *mixing matrix* that specifies who contacts whom [140]. Fig. 3.15 provides a schematic view of this model.

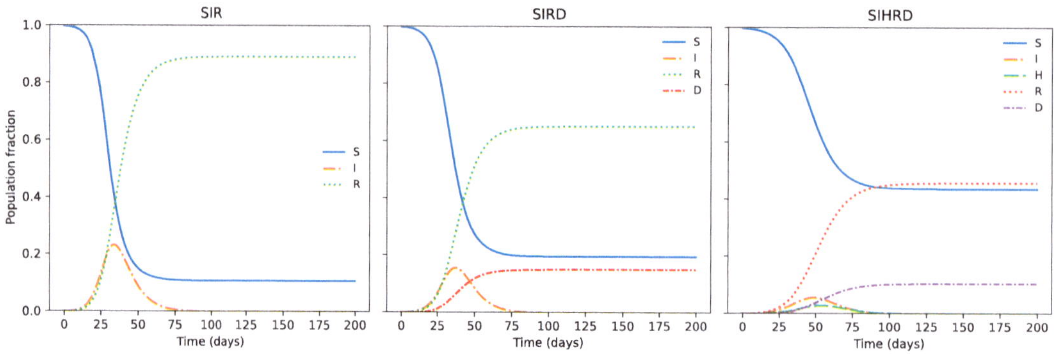

FIGURE 3.14

Comparison of SIR, SIRD, and SIHRD trajectories under identical transmission and recovery parameters.

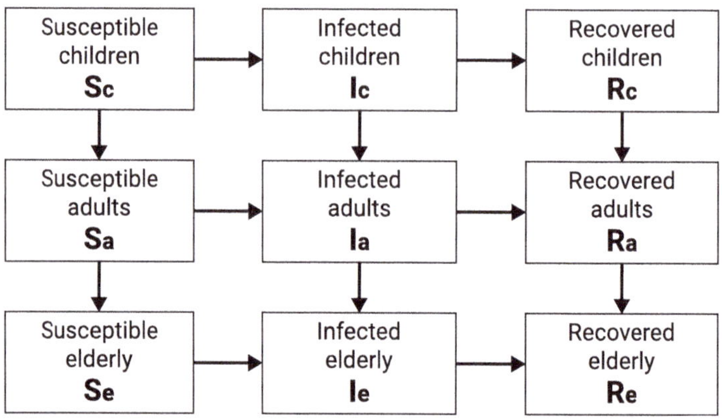

FIGURE 3.15

A schematic view of a multi-group age SIR model. Each group i has its own compartments S_i, I_i, R_i. Groups are coupled through the force of infection $\lambda_i(t)$, which aggregates infectious pressure from all groups j via a mixing/contact matrix C_{ij}.

Formally, let groups be indexed by $i = 1, \ldots, m$, with group sizes $N_i = S_i + I_i + R_i$ and total population $N = \sum_i N_i$. A common frequency-dependent formulation uses a contact (or effective transmission) matrix C_{ij}, where C_{ij} represents the rate of potentially infectious contacts *made by a typical individual in group i toward individuals in group j*. Define the group-specific force of infection as

$$\lambda_i(t) = \underbrace{\sum_{j=1}^{m} \beta C_{ij} \frac{I_j(t)}{N_j}}_{\substack{\text{infectious pressure from group } j \\ \text{weighted by mixing from } i \text{ to } j}}, \tag{3.46}$$

so that susceptible individuals in group i become infected at rate $\lambda_i(t)S_i(t)$. The multi-group SIR dynamics can then be written compactly as

$$\frac{dS_i}{dt} = -\underbrace{\lambda_i(t)S_i}_{\substack{\text{new infections in group } i\\ \text{due to all source groups}}}, \tag{3.47}$$

$$\frac{dI_i}{dt} = \underbrace{\lambda_i(t)S_i}_{\text{inflow from infection}} - \gamma I_i, \tag{3.48}$$

$$\frac{dR_i}{dt} = \gamma I_i. \tag{3.49}$$

More generally, β may be group- or pair-specific (β_{ij}), and additional compartments (e.g., E_i for latency) can be added without changing the coupling idea: heterogeneity enters through the matrix controlling the force of infection.

A key conceptual change is that R_0 is no longer, in general, a simple ratio such as β/γ. Instead, early spread is governed by a *next-generation matrix* K, where K_{ij} is the expected number of new infections in group i caused by a typical infectious individual from group j introduced into an otherwise susceptible population [220,223]. For the model above at the disease-free state ($S_i \approx N_i$), a standard form is

$$K_{ij} = \underbrace{\frac{1}{\gamma}}_{\substack{\text{mean time}\\ \text{infectious}}} \underbrace{\beta C_{ij}}_{\substack{\text{contact/transmission}\\ \text{from } i \text{ to } j}} \underbrace{\frac{N_i}{N_j}}_{\substack{\text{availability of susceptibles}\\ \text{in recipient group}}}, \tag{3.50}$$

and the basic reproduction number is the spectral radius

$$R_0 = \rho(K). \tag{3.51}$$

This representation clearly separates (i) who infects whom (mixing via C_{ij} and any β_{ij}), (ii) how long infections last (via γ), and (iii) how interventions act by altering either contact structure or infectiousness/duration.

Structured models can generate distinct epidemic timing and burden across groups: for example, groups with higher contact intensity or more assortative mixing (such as school-age children) can peak earlier and drive transmission into other groups, while vulnerable groups may peak later but experience higher risk of severe outcomes. They also highlight why targeted interventions can be nonlinear: vaccinating or reducing contacts in a highly connected group can disproportionately reduce $\rho(K)$ and hence overall transmission.

To illustrate how population structure changes both the timing and composition of an outbreak, Fig. 3.16 shows a three-group SIR simulation where each group has its own (S_i, I_i, R_i) trajectories. The figure is styled so that colors encode epidemiological state (S in green, I in red, R in blue), while line style encodes the age group (children solid, adults dashed, older dash-dot).

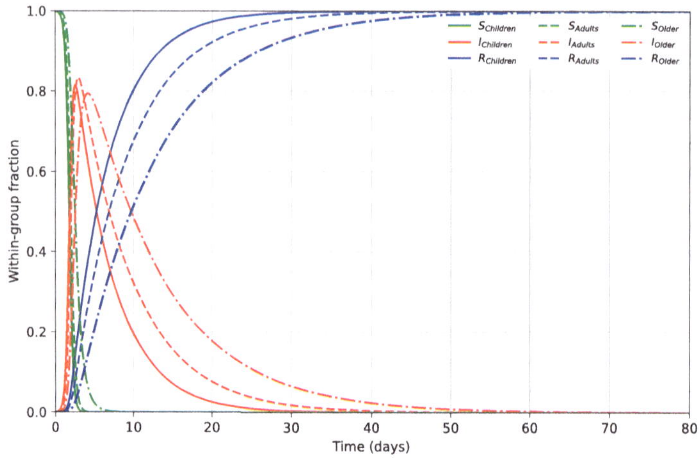

FIGURE 3.16

Three age-group SIR dynamics. The groups are coupled by a mixing matrix C_{ij} through the force of infection $\lambda_i(t) = a_i \beta \sum_j C_{ij} I_j / N_j$.

3.1.7 Vector-borne extensions

For vector-borne diseases (e.g., malaria, dengue, Zika, chikungunya, yellow fever), transmission requires a host-vector-host sequence: vectors acquire infection by biting infectious hosts and later transmit to susceptible hosts while biting [236]. A host-only SIR model cannot represent the vector population, its biting intensity, or its survival, all of which can determine whether infection persists and how strongly seasonal or control-driven changes in vector ecology affect incidence. Ross–Macdonald-type models extend standard compartmental bookkeeping by introducing explicit vector compartments and coupling host and vector infection processes through biting-mediated forces of infection.

Mathematically, let N_h be the host population size and N_v the vector population size. Hosts are modeled with compartments $S_h(t)$, $I_h(t)$, $R_h(t)$, while vectors are modeled with $S_v(t)$, $I_v(t)$. A standard frequency-dependent coupling defines the forces of infection

$$\lambda_h(t) = \underbrace{a}_{\text{biting rate}} \quad \underbrace{b}_{\substack{\text{transmission prob.} \\ \text{vector}\to\text{host}}} \quad \underbrace{\frac{I_v(t)}{N_v}}_{\text{infectious vector fraction}}, \tag{3.52}$$

$$\lambda_v(t) = \underbrace{a}_{\text{biting rate}} \quad \underbrace{c}_{\substack{\text{transmission prob.} \\ \text{host}\to\text{vector}}} \quad \underbrace{\frac{I_h(t)}{N_h}}_{\text{infectious host fraction}}. \tag{3.53}$$

Thus, new host infections occur at rate $\lambda_h(t)S_h(t)$ and new vector infections at rate $\lambda_v(t)S_v(t)$.

A representative Ross–Macdonald system [237] with host recovery rate γ and vector mortality rate μ_v (with births balancing deaths to keep N_v constant) is

$$\frac{dS_h}{dt} = -\underbrace{\lambda_h(t)S_h}_{\substack{\text{new host infections}\\ \text{from infectious vectors}}}, \tag{3.54}$$

$$\frac{dI_h}{dt} = \underbrace{\lambda_h(t)S_h}_{\text{inflow from infection}} - \underbrace{\gamma I_h}_{\text{host recovery}}, \tag{3.55}$$

$$\frac{dR_h}{dt} = \underbrace{\gamma I_h}_{\text{accumulated recoveries}}, \tag{3.56}$$

$$\frac{dS_v}{dt} = \underbrace{\mu_v N_v}_{\substack{\text{vector births}\\ \text{(balance deaths)}}} - \underbrace{\lambda_v(t)S_v}_{\substack{\text{new vector infections}\\ \text{from infectious hosts}}} - \underbrace{\mu_v S_v}_{\text{vector deaths}}, \tag{3.57}$$

$$\frac{dI_v}{dt} = \underbrace{\lambda_v(t)S_v}_{\text{inflow from infection}} - \underbrace{\mu_v I_v}_{\text{vector deaths}}. \tag{3.58}$$

Fig. 3.17 presents a schematic view of this model.

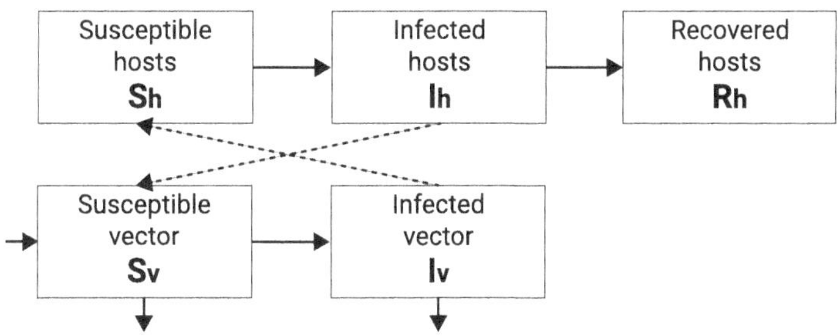

FIGURE 3.17

A schematic view of a Ross–Macdonald-type vector-borne model. Coupling occurs through the forces of infection $\lambda_h(t)$ (vector-to-host) and $\lambda_v(t)$ (host-to-vector), driven by biting rate a and transmission probabilities b, c.

This model makes explicit that sustained transmission depends jointly on host infectious duration ($1/\gamma$), vector lifespan ($1/\mu_v$), biting intensity (a), and the vector-to-host ratio $m = N_v/N_h$. Unlike host-only models, interventions can act on either side of the cycle: e.g., treatment shortens host infectiousness (increasing γ), while insecticides reduce vector lifespan (increasing μ_v), and bed nets reduce effective biting (reducing a). Thus, a key implication is that the reproduction number combines host and vector contributions. In the classic Ross–Macdonald setting (with constant N_h and N_v), one common expression is

$$R_0 = \sqrt{\frac{a^2\, b\, c\, m}{\gamma\, \mu_v}}, \tag{3.59}$$

reflecting that a successful transmission chain requires completion of both legs of the cycle (host→vector and vector→host). This form also immediately clarifies control levers: reducing a or m, or increasing μ_v or γ, can push R_0 below one even without changing host contact behavior [236].

To demonstrate the model behavior, Fig. 3.18 shows simulated host and vector prevalence under a baseline parameter set and under a simple vector-control scenario modeled as increased vector mortality (larger μ_v). Because vector survival enters R_0 directly, modest changes to μ_v can substantially reduce both vector prevalence and downstream host prevalence, illustrating why vector control can be highly effective even when host-side interventions are limited.

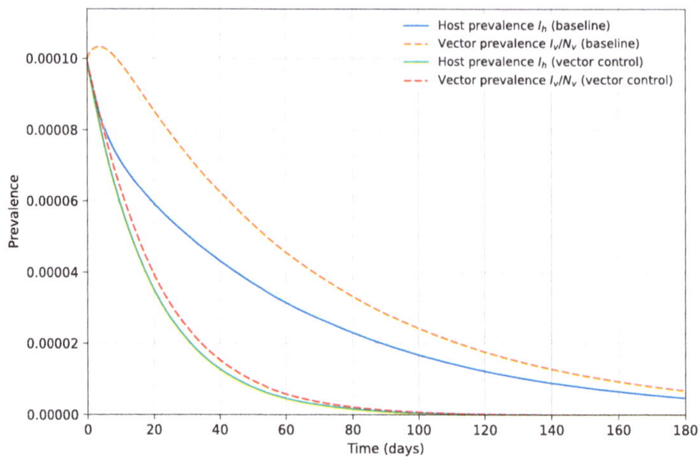

FIGURE 3.18

Example Ross–Macdonald dynamics showing host prevalence I_h and vector prevalence I_v/N_v under baseline conditions and a vector-control scenario (increased vector mortality μ_v). The reduction in vector lifespan lowers R_0 and suppresses both vector infection and subsequent host infection.

3.2 Process extensions

State extensions enrich compartmental models by adding new epidemiological compartments (e.g., latency, waning immunity, demography, structured groups). In contrast, *process extensions* keep the same state variables but change the dynamical and statistical mechanisms that govern how those states evolve and how they are observed. Most importantly, stochastic versions of SIR-type models replace deterministic flows with random events, capturing intrinsic transmission noise, demographic stochasticity in finite populations, and uncertainty in the measurement process.

3.2.1 Time-varying transmission

Assuming a constant transmission rate β is often untenable in real epidemics. Contact patterns change because of policy interventions (e.g., school closures, gathering limits, mask mandates), voluntary behavior change in response to perceived risk (e.g., reduced mobility when incidence rises), and seasonal/environmental context (e.g., indoor crowding and humidity effects that change respiratory virus transmissibility) [238]. A convenient modeling approach is to keep the compartmental state representation but allow transmission to vary over time, replacing β by $\beta(t)$. This decision is more often than not is driven by a lack of ability (or need) to represent these changes in a state-based manner. In an SIR model, a time-varying transmission rate yields

$$\frac{dS}{dt} = -\underbrace{\beta(t)}_{\substack{\text{time-varying}\\ \text{transmission}}} SI, \tag{3.60}$$

$$\frac{dI}{dt} = \underbrace{\beta(t)SI}_{\text{new infections}} - \gamma I, \tag{3.61}$$

$$\frac{dR}{dt} = \gamma I, \tag{3.62}$$

so that all nonstationarity enters through $\beta(t)$. In this setting it is often helpful to distinguish the intrinsic transmissibility encoded by $\beta(t)$ from the depletion of susceptibles encoded by $S(t)$. For example, even if $\beta(t)$ increases seasonally, incidence may still fall if $S(t)$ has been sufficiently depleted. A closely related quantity is the effective reproduction number. In the simplest SIR interpretation, one may write

$$R_t = \frac{\beta(t)}{\gamma}\frac{S(t)}{N}, \tag{3.63}$$

highlighting that R_t depends on both the time-varying contact environment and the evolving susceptible pool. In more structured models (e.g., SEIR or stage-structured systems), translating fitted growth into R_t must respect generation-interval structure, so the mapping between $\beta(t)$ and R_t is model-dependent [232]. Notably, γ is not time-dependent in this model as it is an inherent biological property of the pathogen at the root of the pandemic, which is assume either not to change at all or in an insignificant amount during the pandemic.

Several functional forms for $\beta(t)$ are widely used. First, a piecewise-constant $\beta(t)$ is often the most transparent way to encode policy phases because it matches how interventions are implemented and communicated as restrictions introduced on specific dates, maintained for a period, and then modified or lifted. In this parameterization, the change points $t_1, t_2, \ldots$ represent known (or inferable) intervention times, and the levels $\beta_0, \beta_1, \beta_2, \ldots$ summarize the net effect of everything that changes transmission during each phase, including contact rates (mobility, gatherings), per-contact transmission probability (masking, ventilation), and mixing patterns (school vs workplace contacts). For example, a strict closure period can be represented by $\beta_1 < \beta_0$, while a partial reopening corresponds to an intermediate level β_2 that remains below baseline but above the lockdown phase.

From an inference perspective, piecewise $\beta(t)$ functions act as a low-dimensional surrogate for complex time variation while remaining identifiable from data under reasonable conditions. In practice, change points are frequently anchored to known dates (e.g., school closure, stay-at-home order, reopening), or estimated from the time series as *change-point models*. A common constraint is to require $\beta(t)$ to be continuous in time (e.g., replacing jumps with short ramps) when interventions take effect gradually or compliance changes over several days, but the piecewise-constant form is often a useful first approximation.

Importantly, β_k should not be interpreted as a purely policy effect. It also absorbs contemporaneous changes in testing/reporting, unmodeled importations, seasonality, or missing biological structure (e.g., latent periods) if these are not handled elsewhere. Therefore, when using piecewise $\beta(t)$ for causal narratives, it is good practice to (i) justify change points using external intervention/mobility information, (ii) include an explicit observation model when reporting changes over time, and (iii) quantify uncertainty in both β_k and derived quantities such as R_t.

Second, seasonal forcing is often modeled as a smooth periodic modulation,

$$\beta(t) = \beta_0\left[1 + \epsilon\cos\left(\frac{2\pi t}{T}\right)\right], \tag{3.64}$$

where T is the forcing period (for example $T = 365$ for a year with a step in time of a day) and $\epsilon \in [0, 1)$ controls the amplitude [238]. This specification is a compact way to represent the empirically observed fact that many infections exhibit recurring seasonal patterns even in the absence of large policy changes. Mechanistically, seasonality can reflect changes in contact structure (e.g., school terms, holiday travel, indoor crowding in winter), changes in pathogen viability or host susceptibility driven by climate (humidity, temperature, UV exposure), and seasonal variation in vector abundance for vector-borne diseases. In respiratory infections such as influenza, RSV, and seasonal coronaviruses, the dominant seasonal drivers are often a combination of contact changes and environmental effects; in diseases such as malaria or dengue, seasonal vector dynamics can create strong periodic forcing.

The cosine form has several practical advantages: i) it enforces smoothness, as transmissibility changes gradually rather than through sharp jumps, which can be more realistic when drivers (weather, indoor time, school calendars) evolve continuously; ii) it is low-dimensional and interpretable: β_0 is the mean transmission intensity over the year, while ϵ determines the relative peak-to-trough contrast. Specifically,

$$\beta_{\max} = \beta_0(1+\epsilon), \qquad \beta_{\min} = \beta_0(1-\epsilon), \tag{3.65}$$

so $\epsilon = 0.2$ corresponds to $\pm 20\%$ seasonal variation around the mean, and the peak-to-trough ratio is $(1+\epsilon)/(1-\epsilon)$. The phase (timing) of the seasonal peak can be included by a shift,

$$\beta(t) = \beta_0\left[1 + \epsilon\cos\left(\frac{2\pi(t-\phi)}{T}\right)\right], \tag{3.66}$$

where ϕ sets the calendar date of maximal transmissibility. This extension is often essential when comparing different climates or hemispheres, or when aligning the model to known winter peaks.

Seasonal forcing also has important dynamical consequences. Even when the mean transmission corresponds to $R_0 = \beta_0/\gamma > 1$, the system may alternate between periods with $\beta(t)/\gamma > 1$ (supercritical growth) and periods with $\beta(t)/\gamma < 1$ (subcritical decay). This can produce recurrent waves whose timing is locked to the forcing cycle, and in some regimes, can generate multi-year patterns via resonance between the forcing period and the system's intrinsic time scales (susceptible depletion and replenishment) [238]. Consequently, apparent "waves" in incidence need not imply successive changes in policy or behavior; they may arise as a predictable response to periodic modulation of transmission.

A third option is behavioral feedback, where effective contacts decrease as perceived risk increases. A common stylized form is

$$\beta(t) = \frac{\beta_0}{1 + k\,g(I(t))}, \tag{3.67}$$

where $g(\cdot)$ is a proxy for risk perception and $k \geq 0$ controls responsiveness. The interpretation is that β_0 represents baseline transmission in the absence of heightened caution, while the denominator captures voluntary (or endogenous) reductions in risky contacts when people perceive the epidemic to be severe. When $k = 0$, behavior is insensitive and $\beta(t) \equiv \beta_0$; as k increases, the same level of prevalence induces a larger reduction in transmission. This mechanism is often used as a parsimonious alternative to explicitly modeling policy changes when the primary driver of time variation is voluntary adaptation in response to rising cases or hospital burden.

The choice of $g(\cdot)$ matters and should reflect what individuals plausibly observe. The simplest option is $g(I) = I$, meaning that perceived risk tracks current prevalence. Other common proxies include incidence (new cases per unit time), hospitalizations, deaths, or a smoothed version of these signals to reflect delayed awareness. For example, one can use a moving-average or exponential smoothing, $g(t) = \int_0^\infty w(\tau)\, I(t-\tau)\, d\tau$, to represent that behavior responds to recent history rather than instantaneous prevalence. In applied settings, $g(\cdot)$ may be linked to external covariates such as mobility indices, survey measures of risk perception, or media intensity, but the conceptual role remains the same: it provides an endogenous coupling from epidemic state to transmission.

However, interpretability requires caution. Namely, the mechanism is observationally confounded with many alternatives: an apparent decline in transmission could reflect policy interventions, seasonal effects, changes in testing/ascertainment, or depletion of susceptible individuals, not necessarily voluntary behavior. In addition, using epidemiological outcomes (cases, prevalence) as the driver of behavior introduces a potential *circularity* in inference: if $g(I)$ is constructed from observed case counts, then reporting artifacts can be mistaken for behavioral response. Moreover, the functional form is a modeling choice rather than a law: the saturating response in $\beta_0/(1+kg)$ is convenient and ensures $\beta(t)$ remains positive, but other forms are plausible, such as exponential responses $\beta(t) = \beta_0 \exp(-kg)$, threshold responses (little change until risk exceeds a level), or responses that depend on cumulative burden rather than instantaneous prevalence. Thus, in practice, behavioral feedback models are most defensible when $g(\cdot)$ is tied to measurable drivers (mobility, contact surveys, policy stringency, hospital occupancy) or when sensitivity analyses show that qualitative conclusions are robust to reasonable alternative forms. When such anchoring is unavailable, it is often preferable to treat behavioral feedback as a hypothesis-generating mechanism and to report k as an *effective* parameter summarizing multiple unmodeled processes, rather than as a direct measure of individual-level responsiveness.

Time-varying $\beta(t)$ is powerful but can be dangerously flexible in inference. If $\beta(t)$ is allowed to vary freely (e.g., one parameter per day), it can absorb misspecification elsewhere such as missing compartments, incorrect generation intervals, and unmodeled importations; undermining mechanistic interpretability. Moreover, changes in testing, reporting rates, or delays can masquerade as changes in transmission unless the observation process is modeled explicitly. For this reason, it is often preferable to impose structure on $\beta(t)$ and to report uncertainty in $\beta(t)$ and in derived R_t, while being explicit about the assumed disease natural history [232].

Fig. 3.19 illustrates how different choices of $\beta(t)$ reshape epidemic trajectories even in a basic SIR setting. Relative to constant transmission, a piecewise drop in $\beta(t)$ flattens and delays the epidemic peak (intervention effect), seasonal forcing can create prolonged or multi-peaked dynamics, and behavioral feedback can dampen peaks by reducing transmission during high-prevalence periods. These examples emphasize that repeated waves need not imply new compartments by themselves: they can also arise from time variation in transmission processes.

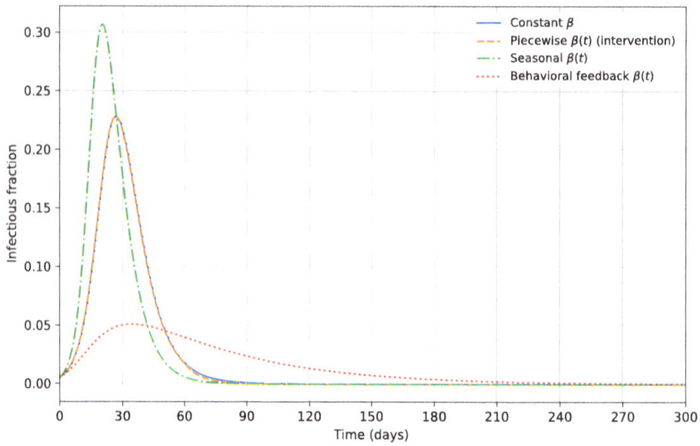

FIGURE 3.19

Illustration of time-varying transmission in an SIR model.

3.2.2 Stochastic models

Deterministic ODE models approximate expected epidemic trajectories under the assumption that populations are effectively infinite and that flows between compartments are smooth [239,240]. In many applications, however, chance events matter. When case counts are small, random timing of infections and recoveries determines whether introductions die out or trigger outbreaks, and even when outbreaks occur, trajectories can differ substantially between realizations. Stochastic compartmental models capture this intrinsic variability (sometimes called "demographic stochasticity" [241]) and provide a natural framework for uncertainty quantification and likelihood-based inference under partial observation [222,242].

A direct stochastic analog of the ODE-represented SIR model is a *continuous-time Markov chain* (CTMC), where the state is $(S(t), I(t), R(t)) \in \{0, 1, \ldots, N\}^3$ and transitions occur as random jumps. In the basic SIR CTMC, the two event types are

$$(S, I, R) \to (S - 1, I + 1, R) \qquad \text{(infection)}, \tag{3.68}$$

$$(S, I, R) \to (S, I - 1, R + 1) \qquad \text{(recovery)}. \tag{3.69}$$

Crucially, the event *rates* match the deterministic hazards:

$$\underbrace{\text{infection rate}}_{\text{jump intensity}} = \underbrace{\beta \frac{SI}{N}}_{\substack{\text{mass action}\\ \text{hazard}}}, \qquad \underbrace{\text{recovery rate}}_{\text{jump intensity}} = \underbrace{\gamma I}_{\substack{\text{per-capita recovery}\\ \text{hazard}}}. \tag{3.70}$$

Thus, the deterministic model can be interpreted as the mean-field limit of a jump process. Simulation is typically performed with Gillespie's stochastic simulation algorithm [243], which generates statistically correct sample paths by repeatedly (1) sampling the time to the next reaction/event and (2) sampling which event happens. At each step compute $a_0 = a_1 + a_2$, draw the next time increment $\tau \sim \text{Exp}(a_0)$, then choose the event type with $\mathbb{P}(\text{infection}) = \frac{a_1}{a_0}$ and $\mathbb{P}(\text{recovery}) = \frac{a_2}{a_0}$. Update (S, I, R) accordingly, advance time by τ, and repeat until $I = 0$ or a final time is reached.

For statistical work, it is often essential to separate the epidemic process from the data-generation mechanism, yielding a Partially Observed Markov Process (POMP). In a POMP formulation, the CTMC (or a stochastic differential equation based approximation) is treated as latent, and observed data are generated through an explicit observation model. For example, if C_t denotes the number of new infections in a reporting interval, one might model reported cases as

$$Y_t \mid C_t \sim \text{NegBin}(\text{mean} = \rho C_t, \text{ dispersion} = \kappa), \tag{3.71}$$

where ρ is a reporting fraction, and κ captures extra-Poisson variation (e.g., heterogeneity in reporting and clustering). More elaborate observation layers can incorporate reporting delays (convolutions), weekday effects, test volume, or measurement error in prevalence [221,222]. This explicit separation is important because changes in reporting can otherwise be mistaken for changes in transmission.

A key qualitative difference from deterministic models is stochastic fade-out: even when $R_0 > 1$, introductions can go extinct with nonzero probability. In early stages, when $S \approx N$ and infections are rare, branching-process approximations often characterize invasion probability and extinction risk. For instance, under a Poisson offspring approximation, the extinction probability q satisfies $q = \exp\{-R_0(1 - q)\}$, so larger R_0 lowers but does not eliminate fade-out probability. Heterogeneous offspring distributions (superspreading) can increase fade-out probability even when the mean R_0 is above one [242].

Fig. 3.20 illustrates these ideas by comparing many CTMC realizations to the deterministic ODE trajectory for the same parameters in a small population. Some realizations die out quickly, while others produce outbreaks of varying size and timing, emphasizing why stochastic models are often more realistic for early outbreak inference, small populations (schools, facilities), and forecasting with uncertainty. The cost is increased computational effort and a stronger dependence on modeling assumptions about what is observed and how.

3.3 The next generation matrix

Recall that in compartmental epidemic models, the basic reproduction number R_0 quantifies the expected number of secondary infections generated by a typical infectious individual introduced into a wholly susceptible population. For models with multiple infected states, such as the extensions presented in this chapter, the computation of R_0 is naturally expressed through the *next-generation matrix* (NGM) [244,245], which generalizes the one infected compartment intuition to a multitype setting. Formally, consider a deterministic continuous-time model written as an autonomous ODE system as follows:

$$\frac{d\mathbf{z}}{dt} = \mathbf{G}(\mathbf{z}), \tag{3.72}$$

such that $\mathbf{z} \in \mathbb{R}^n_{\geq 0}$. Partition the state vector into infected components $\mathbf{x} \in \mathbb{R}^m_{\geq 0}$ and non-infected components $\mathbf{y} \in \mathbb{R}^{n-m}_{\geq 0}$, so that $\mathbf{z} = (\mathbf{x}, \mathbf{y})$. The NGM construction begins by decomposing the infected

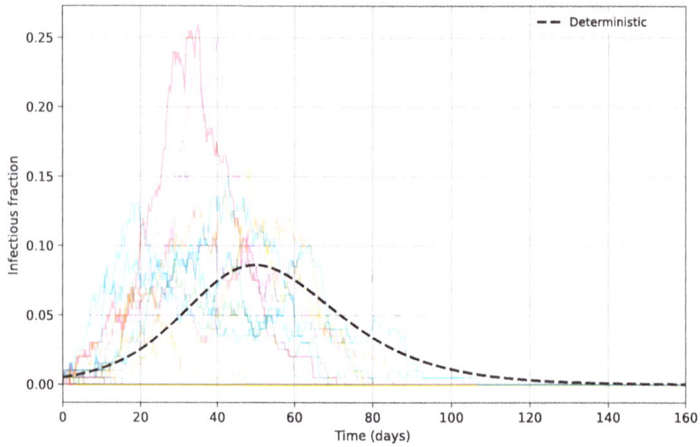

FIGURE 3.20

Stochastic SIR trajectories (CTMC) in a small population compared to the deterministic ODE solution.

subsystem into *new infection* terms and all other terms:

$$\frac{d\mathbf{x}}{dt} = \mathbf{F}(\mathbf{x}, \mathbf{y}) - \mathbf{V}(\mathbf{x}, \mathbf{y}). \tag{3.73}$$

Here, $\mathbf{F}$ collects rates at which new infections enter infected compartments, while $\mathbf{V}$ collects all other transfers (for example, progression between infected states, recovery, isolation, disease-induced death, and natural death). Let $(\mathbf{x}^*, \mathbf{y}^*)$ denote a disease-free equilibrium (DFE), with $\mathbf{x}^* = \mathbf{0}$ and $\mathbf{y}^*$ representing the susceptible-only state.

Linearizing (3.73) around the DFE yields

$$\frac{d\mathbf{x}}{dt} \approx (F - V)\mathbf{x}, \tag{3.74}$$

where

$$F := \left.\frac{\partial \mathbf{F}}{\partial \mathbf{x}}\right|_{(\mathbf{0}, \mathbf{y}^*)}, \qquad V := \left.\frac{\partial \mathbf{V}}{\partial \mathbf{x}}\right|_{(\mathbf{0}, \mathbf{y}^*)}. \tag{3.75}$$

Assuming V is nonsingular (which holds under standard conditions ensuring that infected states eventually leave in the absence of new infection), the NGM is defined as

$$K := F V^{-1}. \tag{3.76}$$

The basic reproduction number is then given by the spectral radius [246]:

$$R_0 := \rho(K), \tag{3.77}$$

i.e., the largest modulus among the eigenvalues of K.

The NGM admits a clear probabilistic interpretation. The entry K_{ij} can be read as the expected number of new infections produced in infected compartment i by an individual introduced into infected compartment j, over the entire duration of infection, under the linearized dynamics at the DFE. In multigroup models, K is often called a "who-acquires-infection-from-whom" (WAIFW) matrix after appropriate scaling, and its dominant eigenvalue characterizes the epidemic threshold. Importantly, under standard regularity and positivity assumptions, the DFE is locally asymptotically stable if $R_0 < 1$ and unstable if $R_0 > 1$. Intuitively, $R_0 < 1$ implies that, on average, each infected introduction produces fewer than one new infection in total, so small outbreaks die out. The eigenvector associated with $\rho(K)$ provides the asymptotic composition of infections across compartments during the early exponential phase, offering additional insight beyond the scalar threshold.

For practical construction, one may follow a systematic procedure:

1. Select infected compartments $\mathbf{x}$ (those that contain infected individuals and can generate future infections, possibly including exposed/latent states depending on modeling conventions).
2. Rewrite the infected dynamics in the form $d\mathbf{x}/dt = \mathbf{F}(\mathbf{x}, \mathbf{y}) - \mathbf{V}(\mathbf{x}, \mathbf{y})$, where $\mathbf{F}$ contains only *new infection inflows* into infected compartments.
3. Evaluate the DFE $(\mathbf{0}, \mathbf{y}^*)$.
4. Compute Jacobians $F = \partial\mathbf{F}/\partial\mathbf{x}$ and $V = \partial\mathbf{V}/\partial\mathbf{x}$ at the DFE.
5. Form $K = FV^{-1}$ and compute $R_0 = \rho(K)$.

Below, we review three examples of NGM generation from the models discussed in this chapter. First, consider the normalized SIR model with frequency-dependent incidence:

$$\dot{S} = -\beta\frac{SI}{N}, \qquad \dot{I} = \beta\frac{SI}{N} - \gamma I, \qquad \dot{R} = \gamma I,$$

where $N = S + I + R$ is constant. The DFE is $(S^*, I^*, R^*) = (N, 0, 0)$. Take the infected vector $\mathbf{x} = (I)$. The new infection term is $F(I, S) = \beta SI/N$ and the remaining term is $V(I) = \gamma I$. Thus,

$$F = \frac{\partial}{\partial I}\left(\beta\frac{SI}{N}\right)\bigg|_{\text{DFE}} = \beta, \qquad V = \frac{\partial}{\partial I}(\gamma I)\bigg|_{\text{DFE}} = \gamma.$$

Hence $K = \beta/\gamma$ (a 1×1 matrix) and $R_0 = \rho(K) = \frac{\beta}{\gamma}$.

A slightly more complex case is SIR with demography, recall that the ODE takes the form (births and natural deaths at rate μ):

$$\dot{S} = \mu N - \beta\frac{SI}{N} - \mu S, \qquad \dot{I} = \beta\frac{SI}{N} - (\gamma + \mu)I, \qquad \dot{R} = \gamma I - \mu R.$$

At the DFE, $S^* = N$ and $I^* = R^* = 0$. Taking $\mathbf{x} = (I)$ again, $F = \beta$, $V = \gamma + \mu$, so $R_0 = \frac{\beta}{\gamma+\mu}$. Going forward to a larger NGM matrix, we consider SIR with a multip;e age groups. Let there be m groups with susceptible and infectious compartments $\{S_i, I_i\}_{i=1}^{m}$, total N_i per group, and group-specific recovery rates γ_i. Suppose new infections in group i arise from infectious individuals in group j via coefficients β_{ij}:

$$\dot{I}_i = \sum_{j=1}^{m} \beta_{ij}\frac{S_i}{N_i}I_j - \gamma_i I_i, \qquad i = 1, \ldots, m.$$

At the DFE, $S_i^* = N_i$. Taking $\mathbf{x} = (I_1, \ldots, I_m)^\top$, we have

$$F_{ij} = \frac{\partial}{\partial I_j}\left(\sum_{k=1}^{m} \beta_{ik}\frac{S_i}{N_i} I_k\right)\Bigg|_{\text{DFE}} = \beta_{ij}, \qquad V = \operatorname{diag}(\gamma_1, \ldots, \gamma_m).$$

Thus $K_{ij} = \frac{\beta_{ij}}{\gamma_j}$ and $R_0 = \rho(K)$. In this setting, the dominant eigenvector of K indicates which groups are most represented in early growth, and K can be visualized as a heatmap (a particularly informative figure for stratified models).

To further demonstrate the usage of NGM, Fig. 3.21 provides an interpretable summary of transmission structure in a stratified (e.g., age- or risk-group) setting via the NGM. The heatmap entries correspond to K_{ij}, the expected number of secondary infections in group i generated by an infectious individual from group j over the course of their infectious period, evaluated at the DFE. The dynamics for each group i are

$$\frac{dS_i}{dt} = -\frac{S_i}{N_i}\sum_{j=1}^{4}\beta_{ij} I_j, \qquad \frac{dI_i}{dt} = \frac{S_i}{N_i}\sum_{j=1}^{4}\beta_{ij} I_j - \gamma_i I_i, \qquad \frac{dR_i}{dt} = \gamma_i I_i,$$

with DFE given by $I_i^* = R_i^* = 0$ and $S_i^* = N_i$ for all i. The transmission coefficients β_{ij} (infections in recipient group i caused by an infectious individual in source group j) and recovery rates γ_i used to generate the heatmap are

$$\beta = \begin{pmatrix} 0.60 & 0.20 & 0.10 & 0.05 \\ 0.15 & 0.55 & 0.20 & 0.10 \\ 0.05 & 0.25 & 0.45 & 0.15 \\ 0.02 & 0.10 & 0.18 & 0.40 \end{pmatrix}, \qquad \gamma = \begin{pmatrix} 0.20 \\ 0.25 \\ 0.30 \\ 0.22 \end{pmatrix}.$$

Under the NGM construction, $F = \beta$, $V = \operatorname{diag}(\gamma_1, \gamma_2, \gamma_3, \gamma_4)$, and hence

$$K = FV^{-1} = \beta\, \operatorname{diag}(\gamma_1^{-1}, \gamma_2^{-1}, \gamma_3^{-1}, \gamma_4^{-1}) = \begin{pmatrix} 3.000 & 0.800 & 0.333 & 0.227 \\ 0.750 & 2.200 & 0.667 & 0.455 \\ 0.250 & 1.000 & 1.500 & 0.682 \\ 0.100 & 0.400 & 0.600 & 1.818 \end{pmatrix}.$$

For these parameters, the spectral radius is $R_0 = \rho(K) \approx 3.863$, and the associated dominant right eigenvector (normalized to sum to one) is approximately

$$\mathbf{v} \approx (0.379, 0.288, 0.200, 0.133)^\top,$$

indicating that, during the early exponential-growth regime near the DFE, infections are most concentrated in groups 1 and 2 for this illustrative contact structure.

Practically, for numerical analysis, it is convenient to implement the Jacobians F and V directly at the DFE and then compute $K = FV^{-1}$ and $R_0 = \rho(K)$. When m is small, one may use explicit formulas; for general models, a robust approach is to compute eigenvalues of K. To this end, Listing 3.2 implements a minimal computational pipeline for the NGM method once the Jacobian matrices F and V have already been derived and evaluated at the DFE. The input matrices are assumed to correspond

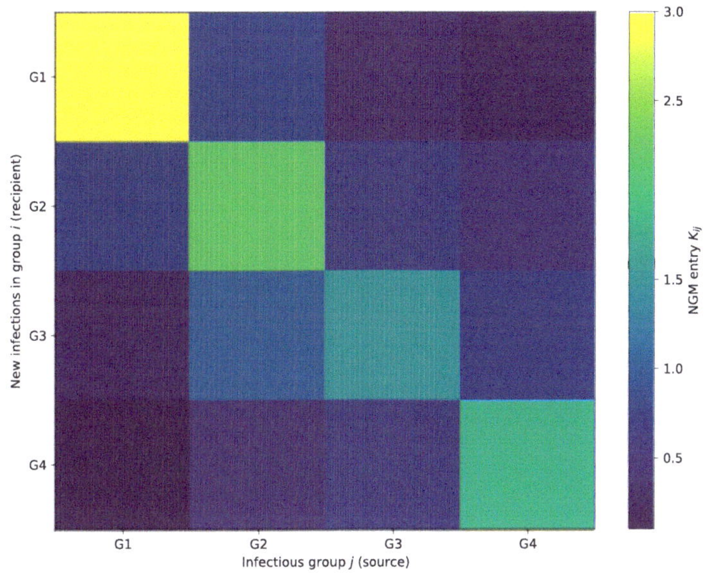

FIGURE 3.21

Heatmap visualization of the next-generation matrix K for a multigroup SIR model. Each entry K_{ij} represents the expected number of new infections in recipient group i caused by a single infectious individual from source group j introduced into an otherwise disease-free population.

to an infected-state vector $\mathbf{x} \in \mathbb{R}^m$, so that $F, V \in \mathbb{R}^{m \times m}$. The function `next_generation_matrix(F,V)` constructs the NGM directly from its defining formula. Internally, the lines `F = np.asarray(F, dtype=float)` and `V = np.asarray(V, dtype=float)` coerce the inputs into floating-point NumPy arrays, which ensures consistent linear-algebra behavior even if the user passes lists or integer arrays. The returned value `F @ np.linalg.inv(V)` uses matrix multiplication (@) to compute F times the matrix inverse of V. In applications, V is typically invertible under standard assumptions, making the construction well-defined. Later, the function `spectral_radius(A)` computes the spectral radius $\rho(A)$, defined as the maximum absolute value among the eigenvalues of A: $\rho(A) = \max_i |\lambda_i(A)|$. It first obtains the (possibly complex) eigenvalues using `np.linalg.eigvals(A)` and then returns the largest modulus with `np.max(np.abs(eigvals))`. This operation is central because, by the NGM theory, the basic reproduction number satisfies $R_0 = \rho(K)$. Finally, `compute_R0(F,V)` provides a convenience wrapper that executes the full workflow.

```
import numpy as np

def next_generation_matrix(F, V):
    """
    Construct the next generation matrix K = F V^{-1}.
    Parameters
    ----------
    F : (m,m) array_like
```

```
        Jacobian of new infection terms at the DFE.
    V : (m,m) array_like
        Jacobian of transition terms at the DFE.
    """
    F = np.asarray(F, dtype=float)
    V = np.asarray(V, dtype=float)
    return F @ np.linalg.inv(V)

def spectral_radius(A):
    """Return the spectral radius rho(A)."""
    eigvals = np.linalg.eigvals(A)
    return np.max(np.abs(eigvals))

def compute_R0(F, V):
    """Compute R0 = rho(F V^{-1})."""
    K = next_generation_matrix(F, V)
    return spectral_radius(K), K
```

Listing 3.2: Python utilities for constructing the next-generation matrix and computing R_0 from Jacobians.

For example, for the 3-group SIR model, F is the transmission matrix (β_{ij}) and $V = \text{diag}(\gamma_i)$. As such, a specific case can take the following format:

```
import numpy as np

# Example with m=3 groups
Beta = np.array([
    [0.6, 0.2, 0.1],
    [0.1, 0.5, 0.2],
    [0.0, 0.2, 0.4]
])

gamma = np.array([0.2, 0.25, 0.3])
F = Beta
V = np.diag(gamma)

R0, K = compute_R0(F, V)
print("K =\n", K)
print("R0 =", R0)
```

Listing 3.3: Python computation of R_0 for a multigroup SIR model from transmission and recovery parameters.

Deepen your knowledge

This pedagogical section provides short exercises designed to build intuition for *process extensions*: time-varying transmission and stochastic epidemic dynamics. Each exercise includes a worked solution and accompanying Python code that can be executed as-is.

Exercise 1: Seasonal forcing and peak-to-trough transmission

Consider seasonal forcing

$$\beta(t) = \beta_0 \left[1 + \epsilon \cos\left(\frac{2\pi t}{T} \right) \right], \qquad 0 \le \epsilon < 1.$$

(a) Compute $\beta_{\max}$ and $\beta_{\min}$. **(b)** Compute the peak-to-trough ratio $\beta_{\max}/\beta_{\min}$. **(c)** If $R_0(t) \approx \beta(t)/\gamma$ near the disease-free state, when is the system locally supercritical ($R_0(t) > 1$)?

Solution

(a) Since $\cos(\cdot) \in [-1, 1]$, we have

$$\beta_{\max} = \beta_0(1+\epsilon), \qquad \beta_{\min} = \beta_0(1-\epsilon).$$

(b) The peak-to-trough ratio is

$$\frac{\beta_{\max}}{\beta_{\min}} = \frac{1+\epsilon}{1-\epsilon}.$$

(c) Near $S \approx N$, $R_0(t) \approx \beta(t)/\gamma$, so $R_0(t) > 1$ iff $\beta(t) > \gamma$. Equivalently,

$$\beta_0 \left[1 + \epsilon \cos\left(\frac{2\pi t}{T} \right) \right] > \gamma,$$

which can be solved for the set of times where the cosine term is large enough. This highlights that even if the *mean* β_0/γ is above one, there can be subcritical seasons (and vice versa).

Python (sanity check plot)

```
import numpy as np
import matplotlib.pyplot as plt
# parameters
beta0 = 0.35
eps   = 0.25
T     = 365.0
gamma = 1/7
# time grid (one year)
t = np.linspace(0, 365, 800)
# seasonal transmission and local threshold
beta_t = beta0 * (1 + eps * np.cos(2*np.pi*t/T))
# plot
plt.figure()
```

```
plt.plot(t, beta_t, label=r'$\beta(t)$')
plt.plot(t, gamma * np.ones_like(t), '--', label=r'$\gamma$ threshold')
plt.xlabel("Time (days)")
plt.ylabel("Rate")
plt.grid(True)
plt.legend()
plt.show()
```

Listing 3.4: Sanity-check plot for seasonal forcing and its threshold relative to the recovery rate.

Exercise 2: Piecewise interventions and peak reduction

In an SIR model with time-varying $\beta(t)$, let

$$\beta(t) = \begin{cases} \beta_0, & t < t_1, \\ \beta_1, & t_1 \le t < t_2, \\ \beta_2, & t \ge t_2, \end{cases}$$

with $\beta_1 < \beta_0$ representing a strong intervention. **(a)** Explain why an intervention can reduce the peak even if it occurs *after* the outbreak begins. **(b)** Using simulation, compare the peak infectious fraction under constant β vs. piecewise $\beta(t)$.

Solution

(a) In SIR, early growth is driven by whether $\beta(t)S(t)/N > \gamma$. An intervention reduces $\beta(t)$, which reduces the force of infection and can push the system closer to (or below) the threshold, slowing growth and lowering the peak. This operates even if susceptibles are not yet depleted; hence timing matters but late interventions can still flatten peaks. (b) The simulation below computes and prints the peak values.

Python (ODE simulation)

```
import numpy as np
import matplotlib.pyplot as plt
from scipy.integrate import solve_ivp

def sir_timevarying(t, y, beta_fun, gamma):
    S, I, R = y
    beta_t = beta_fun(t, S, I, R)
    dSdt = -beta_t * S * I
    dIdt =  beta_t * S * I - gamma * I
    dRdt =  gamma * I
    return [dSdt, dIdt, dRdt]

# parameters
gamma = 1/7
beta0 = 0.35
```

```
# beta(t) options
def beta_constant(t, S, I, R):
    return beta0

def beta_piecewise(t, S, I, R):
    if t < 60:
        mult = 1.0
    elif t < 160:
        mult = 0.45
    else:
        mult = 0.80
    return beta0 * mult

T = 365.0
eps = 0.25
def beta_seasonal(t, S, I, R):
    return beta0 * (1.0 + eps * np.cos(2.0*np.pi*t/T))

k = 18.0
def beta_behavioral(t, S, I, R):
    return beta0 / (1.0 + k*I)

# initial conditions
S0, I0, R0 = 0.995, 0.005, 0.0
y0 = [S0, I0, R0]

# time grid
t_span = (0, 300)
t_eval = np.linspace(*t_span, 1200)

# solve helper
def solve(beta_fun):
    sol = solve_ivp(
        sir_timevarying,
        t_span,
        y0,
        args=(beta_fun, gamma),
        t_eval=t_eval,
        rtol=1e-8,
        atol=1e-10
    )
    return sol.y

# solve models
S_c, I_c, R_c = solve(beta_constant)
S_p, I_p, R_p = solve(beta_piecewise)
S_s, I_s, R_s = solve(beta_seasonal)
```

```
S_b, I_b, R_b = solve(beta_behavioral)

# plot infectious fraction
plt.figure(figsize=(7.8, 4.9))
plt.plot(t_eval, I_c, '-', label=r"Constant $\beta$")
plt.plot(t_eval, I_p, '--', label=r"Piecewise $\beta(t)$ (intervention)")
plt.plot(t_eval, I_s, '-.', label=r"Seasonal $\beta(t)$")
plt.plot(t_eval, I_b, ':', label=r"Behavioral feedback $\beta(t)$")
plt.xlabel("Time (days)")
plt.ylabel("Infectious fraction")
plt.xlim(t_span)
plt.xticks(np.arange(0, 301, 30))
plt.grid(True)
plt.legend()
plt.tight_layout()
plt.show()
```

Listing 3.5: ODE simulation comparing constant, piecewise, seasonal, and behavioral-feedback transmission $\beta(t)$ in an SIR model.

Exercise 3: CTMC SIR and extinction probability

Consider the stochastic SIR CTMC with transitions

$$(S, I, R) \to (S-1, I+1, R) \text{ at rate } \beta\frac{SI}{N}, \qquad (S, I, R) \to (S, I-1, R+1) \text{ at rate } \gamma I.$$

(a) Explain why extinction is possible even when $R_0 = \beta/\gamma > 1$. **(b)** Implement Gillespie simulation and estimate the extinction probability from $I(0) = 1$. **(c)** Compare to the Poisson branching approximation $q = \exp\{-R_0(1-q)\}$.

Solution

(a) Early on, I is small, so randomness in event timing is decisive: recovery can occur before the infection generates secondary cases, terminating the chain. Deterministic models do not represent this discrete chance. (b) Simulate many realizations and count the fraction that hit $I = 0$ before a time horizon. (c) Solve the fixed-point equation numerically and compare.

Python (Gillespie + branching approximation)

```
import numpy as np

def gillespie_sir(beta, gamma, N, S0, I0, R0, t_max, rng):
    t = 0.0
    S, I, R = int(S0), int(I0), int(R0)

    while t < t_max and I > 0:
        rate_inf = beta * S * I / N
```

```
        rate_rec = gamma * I
        rate_tot = rate_inf + rate_rec
        if rate_tot <= 0:
            break

        # time to next event
        dt = rng.exponential(1.0 / rate_tot)
        t += dt
        if t > t_max:
            break

        # which event occurs?
        if rng.random() < rate_inf / rate_tot:
            if S > 0:
                S -= 1
                I += 1
        else:
            I -= 1
            R += 1

    return (I == 0)  # extinct by time t_max?

# parameters
rng = np.random.default_rng(0)
N = 200
gamma = 1/7
beta  = 0.23
t_max = 120

# initial conditions
S0, I0, R0 = N-1, 1, 0

# simulate many realizations
n_sims = 5000
extinct = 0
for _ in range(n_sims):
    extinct += gillespie_sir(beta, gamma, N, S0, I0, R0, t_max, rng)

p_ext = extinct / n_sims
R0_det = beta / gamma

print("R0 =", R0_det)
print("Estimated extinction probability =", p_ext)
```

Listing 3.6: Gillespie simulation for the stochastic SIR CTMC and estimation of early extinction probability.

Exercise 4: Adding an observation model (POMP idea)

Let the latent epidemic be a CTMC (or ODE), but reported cases are noisy. Suppose the true new infections in week t are C_t, and observed cases are

$$Y_t \mid C_t \sim \text{NegBin}(\text{mean} = \rho C_t, \ \text{dispersion} = \kappa).$$

(a) Explain why changing ρ can mimic changes in $\beta(t)$ if ρ is not modeled. **(b)** Simulate weekly observations from an ODE SIR trajectory with fixed ρ and κ. **(c)** Plot C_t and Y_t to see how observation noise obscures the latent signal.

Solution

(a) If only Y_t is seen, a drop in reporting (smaller ρ) looks like a drop in incidence, which could be falsely attributed to reduced transmission. Without an observation model, inference tends to absorb reporting changes into $\beta(t)$. (b,c) The code below computes weekly incidence from an ODE solution and generates negative-binomial noisy observations.

Python (ODE latent + negative binomial observation)

```
import numpy as np
import matplotlib.pyplot as plt
from scipy.integrate import solve_ivp

def sir_rhs(t, y, beta, gamma):
    S, I, R = y
    dSdt = -beta * S * I
    dIdt =  beta * S * I - gamma * I
    dRdt =  gamma * I
    return [dSdt, dIdt, dRdt]

# parameters
beta = 0.35
gamma = 1/7

# initial conditions
S0, I0, R0 = 0.995, 0.005, 0.0
y0 = [S0, I0, R0]

# solve latent ODE
t_span = (0, 160)
t_eval = np.linspace(*t_span, 2000)
sol = solve_ivp(sir_rhs, t_span, y0, args=(beta, gamma), t_eval=t_eval)
S, I, R = sol.y
t = sol.t

# latent weekly infections C_t = \int beta S I dt over each week
lam = beta * S * I
week = 7.0
```

```
edges = np.arange(t_span[0], t_span[1] + week, week)

C = []
for a, b in zip(edges[:-1], edges[1:]):
    mask = (t >= a) & (t < b)
    C.append(np.trapz(lam[mask], t[mask]))
C = np.array(C)

# observation model: Y_t | C_t ~ NegBin(mean = rho*C_t, dispersion = kappa)
rng = np.random.default_rng(1)
rho = 0.35
kappa = 20.0  # larger -> closer to Poisson

mean = rho * C
# Gamma-Poisson mixture for NegBin:
Lambda = rng.gamma(shape=kappa, scale=np.maximum(mean, 1e-12) / kappa)
Y = rng.poisson(Lambda)

# plot latent vs observed
plt.figure()
plt.plot(C, marker='o', label=r"latent weekly infections $C_t$")
plt.plot(Y, marker='s', label=r"reported cases $Y_t$")
plt.xlabel("Week index")
plt.ylabel("Count (arbitrary scale)")
plt.grid(True)
plt.legend()
plt.show()
```

Listing 3.7: ODE latent process with a negative-binomial observation model (POMP-style).

Exercise 5 (optional): When does time-variation become too flexible?

Suppose $\beta(t)$ is allowed to vary freely (e.g., one parameter per day). **(a)** Explain how this can absorb misspecification (missing compartments, wrong delays, or reporting changes). **(b)** Propose two regularization strategies to keep $\beta(t)$ interpretable.

Solution

(a) If the model cannot represent latent periods, reporting delays, or changing ascertainment, a flexible $\beta(t)$ can be tuned to match the observed curve anyway, but the fitted $\beta(t)$ no longer corresponds to genuine transmission changes. (b) Two common strategies are: (i) low-dimensional parameterizations (seasonal sinusoid, few change points), and (ii) smoothness penalties / splines with regularization, ideally anchored to covariates (mobility, policy indices) and paired with an explicit observation model.

CHAPTER

Spatio-temporal compartmental models

4

Spatially explicit compartmental models relax the homogeneous (well-mixed) contact assumption (recall that under the well-mixing assumption, for each point in time, every pair of individuals is equally likely to interact; equivalently, the instantaneous *who-meets-whom* distribution is uniform over the population) by replacing a single, population-level force of infection with a structured contact operator that depends on location and movement. As such, transmission is no longer driven by $I(t)/N(t)$ alone, but by a location-indexed prevalence field together with (i) connectivity constraints (who can contact whom) and (ii) mobility (who is co-located with whom, and when). This perspective unifies a wide family of spatio-temporal models: the main modeling choice is how to encode the operator that maps spatially distributed infectiousness into a hazard of infection for each susceptible individual or subpopulation [247–249].

In this chapter, we focus on three common strategies for building that operator. First, graph-based (discrete-space) extensions represent space as nodes (e.g., regions, cities, facilities), with edges encoding interaction opportunities via travel, commuting, or other mixing channels. The dynamics become a coupled system of compartmental models, typically with an explicit movement matrix (possibly time-varying) that transports individuals or redistributes effective contacts across nodes. This "metapopulation on a network" view is particularly natural when mobility data are intrinsically relational (origin–destination flows), and it has been used from global airline-mediated spread to multiscale commuting–air travel coupling [250–252]. Second, norm-based (also known as distance-based or metric-kernel-based) extensions parameterize interaction strength as a decreasing function of spatial separation (or a generalized distance on a graph), yielding lattice, kernel, or diffusion-like formulations; these models often expose interpretable notions such as spatial wavefronts, invasion speeds, and characteristic spread scales [253,254]. A third approach blends spatial information into an otherwise temporal model by making key rates explicitly data-driven. Rather than indexing compartments by location, one keeps a low-dimensional state (e.g., three states like in the SIR model) and allows parameters (most notably $\beta(t)$) to depend on mobility-derived covariates that proxy changing contact opportunity, co-location, or importation pressure [255–257]. For instance, Berke et al. [258] provide a clear example in a national case study (Andorra) that combines near-complete telecom coverage with high-participation serology data. They consider an SEIR backbone in which is (a) constant (baseline), (b) linked via a monotone (log-linear/exponential) function to a trips-based mobility metric, or (c) linked to a trips metric augmented by an "entrances" term that captures additional seeding from outside the country. To connect latent epidemic dynamics to surveillance records, they include a reporting process in which only a fraction of recoveries is observed after a fixed delay. This approach is prone to over-fitting, and, while providing a promise of simplicity, it is rarely really easily explainable. As such, it will be out of the scope of this book.

Epidemic Modeling. https://doi.org/10.1016/B978-0-44-355525-1.00010-2

4.1 Graphs of locations: metapopulation and network-of-patches models

Location-graph models operationalize these mechanisms by discretizing space into patches (cities, districts, regions) and coupling patch-level compartmental dynamics through a mobility or contact network. The resulting system can be read as a dynamical process on a graph, where nodes host local epidemics, and edges determine seeding, persistence, and synchronization patterns [208,259,260].

Adopting the simplest temporal epidemiological model from the previous chapter, the SIR model, with these spatial extensions, takes the following form. Let locations be indexed by $p = 1, \ldots, P < \infty$, with $(S_p(t), I_p(t), R_p(t))$ denoting the patch-level (i.e., graph's node) compartments and $N_p(t) = S_p(t) + I_p(t) + R_p(t)$ the (possibly time-varying) patch population. We still assume that $\sum_{p \in P} N_p(t) = N$, where N is constant in time. A minimal metapopulation specification introduces a time-constant mobility matrix $M = [m_{pq}]$, where $m_{pq} \geq 0$ is the per-capita movement rate from p to q. A common migration/flux formulation treats movement as continuous exchange occurring in all compartments, yielding

$$\frac{dS_p}{dt} = -\beta_p \frac{S_p I_p}{N_p} + \underbrace{\sum_{q \neq p} (m_{qp} S_q - m_{pq} S_p)}_{\text{movement term}}, \tag{4.1}$$

$$\frac{dI_p}{dt} = \beta_p \frac{S_p I_p}{N_p} - \gamma_p I_p + \underbrace{\sum_{q \neq p} (m_{qp} I_q - m_{pq} I_p)}_{\text{movement term}}, \tag{4.2}$$

$$\frac{dR_p}{dt} = \gamma_p I_p + \underbrace{\sum_{q \neq p} (m_{qp} R_q - m_{pq} R_p)}_{\text{movement term}}. \tag{4.3}$$

When M is balanced (e.g., row-sums match inflow in an appropriate sense), total population is conserved globally; otherwise, $N_p(t)$ varies as the epidemic redistributes people across patches. It is often convenient to interpret M as a continuous-time Markov generator, with $m_{pp} = -\sum_{q \neq p} m_{pq}$, which makes bookkeeping of net fluxes transparent [150].

Fig. 4.1 visualizes the location graph implied by the mobility matrix $M = [m_{pq}]$. Each node represents a patch $p \in \{1, \ldots, P\}$, and each directed edge $p \to q$ indicates a positive per-capita movement rate $m_{pq} > 0$, i.e., the flux of individuals leaving patch p for patch q in the migration/flux formulation. Edge labels report the numerical values of m_{pq} (and edge thickness may be used to emphasize relative magnitude), making the coupling structure explicit. The matrix shown beneath the graph is the same M written in tabular form; reading across row p gives all destinations q reachable from origin p, while reading down column q gives all origins sending flow into q. This graphical view is often the most immediate way to see whether the network is sparse or dense, whether coupling is approximately symmetric, and which patches are most central as sources or sinks of movement, all of which shape seeding patterns and synchronization in metapopulation epidemics.

A complementary mixing/force-of-infection formulation keeps individuals anchored to a "home" patch but allows infectious pressure to be imported through visitation, commuting, or shared venues.

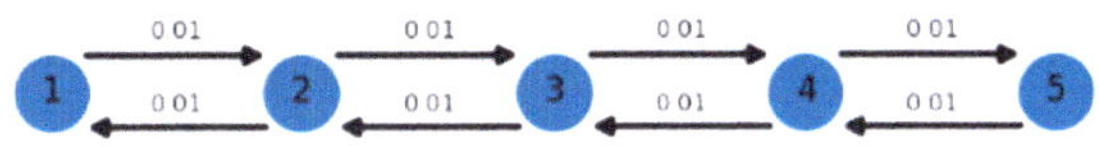

Mobility matrix M (entries m_{pq}: per-capita rate from p to q)

	1	2	3	4	5
1	0.00	0.01	0.00	0.00	0.00
2	0.01	0.00	0.01	0.00	0.00
3	0.00	0.01	0.00	0.01	0.00
4	0.00	0.00	0.01	0.00	0.01
5	0.00	0.00	0.00	0.01	0.00

FIGURE 4.1

Location graph induced by the mobility matrix M. Nodes are patches, and directed edges $p \to q$ indicate movement at rate m_{pq}; the matrix beneath the graph reports the same rates in origin–destination form.

Table 4.1 Two common couplings for SIR dynamics on a location graph. Both relax well-mixing across space, but differ in whether they explicitly relocate individuals or only import infectious pressure.

	Flux / migration	**FOI / mixing**
State variables	(S_p, I_p, R_p) are the "present" counts in patch p	(S_p, I_p, R_p) are typically "home" counts in patch p
Coupling object	Movement rates m_{pq} transporting all compartments	Effective contact rates β_{pq} mixing prevalence across patches
Best suited when	Modeling relocation, migration, long stays; conservation matters	Modeling commuting/visitation, exposure importation, venue sharing
Common pitfall	Interpreting I_p as "residents" when it is "present"	Hiding mechanistic movement assumptions inside β_{pq}

Formally, the operator takes the form:

$$\lambda_p(t) = \sum_{q=1}^{P} \beta_{pq} \frac{I_q(t)}{N_q(t)}, \tag{4.4}$$

where the susceptible equation becomes $dS_p(t)/dt = -\lambda_p(t)S_p$, with corresponding infected and recovered equations as in the usual temporal model but driven by $\lambda_p(t)$. Here β_{pq} summarizes effective cross-patch coupling without explicitly relocating people [145]. Practically, this viewpoint is often closer to what mobility data provide – time spent elsewhere, origin–destination fluxes, co-location proxies. Conceptually, it replaces the *who moved* perspective with *who contributed to exposure* perspective. Table 4.1 provides a simple comparison between these two approaches.

In both formulations, the location graph determines who can infect whom. A patch that would be locally subcritical in isolation (i.e., small population, strong interventions, short infectious period) may nonetheless experience repeated reintroductions that prevent fade-out, a spatial analogue of metapopulation *rescue*, where persistence emerges from the balance between local extinction and recolonization [261]. This mechanism becomes especially prominent when demographic or transmission stochasticity is included: the relevant question is no longer only "does $R_0 > 1$ locally?" but also "how often are sparks imported, and how likely are they to ignite?" [242].

Either way, location graphs introduce competition between at least two time scales: a local epidemic scale set by transmission and recovery, and a mobility scale set by movement/contact coupling. When mobility is weak relative to local growth, outbreaks tend to be asynchronous and spread as a sequence of seeded local epidemics. Empirically, such regimes can manifest as traveling-wave-like patterns and spatial hierarchies in incidence [262]. When mobility is strong, patches synchronize, and the system approaches an effectively well-mixed aggregate where spatial detail mainly affects timing and transient heterogeneity rather than overall attack rate (defined as the proportion of those who became ill after a specified exposure over a specified period of time) [145]. In this context, a particularly useful conceptual tool is that spread is often better described by an *effective distance* induced by weighted travel fluxes than by Euclidean geography. In the effective-distance view, the "geometry" of contagion is set by connectivity, as outbreaks can trace approximately concentric wavefronts in effective-distance space even when geographic maps show irregular patterns. This helps explain why long-range, high-flux edges (e.g., air travel and major commuting corridors) can dominate arrival times, whereas purely geographic adjacency may be a poor predictor [263,264].

From a computational perspective, Listing 4.1 implements a deterministic metapopulation SIR model over $P = 5$ patches by integrating a $3P$-dimensional ODE with `solve_ivp`. After importing `numpy`, `solve_ivp`, and `matplotlib`, the code sets patch-specific transmission and recovery parameters via `beta = np.array([...])` and `gamma = np.array([...])`, and defines a mobility (flux) matrix `M`, where `M[p,q]` is the per-capita rate of movement from patch p to patch q with a zero diagonal (commented as `# mobility rates m_pq (p->q), zero diagonal`). The ODE right-hand side is given by `def rhs(t, y):`, where the stacked state vector `y` is unpacked into patchwise compartments using `S = y[0:P]`, `I = y[P:2*P]`, and `R = y[2*P:3*P]`, and the local patch population is computed as `N = S + I + R`. The next block encodes within-patch SIR dynamics, with incidence and recovery written in vectorized form as `dS = -beta * S * I / np.maximum(N, 1e-12)`, `dI = beta * S * I / np.maximum(N, 1e-12) - gamma * I`, and `dR = gamma * I`; the `np.maximum(N, 1e-12)` safeguard prevents division by zero in the frequency-dependent term. Mobility is then added in flux form (comment `# flux coupling: sum_q!=p (m_qp X_q - m_pq X_p)`) by looping over patches `for p in range(P):` and computing, for each compartment $X \in \{S, I, R\}$, an inflow `np.sum(M[:, p] * X)` and an outflow `np.sum(M[p, :] * X[p])`, then updating derivatives via `dS[p] += inflow_S - outflow_S` (and analogously for `dI` and `dR`); this structure enforces that movement redistributes individuals between patches rather than creating or removing them. The function returns the concatenated derivative vector with `return np.concatenate([dS, dI, dR])`. Initial conditions are set by specifying patch totals `N0 = np.array([...])`, seeding infection in a single patch using `I0 = np.zeros(P); I0[0] = 50.0`, and forming `S0 = N0 - I0`, `R0 = np.zeros(P)`, and the stacked initial state `y0 = np.concatenate([S0, I0, R0])`. The system is integrated on `(0, 200)` with `t_eval=np.linspace(0, 200, 1000)`, the solution is unpacked back into time series arrays `S`, `I`, `R`, and `N = S + I + R`.

```
import numpy as np
from scipy.integrate import solve_ivp
import matplotlib.pyplot as plt

# P patches
P = 5
```

```
beta = np.array([0.30, 0.28, 0.25, 0.32, 0.27])
gamma = np.array([0.10, 0.10, 0.10, 0.10, 0.10])

# mobility rates m_pq (p->q), zero diagonal
M = np.array([
    [0, 0.01, 0.00, 0.00, 0.00],
    [0.01, 0, 0.01, 0.00, 0.00],
    [0.00, 0.01, 0, 0.01, 0.00],
    [0.00, 0.00, 0.01, 0, 0.01],
    [0.00, 0.00, 0.00, 0.01, 0],
], dtype=float)

def rhs(t, y):
    S = y[0:P]
    I = y[P:2*P]
    R = y[2*P:3*P]
    N = S + I + R

    # local infection
    dS = -beta * S * I / np.maximum(N, 1e-12)
    dI =  beta * S * I / np.maximum(N, 1e-12) - gamma * I
    dR =  gamma * I

    # flux coupling: sum_{q!=p} (m_qp X_q - m_pq X_p)
    for p in range(P):
        inflow_S = np.sum(M[:, p] * S) - M[p, p] * S[p]
        outflow_S = np.sum(M[p, :] * S[p]) - M[p, p] * S[p]
        inflow_I = np.sum(M[:, p] * I) - M[p, p] * I[p]
        outflow_I = np.sum(M[p, :] * I[p]) - M[p, p] * I[p]
        inflow_R = np.sum(M[:, p] * R) - M[p, p] * R[p]
        outflow_R = np.sum(M[p, :] * R[p]) - M[p, p] * R[p]

        dS[p] += inflow_S - outflow_S
        dI[p] += inflow_I - outflow_I
        dR[p] += inflow_R - outflow_R

    return np.concatenate([dS, dI, dR])

# initial condition: seed one patch
N0 = np.array([2e5, 1e5, 1.5e5, 1.2e5, 0.8e5], dtype=float)
I0 = np.zeros(P); I0[0] = 50.0
S0 = N0 - I0
R0 = np.zeros(P)

y0 = np.concatenate([S0, I0, R0])
sol = solve_ivp(rhs, (0, 200), y0, t_eval=np.linspace(0, 200, 1000))

S = sol.y[0:P, :]
```

```
I = sol.y[P:2*P, :]
R = sol.y[2*P:3*P, :]
N = S + I + R
```

Listing 4.1: Minimal metapopulation SIR simulation (flux form) and a conservation sanity check.

Moreover, Fig. 4.2 summarizes how the choice of location-coupling structure shapes epidemic potential over time. The top row shows three deliberately distinct 5×5 coupling matrices M (displayed as heatmaps with entries printed in each cell), interpreted here in a force-of-infection mixing form in which the per-patch infectious pressure is

$$\lambda_p(t) = \beta \sum_{q=1}^{P} M_{pq} \frac{I_q(t)}{N},$$

with $\beta = 0.025$ and constant patch size $N = 10\,000$. Starting from a small seed infection in one patch, we simulate a deterministic SIR system forward for 100 discrete time steps (explicit Euler updates with recovery rate $\gamma = 0.14$). At each step t, we compute an effective reproduction index

$$R_0(t) = \frac{\beta}{\gamma} \rho\Big(\text{diag}\Big(\frac{S(t)}{N}\Big)\, a(t)\, M\Big),$$

where $\rho(\cdot)$ denotes the spectral radius, and $\text{diag}(S(t)/N)$ accounts for depletion of susceptibles; the scalar $a(t)$ is an exogenous, time-varying contact-intensity multiplier used to mimic behavioral or policy-driven changes in mixing. The bottom row plots $R_0(t)$ for each matrix, with the horizontal dashed line marking the threshold $R_0(t) = 1$ – used to indicate an outbreak when the line above it and pandemic fading out when the line below it. Across panels, the curves illustrate two key lessons: (i) even with identical biological parameters (β, γ), different network/mixing topologies can produce markedly different initial epidemic potential (e.g., hub-like coupling yields a larger early spectral radius and thus higher initial R_0); and (ii) $R_0(t)$ is not static in spatial systems, because it responds both to interventions/contact modulation (via $a(t)$) and to endogenous susceptible depletion (via $S(t)$), causing transitions from supercritical to subcritical regimes at different times depending on connectivity. In short, the figure highlights that "who mixes with whom" can be as important as β and γ for understanding the timing and strength of metapopulation spread.

Over the classical SIR model that is divided into multiple locations, many natural extensions derive from this formalization without the need to add more dynamics. The most common extension is heterogeneous patch properties, in which we allow patch-specific parameters (e.g., β_p, γ_p) or layered within-patch structure to express differences in density, intervention intensity, and healthcare access [118]. Even with unchanged coupling, heterogeneity can shift which patches act as *sources* (net exporters of infection) versus *sinks* (net importers), which is often central to prioritizing surveillance and control. Another extension is the commuting and multi-timescale mobility, where daily commuting differs from migration: individuals spend time in multiple patches but repeatedly return home. Mechanistically, one can (i) introduce short-term "present-in-q" subcompartments for each home patch, or (ii) derive effective β_{pq} coefficients that approximate time spent away and where transmission occurs [145]. This matters because commuting can create strong directionality in exposure (weekday vs. weekend, day vs. night) and can generate patterns that are not captured by symmetric diffusion-like exchange. Lastly, the movement itself can be temporal to capture seasonality, crises, and policy

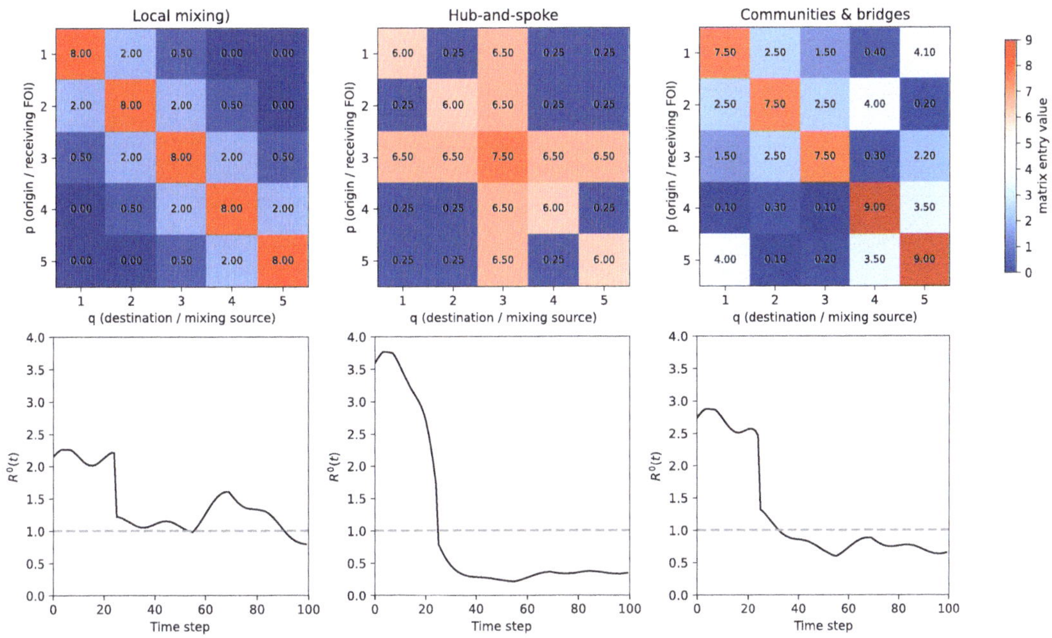

FIGURE 4.2

Top row: three contrasting 5×5 coupling matrices M (entries shown). Bottom row: the corresponding effective reproduction index $R^0(t)$ over 100 time steps computed from the spectral radius of $\mathrm{diag}(S(t)/N)\,a(t)\,M$ with $\beta = 0.025$, $\gamma = 0.14$, and $N = 10{,}000$ per patch; the dashed line marks the threshold $R^0(t) = 1$.

changes. Formally, making $M(t)$ or $\beta_{pq}(t)$ time-varying can capture abrupt connectivity shifts and their implications for importation pressure [265]. In practice, even when a model remains deterministic, time variation in coupling is often essential to avoid attributing behavioral or policy changes to biological parameters.

In order to obtain a better intuition of these extensions, Fig. 4.3 illustrates how epidemic outcomes in a metapopulation are shaped by structural heterogeneity, temporal mixing patterns, and abrupt changes in connectivity. In the top row, both simulations use the same mobility matrix, differing only in local transmission rates. While homogeneous parameters lead to synchronized decay across patches, introducing heterogeneity causes a single high-transmission patch to dominate late-stage dynamics. The middle row contrasts effective mixing patterns between weekdays and weekends. Weekday mobility exhibits strong directional coupling toward a central hub, whereas weekend mixing is largely local. These matrices highlight how commuting induces time-dependent and asymmetric coupling between patches, motivating the use of time-varying mixing matrices $M(t)$ rather than static networks. The bottom row shows the effect of an early connectivity shock, modeling a border closure or mobility restriction at day $t_c = 7$. The intervention sharply reduces global prevalence while affecting patches unevenly: highly connected patches respond immediately, whereas locally driven patches decay more slowly.

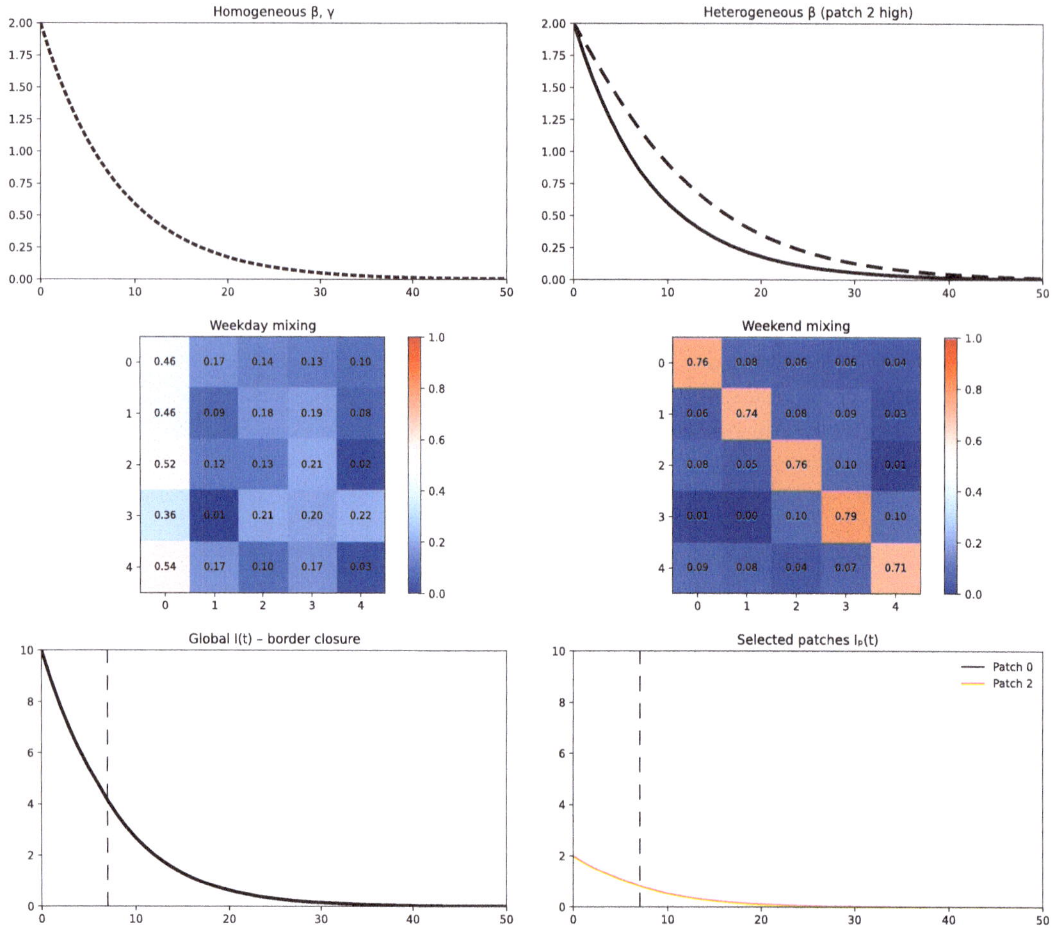

FIGURE 4.3

Metapopulation epidemic dynamics under heterogeneity, time-dependent mixing, and connectivity shocks. **(Top row)** Infection trajectories $I_p(t)$ on the same mobility network. Left: homogeneous transmission and recovery rates across patches. Right: heterogeneous transmission, with one high-density patch acting as a persistent driver. **(Middle row)** Effective mixing matrices. Left: weekday mixing dominated by directional commuting into a central hub. Right: weekend mixing characterized by stronger local (diagonal) interactions. **(Bottom row)** Early connectivity shock (border closure) at day $t_c = 7$. Left: global prevalence $I(t)$. Right: selected patch-level trajectories $I_p(t)$, highlighting differential responses to the intervention.

Motivating the usage of location-graph spatial extensions, here are two recent examples. First, Bognanni et al. [266] develop and estimate a spatial Econ-SIR framework that couples epidemiological dynamics with economic behavior. Agents endogenously adjust activity in response to infection risk and non-pharmaceutical interventions, and the resulting feedback alters transmission and economic outcomes. Importantly for the location-graph perspective, the model is explicitly spatial across

U.S. counties and is disciplined using a panel of high-frequency county-level epidemiological and economic data, including mobility and employment/activity measures. This line of work highlights a broader modeling lesson: once location coupling is present, behavioral responses (and their spatial heterogeneity) can be as important as biological heterogeneity for explaining why some patches surge while others lag. Second, Goel and Sharma [267] propose a mobility-augmented SIR model in which each node represents a location, and infection arises from both *local* transmission within a node and *global* transmission imported via movement from other nodes. Their formulation introduces a social connectivity parameter (capturing how strongly locations are linked) and an explicit mobility term $c(i, j)$ that modulates cross-node infectious pressure. Using call detail records between Estonian counties, they illustrate how changing connectivity or restricting movement from highly connected locations can shift epidemic curves in simulation. While such parsimonious models can be too coarse for accurate forecasting in real time, they provide a clean sandbox for stress-testing spatial pandemic intervention policies and for clarifying which spatial mechanisms a richer model must represent.

4.2 Graphs of individuals

Location graphs treat space as a set of interacting patches; an alternative is to make *individual-to-individual* contact structure explicit. In contact network models, individuals are nodes of a graph $G = (V, E)$ and an edge $(i, j) \in E$ indicates that individuals i and j have a potentially infectious interaction channel, such as direct physical contact, repeated proximity, household co-residence, or any defined "contact opportunity" [268,269]. This shifts the central modeling object from a population-level mixing operator to a graph of "transmission opportunities": heterogeneity is encoded in who is connected to whom, how densely, and in what patterns (hubs, communities, triangles) [270]. There is inherent trade-off between the location-based and infection-based.

Like before, we start with a continuous-time SIR transmission on a fixed contact network. In this context, each node is in one of three states (susceptible, infectious, removed), and infection spreads along edges: an infectious node transmits to each susceptible neighbor at rate τ (note, we do not use the β notation, as it is used for population-level infection dynamics), while recovering (becoming removed) at rate γ. Under these assumptions, the process is a continuous-time Markov chain on the space of node-state configurations, and it reduces to the classical homogeneous-mixing SIR only under strong symmetry/mean-field limits that effectively wash out network structure. A particularly useful quantity in this setting is the *edge transmissibility*: if recovery is exponential, then the probability that infection crosses a given $S - I$ edge before the infectious endpoint recovers is

$$T = \frac{\tau}{\tau + \gamma}. \tag{4.5}$$

This mapping is not merely a convenience. For broad classes of random networks, SIR final outbreak size can be related to bond percolation with bond-occupation probability T, connecting epidemic outcomes directly to connectivity and component structure [271]. Intuitively, in the percolation view, an epidemic explores the "occupied" subset of edges that would succeed in transmitting if challenged; a large outbreak is possible precisely when the occupied graph contains a giant connected component. Fig. 4.4 illustrates transmission on a contact network where nodes are individuals and edges represent potential infectious contact.

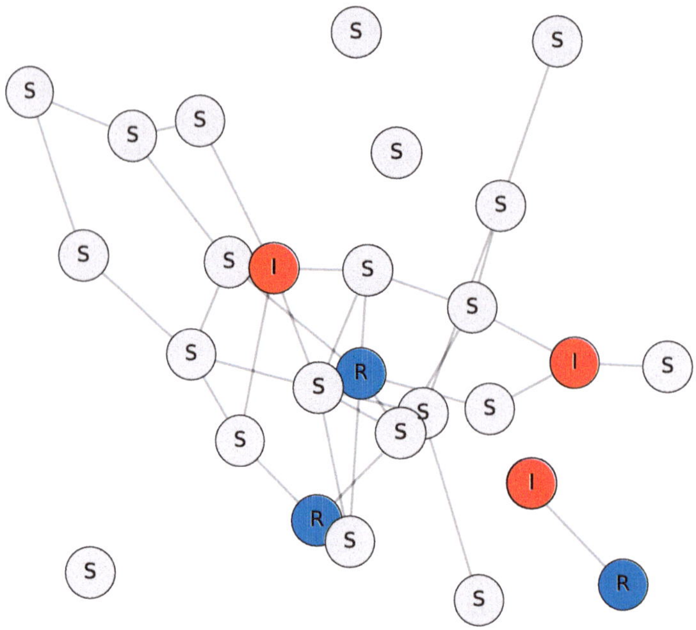

FIGURE 4.4

SIR transmission on a contact network. Nodes are individuals, and edges represent potential infectious contact.

Because transmission follows edges, epidemic thresholds depend strongly on connectivity heterogeneity rather than on population averages alone. For configuration-model networks with degree distribution $P(k)$, the percolation argument yields a simple condition for the possibility of a macroscopic outbreak:

$$T \frac{\langle k^2 \rangle - \langle k \rangle}{\langle k \rangle} > 1, \tag{4.6}$$

where $\langle k \rangle$ and $\langle k^2 \rangle$ are the first two moments[1] of the degree distribution [271]. The factor $(\langle k^2 \rangle - \langle k \rangle)/\langle k \rangle$ is the mean "excess degree" encountered when following a random edge and captures a key network mechanism: infection does not sample nodes uniformly but is biased toward high-degree individuals, because they are more likely to be reached through one of their many edges. In consequence, heavy-tailed degree distributions (many low-degree nodes plus a few hubs) can dramatically lower invasion thresholds and enlarge outbreaks, even when average degree is modest, because hubs create abundant onward transmission opportunities and accelerate early growth [272].

[1] Statistical moments are quantitative measures describing the shape of a data distribution, with the first four revealing its mean (location), variance (spread), skewness (asymmetry), and kurtosis (peakedness/tailedness).

Table 4.2 Representative threshold criteria illustrating how contact structure changes the objects that control epidemic potential.

Setting	Invasion / large-outbreak condition	What controls it?
Well-mixed SIR	$R_0 = \beta/\gamma > 1$	Mean infectious contacts
Network SIR (config.)	$T\,\frac{\langle k^2\rangle - \langle k\rangle}{\langle k\rangle} > 1$	Degree heterogeneity
Network SIS (spectral)	$\tau/\gamma > 1/\rho(A)$	Dense cores / feedback cycles

Table 4.2 summarizes a useful *thresholds-at-a-glance* view that highlights how the notion of epidemic take-off depends on the modeling resolution and, in particular, on what object encodes *who-meets-whom.*

A more dynamic case is the SIS model. For SIS-type dynamics, where recovered individuals return to susceptibility and reinfection is allowed, persistence thresholds are often expressed in spectral terms. A common approximation links invasion to the spectral radius of the adjacency matrix A: writing infection rate τ and recovery rate γ, sustained endemicity typically requires

$$\frac{\tau}{\gamma} > \frac{1}{\rho(A)}, \tag{4.7}$$

highlighting that long-run circulation depends on feedback loops and strongly connected substructures, rather than only on mean degree. This spectral viewpoint is also useful in SIR contexts as a qualitative guide: dense cores and tightly connected communities can act as local amplifiers that shape peak timing and spatially localized surges.

Beyond thresholds, network epidemics admit a hierarchy of tractable approximations that make explicit which structural features matter for growth, peak timing, and final size. Degree-based mean-field models track the fraction of degree-k nodes in each state and thus capture the early concentration of infection among highly connected individuals. More accurate descriptions on locally tree-like graphs are provided by edge-based compartmental models and message-passing formulations, which derive closed systems for quantities such as the probability that a random edge has not yet transmitted, and thereby yield sharp predictions for incidence curves and final size under broad conditions [273,274]. A unifying insight across these methods is that dynamics are governed not only by depletion of susceptible *people* but by depletion of susceptible *edges*: transmission opportunities vanish as edges become "used up" (a susceptible neighbor becomes infected/removed), which changes the effective branching process experienced by the infection. This makes clear why targeted interventions that preferentially remove or protect high-degree nodes can be disproportionately effective: they delete a large fraction of the network's transmission *capacity*, not merely a large fraction of its population.

Real contact networks, however, are rarely tree-like. Clustering (triangles), households, and repeated contacts introduce correlation and "redundancy": an infectious individual may repeatedly attempt transmission along edges leading to neighbors who have already been exposed through shared ties, which can slow spread relative to locally tree-like predictions. Pair-approximation methods and household-structured models provide common ways to incorporate such effects while remaining far more tractable than fully explicit agent-based simulations [270]. In practice, these extensions are essential when the units of close contact (households, classrooms, workplace teams) dominate transmission, because the relevant bottlenecks are then the connections between tightly knit groups rather than within them.

Many practical applications require enriching the edge set beyond a simple unweighted, untyped graph. Contacts differ in duration, intensity, and setting (home, work, school), and representing this heterogeneity can be done by weighting edges, assigning edge types, or using multilayer networks. This allows one to encode, for example, that household edges have high transmissibility while incidental community edges have low transmissibility, and it provides a clean way to represent setting-specific interventions (e.g., closing schools removes or down-weights school-type edges) in a manner that parallels, but is finer-grained than, multi-group mixing matrices. Moreover, contacts are often inherently time-dependent: people meet different sets of individuals across days and contexts, and the temporal ordering of encounters matters because transmission requires a *time-respecting path*. Temporal networks can be represented via time-indexed edge sets $E(t)$ or event lists, and they can either amplify or suppress spread relative to a static aggregation depending on how bursty, repetitive, and synchronized contacts are [275]. Finally, empirical grounding for individual contact graphs can come from contact surveys and sensor-based proximity studies, which reveal strong heterogeneity and setting-specific patterns that guide model structure and calibration.

Fig. 4.5 illustrates why temporal contact networks cannot, in general, be reduced to a static graph without potentially changing the set of feasible transmission routes. In the left panel, contacts occur as time-stamped events: B meets C at $t = 1$, and A meets B later at $t = 2.6$. If infection starts at A at $t = 0$, then A may infect B during their later contact, but B cannot subsequently infect C because the only B–C interaction occurred earlier in time; in other words, the path $A \to B \to C$ is not *time-respecting*. In the right panel, aggregating all contacts into a static network produces edges A–B and B–C, which suggests a connected path from A to C even though that route never exists in the correct temporal ordering. This simple example motivates temporal-network representations (time-indexed edge sets or contact event streams) whenever ordering, burstiness, or repeated contacts are expected to materially affect reachability, growth rates, or intervention impact.

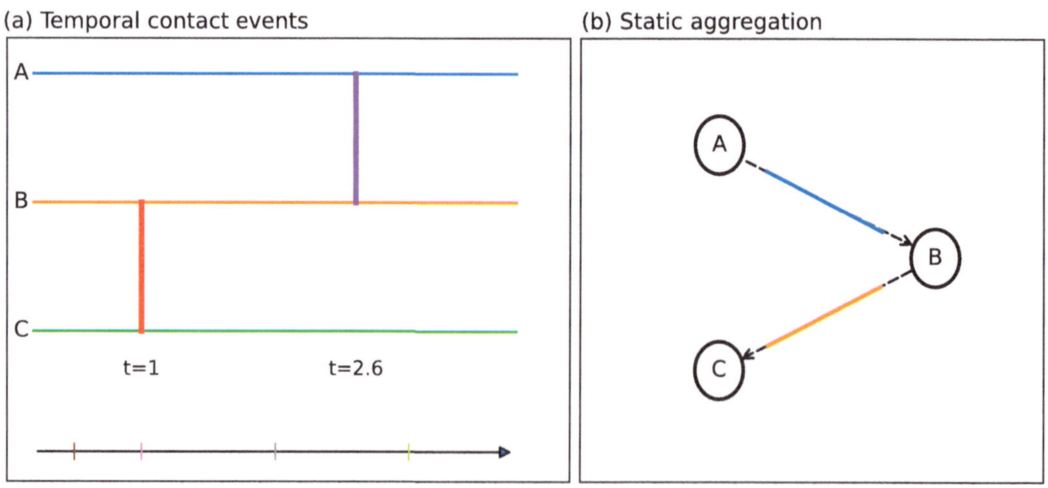

FIGURE 4.5

A transmission path must respect time ordering of contacts; aggregating all edges into a static graph can create spurious paths that never exist in time.

A major conceptual advantage of individual-network models is that many interventions admit an immediate graph interpretation. Vaccination, prophylaxis, or prior immunity can be represented as removal or protection of nodes; distancing measures can remove, rewire, or down-weight certain edge types; and contact tracing and quarantine can be represented as targeted deletion of nodes (isolating infected/contacts) or edges (blocking interactions) [270]. This lens is analytically powerful because it connects policy design directly to topology and to the distribution of transmission opportunities, clarifying why the same biological parameters can produce very different outcomes under different contact structures.

Individual contact networks are therefore particularly compelling when heterogeneity of contacts is central (superspreading opportunities, hubs, households), when the goal is to evaluate targeted interventions, and when fine-scale contact data exist (or plausible synthetic networks can be constructed). They are less compelling when the population is very large and the contact graph is unknown (forcing strong structural assumptions), when long-range spatial mobility and importation dominate dynamics (often better handled by location graphs or hybrid models), or when computational and identifiability constraints preclude inference over high-dimensional network structure. In many applications, a pragmatic compromise is a hybrid: networks within patches (schools, workplaces, households) coupled by mobility between patches, combining individual-level heterogeneity with the spatial importation mechanisms emphasized in metapopulation models [145,150].

4.3 Continuous (norm) spatial domains

Location graphs discretize geography into a finite set of patches; continuous-space models take the opposite approach and represent space as a domain $\Omega \subset \mathbb{R}^d$ (typically $d = 2$ for map-based applications). The central modeling choice is again the contact/movement operator where instead of a $P \times P$ coupling matrix, we specify an operator that transports individuals (or, in a mixing interpretation, transports infectious pressure) across nearby locations. Especially, compartments become spatial densities $S(x,t)$, $I(x,t)$, and $R(x,t)$, and transmission at location x is driven by a local force of infection

$$\lambda(x,t) = \beta(x,t)\frac{I(x,t)}{N(x,t)}, \qquad N(x,t) = S(x,t) + I(x,t) + R(x,t), \tag{4.8}$$

together with a spatial redistribution mechanism that encodes how people move (or how contacts are realized) over the landscape [276,277].

A widely used baseline assumes that mobility is "local" and approximately random, yielding a reaction-diffusion system [278] that formally takes the following partial differential equation form:

$$\frac{\partial S(x,t)}{\partial t} = \underbrace{-\beta(x,t)\frac{S(x,t)I(x,t)}{N}}_{\text{infection (reaction)}} + \underbrace{\nabla\cdot(D_S(x)\nabla S(x,t))}_{\text{mobility (diffusion)}}, \tag{4.9}$$

$$\frac{\partial I(x,t)}{\partial t} = \underbrace{\beta(x,t)\frac{S(x,t)I(x,t)}{N}}_{\text{infection (reaction)}} - \underbrace{\gamma(x,t)I(x,t)}_{\text{recovery (reaction)}} + \underbrace{\nabla\cdot(D_I(x)\nabla I(x,t))}_{\text{mobility (diffusion)}}, \tag{4.10}$$

$$\frac{\partial R(x,t)}{\partial t} = \underbrace{\gamma(x,t)I(x,t)}_{\text{recovery (reaction)}} + \underbrace{\nabla\cdot(D_R(x)\nabla R(x,t))}_{\text{mobility (diffusion)}}, \tag{4.11}$$

where $D_S(x), D_I(x), D_R(x) \geq 0$ are diffusion coefficients (possibly spatially varying). The reaction terms are the usual within-location SIR dynamics, while the diffusion terms arise as the continuum limit of a spatial random walk: individuals execute many small steps, and the net effect is a Laplacian smoothing of densities across nearby locations. Allowing $D_.(x)$ to vary with x provides a direct way to encode heterogeneous mobility (e.g., faster mixing along well-connected corridors and slower mixing across physical barriers). The following three figures are aiming to demonstrate the norm distribution behavior.

Fig. 4.6 provides a term-by-term visual mapping of the reaction–diffusion SIR system in Eqs. (4.9)–(4.11). The *reaction* components act locally at each location x: susceptibles are converted into infectious individuals at rate $\beta(x,t)S(x,t)I(x,t)/N$, and infectious individuals are removed at rate $\gamma(x,t)I(x,t)$. If diffusion were absent, every spatial point would evolve independently as a temporal SIR model driven only by its local state. The *diffusion* components, $\nabla\cdot(D_.(x)\nabla\cdot)$, encode local mobility as a random-walk-like mixing operator: density flows from regions of higher concentration to nearby regions of lower concentration. In flux form, the diffusion term corresponds to a compartment-specific flux

$$J_X(x,t) = -D_X(x)\nabla X(x,t), \qquad X \in \{S, I, R\},$$

so that $\partial_t X = \cdots - \nabla\cdot J_X$. This makes the interpretation explicit: mobility exports mass down gradients, and the divergence aggregates net inflow/outflow at each location. As a result, reaction drives *within-location* epidemic dynamics ($S \to I \to R$), while diffusion drives *between-location* redistribution, producing smooth spatial spread even when infection is initially localized.

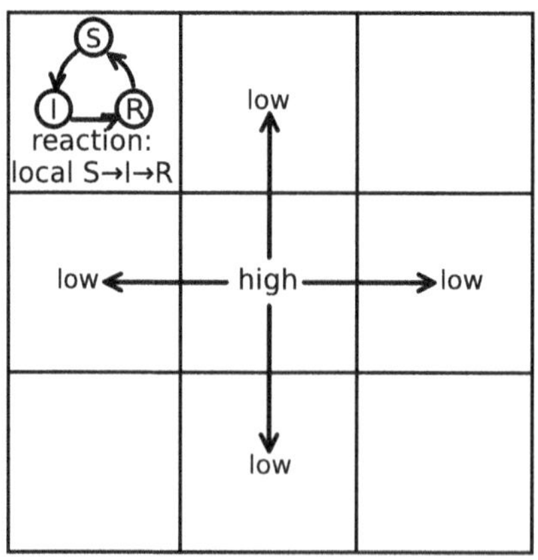

FIGURE 4.6

Reaction–diffusion decomposition of Eqs. (4.9)–(4.11). **Reaction** terms convert compartments locally ($S \to I \to R$) through infection and recovery. **Diffusion** terms represent local random mobility, moving density down spatial gradients as $\nabla\cdot(D\nabla X)$ (equivalently $-\nabla\cdot J_X$ with $J_X = -D_X\nabla X$).

Fig. 4.7 explains why diffusion is a natural baseline operator for local mobility. In a microscopic picture, individuals execute many short, approximately unbiased steps (a spatial random walk). When the population is observed at a coarser spatiotemporal resolution – large enough that many steps occur between observations – the discrete walk converges to a smooth density evolution governed by a diffusion (heat) equation. In this continuum limit, the effective diffusivity D scales with the typical step length ℓ and time step Δt as $D \propto \ell^2/\Delta t$ (up to a dimension-dependent constant). This scaling highlights the modeling meaning of $D_X(x)$ in Eqs. (4.9)–(4.11): it summarizes the intensity of short-range movement for compartment X and, when spatially varying, it encodes that some places support faster mixing than others. Diffusion is therefore best interpreted not as a literal physical process, but as a parsimonious approximation to frequent local mobility.

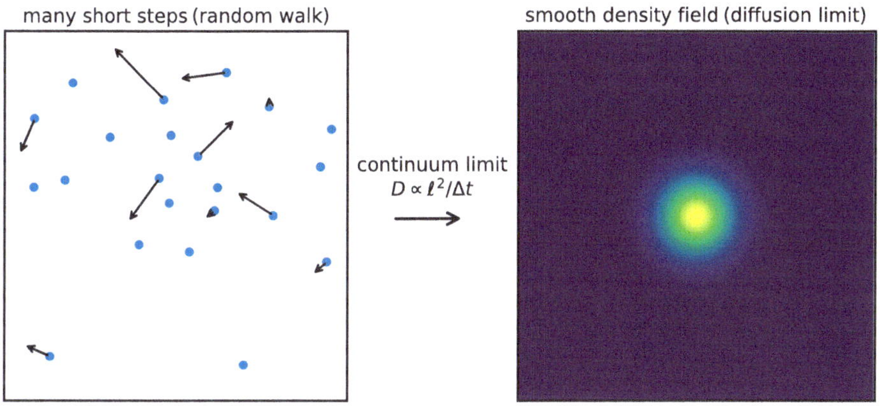

FIGURE 4.7

Diffusion as the continuum limit of local mobility. Left: individuals take many short steps (random walk) with characteristic step length ℓ over time scale Δt. Right: at coarse resolution, the aggregate is well described by a smooth density field satisfying a diffusion equation (e.g., $\partial_t u = D\Delta u$), with $D \propto \ell^2/\Delta t$.

Fig. 4.8 illustrates the qualitative spatiotemporal behavior implied by reaction–diffusion coupling. The simulation is initialized with a small, localized infection seed in an otherwise susceptible domain and then evolved forward under reaction–diffusion dynamics. For this illustrative run, we discretize a two-dimensional domain on a regular grid and impose a no-flux (Neumann) boundary condition at the edges; reaction is implemented via the standard incidence term $\beta S(x,t)I(x,t)/N$ and recovery $\gamma I(x,t)$, while diffusion is implemented with a discrete Laplacian operator. We then visualize snapshots of the infectious density field $I(x,t)$ at four time indices. The resulting sequence highlights the two coupled mechanisms: local reaction amplifies infection where it is present, while diffusion exports infection to neighboring locations, producing an expanding affected region. Visually, this yields smooth prevalence gradients and a recognizable *invasion front* separating largely susceptible areas (low I) from areas where infection has established (higher I). This geometry is precisely what motivates PDE models when spread is dominated by short-range mobility and spatially local mixing.

In continuous-space models, the boundary $\partial\Omega$ is not a technical afterthought: it is the place where we specify how the modeled region exchanges people (or, in a mixing interpretation, infectious pressure) with the outside world. A convenient way to see this is to rewrite diffusion in flux form. For

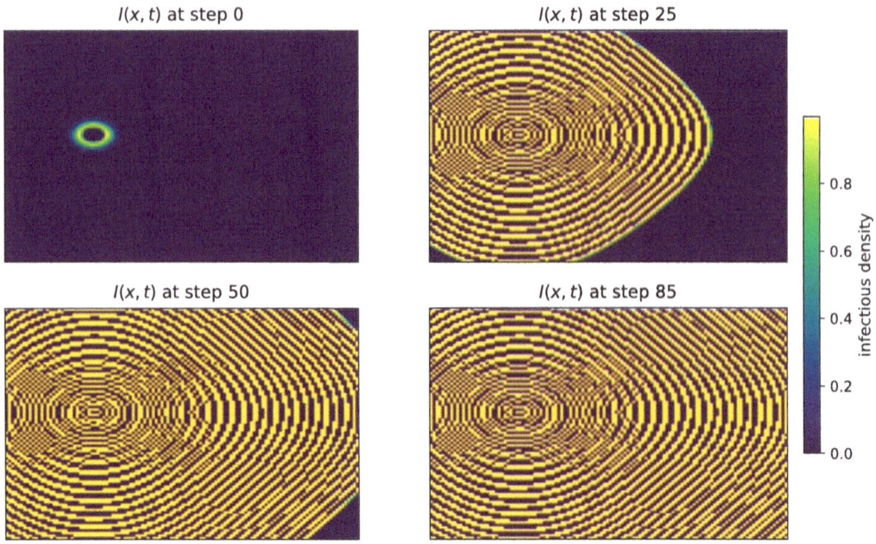

FIGURE 4.8

Spatiotemporal spread under a reaction–diffusion SIR model (illustrative simulation). Starting from a localized seed, infection grows locally through reaction and spreads spatially through diffusion, producing a smoothly expanding infected region and an invasion-front-like pattern in $I(x,t)$. The panels show $I(x,t)$ at four time indices from the same run under no-flux boundaries and homogeneous parameters (see text for running details).

a compartment density $X(x,t)$ with diffusion coefficient $D_X(x)$, define the mobility flux

$$J_X(x,t) := -D_X(x)\nabla X(x,t),$$

so that the diffusion operator becomes a conservation law,

$$\frac{\partial X}{\partial t} = \cdots - \nabla \cdot J_X.$$

The boundary condition specifies what happens to the *normal component* of this flux at the border. Let $n(x)$ denote the outward unit normal on $\partial\Omega$. Then $J_X \cdot n$ is the net rate at which density leaves the domain through the boundary at location $x \in \partial\Omega$. Choosing boundary conditions is therefore equivalent to choosing an epidemiological hypothesis about border permeability, external reservoirs, and importation/exportation. Common choices include Neumann, Dirichlet, and Robin [276]. First, Neumann (no-flux / reflecting boundary) imposes

$$(D_X \nabla X) \cdot n = 0 \qquad \text{on } \partial\Omega,$$

equivalently, $J_X \cdot n = 0$. This means there is *no net mobility flow* across the boundary: individuals do not leave the modeled region and no individuals enter through the border. In diffusion-driven problems, Neumann boundaries are "reflecting": gradients are flattened at the boundary because the solution cannot "spill out" of the domain. In epidemic terms, this is appropriate for an isolated region (or for a modeling window where cross-border movement is negligible at the spatial scale of interest). A practical consequence is that diffusion alone conserves the total mass $\int_\Omega X(x,t)\,dx$ (up to numerical error), so changes in $\int_\Omega S$, $\int_\Omega I$, $\int_\Omega R$ are driven only by the reaction terms. Second, Dirichlet (fixed value / absorbing or externally pinned boundary) prescribes the boundary value:

$$X(x,t) = \bar{X}(x,t) \qquad \text{on } \partial\Omega,$$

where $\bar{X}$ is an externally imposed condition. When $\bar{X} \equiv 0$, the boundary is "absorbing": diffusing density that reaches the boundary is effectively removed, which mimics strong outflow or a sink outside the domain. More generally, specifying $\bar{X}(x,t) > 0$ can represent an *external reservoir* that maintains a fixed level at the border (e.g., persistent infection pressure on the boundary due to surrounding regions). Epidemiologically, Dirichlet boundaries can thus encode constant importation pressure (if $\bar{I} > 0$), constant susceptibility at the boundary, or any fixed external state. Unlike Neumann boundaries, Dirichlet boundaries typically do not conserve total mass, since the boundary can inject or remove density depending on $\bar{X}$. Finally, Robin (mixed / leaky boundary) interpolates between Neumann and Dirichlet by linking boundary flux to boundary value, formalized as follows:

$$(D_X \nabla X)\cdot n + \kappa X = \kappa X_{\text{ext}}(x,t) \qquad \text{on } \partial\Omega,$$

for some leakage/exchange rate $\kappa \geq 0$ and (optionally) an external state X_{ext}. In the common special case $X_{\text{ext}} \equiv 0$, this becomes

$$(D_X \nabla X)\cdot n = -\kappa X,$$

meaning that net outflow across the boundary is proportional to how much density is present near the boundary. Small κ yields an almost reflecting boundary (close to Neumann), while large κ yields a strongly absorbing boundary (approaching Dirichlet with $X = 0$). Epidemiologically, Robin conditions provide a simple, interpretable model of partial border permeability, where individuals can leave (or enter) but at a rate controlled by κ, rather than being perfectly blocked or perfectly pinned.

These choices materially affect invasion, persistence, and inference. For example, a Dirichlet boundary with $\bar{I} > 0$ can sustain repeated re-seeding at the border even when the interior is locally subcritical, whereas a Neumann boundary removes that mechanism entirely. Likewise, fitting spatial parameters without accounting for boundary exchange can lead to misattribution: a leaky boundary (mass loss) may be incorrectly explained by a lower transmission rate or higher recovery if the model assumes no-flux isolation. Fig. 4.9 visualizes the qualitative differences between the three boundary types for the same diffusion process.

Moreover, Reaction-diffusion epidemic models connect directly to classical results on traveling waves [279,280]. The canonical reference point is the Fisher-Kolmogorov-Petrovsky-Piskunov (Fisher–KPP) equation [281]:

$$\frac{\partial u}{\partial t} = D\,\Delta u + r\,u(1-u), \tag{4.12}$$

where $u(x,t) \in [0,1]$ is a normalized "invading" quantity, $D > 0$ is a diffusivity, and $r > 0$ is the early exponential growth rate. In one spatial dimension (and, more generally, under mild conditions which is out of the scope of this book), Eq. (4.12) admits traveling-wave solutions $u(x,t) = U(x - ct)$ that

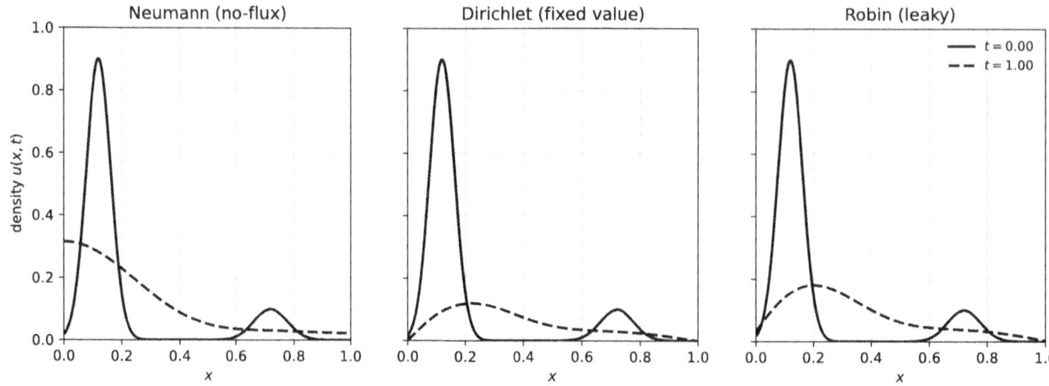

FIGURE 4.9

Illustrative effect of boundary conditions on a diffusing density $u(x,t)$ in a one-dimensional domain. **Neumann (no-flux)** boundaries are reflecting and conserve mass under diffusion. **Dirichlet (fixed value)** boundaries (here set to $u = 0$) are absorbing and remove mass. **Robin (leaky)** boundaries interpolate between the two by allowing partial outflow proportional to the boundary value. All panels start from the same initial condition and show several time snapshots.

connect $u \approx 1$ behind the front to $u \approx 0$ ahead of it [282,283]. A key quantitative prediction is that the invasion speed is set by diffusion and early growth: for the KPP nonlinearity, the minimal wave speed is

$$c^* = 2\sqrt{Dr}. \tag{4.13}$$

This formula makes transparent why invasion is jointly controlled by movement and reproduction potential: increasing either D (more rapid local mobility) or r (faster local amplification) increases the speed at which the wavefront advances. For epidemic models, the Fisher-KPP logic often emerges from early-stage linearization. Consider a reaction-diffusion SIR system in which (locally) infection grows as $\beta(x,t)\,SI/N$ and decays as $\gamma(x,t)I$, and (spatially) infectious density is transported by diffusion. In the early phase, $S(x,t) \approx S_0(x)$ is approximately constant, and one obtains a leading-order approximation of the form

$$\frac{\partial I}{\partial t} \approx D_I\,\Delta I + r(x)\,I, \qquad r(x) = \beta(x,t)\,\frac{S_0(x)}{N(x,0)} - \gamma(x,t), \tag{4.14}$$

so that $r(x)$ plays the role of an intrinsic local growth rate. When coefficients are homogeneous (or slowly varying), invasion resembles a coherent front whose speed is governed by a balance between amplification (reaction) and export (diffusion). In two dimensions, this corresponds to approximately expanding level sets of $I(x,t)$: in homogeneous isotropic settings, fronts are roughly concentric; in heterogeneous landscapes, the front propagates faster in regions with larger effective $r(x)$ or larger mobility (larger $D(x)$), producing anisotropic, channel-like spread.

Fig. 4.10 illustrates these ideas in a minimal setting. Panels (a)–(c) show a one-dimensional Fisher–KPP wave initiated from a localized seed under reflecting (no-flux) boundaries. Panel (b) tracks the position of the $u = 0.5$ level set over time and compares the empirically estimated speed with the theoretical value $c^* = 2\sqrt{Dr}$. The figure demonstrates the core mechanism: local growth builds the wave behind the front, while diffusion continually exports mass into the leading edge where growth can take hold, producing a persistent, coherent invasion front. A second lesson from spatial epidemic theory is that invasion patterns can be highly sensitive to the *tails* of the movement operator. Pure diffusion corresponds to many short steps and thin-tailed displacement statistics. When movement includes rare long-range jumps, diffusion can underestimate arrival speed because distant seeding events establish new foci ahead of the main front. One way to represent such jump processes is a nonlocal dispersal operator with kernel K (an integro-differential analogue of diffusion). Panel (d) of Fig. 4.10 compares a thin-tailed (Gaussian) kernel with a fat-tailed (Cauchy-like) kernel: the fat-tailed kernel yields earlier and faster front advance in this illustrative run, reflecting the ability of rare long jumps to dominate arrival times. This is precisely the regime in which purely diffusive PDEs become a strong hypothesis rather than a neutral baseline [248,284].

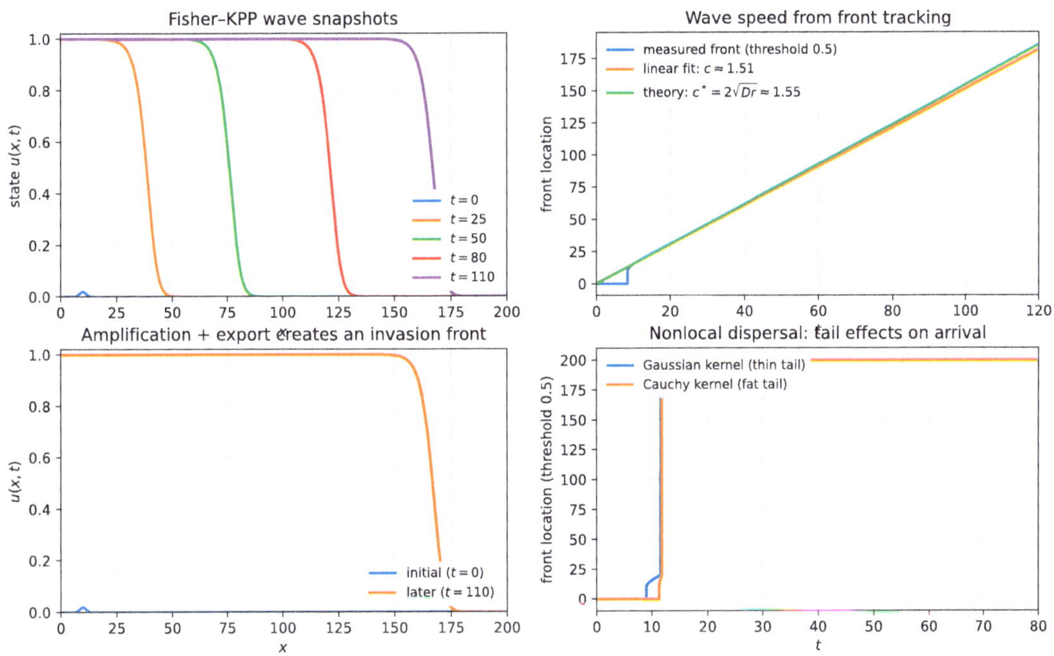

FIGURE 4.10

Wavefronts and invasion geometry (illustrative). **(a)** Fisher–KPP wave snapshots $u(x,t)$ from a localized seed. **(b)** Front tracking using the $u = 0.5$ level set: measured front location over time, a linear fit after transient, and the theoretical KPP prediction $c^* = 2\sqrt{Dr}$. **(c)** Overlay of the initial and late profiles highlighting amplification behind the front and export into the leading edge. **(d)** Tail sensitivity under nonlocal dispersal: front locations for a thin-tailed Gaussian kernel versus a fat-tailed Cauchy-like kernel, showing earlier advance when rare long jumps are present. Running details: 1D domain $x \in [0, 200]$ with reflecting (Neumann) boundaries; Fisher–KPP parameters $D = 1.0$, $r = 0.6$; the front is defined by the largest x where $u(x,t) \geq 0.5$; the speed fit is performed over $t \in [30, 110]$. Nonlocal runs use $u_t = ru(1-u) + m(K * u - u)$ with $m = 1$ and kernels normalized to sum to one.

Common extensions of norm-based spatial epidemic models start from the observation that the simplest reaction–diffusion formulation is best interpreted as a conservation law with a spatial flux. A first extension replaces purely diffusive mixing by a combination of diffusion and directed motion (advection, drift), reflecting the fact that human mobility is often organized along transportation corridors, commuting gradients, and migration streams rather than being a symmetric random walk. In the infectious equation, this is commonly written as

$$\frac{\partial I}{\partial t} = \beta(x,t)\frac{SI}{N} - \gamma(x,t)\,I - \nabla\cdot\big(v(x,t)\,I\big) + \nabla\cdot\big(D_I(x)\nabla I\big), \tag{4.15}$$

where the advective term $-\nabla\cdot(vI)$ produces asymmetric spread and corridor-driven invasion patterns [276,277]. Beyond the qualitative statement "advection makes things directional", Eq. (4.15) clarifies three modeling choices that matter in practice. First, $v(x,t)$ can be interpreted as a velocity field transporting infectious individuals (or infectious presence) and can be time-varying to encode daily commuting, seasonal migration, or disruption during interventions; if one expects a return flow (e.g., home-work-home), v may naturally oscillate in time, or it may be derived from a potential capturing *attractiveness* of destinations. Second, if D_I is allowed to be a tensor rather than a scalar, then diffusion itself can encode directional ease of movement (e.g., faster along roads than across them), and the distinction between directionality from v and directionality from anisotropic D_I becomes a matter of whether movement is coherent (transport-like) or dispersive (spread-like). Third, because advection can strongly steepen fronts, numerical discretization often requires advection-aware schemes (e.g., upwinding) to avoid spurious oscillations; conceptually this is the same reason why directed mobility can yield sharp, fast-moving invasion along corridors, compared to the smoother wavefronts of pure diffusion [276,277]. The flux form also makes it transparent how to impose epidemiologically meaningful boundary conditions: a no net movement across the boundary assumption corresponds to $(vI - D_I\nabla I)\cdot n = 0$ on $\partial\Omega$, whereas an open boundary might instead prescribe inflow/outflow rates (Robin-type conditions) to represent cross-border travel.

A second, pervasive extension is to allow the landscape to be heterogeneous by making transmission, recovery, and mobility vary over space (and often time). Writing $\beta(x,t)$ and $\gamma(x,t)$ acknowledges that contact patterns and clinical course are not spatially uniform: urban versus rural differences in contact rates, geographically structured seasonality and environmental suitability, and spatially varying intervention intensity all translate naturally into spatially varying reaction terms. Allowing $D_{\cdot}(x)$ (and/or a tensor $D_{\cdot}(x)$) to vary provides a direct way to represent mobility constraints such as mountains or water (low conductance), administrative boundaries (reduced cross-boundary mixing), and high-conductance corridors such as roads and rail (enhanced conductance in preferred directions). The important point is not merely that coefficients "can vary," but that such variation interacts with front propagation in qualitatively distinctive ways: low-mobility or low-transmission regions can slow, bend, or even pin wavefronts; corridor-like regions can act as invasion highways; and patches with locally favorable conditions (high β/γ and/or reduced outward mobility) can form spatial refugia that maintain infection even when surrounding areas fade out [276,284]. In this sense, heterogeneity is not just a nuisance parameterization; it is a mechanism for pattern formation and persistence. It also provides a bridge to inference, because $\beta(x,t)$ can be linked to covariates (population density, school calendars, climate proxies, intervention timing), and $D(x)$ or $v(x,t)$ can be linked to mobility data (commuting flows, transit accessibility), making the model interpretable as a structured regression embedded inside a mechanistic dynamical system.

A third extension addresses a limitation of diffusion itself. Diffusion is a good approximation when movement consists of many small, roughly symmetric steps, but it cannot represent occasional long-range jumps (inter-city commuting, air travel, mass gatherings) that can dominate modern contagion. A common alternative replaces local diffusion by nonlocal dispersal via a movement kernel $K(x, y)$, yielding an integro-differential formulation such as

$$\frac{\partial I(x,t)}{\partial t} = \beta(x,t)\frac{S(x,t)\,I(x,t)}{N(x,t)} - \gamma(x,t)\,I(x,t) + \int_{\Omega} K(x,y)\,I(y,t)\,dy - \left(\int_{\Omega} K(y,x)\,dy\right) I(x,t). \quad (4.16)$$

Here, $K(x, y)$ is naturally interpreted as a rate kernel for movement from y to x: the first integral aggregates inflow into x from all origins y, while the second term subtracts total outflow from x to all destinations, ensuring mass balance (the net nonlocal contribution integrates to zero over Ω under mild conditions) [284]. This formulation is particularly helpful because it makes explicit how the tail behavior of movement shapes epidemic speed and pattern. If $K(x, y)$ is concentrated near $x = y$ (short-range movement with finite variance), then Eq. (4.16) admits a diffusion approximation in which the integral operator reduces to a Laplacian-like term with an effective diffusion coefficient; conversely, if K has heavy tails (rare but consequential long jumps), invasion can accelerate and leapfrog, producing satellite outbreaks well ahead of the main front [248]. In practice, the same mathematical object provides a continuum analogue of network coupling: discretizing space into cells turns the integrals into sums, and the operator becomes a weighted graph Laplacian with weights determined by the kernel, which clarifies why kernel-based models "interpolate" between partial differential equation-like local spread and metapopulation-style long-range coupling.

Finally, many applied spatio-temporal epidemic models are best described as hybrid couplings, where geographically smooth within-region spread (captured by diffusion/advection in continuous space) coexists with sparse nonlocal dissemination (captured by kernel tails, explicit travel matrices, or stochastic importations). This matches a recurring empirical signature: local spread is coherent and wave-like, while nonlocal spread appears as discontinuous seeding that establishes new foci which then expand locally [145,263]. Operator-wise, one can think of combining a local differential operator (advection–diffusion, possibly with heterogeneous coefficients) with a nonlocal term that is either continuous (an integral kernel), discrete (coupling between a set of hubs or administrative regions), or event-like (a point-source importation process). The hybrid perspective is often more than a computational convenience: it encodes multi-scale structure explicitly (neighborhood mixing versus inter-city travel), allows different data streams to inform different components (mobility traces informing the nonlocal coupling, spatial covariates informing $\beta(x, t)$, and local surveillance informing the reaction terms), and provides a clear narrative interpretation for observed patterns such as corridor-driven spread, occasional long-distance jumps, and persistence in refugia [145,263].

Table 4.3 summarizes these extensions, highlights a distinct extension, beginning with the additional operator or term introduced into the governing equations, and then interpreting that term in terms of real-world disease spread mechanisms.

Fig. 4.11 illustrates how transport mechanisms and spatial heterogeneity shape epidemic spread. In the absence of directed motion, pure diffusion yields a radially symmetric infection profile, reflecting isotropic local mixing. When advection is included, the infection is displaced along the direction of flow, generating an asymmetric distribution and a tilted front. Introducing spatially varying transmission and mobility parameters further modifies the dynamics: regions of elevated transmission accelerate local spread, while reduced mobility can slow or pin the wavefront, leading to pronounced bending and

Table 4.3 Comparison of several common extensions of norm-based spatial epidemic models.

	Advection	Heterogeneous coefficients	Nonlocal dispersal	Hybrid models
Mathematical term	$-\nabla \cdot (vI)$	$\beta(x), \gamma(x), D(x)$	$\int K(x, y)I(y)dy$	PDE + graph/kernel
Interpretation	Directed mobility, commuting	Spatial variation in risk and mobility	Long-range jumps	Local spread + seeding
Data sources	Travel surveys, mobile phone data	Demography, climate, interventions	Travel matrices, gravity models	Surveillance + mobility data
Common pitfalls	Confounding with anisotropic D	Overfitting, identifiability	Kernel tail vs. $\beta(t)$	Scale mismatch

spatial irregularities. Finally, the movement kernels shown in the bottom panel emphasize the role of dispersal structure. While short-range kernels support classical diffusive behavior, heavy-tailed kernels permit rare long-distance jumps, which can undermine wavefront pinning and enable rapid invasion across heterogeneous landscapes.

Up to this point, the norm-based (metric-kernel / partial differential equation representation) view has treated space as continuous, so that infection and mobility are governed by operators defined directly on Ω (diffusion, advection, or nonlocal kernels). In many applications, however, one eventually wants to replace this continuous operator with a location graph [285]. The motivation is pragmatic – graph models are typically cheaper to simulate and calibrate, align naturally with many available data products, and provide a direct handle for policy questions phrased in terms of regions and travel restrictions. This is especially appealing when (i) the domain is large; (ii) one needs to run the model many times (forecast ensembles, optimization, uncertainty quantification); or (iii) the relevant connectivity is better described by transport links than by Euclidean distance.

The transformation is not trivial, and it is not merely a matter of choosing a grid. A key difficulty is that a norm-based model encodes infinitely many possible locations and interactions, while a location graph assumes a finite set of nodes with within-node well-mixing. In general, there is no closed-form, analytical mapping that preserves the epidemic trajectories exactly. Simply put, constructing an analytical transformation from a norm-based spatial representation to a graph-based representation that is dynamically equivalent is not possible for any epidemiological model. The proof is simple, let us assume a SIR model in which the pairwise infection rate between two agents depends on their instantaneous distance,

$$\beta_{i,j}(t) = \frac{1}{\|L_i(t) - L_j(t)\|},$$

and the walk dynamics update each agent's location by averaging the locations of all other agents. Formally, the i_{th} agent in a population (P) location at time t is denoted by L_i^t and with the update rule: $L_i^{t+1} = L_i^t + \delta \sum_{j \in P}(L_j^t)/|P|$ where $\delta > 0$ is the step size. Under these dynamics, the smallest interaction-relevant distance can shrink arbitrarily (converging to zero), so there is no fixed discretization size that is guaranteed to resolve the interaction scale over time. In other words, a naive "mesh graph" construction is not guaranteed to yield an isomorphic graph model unless one allows the graph resolution to become arbitrarily fine, which defeats the computational motivation. This is why the

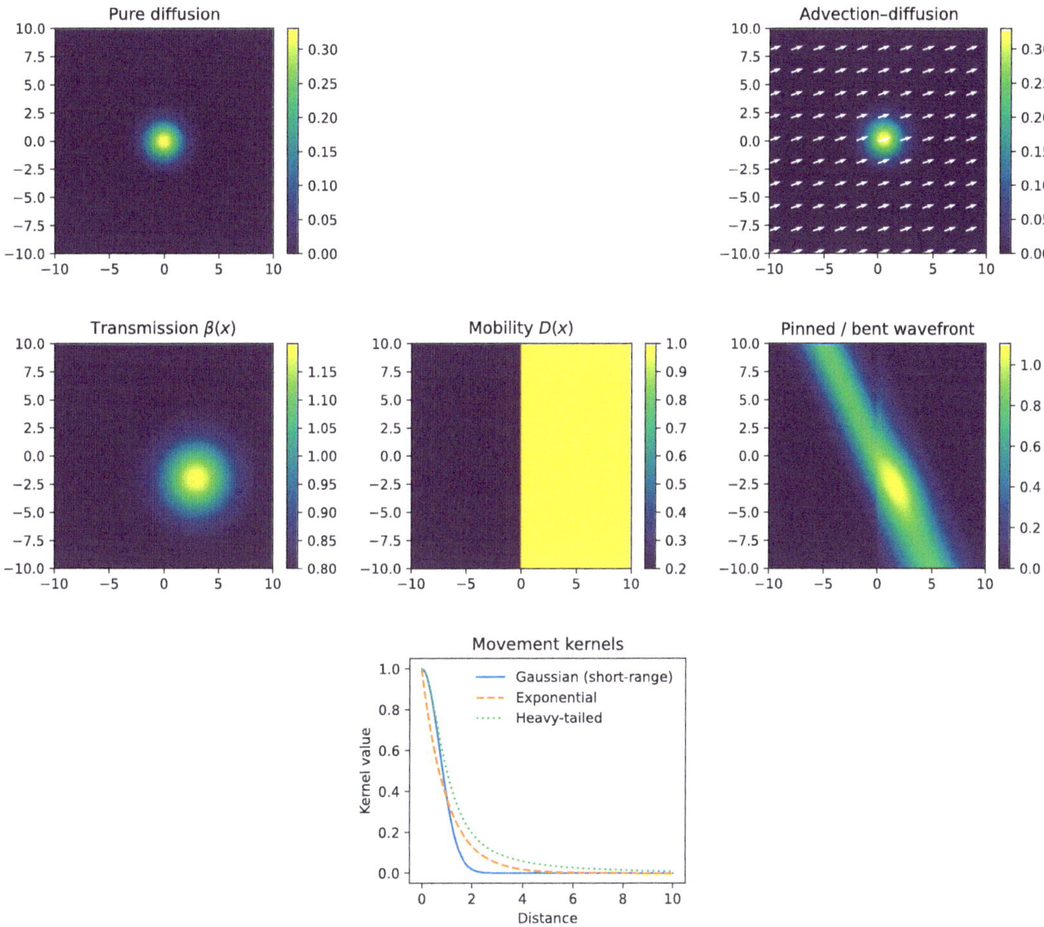

FIGURE 4.11

Effects of spatial transport, heterogeneity, and movement kernels on epidemic spread. **Top row:** Pure diffusion produces a symmetric, isotropic smoothing of the initial infection, while advection–diffusion introduces a directional bias aligned with the imposed velocity field. **Middle row:** Spatial heterogeneity in transmission $\beta(x)$ and mobility $D(x)$ deforms the propagating infection front, producing pinning and bending where transmission is enhanced or mobility is reduced. **Bottom row:** Representative movement kernels illustrate the contrast between short-range (Gaussian), intermediate (exponential), and heavy-tailed dispersal, highlighting how long-range movements can fundamentally alter front propagation dynamics.

transformation problem must be treated as a *numerical approximation* problem rather than an analytical conversion.

A useful way to frame the conversion is: we want a coarse-grained graph that preserves the epidemiologically relevant consequences of the norm-based operator, not the operator itself. As a result, suppose the norm-based model induces through its movement and contact rules a set of infection and recovery trajectories or compartment time series $A_n(t)$. We seek a location-graph model with trajectories $A_g(t)$ that are close in an appropriate metric while keeping the graph small. This task can be formalized as a constrained optimization problem of the form

$$\min_{|V|} \min \|A_n - A_g\|, \tag{4.17}$$

highlighting that there is an explicit trade-off between graph complexity, reflected by the number of nodes, and dynamical fidelity, reflected by the distance between A_n and A_g. The hard part is that both (i) the partition of space into nodes and (ii) the induced mobility/contact coupling must be chosen so that the graph reproduces the continuous model's effective co-location and flow patterns.

A simple outline of a practical conversion pipeline is the following.

(1) Choose a spatial partition (nodes). Pick a finite set of regions $\{\Omega_p\}_{p=1}^{P}$ whose union approximates Ω (possibly excluding obstacles). This can be done by administrative boundaries, a uniform grid, an adaptive partition (e.g., quadtree), or data-driven clustering of trajectories/visitation patterns. The goal is to ensure that within each Ω_p, individuals are "close enough" and mix fast enough that a within-node well-mixing approximation is reasonable on the infection time scale.

(2) Aggregate continuous states to node states. Define node-level compartments as spatial integrals (or sums for agent data),

$$S_p(t) = \int_{\Omega_p} S(x,t)\,dx, \quad I_p(t) = \int_{\Omega_p} I(x,t)\,dx, \quad R_p(t) = \int_{\Omega_p} R(x,t)\,dx,$$

with $N_p(t) = S_p(t) + I_p(t) + R_p(t)$. This step fixes what the graph model will consider the "state" of a node.

(3) Approximate mobility coupling (edges / movement). If the norm-based model uses diffusion/advection, one can approximate inter-node movement in two broad ways: (i) *flux-based*: estimate net flows across shared boundaries and convert them into rates m_{pq} so that the graph implements the same conservation-of-mass logic as a finite-volume discretization; or (ii) *trajectory-based*: if agent traces or empirical mobility data are available, estimate a transition kernel directly as

$$m_{pq} \propto \Pr\{\text{agent in } \Omega_p \text{ at } t \rightarrow \Omega_q \text{ at } t + \Delta t\},$$

possibly time-varying and stratified by day-type or covariates. Learning this *walk dynamics approximation* is a central component: the graph must reproduce how agents redistribute across regions, not just where regions are placed.

(4) Approximate cross-node exposure (FOI coupling). If the norm-based model uses a distance kernel or nonlocal dispersal, one can derive an effective mixing matrix by averaging the kernel over regions:

$$\beta_{pq} = \frac{1}{|\Omega_p|} \int_{\Omega_p} \int_{\Omega_q} \beta(x,y)\,dy\,dx,$$

or, when the kernel depends only on distance $K(\|x-y\|)$,

$$\beta_{pq} = \frac{\beta}{|\Omega_p|} \int_{\Omega_p} \int_{\Omega_q} K(\|x-y\|)\,dy\,dx.$$

This produces a location-graph force-of-infection operator

$$\lambda_p(t) = \sum_{q=1}^{P} \beta_{pq} \frac{I_q(t)}{N_q(t)},$$

which is the direct analogue of the norm-based infectious pressure integral, but expressed on a finite graph. Importantly, the approximation quality depends strongly on whether the within-region heterogeneity (gradients inside Ω_p) is small enough to be ignored.

(5) Validate and (if needed) iteratively refine. Because no analytical mapping exists in general, the graph must be treated as a fitted surrogate. One typically checks whether the graph model matches targeted summaries of the norm-based dynamics: compartment trajectories, arrival times across regions, peak timing, or the effective reproduction index induced by the coupling. If the mismatch is large, refinement can proceed by (i) increasing $|V|$ in high-activity areas, (ii) changing the partition to better align with movement corridors, or (iii) re-estimating the walk and mixing operators. This iterative, numerical view is exactly the stance advocated in the norm-to-graph transformation framework, where the graph search (partition) and walk-dynamics learning (mobility approximation) are treated as distinct but coupled approximation problems.

Deepen your knowledge

Exercise 1 (Location-based spatio-temporal risk)

You observe case events $\{(s_i, t_i)\}_{i=1}^{n}$ where $s_i \in \mathbb{R}^2$ is the (projected) location and t_i is time. Consider the location-based risk score

$$r(s, t) = \sum_{i=1}^{n} \exp\Big(-\frac{\|s - s_i\|_2^2}{2\sigma^2}\Big) \exp\Big(-\frac{t - t_i}{\tau}\Big) \mathbf{1}\{t \geq t_i\}.$$

(a) Interpret σ and τ. (b) Give two reasons to prefer a projected coordinate system (meters) over latitude/longitude for this model.

Solution. (a) σ controls *spatial reach*: larger σ spreads influence farther in space. τ controls *temporal memory*: larger τ makes past cases matter longer (slower decay). (b) (i) The Gaussian kernel uses Euclidean distances; on lat/lon this is distorted and scale varies with latitude. (ii) Using meters makes σ interpretable (e.g., "500m influence radius") and numerically stable.

Exercise 2 (Individuals infection graphs)

An *infection graph* is a directed graph $G = (V, E)$ where V are individuals, and an edge $(i \to j) \in E$ means individual i infected j. Assume each infection has a time stamp T_j and each directed edge has transmission time $T_{i \to j} = T_j$.

(a) Explain how to build a *time-respecting* infection graph from contact events (i, j, t) (who met whom and when), assuming a fixed infectious window of length Δ after infection. (b) Define an estimator of the individual reproduction number R_i from the infection graph, and a population-level R.

Solution. (a) If j is infected at time T_j, then any infector i must satisfy: (i) i was infected earlier ($T_i < T_j$); (ii) there exists at least one contact event (i, j, t) with $t \in [T_i, T_i + \Delta]$ and $t \leq T_j$; and (iii) choose the most plausible i (e.g., the latest qualifying contact before T_j, or probabilistically if multiple candidates exist). Add edge $i \to j$. This enforces *causality* and an infectious period. (b) A simple estimator is out-degree: $R_i = \text{outdeg}(i)$. A population-level reproduction estimate is

$$R = \frac{1}{|V_I|} \sum_{i \in V_I} \text{outdeg}(i),$$

where V_I are infected individuals (or restrict to those whose infectious periods are fully observed to reduce right-censoring).

Exercise 3 (Norm-based stability in spatio-temporal models)

Let $y_t \in \mathbb{R}^m$ be incidence across m regions and consider a linear spatio-temporal autoregressive model

$$y_t = A\, y_{t-1} + \varepsilon_t, \qquad A = \alpha I + \beta W,$$

where W is a row-stochastic spatial weights matrix (nonnegative rows summing to 1).

(a) State the stationarity/stability condition in terms of the spectral radius $\rho(A)$. (b) Give a *sufficient* condition using an induced matrix norm $\|\cdot\|$. (c) Using $\|W\|_\infty = 1$ for row-stochastic W, provide an easy-to-check bound involving $|\alpha|$ and $|\beta|$.

Solution. (a) Stability requires $\rho(A) < 1$. (b) Since $\rho(A) \leq \|A\|$ for any induced norm, a sufficient condition is $\|A\| < 1$. (c) With the induced ∞-norm, $\|A\|_\infty = \|\alpha I + \beta W\|_\infty \leq |\alpha| \|I\|_\infty + |\beta| \|W\|_\infty = |\alpha| + |\beta|$. So a simple sufficient condition is

$$|\alpha| + |\beta| < 1.$$

Exercise 4 (Location-based risk surface in Python)

Implement the risk score $r(s,t)$ from Exercise 1 and compute risks at two query points.

```
import numpy as np

def risk_score(pts_xy, pts_t, query_xy, query_t, sigma, tau):
    """
    pts_xy: (n,2) case locations (meters)
    pts_t : (n,)  case times (same units as query t)
    query_xy: (q,2) query locations
    query_t : scalar query time
    """

    pts_xy = np.asarray(pts_xy, dtype=float)
    pts_t  = np.asarray(pts_t, dtype=float)
    query_xy = np.asarray(query_xy, dtype=float)

    # Only past events contribute
    mask = pts_t <= query_t
    xy = pts_xy[mask]
    tt = pts_t[mask]
```

```
    # Spatial squared distances: (q,n_past)
    d2 = np.sum((query_xy[:, None, :] - xy[None, :, :])**2, axis=2)
    K_space = np.exp(-d2 / (2.0 * sigma**2))

    # Temporal decay: (n_past,)
    K_time = np.exp(-(query_t - tt) / tau)

    # Risk per query: (q,)
    return K_space @ K_time

# Example data: 4 cases with (x,y) in meters and times in days
pts_xy = np.array([[0,0], [200, 50], [400, 0], [150, 300]])
pts_t  = np.array([0.0, 2.0, 3.0, 7.0])

sigma = 250.0  # meters
tau   = 5.0    # days

query_xy = np.array([[100, 0], [500, 0]])
query_t  = 6.0

r = risk_score(pts_xy, pts_t, query_xy, query_t, sigma, tau)
print("risk =", r)
```

Listing 4.2: Python computation of a location-based spatio-temporal risk score with Gaussian spatial kernel and exponential temporal decay.

Solution. The function implements exactly

$$r(s,t) = \sum_i \exp(-\|s - s_i\|_2^2/(2\sigma^2)) \exp(-(t - t_i)/\tau) \mathbf{1}\{t \geq t_i\}.$$

Interpreting the output: larger values indicate higher proximity (in space and time) to past cases.

Exercise 5 (Individuals infection graph from temporal contacts)

You are given temporal contacts (i, j, t) and infection times T_k for each individual k (or ∞ if never infected). Build a directed infection graph by assigning each infected j an infector i, chosen as the *latest* contact before T_j such that $t \in [T_i, T_i + \Delta]$.

```
import numpy as np

def build_infection_graph(contacts, T, Delta):
    """
    contacts: list of (i, j, t) with i!=j
    T: array of infection times (np.inf if never infected)
    Delta: infectious window length
    Returns: list of directed edges (i, j) and an array R_i (out degree)
    """
    T = np.asarray(T, dtype=float)
```

```
    n = len(T)

    edges = []
    # Sort contacts by time to make "latest before T_j" easy to implement
    contacts_sorted = sorted(contacts, key=lambda x: x[2])

    for j in range(n):
        if not np.isfinite(T[j]):
            continue  # never infected
        best_i = None
        best_t = -np.inf

        for (i, jj, t) in contacts_sorted:
            if jj != j:
                continue
            if t > T[j]:
                break  # contacts are sorted; no need to keep scanning
            if not np.isfinite(T[i]) or T[i] >= T[j]:
                continue  # i must be infected earlier
            if (T[i] <= t) and (t <= T[i] + Delta):
                # candidate infector; keep the latest contact
                if t > best_t:
                    best_t = t
                    best_i = i

        if best_i is not None:
            edges.append((best_i, j))

    # Out-degree as individual reproduction number proxy
    R = np.zeros(n, dtype=int)
    for (i, j) in edges:
        R[i] += 1
    return edges, R

# Example with 5 individuals (0..4)
T = np.array([0.0, 2.0, 4.0, np.inf, 6.0])  # infection times
Delta = 3.0

contacts = [
    (0, 1, 1.8),
    (1, 2, 3.9),
    (0, 2, 2.5),  # earlier than 3.9, but still a candidate
    (2, 4, 5.5),
    (1, 4, 4.5),  # not valid if outside infectious window
]

edges, R = build_infection_graph(contacts, T, Delta)
print("edges =", edges)
```

```
print("R_i =", R, "  R_bar =", R[np.isfinite(T)].mean())
```

Listing 4.3: Python construction of a time-respecting infection graph from temporal contacts and infection times.

Solution. The algorithm enforces causality ($T_i < T_j$) and an infectious window ($t \in [T_i, T_i + \Delta]$), then selects the latest qualifying contact as the infector. The resulting edges define the infection graph; R_i is computed as out-degree.

Exercise 6 (Norm-based bounds in code: spectral radius vs. induced norms)

Given a spatio-temporal transmission operator A, compute: (i) the spectral radius $\rho(A)$, (ii) induced norms $\|A\|_2$ and $\|A\|_\infty$, and verify $\rho(A) \le \|A\|_2$ and $\rho(A) \le \|A\|_\infty$ numerically.

```
import numpy as np

def spectral_radius(A):
    eigvals = np.linalg.eigvals(A)
    return np.max(np.abs(eigvals))

# Example: A = alpha I + beta W with row stochastic W
W = np.array([
    [0.6, 0.4, 0.0],
    [0.2, 0.5, 0.3],
    [0.1, 0.1, 0.8]
], dtype=float)

alpha = 0.2
beta  = 0.7
A = alpha * np.eye(W.shape[0]) + beta * W

rho = spectral_radius(A)
n2  = np.linalg.norm(A, 2)        # induced 2 norm (spectral norm)
ninf = np.linalg.norm(A, np.inf) # induced infinity norm (max row sum)

print("rho(A)    =", rho)
print("||A||_2  =", n2)
print("||A||_inf=", ninf)
print("rho<=||A||_2  ?", rho <= n2 + 1e-12)
print("rho<=||A||_inf?", rho <= ninf + 1e-12)

# Easy to check sufficient condition from Exercise 3:
print("|alpha|+|beta| =", abs(alpha) + abs(beta))
```

Listing 4.4: Python verification of norm-based bounds for spatio-temporal stability: spectral radius vs. induced norms.

Solution. For any induced norm, $\rho(A) \le \|A\|$ holds, so the printed booleans should evaluate to `True` up to numerical tolerance. If additionally $\|A\| < 1$ (or the sufficient bound $|\alpha| + |\beta| < 1$ holds for row-stochastic W under $\|\cdot\|_\infty$), the system is guaranteed stable.

CHAPTER

Multi-strain and evolving pandemics

5

Compartmental models, SIR and its temporal and spatial extensions, treat the pathogen as a single, fixed "type" and typically collapse immunity into a small number of states. This abstraction is often adequate for short outbreaks dominated by one lineage, but it becomes strained when epidemiological dynamics unfold on the same timescale as antigenic change, immune escape, importation of divergent lineages, or the ecological coexistence of many serotypes [286–289]. In such settings, the epidemic is not merely the spread of infection through a population, but rather it is a dynamical process on a shifting landscape of susceptibility shaped by prior exposures and interventions [290–293].

Multi-strain structure arises whenever infection- or vaccine-induced immunity is incomplete, wanes, or is only partially cross-protective [294]. The resulting frequency-dependent feedback can generate strain replacement, long-term coexistence, and complex oscillations even when strains are identical except for antigenic identity [295–297]. A central modeling challenge is that host immune memory is inherently high-dimensional including individuals differ in whether they have been infected, in which strains they have seen, and how those exposures map to protection against current strains [298,299].

Pandemics further complicate this picture because strain diversity is not necessarily pre-existing; it can be (and usually does) created during spread via mutation, recombination, or reassortment, with selection acting through transmission opportunities and immune evasion. Influenza provides a canonical example in which ecological forcing and immune selection shape antigenic drift and lineage turnover [300–303]. More recently, SARS-CoV-2 has illustrated how shifts in population immunity can be followed by the emergence and global spread of variants carrying suites of mutations that affect transmissibility and antigenicity [304–307].

A second challenge is representational. Even before introducing evolution, how should "strain space" be encoded? At one extreme are strain-indexed compartmental models with explicit cross-immunity matrices, which are transparent but can scale poorly with the number of strains [295,297]. At the other extreme are formulations that summarize immune history by *status* variables or symmetries, enabling tractable analysis and simulation for many strains while retaining key immunological interactions. Between these lie models that embed strains in continuous antigenic space, where antigenic distance induces graded cross-protection and supports phenomena such as traveling antigenic waves and lineage clustering [299]. A particularly promising direction is to integrate genotype, antigenic phenotype, and incidence into unified inferential frameworks [308,309]. For influenza, models that jointly represent molecular evolution and antigenic drift have clarified how year-to-year antigenic change relates to incidence patterns and lineage competition [310,311]. These developments motivate a view of surveillance data as partial projections of one coupled dynamical system.

In this chapter, we develop a modeling toolkit for multi-strain and evolving pandemics. Building on the foundations of the SIR model and its extensions, we introduce strain-structured transmission models with cross-immunity. Next, we incorporate mutation and invasion to represent evolutionary emergence

Epidemic Modeling. https://doi.org/10.1016/B978-0-44-355525-1.00011-4

and replacement. We connect discrete-strain models to continuous antigenic-space formulations, and discuss when simplifications (e.g., status-based reductions) preserve the phenomena of interest. We then turn to inference, emphasizing how genomic and antigenic data constrain transmission and selection. Finally, we discuss implications for forecasting and control when the pathogen adapts in response to the immunity and interventions we create [299,309].

5.1 Parallel-strains

We begin with the simplest multi-strain generalization of multiple lineages circulating in parallel, each with its own transmission and clearance parameters, but we treat post-infection immunity as fully protective against all circulating strains. This is an intentionally strong simplification, as this is rarely a biologically realistic representation. Nevertheless, it provides a clean baseline for understanding competition for susceptibles and the conditions under which one strain replaces another.

Formally, let $k \in \{1, \ldots, K\}$ index strains. We track a single susceptible pool $S(t)$ and a single removed/immune class $R(t)$, while keeping strain-specific infectious compartments $I_k(t)$. The total population size is

$$N(t) = S(t) + \sum_{k=1}^{K} \big(I_k(t)\big) + R(t).$$

Each strain generates a force of infection

$$\lambda_k(t) = \beta_k \frac{I_k(t)}{N(t)}, \qquad \lambda(t) = \sum_{k=1}^{K} \lambda_k(t),$$

so susceptible hosts become infected by *one of the strains* at a total rate $\lambda(t)S(t)$, and infections are attributed to strain k in proportion to $\lambda_k(t)$. Thus, a K-strain SIR system is then

$$\frac{dS(t)}{dt} = -\underbrace{\sum_{k=1}^{K} \beta_k \frac{S(t)I_k(t)}{N}}_{\text{infection by any strain}}, \tag{5.1}$$

$$\frac{dI_k}{dt} = \underbrace{\beta_k \frac{S(t)I_k(t)}{N}}_{\text{infected with strain}} \underbrace{-\gamma_k I_k(t)}_{\text{onset of infectiousness and recovery}}, \qquad k = 1, \ldots, K, \tag{5.2}$$

$$\frac{dR(t)}{dt} = \underbrace{\sum_{k=1}^{K} \gamma_k I_k(t)}_{\text{recovery from any strain}}. \tag{5.3}$$

Fig. 5.1 provides a schematic view of the model.

Notably, Eqs. (5.1)–(5.3) encode competition through susceptible depletion as strains interact only because they infect hosts from the same susceptible pool. This structure is often used as a baseline for "strain replacement" arguments. Namely, if a new strain has a higher growth advantage when rare, it can

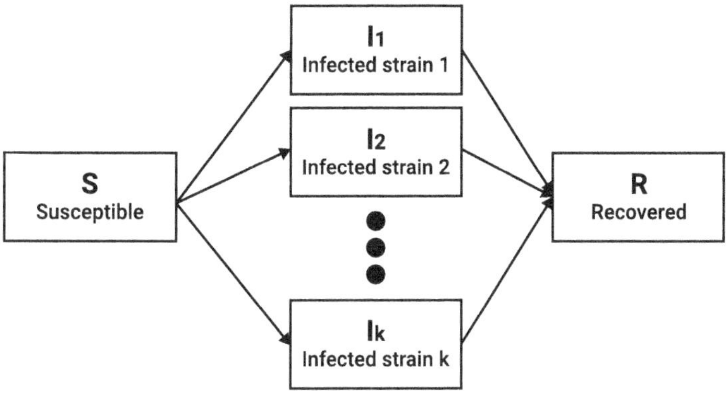

FIGURE 5.1

Schematic view of a K-strain SIR model with a shared susceptible pool.

rise and reduce the susceptible pool in a way that suppresses other strains. This simplification, which places all recovered individuals into a single R class, inherently assumes that infection by any strain grants complete protection against all strains. However, in practice, this is rarely exact for antigenically diverse pathogens. For example, a canonical early example is the analysis of cocirculating antigenic types in SIR-like systems by Andreasen and colleagues, who showed that even simple multi-strain structures can generate replacement and oscillatory dynamics once immune interactions are added [295]. The parallel-strain model above can be viewed as the "null" case in which the only coupling is through the shared susceptible pool; subsequent sections introduce the immunological coupling mechanisms that drive richer behavior [296,297].

To illustrate the dynamics, Fig. 5.2 shows a K-strain parallel SEIR model, where each strain k has its own exposed and infectious compartments (E_k, I_k), but all strains draw infections from the same susceptible pool S. In this formulation, the strains interact only indirectly, via competition for susceptibles:

$$\frac{dS(t)}{dt} = -S(t)\sum_{k=1}^{K} \beta_k I_k(t), \qquad \frac{dI_k(t)}{dt} = \delta_k E_k(t) - \gamma_k I_k(t).$$

Because recovery is pooled into a single R class, this baseline model corresponds to the strong assumption that infection by any strain moves individuals into the same "removed/immune" state. Even under this simplifying assumption, adding strains changes the epidemic trajectory because each additional lineage can generate infections while simultaneously reducing $S(t)$, thereby lowering $R_{e,k}(t) = (\beta_k/\gamma_k)\, S(t)$ for all strains. In the code that generates Fig. 5.2, strains 1–2 keep identical parameters across panels, while strains 3–4 are added (with small initial seeds) to create new infectious curves and demonstrate how co-circulation reshapes incidence even without explicit cross-immunity. In order to generate Fig. 5.2, we used the same parameter values for strains 1–2 in all panels, and introduced additional strains (3 and 4) only in the $K = 3$ and $K = 4$ panels. The model is simulated in proportions (so $N = 1$), and the initial susceptible fraction is computed as $S(0) = 1 - \sum_k (E_k(0) + I_k(0)) - R(0)$, see Table 5.1.

Table 5.1 Parameter values used in the parallel-strain SEIR simulations. Strains 1–2 are present in all panels; strain 3 is added in the $K = 3$ panel; strains 3–4 are added in the $K = 4$ panel.

Strain	β_k	δ_k	γ_k	$R_{0,k} = \beta_k/\gamma_k$	$I_k(0)$	$E_k(0)$
1	0.85	1/3	1/6	5.10	2×10^{-4}	0
2	0.65	1/4	1/7	4.55	1×10^{-4}	0
3	0.95	1/2.5	1/5	4.75	7×10^{-5}	0
4	0.75	1/5	1/8	6.00	5×10^{-5}	0

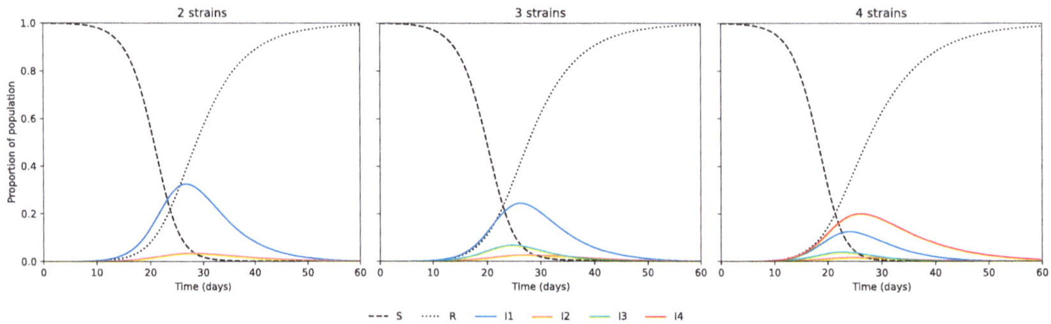

FIGURE 5.2

Parallel-strain SEIR dynamics with a shared susceptible pool for $K = 2, 3, 4$ strains (left to right).

Listing 5.1 gives the computational core used to generate the three-panel figure: it defines the K-strain parallel SEIR right-hand side, constructs the initial condition vector for a chosen K, solves the ODE, and extracts the trajectories.

```
def rhs_parallel_seir(t, y, beta, delta, gamma):
    # y = [S, E_1..E_K, I_1..I_K, R] in proportions (N=1)
    K = len(beta)
    S = y[0]
    E = y[1:1+K]
    I = y[1+K:1+2*K]

    # Total force of infection (frequency dependent with N=1)
    lam = np.dot(beta, I)

    dS = -S * lam
    dE = beta * S * I - delta * E
    dI = delta * E - gamma * I
    dR = np.dot(gamma, I)

    return np.concatenate(([dS], dE, dI, [dR]))
```

```python

def simulate_parallel_seir(K, params_all, init_all, t_eval):
    # Keep strains 1..K (strains 1-2 unchanged across panels; higher strains appear only when
        K increases)
    beta  = np.array(params_all["beta"][:K], dtype=float)
    delta = np.array(params_all["delta"][:K], dtype=float)
    gamma = np.array(params_all["gamma"][:K], dtype=float)

    E0 = np.array(init_all["E0"][:K], dtype=float)
    I0 = np.array(init_all["I0"][:K], dtype=float)
    R0 = float(init_all.get("R0", 0.0))

    # Proportions: N=1, so S0 is whatever remains
    S0 = 1.0 - (E0.sum() + I0.sum() + R0)
    y0 = np.concatenate(([S0], E0, I0, [R0]))

    sol = solve_ivp(
        fun=lambda t, y: rhs_parallel_seir(t, y, beta, delta, gamma),
        t_span=(t_eval[0], t_eval[-1]),
        y0=y0,
        t_eval=t_eval,
        rtol=1e-7, atol=1e-10
    )

    # Return the pieces typically plotted
    S = sol.y[0, :]
    E = sol.y[1:1+K, :]
    I = sol.y[1+K:1+2*K, :]
    R = sol.y[-1, :]

    return sol.t, S, E, I, R
```

Listing 5.1: Computational core for simulating the parallel-strain SEIR model (no imports, no printing).

5.2 Multi-strain with reinfection

A more realistic multi-strain model considers a partial immunity between strains and therefore allows reinfection between strains. This extension can occur for both SIR and SIS type models, where SIR extended models allow reinfection once per strain, while SIS extended models allow multiple reinfections for each strain.

Considering the SIR extended version, formally, we consider a closed population of constant size N and a pathogen that can appear in m distinct strains, indexed by $M := \{1, \ldots, m\}$. Strain diversity may be interpreted as arising endogenously (e.g., strains $2, \ldots, m$ are mutants derived from an ancestral lineage during transmission) or as pre-existing diversity introduced by importation or ecological coexistence. In both interpretations, we assume that an individual can be infected by at most one strain at a time, but may be infected repeatedly over the course of the outbreak because immunity is strain-specific. The key modeling goal is therefore to represent multi-strain transmission while allowing reinfection by strains not yet experienced.

To make reinfection explicit, we track immune history. Let $P(M)$ denote the power set of M. For each immune-history set $J \in P(M)$, we define $R_J(t)$ as the individuals who are not currently infectious and have previously recovered from exactly the strains in J, and $R_J I_i(t)$ as the individuals with immune history J who are currently infectious with strain $i \in M$, where necessarily $i \notin J$. For this reason, the compartment $R_\emptyset$ corresponds to individuals with no prior infections and thus susceptible to all strains; as such, it plays the role of the susceptible class (i.e., $S(t)$) in a single-strain SIR model.

Population conservation is enforced by

$$N = \sum_{i \in M} \sum_{J \in P(M \setminus \{i\})} R_J I_i(t) + \sum_{J \in P(M)} R_J(t), \tag{5.4}$$

where the restriction $J \in P(M \setminus \{i\})$ reflects that an individual currently infectious with strain i cannot already have recovered from i. Fig. 5.3 presents a schematic view of this model.

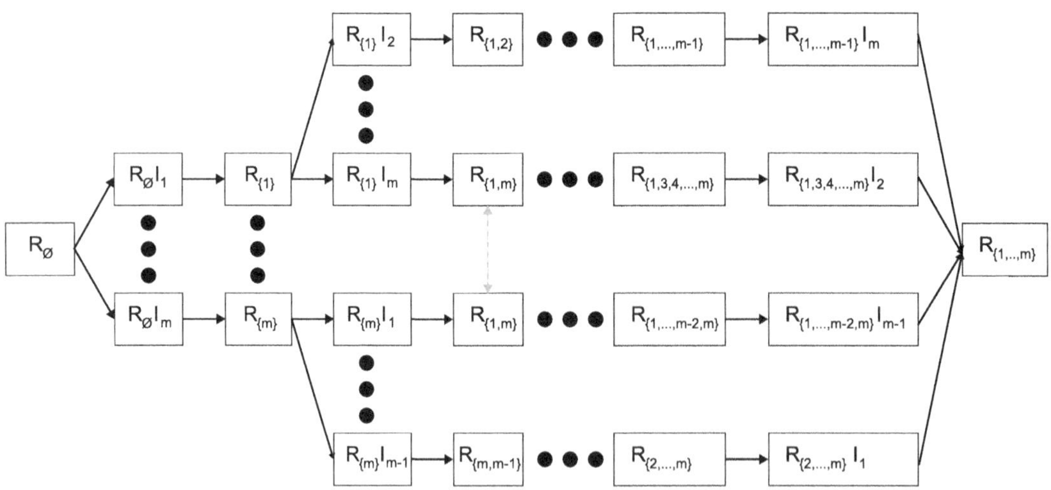

FIGURE 5.3

Schematic view of multi-strain progression with reinfection. Individuals in R_J are immune to strains in J but remain susceptible to strains in $M \setminus J$. Infection by strain $i \notin J$ moves $R_J \to R_J I_i$. Infectious individuals $R_J I_i$ recover into $R_{J \cup \{i\}}$.

Individuals in R_J are protected against strains in J and remain susceptible to strains in $M \setminus J$. If they are exposed to a strain $i \in M \setminus J$, they transition to $R_J I_i$. We allow immune history to modify both the risk of infection and the clinical course, using: $\beta_{J,i}$ for the transmission rate into infection with strain i given history J, $\gamma_{J,i}$ for the removal rate from $R_J I_i$, and $\phi_{J,i} \in [0, 1]$ for the probability that removal corresponds to recovery (so $1 - \phi_{J,i}$ corresponds to death).

With these definitions, the infectious-with-history compartments evolve, for each strain $i \in M$ and each history set $J \in P(M)$ with $i \notin J$, as

$$\frac{dR_J I_i(t)}{dt} = -\gamma_{J,i}\, R_J I_i(t) + \beta_{J,i}\, R_J(t) \sum_{K \in P(M): i \notin K} R_K I_i(t). \tag{5.5}$$

The gain term collects infectious contacts from all individuals currently carrying strain i, independent of their own history K, subject to the constraint $i \notin K$. In this formulation, the recipient's history J controls susceptibility (through $\beta_{J,i}$), while the current strain i controls who can transmit.

Recovered compartments change because individuals clear infections (and update their immune history) and because they can be reinfected by strains not yet seen. For any $J \in P(M)$,

$$\frac{dR_J(t)}{dt} = \sum_{i \in J} \gamma_{J\setminus\{i\},i}\, \phi_{J\setminus\{i\},i}\, R_{J\setminus\{i\}} I_i(t) - \sum_{i \in M\setminus J} \beta_{J,i}\, R_J(t) \sum_{K \in P(M): i \notin K} R_K I_i(t). \tag{5.6}$$

The first sum accounts for recoveries into R_J: to arrive in R_J, an individual must have been infectious with some $i \in J$ while previously having history $J \setminus \{i\}$, and then recover. The second sum is the reinfection outflow: individuals in R_J remain susceptible to every strain not included in J.

Eqs. (5.5)–(5.6) define a strain-structured SIR-type model in which reinfection is driven by immune history. The price of representing reinfection faithfully is that immune memory is encoded by subsets $J \subseteq M$, and the number of compartments grows rapidly with m. Later sections discuss reductions (for example, grouping histories into lower-dimensional "status" variables or embedding strains in antigenic space) that preserve the reinfection mechanism while improving tractability.

To simplify the notation, we define the total prevalence of strain i across all immune histories as

$$I_i^{\text{tot}}(t) = \sum_{K \in P(M): i \notin K} R_K I_i(t). \tag{5.7}$$

Thus, the full model takes the form:

$$\frac{dR_J I_i(t))}{dt} = \underbrace{\beta_{J,i}\, R_J\, I_i^{\text{tot}}(t)}_{\text{reinfection of history } J \text{ by strain } i} - \underbrace{\gamma_{J,i}\, R_J I_i(t)}_{\text{exit from infection (recovery or death)}}, \quad i \notin J, \tag{5.8}$$

$$\frac{dR_J(t)}{dt} = \underbrace{\sum_{i \in J} \gamma_{J\setminus\{i\},i}\, \phi_{J\setminus\{i\},i}\, R_{J\setminus\{i\}} I_i(t)}_{\text{recovery updates immune history into } J} - \underbrace{\sum_{i \in M\setminus J} \beta_{J,i}\, R_J\, I_i^{\text{tot}}(t)}_{\text{reinfection by strains not yet in } J}. \tag{5.9}$$

Computationally, Listing 5.2 shows the computational core for simulating Eqs. (5.8)–(5.8). We encode immune histories $J \subseteq M$ as bitmasks $J \in \{0, \ldots, 2^m - 1\}$. The state vector is ordered as

$$y(t) = \left[R_0, \ldots, R_{2^m-1},\ I_{1,0}, \ldots, I_{1,2^m-1},\ \ldots,\ I_{m,0}, \ldots, I_{m,2^m-1},\ D\right],$$

where $I_{i,J}$ represents $R_J I_i$. Entries with $i \in J$ are invalid and are kept at zero; the code enforces zero dynamics for them.

```
def rhs_multistrain_reinfection(t, y, m, beta, gamma, phi):
    nS = 1 << m  # number of immune history subsets = 2^m

    # Unpack state
    R = y[:nS]                              # R[J]
    I = y[nS:nS + m * nS].reshape(m, nS)        # I[i, J]   R J I i
    D = y[-1]

    # Validity mask: valid[i, J] is True iff i not in J
    valid = np.empty((m, nS), dtype=bool)
    for i in range(m):
        bit = 1 << i
        valid[i, :] = (np.arange(nS) & bit) == 0

    # Total prevalence of each strain i across admissible histories
    I_tot = np.sum(I * valid, axis=1)           # shape (m,)

    # Infectious derivatives
    dI = np.zeros_like(I)
    for i in range(m):
        # Vectorized over histories J
        # dI[i,J] = -gamma[J,i]*I[i,J] + beta[J,i]*R[J]*I_tot[i] for valid histories
        gi = gamma[:, i]
        bi = beta[:, i]
        dI[i, :] = valid[i, :] * (-gi * I[i, :] + bi * R * I_tot[i])

    # Recovered derivatives
    dR = np.zeros_like(R)
    for J in range(nS):
        inflow = 0.0
        outflow = 0.0

        for i in range(m):
            bit = 1 << i

            # inflow into R[J] from recovering infections that add strain i to history
            if (J & bit) != 0:
                J_prev = J & ~bit
                inflow += gamma[J_prev, i] * phi[J_prev, i] * I[i, J_prev]

            # outflow from R[J] due to reinfection by strains not yet in J
            else:
                outflow += beta[J, i] * R[J] * I_tot[i]

        dR[J] = inflow - outflow

    # Death derivative
```

```
    # dD = sum_{i,J valid} gamma[J,i]*(1-phi[J,i])*I[i,J]
    dD = 0.0
    for i in range(m):
        dD += np.sum(valid[i, :] * gamma[:, i] * (1.0 - phi[:, i]) * I[i, :])

    # Pack derivative
    return np.concatenate([dR, dI.reshape(-1), np.array([dD])])


def simulate_multistrain_reinfection(m, beta, gamma, phi, R0, I0, D0, t_eval):
    nS = 1 << m
    y0 = np.concatenate([np.array(R0, dtype=float),
                         np.array(I0, dtype=float).reshape(m * nS),
                         np.array([float(D0)])])

    sol = solve_ivp(
        fun=lambda t, y: rhs_multistrain_reinfection(t, y, m, beta, gamma, phi),
        t_span=(t_eval[0], t_eval[-1]),
        y0=y0,
        t_eval=t_eval,
        rtol=1e-7, atol=1e-10
    )

    R = sol.y[:nS, :]
    I = sol.y[nS:nS + m * nS, :].reshape(m, nS, -1)
    D = sol.y[-1, :]
    return sol.t, R, I, D
```

Listing 5.2: Computational core for the multi-strain reinfection model (no imports, no printing). Arrays `beta`, `gamma`, `phi` are shaped $(2^m, m)$.

Fig. 5.4 illustrates how explicit immune history induces multi-wave behavior even without allowing simultaneous co-infection. Individuals begin in the no-history class $R_\emptyset(t)$ $(= S(t))$. Infection by strain i moves them to an infectious-with-history state $R_J I_i(t)$, and upon leaving infection they either recover into $R_{J\cup\{i\}}(t)$ or die into $D(t)$. Reinfection is possible because membership in $R_J(t)$ confers immunity only to strains in J; individuals remain susceptible to any strain not yet experienced ($M \setminus J$). As a result, when additional strains are present (middle and right panels), a population that has largely depleted $R_\emptyset(t)$ can still sustain later epidemics driven by infections in partially immune individuals. The strain-level parameters used in the simulations were constant across immune histories J (so $\beta_{J,i} \equiv \beta_i$, $\gamma_{J,i} \equiv \gamma_i$, and $\phi_{J,i} \equiv \phi_i$), with values: $(\beta_1, \gamma_1, \phi_1) = (0.85,\ 1/6,\ 0.995)$, $(\beta_2, \gamma_2, \phi_2) = (0.65,\ 1/7,\ 0.994)$, $(\beta_3, \gamma_3, \phi_3) = (0.55,\ 1/5,\ 0.992)$, $(\beta_4, \gamma_4, \phi_4) = (0.75,\ 1/8,\ 0.993)$.

5.3 Mutation-driven emergence of new strains

So far, the strain set $M = \{1, \ldots, m\}$ has been treated as fixed over time. Strains circulate "in parallel" and interact only through the depletion of susceptible (possibly structured by immune history). During

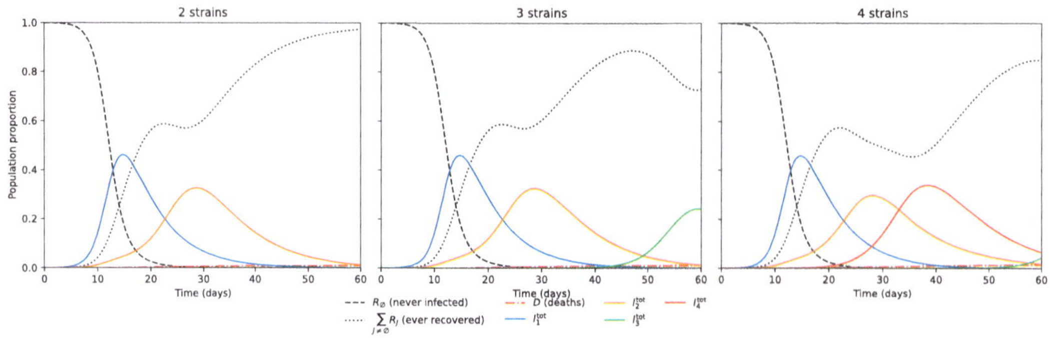

FIGURE 5.4

Multi-strain compartmental dynamics with reinfection for $m = 2, 3, 4$ strains (left to right).

pandemics, however, strains can also be created during transmission via mutation, recombination, or reassortment [292,312,313]. In compartmental models, this evolutionary layer is commonly represented by allowing infections caused by one strain to sometimes found infections of another strain. This turns the force of infection into a mutational mixture over donor strains. Formally, let $q_{i\to j} \in [0, 1]$ denote the probability that a transmission event originating from a donor infected with strain i results in the recipient being infected with strain j, with $\sum_{j\in M} q_{i\to j} = 1$ for each i. The matrix $Q = (q_{i\to j})$ may be interpreted as a coarse "genotype-to-genotype" mutation kernel per transmission generation. Using the immune-history notation from the previous section, define the total prevalence of strain i,

$$I_i^{\text{tot}}(t) = \sum_{K: i\notin K} R_K I_i(t), \tag{5.10}$$

and define the mutationally mixed infection pressure that produces infections of type j in hosts with history J,

$$\Lambda_{J\to j}(t) = \sum_{i\in M} \beta_{J,i}\, q_{i\to j}\, I_i^{\text{tot}}(t). \tag{5.11}$$

Here, $\beta_{J,i}$ is the (history-dependent) per-capita infection rate experienced by individuals in R_J due to contact with infectious individuals carrying donor strain i; mutation then allocates these transmission events to an offspring strain j.

With this notation, the multi-strain reinfection model with mutation can be written compactly as

$$\frac{d(R_J I_j)}{dt} = \underbrace{\mathbb{1}[j\notin J]\, R_J(t)\, \Lambda_{J\to j}(t)}_{\substack{\text{new infections of strain } j\\ \text{generated by all donor strains via } Q}} - \underbrace{\gamma_{J,j}\, R_J I_j(t)}_{\text{exit from infectiousness (recovery or death)}}, \tag{5.12}$$

$$\frac{dR_J}{dt} = \underbrace{\sum_{i\in J} \gamma_{J\setminus\{i\},i}\, \phi_{J\setminus\{i\},i}\, R_{J\setminus\{i\}} I_i(t)}_{\substack{\text{recoveries into history } J\\ \text{after infection with strains in } J}} - \underbrace{\sum_{j\in M\setminus J} R_J(t)\, \Lambda_{J\to j}(t)}_{\substack{\text{all infections leaving } R_J\\ \text{into not-yet-seen strains}}}, \tag{5.13}$$

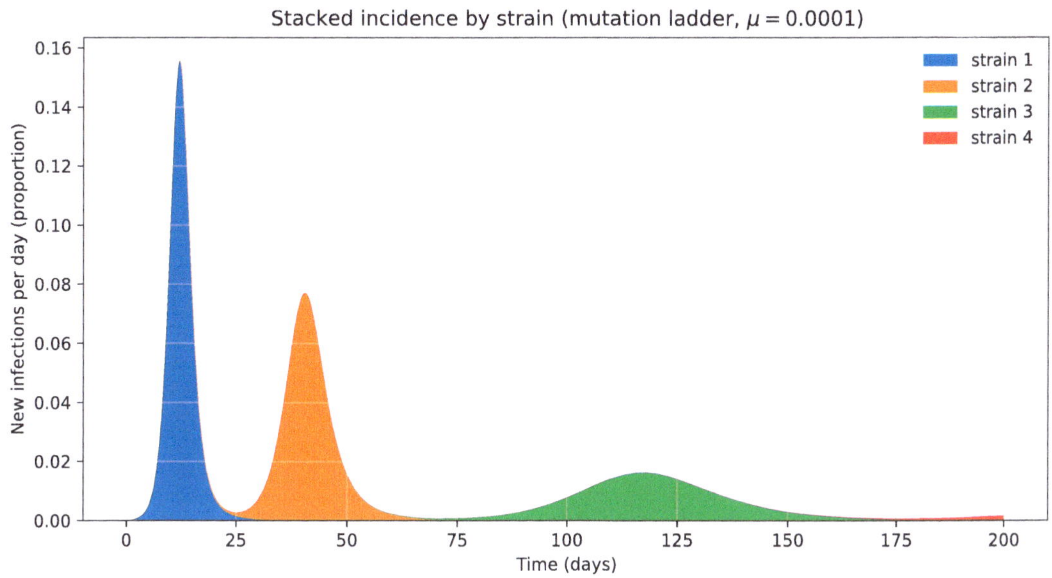

FIGURE 5.5

Stacked incidence by strain under mutation-driven emergence ($m = 4$, mutation ladder kernel, $\mu = 10^{-4}$).

$$\frac{dD}{dt} = \underbrace{\sum_{J \subseteq M} \sum_{j \in M \setminus J} \gamma_{J,j} \left(1 - \phi_{J,j}\right) R_J I_j(t)}_{\text{deaths upon leaving infection}}. \tag{5.14}$$

The indicator $\mathbb{1}[j \notin J]$ enforces the simplifying assumption of perfect homologous immunity (no infection by strains already in the history set J). Notably, this can be relaxed to partial cross-immunity by allowing infections into $j \in J$ with reduced rates, however, it is out of the scope of this chapter.

To make the mutation mechanism concrete, we conclude this section with three simulation figures generated from the mutation-at-transmission model (Eqs. (5.12)–(5.14)) using a sequential "mutation ladder" kernel Q, in which most transmissions preserve the donor strain while a small fraction produce the next strain. Unless stated otherwise, we simulate $m = 4$ strains, use the same history-independent parameters across immune histories ($\beta_{J,i} \equiv \beta_i$, $\gamma_{J,i} \equiv \gamma_i$, $\phi_{J,i} \equiv \phi_i$), and seed infection only in strain 1 (mutation generates the remaining strains). The strain-level values used in the code are $(\beta_1, \gamma_1, \phi_1) = (0.85, 1/6, 0.995)$, $(\beta_2, \gamma_2, \phi_2) = (0.55, 1/6, 0.995)$, $(\beta_3, \gamma_3, \phi_3) = (0.28, 1/6, 0.995)$, $(\beta_4, \gamma_4, \phi_4) = (0.23, 1/6, 0.995)$, with initial condition $R_\emptyset(0) = 1 - I_1^{\text{tot}}(0)$, $I_1^{\text{tot}}(0) = 2 \times 10^{-4}$, and $I_j^{\text{tot}}(0) = 0$ for $j \geq 2$, and $D(0) = 0$.

Fig. 5.5 visualizes variant turnover as a decomposition of incidence into strain-attributed components. In this model, infections are not "born" as a fixed strain: rather, each transmission event has a donor strain i and an offspring strain j determined by the kernel Q. As a result, later strains can rise even when they are absent initially, provided the system generates sufficient mutation flux from earlier strains. The stacked representation makes this causal chain explicit: early epidemic growth is

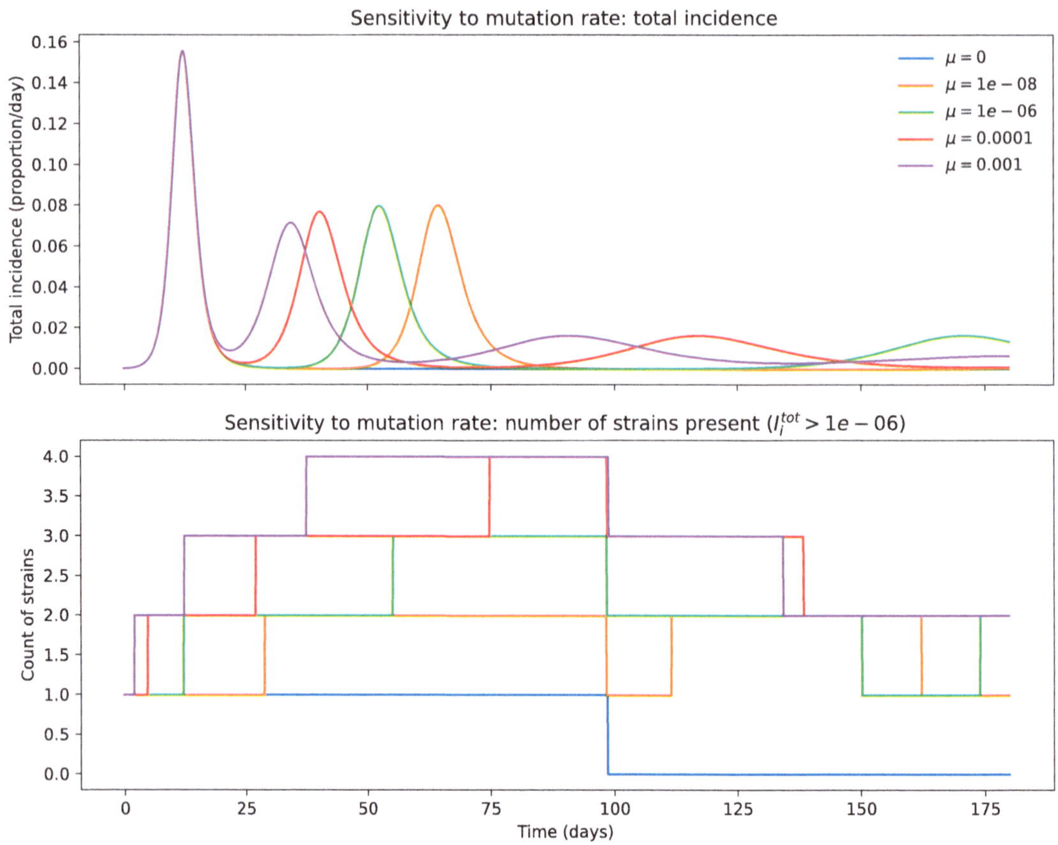

FIGURE 5.6

Sensitivity of mutation-driven dynamics to the mutation probability μ (mutation ladder kernel, $m = 4$). Top: total incidence $\sum_j \text{inc}_j(t)$. Bottom: the number of strains currently "present" (defined in the code by $I_j^{\text{tot}}(t) > 10^{-6}$). Increasing μ accelerates the appearance of new strains, tends to increase transient diversity, and can introduce additional waves through infections in individuals whose immune history does not include the newly emergent strains.

dominated by strain 1, after which mutation-generated strains contribute an increasing fraction of new cases.

Fig. 5.6 shows how a single evolutionary control knob reshapes epidemic trajectories. When $\mu = 0$, no new strains are created, and the system behaves like a single-strain outbreak seeded in strain 1. As μ increases, new strains appear earlier and in larger numbers, which can alter the timing and amplitude of incidence peaks. The bottom panel provides a simple, operational view of diversity: the number of strains whose total prevalence exceeds a small threshold.

Fig. 5.7 summarizes the same mechanism across mutation rates by converting trajectories into two interpretable outcome measures. First, we count the number of distinct incidence peaks (waves) in

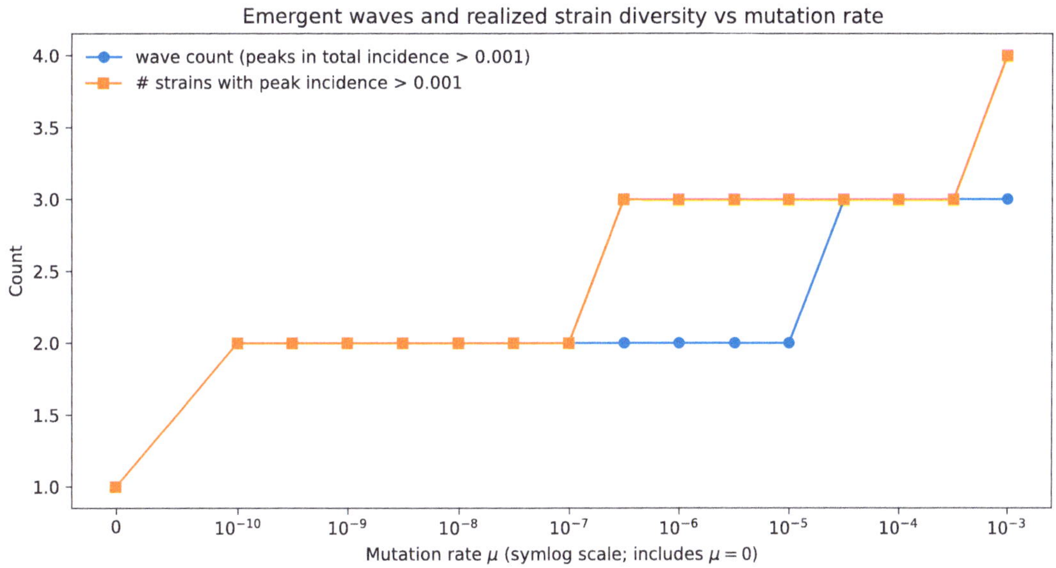

FIGURE 5.7

Emergent wave count and realized diversity versus mutation probability μ (mutation ladder kernel, $m = 4$). Circles show the number of peaks in total incidence above a fixed floor (as detected by a peak-finding rule in the code), while squares show the number of strains whose *peak* incidence exceeds a threshold within the simulated horizon. Both measures increase with μ over a range of values, illustrating how mutation can induce qualitatively richer multi-wave dynamics even in a deterministic compartmental approximation.

the total incidence curve. Second, we count how many strains become epidemiologically meaningful, operationalized as achieving a sufficiently large peak incidence at some point during the simulation. Both measures are crude but useful: they translate high-dimensional multi-strain dynamics into quantities that can be compared across scenarios, and they highlight an important qualitative transition. Below some mutation rates, the dynamics remain close to a single-strain epidemic, whereas above them, the system can produce repeated waves and sustained turnover as newly created strains exploit remaining susceptibility in immune-history space.

Deepen your knowledge

Exercise 1 (How many compartments does immune history create?)

Let $M = \{1, \ldots, m\}$ be the strain set and let immune history be represented by subsets $J \subseteq M$. Consider the reinfection model with compartments R_J, $R_J I_i$ (only for $i \notin J$), and D.

(a) How many R_J compartments are there as a function of m?

(b) How many infectious-with-history compartments $R_J I_i$ are there in total?

(c) Compare this to a "parallel strains" SIR/SEIR model with m strains and no immune-history structure.

Solution. (a) There is one recovered-history compartment for each subset $J \subseteq M$, hence 2^m compartments R_J.

(b) Fix a strain i. The number of histories that do *not* contain i is 2^{m-1}, so there are 2^{m-1} valid compartments of the form $R_J I_i$. Summing over $i \in M$ gives $m\,2^{m-1}$ infectious compartments.

(c) A parallel-strain SIR has S plus m infectious compartments and one R: total $m+2$. A parallel-strain SEIR has S, m exposed, m infectious, and R: total $2m+2$. The immune-history model scales exponentially in m, which is why reductions are often needed.

Exercise 2 (Strain-specific effective susceptibility and invasion)

In the immune-history reinfection model, define the strain-specific effective susceptible mass

$$S_i^{\text{eff}}(t) = \sum_{J:i \notin J} R_J(t).$$

Assume for simplicity that $\beta_{J,i} \equiv \beta_i$ and $\gamma_{J,i} \equiv \gamma_i$ do not depend on J.

(a) Give a natural approximation for $R_{e,i}(t)$ in terms of $S_i^{\text{eff}}(t)$, β_i, and γ_i.

(b) Interpret $S_i^{\text{eff}}(t)$ in words and explain how it differs from the single-strain susceptible fraction $S(t)$.

Solution. (a) A natural approximation is

$$R_{e,i}(t) \approx \frac{\beta_i}{\gamma_i} S_i^{\text{eff}}(t),$$

because only individuals whose histories do not include i are susceptible to infection by i.

(b) $S_i^{\text{eff}}(t)$ is the fraction of the population that has *not yet acquired immunity to strain i* (i.e., has not recovered from i). Unlike a single $S(t)$, which collapses all immunity into one number, $S_i^{\text{eff}}(t)$ is strain-specific and can remain large for newly emergent strains even when the never-infected class $R_\emptyset(t)$ is small.

Exercise 3 (Mutation kernel sanity check)

In the mutation-at-transmission extension, donor strain i produces offspring strain j with probability $q_{i\to j}$, where each donor row sums to one: $\sum_j q_{i\to j} = 1$. Define the mixed infection pressure into offspring strain j (for a given history J) as

$$\Lambda_{J\to j}(t) = \sum_{i \in M} \beta_{J,i}\, q_{i\to j}\, I_i^{\text{tot}}(t).$$

(a) Show that if infection into *all* strains $j \in M$ were permitted from R_J (ignoring the restriction $j \notin J$), then the total pressure $\sum_{j\in M} \Lambda_{J\to j}(t)$ does not depend on Q.

(b) Explain why, once we enforce immunity (infection only into $j \notin J$), Q can change the *total* outflow from R_J, not just its allocation across strains.

Solution. (a) Using row-stochasticity,

$$\sum_{j\in M} \Lambda_{J\to j}(t) = \sum_{j\in M}\sum_{i\in M} \beta_{J,i}\, q_{i\to j}\, I_i^{\text{tot}}(t) = \sum_{i\in M} \beta_{J,i}\, I_i^{\text{tot}}(t) \sum_{j\in M} q_{i\to j} = \sum_{i\in M} \beta_{J,i}\, I_i^{\text{tot}}(t).$$

So Q only redistributes infections across offspring strains, leaving the total unchanged.

(b) With immunity constraints, infections into $j \in J$ are disallowed (or reduced). Then parts of $\sum_j \Lambda_{J \to j}$ may be "blocked" by history J, so the effective outflow $\sum_{j \notin J} \Lambda_{J \to j}$ depends on how Q places offspring mass on allowed vs. blocked strains.

Exercise 4 (Waiting time to a successful mutant)

Suppose strain i is currently prevalent and produces candidate mutants of strain j at per-transmission probability μ (e.g., $q_{i \to j} = \mu$ in a ladder kernel). Let $C_i(t)$ denote the instantaneous rate of transmissions generated by strain i (new infections per unit time attributable to donor i). A newly created strain-j lineage establishes with probability $p_{\text{est}}(t)$.

(a) Under a "rare successful mutant" approximation, give the instantaneous rate $\lambda_{\text{succ}}(t)$ at which successful strain-j lineages appear.

(b) If $C_i(t) \approx C$ and $p_{\text{est}}(t) \approx p$ are roughly constant over a window, approximate the expected waiting time to the first successful mutant.

(c) Using a branching-process heuristic, give a common approximation for p_{est} in terms of $R_{e,j}$ when $R_{e,j} > 1$.

Solution. (a) Successful mutants appear at rate $\lambda_{\text{succ}}(t) \approx \mu\, C_i(t)\, p_{\text{est}}(t)$.

(b) For constant C and p, the waiting time is approximately exponential with mean

$$\mathbb{E}[T] \approx \frac{1}{\mu C p}.$$

(c) A common heuristic is $p_{\text{est}} \approx 1 - \frac{1}{R_{e,j}}$ for $R_{e,j} > 1$ (and $p_{\text{est}} \approx 0$ if $R_{e,j} \leq 1$). This connects establishment to the available strain-specific susceptible mass.

Exercise 5 (Measuring strain diversity over time)

Let $I_i^{\text{tot}}(t)$ be total prevalence of strain i. Define strain frequencies among infectious individuals as

$$p_i(t) = \frac{I_i^{\text{tot}}(t)}{\sum_{\ell \in M} I_\ell^{\text{tot}}(t)} \quad \text{(when the denominator is nonzero).}$$

(a) Define the Shannon diversity $H(t)$ and the "effective number of strains" $N_{\text{eff}}(t)$.

(b) What are the minimum and maximum possible values of $H(t)$ for m strains?

Solution. (a) Shannon diversity is

$$H(t) = -\sum_{i \in M} p_i(t) \log p_i(t), \qquad N_{\text{eff}}(t) = \exp(H(t)).$$

N_{eff} is interpretable as the number of equally common strains that would give the same entropy.

(b) $H(t) = 0$ when one strain dominates completely ($p_i = 1$ for some i). The maximum is $\log m$, achieved when all strains are equally frequent ($p_i = 1/m$).

Exercise 6 (Wave counting and diversity in Python)

You simulate the mutation model and obtain arrays `I_tot` of shape (m, T) and `inc` of shape (m, T), where `I_tot[i,t]` $\approx I_{i+1}^{\text{tot}}(t)$ and `inc[i,t]` $\approx \text{inc}_{i+1}(t)$.

(a) Write code to compute $H(t)$ and $N_{\text{eff}}(t)$ from `I_tot`.

(b) Write code to count waves as the number of peaks in total incidence $\sum_i \mathrm{inc}_i(t)$ above a threshold.

```
import numpy as np
from scipy.signal import find_peaks

def shannon_diversity(I_tot, eps=1e-18):
    """
    I_tot: (m,T) array of total prevalence by strain
    Returns: H(t), N_eff(t)
    """
    It = np.asarray(I_tot, dtype=float)
    denom = It.sum(axis=0) + eps
    p = It / denom
    H = -np.sum(p * np.log(p + eps), axis=0)
    N_eff = np.exp(H)
    return H, N_eff

def count_waves_from_incidence(inc_total, height=1e-3):
    """
    inc_total: (T,) total incidence time series
    Returns: integer wave count (peaks above 'height')
    """
    x = np.asarray(inc_total, dtype=float)
    peaks, _ = find_peaks(
        x,
        height=height,
        prominence=0.01 * x.max(),
        distance=40
    )
    return len(peaks)

# Example usage:
# H, N_eff = shannon_diversity(I_tot)
# waves = count_waves_from_incidence(inc.sum(axis=0), height=1e-3)
```

Listing 5.3: Computing diversity and wave counts from simulated multi-strain mutation dynamics.

Solution. (a) The function `shannon_diversity` computes $p_i(t)$, then $H(t)$, and $N_{\mathrm{eff}}(t)$.

(b) The function `count_waves_from_incidence` applies a simple peak-finding rule; the parameters (height, prominence, distance) should be chosen to match the time scale and noise level of the simulated or observed data.

CHAPTER

Introduction to multi-agent simulations

6

In the design of systems that are complex, dynamic, and stochastic, simulation is widely regarded as one of the most effective technologies for supporting design activities and for informing strategic and tactical decision making [314–319]. A simulation-based model specifies a set of rules that governs how a system evolves over time as a function of its current (and past) state(s). In contrast to analytical models, which are typically "solved", a simulation model is executed, enabling the user to observe state trajectories at any desired moment. This emphasis on (spatio-)temporal evolution offers insight into underlying system dynamics, rather than merely mapping particular inputs to corresponding outputs [320]. Accordingly, simulation should be understood not as a decision-making instrument, but as a decision-support technology that facilitates better-informed choices.

As the famous aphorism stated by the statistician George E. P. Box, "All models are wrong, but some are useful" [321]. Indeed, since real-world systems are inherently complex and not fully observable, any simulation model necessarily constitutes an approximation of its target system. The central craft of simulation modeling, therefore, lies in abstraction and simplification such that the model should include only those characteristics that are essential for the study's objectives and for meaningful analysis of the system under investigation [322].

Simulation is commonly employed either to deepen understanding of how a target system operates or to anticipate aspects of its performance. It may be viewed as an "artificial white room" in which hypotheses, policies, and practices can be explored without disrupting the everyday functioning of the focal organization. If an appropriate theory has been formulated for the target system and faithfully encoded in a computational model, simulation can assist in addressing questions such as the following:

- What forms of behavior arise under arbitrarily specified parameter combinations and initial conditions?
- What behavior is the target system likely to exhibit in the future?
- What state is the target system likely to reach in the future?

The level of accuracy required of a simulation model depends strongly on the questions being posed. Addressing the first question primarily demands an explanatory model, which typically tolerates lower precision in input data. By contrast, the latter two questions require a predictive model and thus depend on highly accurate inputs to produce credible outputs. Even then, such predictions are usually best interpreted as indications of trends and tendencies rather than as exact, absolute forecasts of system performance, especially in the epidemiological context [323–325].

Roughly, there are three principal simulation paradigms to investigate and explain the behavior of organizational systems, each suited to a distinct class of problems: Discrete-Event Simulation (DES) [326], System Dynamics (SD) [327], and Agent-Based Simulation (ABS) [328]. In DES, a system is

Epidemic Modeling. https://doi.org/10.1016/B978-0-44-355525-1.00012-6

represented as a collection of entities that move through processes and evolve over time in response to resource constraints and the occurrence of events. The simulator advances the model by managing a time-ordered event list. System Dynamics adopts a top-down perspective, characterizing system evolution through a small set of key state variables that capture dominant behaviors. These variables are linked via coupled differential (or difference, in discrete cases) equations, enabling analysis at an aggregate level. SD is most appropriate when individual actors need not be modeled in detail and when system knowledge is available in aggregated form, as in many population, ecological, and economic applications [329–331]. ABS, by contrast, is a bottom-up approach in which the researcher specifies decision-making and behaviors rules for individual actors at the micro level. Macro-level patterns then arise endogenously from agents' actions and from their interactions with one another and with the environment. While DES and SD are long-established and widely deployed in both academic and industrial contexts, ABS has historically been used predominantly as a research methodology, particularly in fields such as the social sciences, economics, ecology (where it is often termed individual-based modeling), and political science [332–334].

Although computational simulation has been used extensively since the 1960s [335,336], ABS rose to prominence only in the early 1990s [316]. Since then, it has become a well-established approach and has increasingly been adopted in applied settings [328,337]. Its intellectual lineage is difficult to trace precisely, in part because the field lacks a single, universally accepted definition of what constitutes an "agent", leading to different accounts of where agent-based modeling begins. From an epidemiological perspective, many of the core ideas underpinning ABS can be seen in early work on distributed, local-interaction models. One foundational thread is John von Neumann's cellular automata in the 1940s [338]: systems composed of discrete cells that transition among predefined states according to local rules and neighborhood histories. Such rule-based, spatially explicit dynamics closely resemble later epidemic models in which infection status changes depend on proximity, contact, and local mixing patterns. Other influential early contributions – though developed in different substantive domains – demonstrated how simple micro-level rules can generate macro-level regularities, including Schelling's segregation dynamics and Axelrod's tournament studies of cooperation [339]. These examples foreshadow a key epidemiological motivation for ABS: population-level disease patterns can emerge from individual contact behavior, mobility, and heterogeneity rather than being imposed by aggregate assumptions.

Early "agent-type" modeling efforts can be traced to work in the 1960s on individual decision behavior [340], and by the 1970s and 1980s, agent-based approaches were increasingly used in the social sciences to study networks, population dynamics, and political processes, alongside proposals to integrate distributed artificial intelligence and multi-actor systems into mainstream modeling. In parallel, developments in artificial life, such as Reynolds' mid-1980s work on flocking, helped popularize the idea that realistic collective dynamics can arise from relatively simple behavioral rules operating at the level of individuals [341].

In modern epidemiological modeling, ABS is particularly valuable because it represents individuals explicitly and allows them to differ in clinically and socially meaningful ways [342–344]: susceptibility and infectiousness, symptom severity, contact rates, movement patterns, household and workplace structure, healthcare-seeking behavior, and adherence to interventions (e.g., isolation, masking, vaccination, or prophylaxis). This bottom-up framing is especially useful when individual variability cannot be neglected, such as when transmission is driven by heterogeneous contacts, clustered mixing, super-spreading contexts, behavioral responses to risk communication, or policy measures that operate through individual decisions. Macro-level outcomes such as incidence curves, spatial waves, healthcare demand, or inequality in burden, can then be understood as emergent consequences of micro-level interactions among agents and between agents and their environment. A further advantage, central to public-health decision support, is that ABS can naturally encode explicit mechanisms for interaction – within households, schools, workplaces, clinics, and community settings – rather than treating people as homogeneous *compartments* or deterministic resources. This is often crucial when evaluating targeted interventions (e.g., school closures, test-and-trace strategies, vaccine privatization, mobility restrictions) whose effectiveness depends on who interacts with whom, where, and under what behavioral constraints. In this sense, ABS offers a modeling framework well suited to systems driven by interaction structure and behavioral feedback, and it can reveal apparently complex population-level dynamics even when individual agents follow comparatively simple rules. At the same time, ABS entails notable challenges. Because interactions among agents, and between agents and their environments, must be specified explicitly, model design can be substantially more complex than in more aggregate approaches [345]. The resulting models can also be computationally demanding, particularly when representing large populations, fine-grained mobility, or detailed health states [346]. Finally, for epidemiological applications in particular, credibility depends heavily on data for contact patterns, behavior, and intervention compliance, as well as careful calibration and uncertainty analysis; without these, fine-grained realism may be more apparent than real. Fig. 6.1 presents a schematic view of when using ABS.

In this chapter, we cover ABS as a computational medium in general, and its meaning for epidemiological modeling, in particular. First, we show that a broad and natural class of ABS formalisms is *Turing complete* – an ABS can emulate any algorithm (equivalently, any Turing machine) given enough time and memory. Second, we explain what this implies for performance analysis: in the fully general case, runtime and memory usage inherit the fundamental limits of program analysis (e.g., the halting problem [347]). Third, we move from the general theory to simple epidemiological ABS and derive concrete runtime/memory scalings for common SIR/SEIR-style constructions.

6.1 Agent-based simulation is Turing complete

First, we need to define Turing completeness. Here, a system is Turing complete if, given enough time and unbounded memory, it can implement any computation that a Turing machine (i.e., modern computer) can. Simply put, a computational system is Turing complete if it can express any algorithm, not just a fixed family of behaviors [348]. In practice, that usually requires (i) some form of conditional branching, (ii) the ability to repeat actions (iteration/recursion), and (iii) access to an unbounded storage resource. It is desirable not because most models need arbitrary computation, but because it guarantees expressive adequacy: whatever mechanism one needs for decision rules, learning/adaptation, bookkeeping, dynamic data structures, or embedding other models (ODE solvers, optimization routines, inference), the modeling framework can represent it without hitting a hard ceiling. For agent-based simulation specifically, Turing completeness reassures that ABS is not merely a stylized dynamical system class but rather can encode complex protocols, institutional rules, and adaptive strategies. The trade-off is that universality comes with theoretical limits, and many global properties become undecidable in full generality, so good practice relies

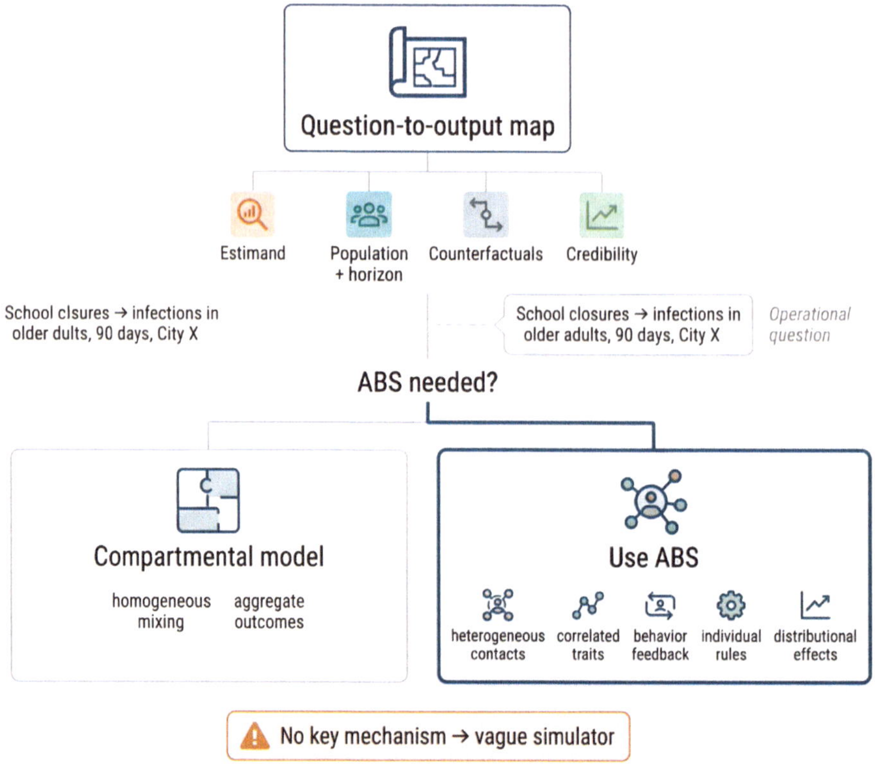

FIGURE 6.1

A schematic view of when using ABS.

on restricted model classes and empirical/parametric performance reasoning rather than universal guarantees.

Following this definition, from the technical side, to reason about computational power and resource use, one first needs to describe an abstract ABS in terms of: (i) a (finite or countably infinite) set of agents A_t at simulation time t where each agent $a \in A_t$ has a finite internal state $\sigma_t(a) \in \Sigma$ (e.g., health state, timers, attributes), a neighborhood/interaction structure (spatial grid, contact network, or message channels) determining which agents can affect which, and a scheduling discipline that defines how updates occur: synchronous (all agents compute next state from the previous global state) or asynchronous (agents update in some sequence, possibly randomized) [314,315].

To this end, the number of agents is fixed to N and each agent has a finite state; then the global state space is finite ($|\Sigma|^N$ times any finite environment state). Such a system must eventually repeat a previous global state, after which it cycles. That is powerful for modeling, but it is not sufficient for Turing completeness. Universality requires an unbounded resource in principle. E.g., unbounded time and an unbounded supply of agents or an unbounded environment size (the analogue of an infinite tape). This is the same assumption used when calling conventional programming languages Turing-complete.

In a more general manner, one can show ABS is Turing complete using reduction to a cellular automaton (CA) [349]. Namely, a one-dimensional CA consists of cells in a line, each holding a small state (often one bit), updated synchronously from a fixed-radius neighborhood rule. Stephen Wolfram conjectured, and Matthew Cook proved, that the elementary CA known as *Rule 110* is capable of universal computation [350,351]. Concretely, Rule 110 can simulate a universal system (via cyclic tag systems), and therefore can emulate any Turing machine given an appropriate initial configuration [351].

Therefore, we state that ABS can be Turing complete via CA emulation. Formally, let us consider an ABS with the following properties:

1. Agents occupy positions on an unbounded 1D lattice indexed by $i \in \mathbb{Z}$ (or, equivalently, the model may create new agents to extend the lattice as needed).
2. Each agent has a binary internal state $\sigma(i) \in \{0, 1\}$.
3. Updates are synchronous: $\sigma_{t+1}(i)$ is a function only of $\sigma_t(i-1), \sigma_t(i), \sigma_t(i+1)$.

Then this ABS class is Turing complete.

The proof is quite simple. Under the stated assumptions, the ABS is exactly a cellular automaton since each agent is a CA cell, and the scheduler enforces synchronous global steps. Choose the local update rule to match Rule 110. Then the ABS can simulate Rule 110 for any initial configuration. Since Rule 110 is universal [351], the ABS can emulate an arbitrary Turing machine (encode the machine and input into the initial configuration, and interpret the CA evolution). Therefore, the ABS is Turing complete.

6.2 Runtime and memory analysis

Because ABS can emulate arbitrary programs, the fully general performance analysis problem is as hard as program analysis for general computation since given an arbitrary ABS specification and initial state, there is no algorithm that can always decide whether the simulation will halt (reach a designated stopping condition) unless you restrict the model class (e.g., bounded agents and bounded environment) [347]. If your ABS can encode a Turing machine, it inherits the halting problem.

Nevertheless, while worst-case analysis is unavoidably "as hard as programming", most scientific ABMs have regular structure. For these, we can provide meaningful bounds in terms of (i) number of agents at time t ($N_t = |A_t|$), (ii) number of interaction edges (E_t), (iii) number of spatial sites/cells (if the environment is a grid, M), (iv) average cost of an agent update (c_{upd}), and (v) number of discrete time steps executed (T). To this end, a typical synchronous ABS with per-agent constant-size neighborhoods satisfies:

$$\text{Time} \in \Theta\left(\sum_{t=0}^{T-1} N_t \, c_{\text{upd}}\right), \tag{6.1}$$

$$\text{Memory} \in \Theta\left(\max_{t \le T} N_t \cdot b + \text{environment}\right), \tag{6.2}$$

where b is the number of bytes (or bits) per agent state record. The hard part is that N_t itself can be data-dependent (birth/death, dynamic network formation), and for universal ABSs it can encode arbitrary computation.

In this book's epidemiological modeling context, agents usually represent people (or animals), and the state is typically from a small compartment set (S, I, R) plus timers for the temporal compartmental-based dynamics. We focus on the computational cost of contact generation and transmission evaluation. Hence, let us consider a deterministic SIR model under homogeneous mixing. ABS representation of this SIR model replaces mass-action terms with explicit contacts between individuals as follows:

- Each agent a has health state $H(a) \in \{S, I, R\}$ (or $\{S, E, I, R\}$).
- Infectious agents generate contacts with susceptible agents according to a mixing rule.
- Each contact transmits with probability p (or with hazard β in continuous time).
- Recovery occurs after a duration or with probability per unit time.

This is conceptually straightforward; computationally, the dominant cost is how contacts are enumerated.

One options is a discrete-time, well-mixed SIR with naive pairwise checking. Under these settings, at each time step, evaluate every pair (a, b) and allow transmission if a is infectious, b is susceptible, and a Bernoulli (p) trial succeeds. There are $\binom{N}{2} \in \Theta(N^2)$ pairs. If pair evaluation is $O(1)$, per-step time is $\Theta(N^2)$ and total time is $\Theta(T\,N^2)$. In addition, agent states require $\Theta(N)$ memory; if no explicit network is stored, additional memory is constant. Notably, this implementation is almost never necessary, but it is a useful worst-case baseline to start with our exploration.

A significantly better option is a discrete-time, well-mixed SIR with k random contacts per infectious agent. Here, each infectious agent samples k contacts uniformly at random (with or without replacement) each step, and attempts transmission on those contacts. For the computational time, let I_t be the number infectious at time t. Then, per-step contact attempts are $k\,I_t$, so total time is $\Theta\left(\sum_{t=0}^{T-1} k\,I_t\right) \subseteq O(T\,k\,N)$. If k is constant (common in stylized ABMs), time is linear in population size per step with the same memory consumption – an improvement in one order of magnitude. It is important to note that, under appropriate scaling of k and p with N, and ignoring correlations, this agent-based scheme approximates mass-action incidence (for more details, see [150]).

Extending to the spatio-temporal SIR model with a contact network, a static graph $G = (V, E)$ defines who can meet. Following the classical definition, infectious agents attempt transmission along incident edges at each step (or at some rate). In here, a straightforward implementation iterates over edges adjacent to infectious vertices:

$$\Theta\left(\sum_{a \in I_t} \deg(a)\right) \leq \Theta(|E|).$$

Thus, total time is $\Theta\left(\sum_t \sum_{a \in I_t} \deg(a)\right)$, and in the worst case, $O(T|E|)$. Moreover, the to store the agent states one needs $\Theta(N)$ and to store adjacency, one needs $\Theta(|E|)$, with a total of $\Theta(N + |E|)$. This implementation is well-used in practice [352].

Taken jointly, for epidemiological ABMs, it is usually meaningful to report:

- Population size N and peak infectious $I_{\max}$.

- Contact structure (well-mixed with k contacts, network with $|E|$, or spatial with M and neighborhood radius).
- Scheduler type (synchronous step, random sequential, or event-driven).
- The dominant loop(s) and their asymptotic scalings (e.g., $O(TkI_{\max})$ or $O(T|E|)$ or $O(K \log N)$).
- Peak memory in terms of N, $|E|$, M, and event-queue size.

These parameters capture most of what determines feasibility, even though the fully general ABS performance problem cannot be solved in closed form for all possible models.

6.3 Building agent-based simulation

The power of ABS to simulate the emergence of complexity comes with a distinct set of methodological burdens. Because ABS are software artifacts rather than pure mathematical abstractions, from a practical perspective, their behavior is inextricably linked to the underlying architectural decisions made during their construction. The representation of time, the scheduling of agent actions, the topology of the environment, and the management of stochasticity are not merely technical implementation details; they are ontological commitments that define the physics of the artificial world. A decision to model time as discrete steps rather than continuous events can introduce synchronization artifacts that destroy the validity of biological population models. Similarly, the choice between synchronous and asynchronous updating schemes can fundamentally alter the stability of cooperative regimes in game-theoretic simulations, turning a robust evolutionary strategy into a fragile one.

In this section, we analyze the structural pillars of ABS, examining the temporal dimension, contrasting the computational efficiency of discrete time stepping with the physical realism of stochastic event-driven methods like the Gillespie algorithm. We dissect the mechanics of interaction topologies, from the rigid geometry of Von Neumann grids to the dynamic fluidity of temporal networks. We explore the stochastic engine, analyzing how pseudo-random number generation and variance reduction techniques determine the reproducibility and statistical power of simulation results. Finally, we analyze the phenomenon of emergence itself, moving beyond qualitative descriptions to rigorous, information-theoretic metrics of causal emergence.

6.3.1 The temporal dimension

The representation of time is the foundational substrate of any dynamic simulation. In the physical world, time is perceived as a continuous, flowing variable where events can occur at any infinitesimal moment. In the digital realm of the computer, however, time must be discretized or serialized to be processed [353]. The mapping of physical time onto computational time dictates not only the resolution and accuracy of the model but also its computational efficiency and its ability to capture causal dependencies. The most prevalent approach in social and ecological agent-based models is Discrete Time Stepping (DTS), often referred to as "time-driven" simulation. In this regime, the simulation advances in fixed increments (Δt). At each "tick" of the clock, the system pauses, agents assess the state of the world, execute their behaviors, and the clock advances to $t + \Delta t$. The dominance of DTS stems from its conceptual alignment with the dynamics "natural" cycles. Budgetary cycles, agricultural harvests, and daily commuting patterns naturally lend themselves to a synchronized, stepped progression. For a modeler simulating the daily operations of a supply chain or the annual migration of a species, the fixed time step provides a logical and intuitive framework for organizing agent activities.

However, despite its intuitive appeal, DTS introduces "discretization error", a phenomenon where the granularity of the time step obscures the true causal ordering of events. Namely, in a continuous reality, it is impossible for two agents to arrive at a resource at the exact same instant; one will always precede the other by some fraction of a second. In a discrete model with a coarse Δt, multiple agents may attempt to access a resource simultaneously, forcing the modeler to implement arbitrary "tie-breaking" rules (e.g., random selection or ID-based priority) that have no basis in the physical system being modeled. Furthermore, DTS suffers from the "empty step" inefficiency. In systems characterized by "bursty" dynamics, such as a market that is quiet for hours and then explodes with trading activity, a fixed time step must be small enough to capture the rapid activity. As such, during the quiet periods, the simulation wastes computational cycles calculating updates for a system where nothing is changing.

To address the limitations of fixed time steps, Discrete Event Simulation (DES) treats time as a continuous variable. In this paradigm, the simulation is driven not by a clock that ticks regardless of activity, but by a queue of events ordered by their scheduled execution time. The system state remains frozen between events; the simulation engine simply skips from the timestamp of one event to the timestamp of the next, skipping over periods of inactivity entirely. This approach is fundamentally different in philosophy since DTS resembles an orchestra following a conducted score, where every instrument plays at specific beats, while DES resembles a jazz ensemble where musicians respond dynamically to the timing of others' improvisations. This makes DES particularly superior for systems involving queuing, logistics, or chemical kinetics, where the exact timing of an arrival or reaction is the critical variable determining system performance.

In such systems, when modeling the reaction of molecules within a cell, assuming a fixed Δt violates the continuous probabilistic nature of some processes, such as molecular collisions [354]. The Gillespie Algorithm [355], also known as the Stochastic Simulation Algorithm (SSA), provides a mathematically exact method for simulating these trajectories without the artificiality of time discretization. The Gillespie algorithm is derived from a single master equation and relies on the premise that, for a well-mixed system, the time between reactions is a random variable drawn from an exponential distribution. This algorithm follows three conceptual steps. First, the algorithm calculates the "propensity" (a_j) for every possible reaction channel in the system. The propensity represents the probability rate of that reaction occurring given the current number of reactant agents. Second, the time until the next reaction (τ) is not fixed but is generated stochastically: $\tau = (1/a_0)\ln(1/r_1)$, where a_0 is the sum of all propensities and r_1 is a random number. This explicitly models the "waiting time" for an event. Finally, a second random number determines which specific event occurs, weighted by their relative propensities.

The Gillespie algorithm is "exact" in that it produces a trajectory that is a statistically correct realization of the master equation. Nevertheless, because it simulates every single reaction event, it becomes computationally prohibitive as the number of agents increases. To bridge this gap, the Tau–Leaping, an approximate method that advances time by a larger step (τ) and estimates how many reactions of each type would fire during that interval, assuming the propensities remain relatively constant [356]. This hybrid approach effectively reintroduces a variable time step to balance the precision of SSA with the efficiency of DTS.

Table 6.1 Comparison of DTS, SSA (Gillespie), and Tau-Leaping simulation approaches.

Feature	Discrete Time Stepping (DTS)	Discrete Event / Gillespie (SSA)	Tau-Leaping (Approx. SSA)
Time Progression	Fixed Δt increments	Variable, precise event timestamps	Variable, approximate intervals
Computational Driver	Loop over all agents per step	Sorted event queue (priority heap)	Adaptive step size calculation
Causal Resolution	Coarse; simultaneous events forced	Fine; explicit sequential ordering	Intermediate; bounded error
Efficiency	High for dense, synchronous activity	High for sparse, bursty activity	Balanced for intermediate scales
Typical Domain	Social sciences, economics (macro)	Chemistry, logistics, queuing	Epidemiology, systems biology

Table 6.1 compares three simulation paradigms – DTS, Gillespie's SSA, and Tau-Leaping – across how time advances, what computation drives each method, how well they resolve causality, when they are most efficient, and the kinds of application domains where they are commonly used.

6.3.2 The synchronization problem

Once the progression of time is established, the modeler must address the problem of concurrency. Real-world agents act simultaneously in parallel. Standard computer processors execute instructions sequentially. The method used to serialize agent actions (the update schedule) is a notorious source of simulation artifacts that can generate spurious phenomena or mask real dynamics.

In a synchronous update regime, the state of the entire system at time $t + 1$ is calculated based only on the state of the system at time t. Conceptually, all agents observe the world, decide on an action, and then execute those actions simultaneously. This is often implemented using a double-buffering technique (a current-state array and a next-state array) to ensure that no agent sees the updated state of a neighbor within the same tick. Synchronous updating is theoretically attractive because it is deterministic and fair – no agent has a "first-mover" advantage. However, it is prone to generating spurious oscillations. A classic example is found in biological predator-prey models on a grid. If a group of predators simultaneously perceives a prey item in a neighboring cell, they may all move toward it in the same time step. This results in an unnatural "clumping" of predators and can lead to a localized extinction of prey that would not occur if the predators moved sequentially. Similarly, in the "Spatial Prisoner's Dilemma," synchronous updating allows cooperators to form stable, crystalline fractal patterns that resist invasion by defectors. However, Huberman and Glance (1993) demonstrated that this stability is largely an artifact of the synchronization [357]. When the update rule is switched to asynchronous (sequential), these intricate spatial structures break down, and the system collapses into a defector-dominated state much more rapidly. This reveals that the "emergence" of cooperation in the synchronous model was not a robust property of the agent interactions, but a fragile consequence of the artificial global clock [358].

As an alternative, asynchronous updating mimics continuous time by allowing agents to act one by one. The state of the system is updated immediately after each agent's action, meaning that an agent acting later in the sequence perceives the consequences of the actions taken by agents earlier in the sequence. While this eliminates synchronization artifacts, it introduces path dependence. The dynamics of the system become sensitive to the order in which agents are updated. There are three main types: Random Order (Random Permutation), Fixed Order, and Round-Robin. First, random order is considered the most robust and widely recommended approach. At each time step, the list of agents is shuffled, and they act in this randomized sequence. This prevents any single agent from accumulating a systematic advantage over time and averages out the first-mover effects. Second, in the fixed order method, agents are updated in a static sequence (e.g., by Agent ID: 1, 2, 3...). While computationally faster (avoiding the shuffle overhead), this introduces severe bias. If agents are competing for limited food, Agent 1 effectively has first priority in every single time step, leading to an artificial hierarchy where lower-ID agents prosper, and higher-ID agents starve. Lastly, the Round-Robin method is used in specific contexts like financial market simulations, where agents might act in a structured sequence to simulate the specific mechanics of an order book or a turn-based negotiation.

Importantly, the synchronization problem becomes acute when scaling ABMs to High-Performance Computing (HPC) environments. When agents are distributed across multiple processor cores (e.g., using MPI or CUDA), the simultaneous execution becomes literal, reintroducing the risk of race conditions. Technically, a race condition occurs when two agents on different cores attempt to modify the same memory address (e.g., consuming the same food patch) at the same time. The outcome depends on the microscopic timing of the threads, leading to non-deterministic and potentially erroneous behavior (e.g., the food patch acts as if it has double capacity). In order to manage this in spatial models, parallel ABMs employ domain decomposition with ghost zones (or halo regions). The simulation space is divided into chunks assigned to different processors. The boundaries of these chunks are duplicated as "ghosts" on neighboring processors. In the synchronization phase, agents act within their local partition. Agents near the boundary interact with the "ghost" copies of neighbors. In the communication phase, occurring at the end of the synchronization step, the processors exchange data to synchronize the state of the ghost zones with the real agents. This "Bulk Synchronous Parallel" (BSP) approach enforces a structured synchrony that prevents causality errors but imposes a communication overhead that can bottleneck performance.

6.3.3 The geometry of interaction

The "environment" in an ABS is not merely a passive container; it is the topology that constrains interaction. The geometry of this space (whether a 2D grid, a continuous plane, or a social network) dictates how information diffuses, how diseases spread, and how clusters form.

Most simply, in cellular automata and grid-based ABS, the definition of "neighborhood" fundamentally shapes the propagation of influence. For example, Von Neumann neighborhood formalization consists of the four orthogonal cells (North, South, East, West). Diffusion in this topology tends to expand in diamond shapes. Comparably, Moore neighborhood includes the eight surrounding cells (adding diagonals). Diffusion expands in square shapes and occurs more rapidly due to the increased connectivity. This decision can influence the simulation's conclusion. For instance, in a Prisoner's Dilemma simulation, Moore neighborhoods often support higher levels of cooperation than Von Neumann neighborhoods. The larger cluster size (8 neighbors vs. 4) provides a more robust buffer for cooperators to cluster together and protect themselves from exploitation by defectors.

More complex, in many social (and epidemiological) simulations, physical distance is less relevant than social distance. Agents are connected via networks, and the structure of these networks determines the system's dynamic properties. Computationally, one can divide these connection graphs into small-world and scale-free networks. Small-world networks, also known as *Watts-Strogatz* networks [359], characterized by high local clustering (cliques) and short average path lengths (shortcuts). These networks are highly efficient for signal propagation and synchronization. Scale-Free Networks, also known as *Barabási-Albert* networks [360], characterized by a power-law degree distribution, meaning a small number of hub nodes have massive connectivity while the vast majority have very few links. These networks are robust to random failure but extremely fragile to targeted attacks on hubs. The topology dramatically alters diffusion processes, such as infection spread. In the Bass diffusion model of product adoption, scale-free networks typically show a faster take-off because if a hub adopts the innovation, they instantly expose a massive number of followers to it [361]. However, for complex contagions – behaviors that require reinforcement from multiple peers before adoption (e.g., joining a risky protest) – highly clustered Small-World networks can be more effective. In a scale-free network, a hub might be connected to many people, but those people are not connected to each other, preventing the local reinforcement loops needed to trigger the complex behavior. A further extenstion comes naturally from the fact that real-world networks are not static; friendships form and dissolve over time. Temporal Networks model these dynamic topologies, where edges appear and disappear. Integrating dynamic networks into ABM allows for the modeling of "co-evolutionary" dynamics: agents influence the network (by breaking ties with dissimilar others), and the network influences the agents (by exposing them to new ideas). This feedback loop is essential for realistic models of polarization and echo chambers.

Another approach is continuous-space spatial models. For models requiring high physical fidelity, continuous space ($\mathbb{R}^2$ or $\mathbb{R}^3$) is preferred over discrete grids. However, continuous space introduces a significant computational hurdle – neighbor detection [362,363]. In a naive implementation, checking if any agent is within interaction radius r of any other agent requires measuring the distance between every pair, resulting in a computational complexity of $O(N^2)$ where N is the number of agents. For a simulation with 10 000 agents, this means 100 million checks per time step, rendering the model unusable. The solution to this challenge is spatial partitioning. A commonly used algorithm for this task is the Quadtree [364]. This algorithm recursively divides the 2D space into four quadrants. If a quadrant contains too many agents, it subdivides again. When an agent searches for neighbors, the algorithm traverses the tree. If a node's bounding box is further than distance r, the entire node (and all agents within it) is pruned from the search. This reduces the complexity from $O(N^2)$ to $O(N \log N)$. In these settings, the 10 000 agent example requires only 132 877 (or 0.13M) checks, three orders of magnitude less. Another option is the cell list, also known as the Bin-Lattice spatial subdivision [365], where the continuous space is overlaid with a virtual grid where the cell size equals the interaction radius r. Agents are mapped to these cells. To find neighbors, an agent only needs to search the agents in its own cell and the 8 adjacent cells. This approach effectively reduces neighbor detection to $O(N)$ but is less accurate.

6.3.4 Managing randomness and reproducibility

Stochasticity is intrinsic to ABS. It captures the heterogeneity of populations, the uncertainty of decision-making, and the inherent variability of natural processes. However, "randomness" in a scientific simulation is a carefully engineered artifact. It must be statistically high-quality, computationally efficient, and, most importantly, reproducible. It is important to recall that computers

(Turing machines in the general sense) are deterministic machines. They cannot generate true randomness and instead, they use Pseudo-Random Number Generators (PRNGs) to produce sequences of numbers that approximate the statistical properties of randomness [366]. As stated already in this book, a cardinal rule of scientific computing is that a simulation run with the same parameters and the same *seed* must produce the exact same results. This reproducibility is essential for debugging (isolating whether a strange behavior is a bug or a feature) and for sensitivity analysis. If a modeler changes a policy parameter and the outcome changes, they must be certain the change is due to the policy and not just a different roll of the dice.

Thus, achieving reproducibility becomes exponentially more difficult in parallel environments. If a simulation is running on 100 processor cores, the order in which agents request random numbers is non-deterministic due to the race conditions. If all agents draw from a single central PRNG, the sequence of numbers assigned to Agent X will vary depending on when Agent Y makes its request. To solve this, parallel ABS employs Strided Random Number Generation or independent streams, where each processor or agent is assigned a fixed sub-stream of the PRNG sequence. For example, Agent 1 uses random numbers at indices $1, 101, 201\ldots$; Agent 2 uses $2, 102, 202\ldots$. This ensures that Agent 1 always receives the same sequence of random numbers regardless of the execution speed of Agent 2.

Because stochastic ABS are inherently *noisy*, distinguishing a signal (the effect of a parameter change) from the background noise requires large sample sizes. Variance Reduction Techniques (VRT), borrowed from Monte Carlo physics, are used to increase the statistical power of simulations without simply running more of them. One option is the Common Random Numbers (CRN). This is the most effective VRT for comparative studies. When comparing Scenario A (Status Quo) and Scenario B (Intervention), the modeler uses the same random seeds for both. This synchronizes the stochastic events (e.g., the same agents get sick at the same time in both worlds), ensuring that the observed difference in outcomes is due solely to the intervention and not to random fluctuation. Another option is Antithetic Variates which involves running pairs of simulations. In the first run, random numbers U are used. In the second, perfectly negatively correlated numbers $1 - U$ are used. The average of these two runs has a lower variance than the average of two independent runs because the negative correlation cancels out extreme deviations. Finally, the importance sampling is often used to study rare events (e.g., black swan market crashes). The simulation is biased to make these events occur more frequently, and the results are then mathematically re-weighted to correct for the bias, providing a precise estimate of the rare event's probability without running millions of empty simulations.

6.3.5 Beyond simple rules

While ABS has often relied on *reactive* agents with simple stimulus-response behaviors, modern simulations increasingly employ *cognitive* agents capable of deliberation, planning, and learning. Probably the most basic reactive ABS type is rooted in the philosophy of Michael Bratman, the Belief-Desire-Intention (BDI) architecture provides a framework for rational agency [367]. Formally, a BDI agent maintains an explicit representation of its *beliefs* B about the world (possibly incomplete or uncertain), a set of *desires* D encoding objectives or preferred world states, and a current set of *intentions* I that commit the agent to particular courses of action. Deliberation can be viewed as a mapping that filters desires into intentions given the agent's beliefs and constraints, while means-end reasoning selects or synthesizes plans that are expected to achieve the chosen intentions. In practical implementations, this typically corresponds to (i) perception updates that revise B, (ii) an option-generation step that proposes candidate goals from D, (iii) an intention-selection step that yields $I \subseteq D$, and (iv) plan selection and execution, where plans are triggered and monitored until completion, failure, or revision. This separation is useful in simulation because it makes explicit *why* an agent acts (desires), *what* it is currently committed to (intentions), and *how* its internal model of the environment evolves (beliefs), enabling richer phenomena such as goal switching, resource-bounded planning, and context-dependent behavior that cannot be captured by purely reactive rules.

BDI agents are often described as "bounded rational": they do not compute globally optimal actions over an unbounded horizon, yet they remain capable of complex teleological (goal-directed) behavior. In particular, they can form explicit plans, monitor their execution, and revise or abandon them when circumstances change. For example, an agent may reason that "I intended to farm, but it is raining, so I will repair my tools instead", thereby switching intentions in light of updated beliefs and altered action affordances. This commitment-based representation of agency makes BDI especially attractive for social simulations in which human decision-making, norm compliance, and the interplay between private goals and public obligations are central. Because intentions persist over time, BDI agents can exhibit temporally extended behavior such as promise-keeping, obligation discharge, and strategic defection, while still remaining sensitive to new information and unexpected events through belief revision and intention reconsideration. However, these benefits come with substantial computational overhead. Maintaining symbolic belief bases, running deliberation cycles, performing plan selection, and handling continual monitoring and replanning are all more expensive than evaluating simple stimulus-response rules. As a result, large-scale simulations with many thousands or millions of agents may find full BDI reasoning prohibitively costly, motivating compromises such as (i) hybrid models where only a subset of agents (or only certain decision points) invoke BDI deliberation, (ii) simplified plan libraries and shallow reconsideration policies, or (iii) meta-models that approximate cognitive reasoning with learned or aggregated behavioral rules. Consequently, while BDI architectures provide interpretability and behavioral richness, they typically reduce the feasible agent count and time resolution relative to purely reactive architectures.

Moving beyond static rules, modern agents often incorporate Reinforcement Learning (RL) [368,369]. In this framework, agents are not told how to behave; instead, they are given a goal encoded as a reward signal and must learn an effective policy through trial-and-error interaction with the environment. Formally, an agent learns a policy $\pi(a \mid s)$ that maximizes the expected discounted return $\mathbb{E}\left[\sum_{t=0}^{\infty} \gamma^t r_t\right]$, where r_t is the reward at time t and $\gamma \in (0, 1)$ is a discount factor. Within ABS, RL can be embedded at the micro-level (each agent learns its own behavior) or at the macro-level (a controller learns policy levers such as intervention intensity, resource allocation, or mobility restrictions). A common design pattern is to build *hybrid agents*: a hand-crafted decision scaffold (e.g., goals, norms, safety constraints, or BDI-like deliberation) that defines admissible actions, paired with an RL component that learns within those bounds. This preserves interpretability and constraint compliance while still enabling adaptation to non-stationary environments and heterogeneous interactions.

In particular, when multiple adaptive agents learn simultaneously, each agent's environment becomes partially defined by other agents' changing policies. Multi-Agent Reinforcement Learning (MARL) addresses this coupling via mechanisms such as centralized training with decentralized execution, self-play, and population-based training. In a hybrid threat simulation, for instance, attacker agents may learn to exploit vulnerabilities in a network while defender agents learn patching and monitoring strategies, resulting in an evolutionary "arms race" in which both sides continuously adapt. For example, one can couple spatio-temporal SIR model with a dual-use healthcare administration problem and train a deep RL policy (e.g., Deep Q-Learning [370]) to allocate limited care capacity between war injuries and infectious disease cases [371]. Another, more botanical example, can be of a spread of coffee leaf rust as a botanical pandemic with intertwined epidemiological and economic dynamics, and employ an agent-based simulator with a deep RL agent to obtain near-optimal management policies under resource constraints [372,373]. Beyond transmission models, RL can also support downstream operational decisions such as hospital staff and resource allocation in stochastic settings [374]. Together, these examples emphasize a broader methodological point: RL enables policy search in complex, stochastic epidemic environments where analytic optimization is often intractable, but it also raises new challenges-training cost, reward misspecification, and the risk of learned policies exploiting simulator artifacts. Indeed, RL-based agents can be computationally demanding because they may require many simulation rollouts for training and extensive reward shaping to align with public-health objectives. In large-scale ABS, a practical compromise is to train RL policies offline and then deploy the learned policy as a fast, fixed controller during scenario runs, or to restrict learning to a small subset of *decision-maker* agents while the majority remain reactive.

The most recent and arguably most radical shift in agent architecture is the integration of *Large Language Models* (LLMs) [375,376]. Instead of coding rigid logical rules (e.g., "if $x > 5$ then y"), the modeler imbues an agent with a natural-language persona (e.g., "You are a conservative farmer in 18th century France...") and provides the current simulation state as a textual prompt. The agent's next action is then generated as conditional text, often mediated by a lightweight scaffold that (i) formats observations, (ii) retrieves relevant memories, (iii) requests a plan or decision from the LLM, and (iv) translates the output into structured actions executable by the simulator. In practice, modern *LLM agents* frequently combine prompting with additional components-memory (logs + retrieval), reflection/summarization, and tool use to maintain coherence across long horizons and to support multi-step reasoning [377]. The advantage of LLMs is that they enable open-ended, high-bandwidth social interaction: agents can converse, negotiate, justify decisions, and flexibly interpret novel situations using broad background knowledge that is difficult to hard-code. This makes them especially appealing for simulations of institutions, norms, and communication processes, where language is not merely an output channel but a causal mechanism of coordination and conflict. On the other hand, the cost of LLM inference is substantial: per-decision generation can be orders of magnitude slower than evaluating reactive rules, and scaling to many agents requires aggressive engineering (batching, caching, distillation, or hierarchical control) [378,379]. Moreover, LLM agents are partially opaque: they may hallucinate facts, drift in persona, or produce inconsistent decisions under small prompt perturbations, threatening reproducibility and transparency. These risks motivate *guardrails* such as constrained action schemas, deterministic decoding settings (e.g., low temperature), prompt/version control, output validators, and repeated-run reporting (multiple seeds / Monte Carlo) to quantify variability rather than hiding it [380]. For example, Park et al. demonstrate *generative agents* in a sandbox town, where agents generate daily plans, remember past experiences, and initiate conversations that produce emergent social behavior [377]. At larger scales, GenSim proposes a general social simulation platform with error-correction mechanisms and reports simulations with up to 10^5 agents [378], while AgentSociety presents a large-scale social simulator intended as a testbed for computational social experiments (e.g., polarization and policy interventions) [379]. Complementing these platform efforts, Ashery et al. study how populations of LLM agents can self-organize to form shared conventions and exhibit collective biases through repeated decentralized interactions [381].

6.4 Epidemiological use-cases

This section presents two representative case studies that illustrate how ABS capture epidemiological dynamics that are inaccessible to compartmental approaches. The first focuses on influenza-like illness (ILI) and emphasizes the role of relatively static social structure in shaping transmission, while the second examines COVID-19, where behavioral adaptation, logistical constraints, and heterogeneity dominate the dynamics.

6.4.1 Influenza-like illness and structural dynamics

The first case study examines Influenza-Like Illness using the Framework for Reconstructing Epidemiological Dynamics (FRED), a large-scale ABS designed to couple disease transmission with realistic social and physical infrastructure [382]. Rather than representing populations as homogeneous compartments, FRED constructs a detailed synthetic population in which every individual in a geographic region is represented as a distinct agent derived from U.S. Census data. Agents possess socio-demographic attributes such as age, household membership, employment or school enrollment, and income, and are embedded in explicit spatial locations corresponding to real households, schools, and workplaces. Daily activity schedules are assigned based on demographic role, with children attending specific schools and adults commuting to specific workplaces using gravity models and administrative boundaries. Critically, FRED explicitly models high-density *group quarters* such as dormitories, prisons, military bases, and nursing homes, which act as accelerants of transmission and are often missed in coarser-grained models.

Within this rich social scaffold, influenza transmission is typically modeled using a spatio-temporal discrete-time SEI^sI^aR, with parameters calibrated to historical pandemics and location graph-based spatial component. Disease progression is heterogeneous: the latent and infectious periods are drawn from distributions rather than fixed values, and infectious agents are stratified into symptomatic and asymptomatic classes. This distinction is crucial because only symptomatic agents are subject to detection, withdrawal from work or school, and isolation. Behavioral withdrawal is explicitly modeled as a redistribution of contacts across layers, reducing contacts in workplaces or schools while increasing contacts within households. Time advances in daily steps, with distinct weekday and weekend schedules, and the model supports inter-city travel that allows infection to propagate across regions, producing spatially realistic diffusion patterns.

Baseline epidemic simulations in FRED reveal dynamics that diverge sharply from ODE predictions. For example, the basic reproduction number R_0 is not imposed but emerges from realized transmission events and is calibrated ex post by tuning per-contact transmissibility. For pandemic influenza scenarios with R_0 in the range of 1.2–1.8, FRED predicts clinical attack rates of roughly 25–35%, with total (including asymptomatic) infection rates reaching 30–50%. These outcomes are highly sensitive to small changes in transmissibility or infectious duration, reflecting nonlinear amplification through the contact network. A robust and recurring finding is the disproportionate role of school-aged children. Schools form the most densely connected layer in the synthetic society, and epidemics are typically led by the school cohort, with prevalence peaking among children weeks before adults. Children act as vectors, introducing infection into households, which then propagates to adults and the elderly. Consequently, households with school-aged children form high-risk network nodes, while childless households remain relatively insulated until community transmission becomes widespread.

FRED also captures the spatial propagation of influenza as a wave rather than as instantaneous mixing. Initial outbreaks emerge as localized clusters around seed locations, such as specific schools or workplaces. Transmission then diffuses outward via commuter flows, following transportation corridors and producing geographically coherent wavefronts. This spatial realism enables the evaluation of geographically targeted interventions, such as county-level closures or travel restrictions, and highlights mechanisms that are invisible to well-mixed models.

Beyond descriptive analysis, FRED is explicitly designed to be used as an experimental platform, enabling researchers and practitioners to reproduce, modify, and extend influenza scenarios. Technically, FRED is implemented as a discrete-time, agent-based simulator written primarily in C++ for performance, with Python and shell-based tooling commonly used for experiment orchestration, parameter sweeps, and post-processing. Each simulation run is driven by a configuration file specifying the geographic region, disease natural history parameters, contact layer weights, and intervention triggers. This design allows users to systematically explore counterfactual scenarios by altering only a small subset of parameters while keeping the underlying synthetic population fixed.

At a practical level, users interact with FRED by defining (i) a population dataset, (ii) a disease model, and (iii) a set of policies. For example, transmissibility is specified as a per-contact probability that is internally modulated by contact duration and layer type (household, school, workplace, neighborhood). Intervention logic is implemented as rule-based switches that activate or deactivate layers or modify agent behavior when epidemiological thresholds are crossed. This makes FRED particularly suitable for policy experiments in which the timing and intensity of interventions are endogenous rather than pre-scripted.

To illustrate how one might prototype a simplified FRED-like experiment outside the full framework, consider the following minimal Python pseudo-code sketch of a daily, layer-based influenza simulation using a synthetic population. While not a replacement for FRED, this example mirrors its core logic:

```
import numpy as np

# Agent states:
# 0 = Susceptible, 1 = Exposed, 2 = Infectious (Symptomatic),
# 3 = Infectious (Asymptomatic), 4 = Recovered
N = 10000
state = np.zeros(N, dtype=int)

# Epidemiological parameters
p_transmit = 0.03
symptomatic_rate = 0.67
latent_period = (1, 3)        # days
infectious_period = (3, 6)  # days

# Contact layers (expected contacts per day)
layers = {
    "household": 3,
    "school_work": 8,
    "community": 2
}
```

```python

# Track time spent in current disease state
days_in_state = np.zeros(N, dtype=int)

def step(state, days_in_state):
    new_state = state.copy()
    new_days = days_in_state + 1

    infectious = np.where((state == 2) | (state == 3))[0]

    # Transmission
    for i in infectious:
        for layer, mean_contacts in layers.items():
            n_contacts = np.random.poisson(mean_contacts)
            for _ in range(n_contacts):
                j = np.random.randint(0, N)
                if state[j] == 0 and np.random.rand() < p_transmit:
                    new_state[j] = 1
                    new_days[j] = 0

    # Disease progression
    for i in range(N):
        if state[i] == 1 and new_days[i] >= np.random.randint(*latent_period):
            new_state[i] = 2 if np.random.rand() < symptomatic_rate else 3
            new_days[i] = 0
        elif state[i] in (2, 3) and new_days[i] >= np.random.randint(*infectious_period):
            new_state[i] = 4
            new_days[i] = 0

    return new_state, new_days

# Seed initial infections
initial_cases = np.random.choice(N, 10, replace=False)
state[initial_cases] = 2
days_in_state[initial_cases] = 0

# Run simulation
T = 160
history = []

for day in range(T):
    state, days_in_state = step(state, days_in_state)
    counts = np.bincount(state, minlength=5)
    history.append(counts)

history = np.array(history)
```

Listing 6.1: Simplified Python implementation of a layer-based agent simulation inspired by FRED.

At initialization, the model constructs a population of N agents. The epidemiological state of each agent is encoded as an integer, corresponding to a five-state disease model: Susceptible (0), Exposed (1), Infectious–Symptomatic (2), Infectious–Asymptomatic (3), and Recovered (4). This representation allows efficient vectorized operations and mirrors the extended SEI^sI^aR formulation. A parallel array tracks the number of days each agent has spent in its current disease state, enabling heterogeneous, duration-based transitions rather than fixed-time progression – this will be used later for analysis of the simulation. Next, for each infectious agent, the model samples a Poisson-distributed number of contacts per layer and randomly selects interaction partners. If a contacted agent is susceptible, transmission occurs with probability p_{transmit} (see line 37). Although contacts are sampled uniformly at random here, the same logic extends naturally to structured contact graphs based on households, schools, and workplaces. Moreover, disease progression is handled stochastically where exposed agents transition to an infectious state once their time in the latent state exceeds a randomly drawn latent period (see line 43), capturing biological heterogeneity. Upon becoming infectious, agents are assigned to symptomatic or asymptomatic classes according to the symptomatic rate parameter (see line 44). Infectious agents then recover after a randomly drawn infectious period, at which point they transition to the recovered state and no longer participate in transmission (see lines 46–48). This explicit accounting of state durations approximates the generation-interval dynamics that drive epidemic growth. The simulation is seeded by introducing a small number of initially infectious agents, after which the system evolves in discrete daily time steps. At each step, the model updates transmission events and disease progression and records the population-level state counts.

Despite its simplicity, this code captures several essential features of ABS based epidemiological modeling: layered contacts, stochastic transmission, heterogeneous disease courses, and emergent macroscopic behavior. One can incrementally enrich the model by introducing age stratification, symptomatic withdrawal from school or work layers, spatially explicit households and workplaces, or intervention logic such as school closures and targeted prophylaxis. With this objective in mind, for researchers wishing to use FRED directly, typical workflows involve running ensembles of simulations (e.g., 50–500 stochastic replicates per scenario) and summarizing outcomes such as peak incidence, attack rates, and timing of epidemic waves. Because R_0 is emergent rather than imposed, calibration is performed iteratively: transmissibility is adjusted until the realized secondary infection counts match target values. Sensitivity analysis is then conducted by perturbing parameters such as latent-period distributions or symptomatic rates, revealing nonlinear effects that are robust across different synthetic populations.

6.4.2 COVID-19 – behavioral feedback and logistic constraints

The second case study addresses COVID-19, a pandemic whose dynamics exposed the limits of static structural models. SARS-CoV-2 transmission is characterized by superspreading, substantial pre-symptomatic and asymptomatic transmission, and strong feedback between disease prevalence and human behavior [383]. To capture these features, we turn to Covasim, a COVID-19-specific agent-based simulator that extends traditional ABMs with detailed intra-host viral dynamics and endogenous behavioral feedback. In Covasim, infectiousness is a continuous function of viral load rather than a binary state, with agents generating individualized viral trajectories that govern their per-contact transmissibility over time. Disease progression is represented with fine granularity, including presymptomatic, symptomatic, asymptomatic, severe, and critical states, with age-dependent probabilities calibrated to clinical data.

A defining innovation of Covasim is the explicit modeling of behavioral adaptation. Agents dynamically reduce their contact rates in response to perceived risk, which may be tied to reported cases or deaths, effectively lowering transmission as fear increases following the BDI framework. At the same time, adherence fatigue erodes compliance over longer horizons, allowing contact rates to rebound even in the presence of ongoing mandates. Some model variants further allow agents to adapt based on infections observed in their local network, producing a complex adaptive system in which psychological feedback dampens or amplifies epidemic waves. These behavioral loops transform transmission dynamics from a static process into a co-evolving system of biology and behavior.

Covasim also treats testing, tracing, and isolation not as parameters but as explicit logistical processes. Symptomatic agents seek tests probabilistically, constrained by finite testing capacity. Results are returned after a delay, and only then can contact tracing be initiated over a specified retrospective window. Identified contacts enter quarantine, which reduces – but does not necessarily eliminate – their contacts, allowing for imperfect compliance. When calibrated to early COVID-19 data from Seattle/King County, including high-resolution mobility data, the model reproduces observed case and death trajectories and reveals the characteristic "Hammer and Dance" pattern: explosive early growth with $R_0 \approx 2.5$–3.0, a sharp suppression following lockdowns, and persistent residual transmission within essential workplaces and long-term care facilities.

Technically, a defining feature of COVID-19 ABS is that infectiousness is not binary but varies continuously over time as a function of viral load. Listing 6.2 shows a simplified Python implementation of agent-specific viral trajectories and their translation into per-contact transmission probabilities.

```
import numpy as np

def viral_load(t, t_peak=4.0, decay=0.8):
    """
    Log-normal-like viral load curve.
    """
    if t < t_peak:
        return np.exp(t - t_peak)
    else:
        return np.exp(-(t - t_peak) * decay)

def transmissibility(t, beta_max=0.15):
    """
    Transmission probability proportional to viral load.
    """
    return beta_max * viral_load(t)

# Example: Infectiousness over 14 days
days = np.arange(0, 14)
beta_t = [transmissibility(d) for d in days]
```

Listing 6.2: Python sketch of agent-specific viral load dynamics and time-varying transmissibility.

Table 6.2 Key parameters in Test–Trace–Quarantine (TTQ) modeling.

Parameter	Typical Range	Interpretation
P_{test}	0.1–0.6	Probability a symptomatic agent seeks testing
Testing capacity	10^2–10^4/day	Maximum tests processed per day
Result delay D	1–5 days	Time from test to result
Tracing yield	50–90%	Fraction of contacts successfully traced
Quarantine compliance	60–100%	Contact reduction during quarantine

This formulation captures pre-symptomatic transmission by allowing high transmissibility before symptom onset. In full-scale models, parameters such as peak timing and decay rate are age-dependent and calibrated to empirical viral shedding data.

In addition, COVID-19 transmission is strongly shaped by behavioral adaptation. Thus, Listing 6.3 illustrates how perceived risk can dynamically reduce contact rates, while adherence fatigue erodes compliance over time.

```
def effective_contacts(base_contacts, cases_today, fatigue, fear_scale=0.02):
    """
    Adjust contact rates based on perceived risk and fatigue.
    """
    fear_response = np.exp(-fear_scale * cases_today)
    fatigue_factor = max(0.5, 1.0 - fatigue)
    return base_contacts * fear_response * fatigue_factor

# Example usage
base_contacts = 10
daily_cases = 500
fatigue = 0.3   # accumulated over time

contacts_today = effective_contacts(base_contacts, daily_cases, fatigue)
```

Listing 6.3: Behavioral feedback loop with fear-based distancing and adherence fatigue.

Such feedback loops generate endogenous waves: declining cases lead to relaxed behavior, which in turn fuels resurgence – an effect that cannot be captured by static transmission parameters.

Furthermore, unlike simple parameterized reductions in transmission, COVID-19 ABMs explicitly simulate the testing and tracing pipeline. Table 6.2 summarizes key logistical parameters typically used in Covasim-style models.

Small changes in these parameters can move the system across a control threshold, producing qualitatively different epidemic trajectories. For instance, one of the most striking findings from COVID-19 ABMs is the extreme sensitivity to testing delays. Fig. 6.2 provides a schematic visualization commonly used to explain this effect. This visualization underscores why ABS predict a sharp transition between suppression and failure as delays cross the generation-interval timescale.

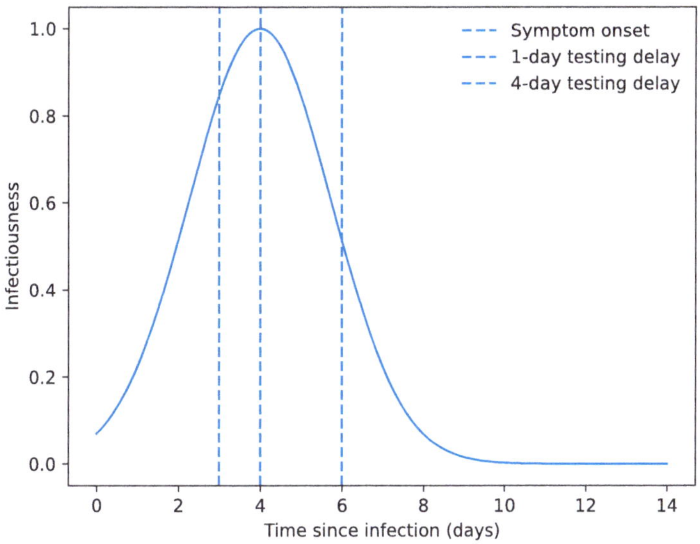

FIGURE 6.2

Effect of testing delays on COVID-19 control. The infectiousness profile peaks near symptom onset, implying that isolation after several days of delay occurs too late to interrupt most transmission events.

6.5 Designing your agent-based simulation

This section is a practical *mini-guide* for designing an epidemiological ABS that is fit-for-purpose, interpretable, and reproducible. The workflow below is presented linearly, but in practice, it is iterative: early choices about contact structure or disease natural history often force revisions in the agent definition, initialization, and outputs. A useful mental model is that every design decision should be traceable back to the research question, and every parameter should have an intended observable consequence in the outputs. Fig. 6.3 provides a schematic view of the components in ABS design and the connections between them.

The first deliverable is not a prototype but a *question-to-output map*. You should articulate (i) the estimand (what quantity you want), (ii) the target population and time horizon, (iii) the counterfactual(s) you want to compare, and (iv) the acceptance criteria for credibility (e.g., which historical patterns the model must reproduce). Good research questions are operational: "How much did school closures reduce cumulative infections among adults over 90 days in City X under observed vaccination coverage?" is better than "How do schools affect transmission?". At this stage, you should also decide whether an ABS is warranted. If the question can be answered with a homogeneous mixing assumption and aggregate outcomes (final size, peak time, R_0), a compartmental model is often more transparent and computationally efficient. ABS becomes valuable when you need one (or more) of the following: heterogeneous contacts (households, workplaces, schools), correlated attributes (age $\times$ occupation $\times$ household size), nonlinear behavioral feedback, policy rules that act on individuals, or distributional outcomes (e.g., inequities across neighborhoods). If you cannot name which of these mechanisms is

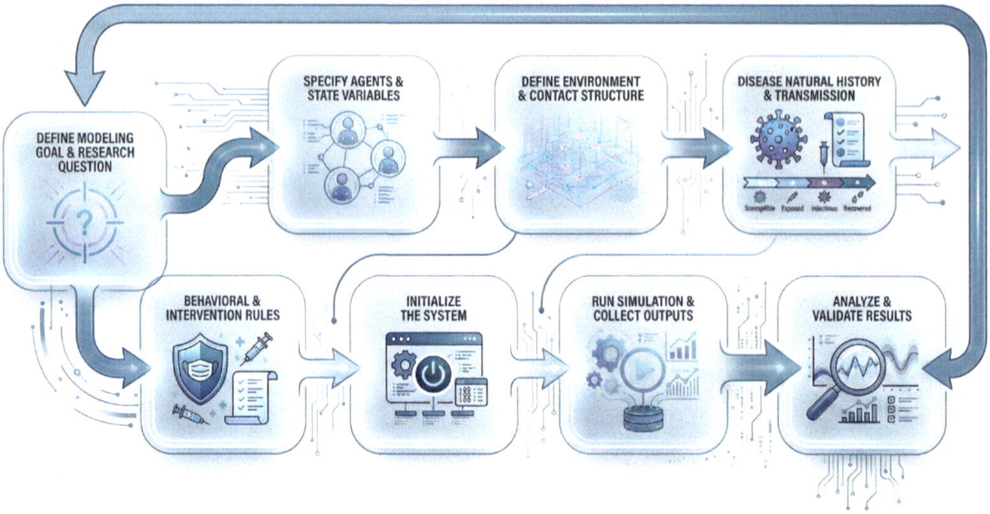

FIGURE 6.3

A schematic view of the components in ABS design and the connections between them.

essential to your estimand, you are at risk of building an elaborate simulator that answers a vague question.

A practical tool is a one-page *model charter*:

- **Primary outcomes:** daily incidence, attack rate by age, hospital occupancy, infections averted.
- **Mechanisms required:** household clustering, age-specific susceptibility, testing-and-isolation.
- **Decision context:** exploratory (mechanism understanding), predictive (forecasting), or evaluative (policy comparison).
- **Scope boundaries:** explicitly list what you will *not* model (e.g., within-hospital transmission).

This document prevents Bonini-style drift (attributed to Charles Bonini, captures the tension between realism and usability in modeling complex systems, as a model incorporates more detail to increase accuracy, it often becomes harder to interpret and less practically useful), where additional realism is continuously added without improving decision relevance.

For the second step, specifying agents and their state variables, we consider agents as the minimal entities needed to realize the mechanisms in the first step. In infectious disease ABS, agents are commonly people, but they can also be households, facilities, or *patches* (spatial units) if that better matches the question. A useful decision rule is: choose the coarsest agent definition that still expresses the needed heterogeneity. For example, if the key mechanism is household transmission, modeling households as agents with internal composition may be sufficient; if workplace closure policies depend on individual occupation, you need individuals. To this end, agent state variables should be divided into: (i) **static attributes** (age, household id, comorbidity category), (ii) **dynamic epidemiological**

Table 6.3 Common modeling choices for agent design and when to use them.

Decision	Option A (prefer when)	Option B (prefer when)
Agent unit	**Individuals** (interventions target persons; mixing depends on age/occupation; distributional outcomes)	**Groups/households** (focus on clustering; limited individual-level policy; speed is critical)
State detail	**Few states** (SEIR; interpretability; sparse data)	**Many states** (testing/isolation, multiple strains, care pathways; data exist to support)
Time handling	**Discrete time steps** (daily policies, simpler implementation, many ABS frameworks)	**Event-driven** (need exact timing; heterogeneous dwell times; high temporal resolution)
Behavior	**Exogenous** (fixed compliance trajectories; limited behavioral evidence)	**Endogenous feedback** (risk perception, fatigue, adaptive contacts; question requires it)

state (SEIR stage, days-since-infection), and (iii) **behavioral/decision state** (risk perception, adherence fatigue, vaccination intent). Each variable must justify itself by changing at least one output in a way that is relevant to the estimand. If you cannot describe the causal pathway from attribute → contact/transmission/progression → output, omit it. Table 6.3 summarizes several agent states modeling choices and alternatives.

As a technical comment, a minimal implementation pattern is to define an agent record and keep all state transitions explicit. Even in object-oriented ABS, consider storing states in arrays or dataframes for speed and easier auditing, as follows:

```
# Each agent i has:
age[i]                  # static
household_id[i]         # static
workplace_id[i]         # static or None

epi_state[i]            # 'S','E','I','R'
t_in_state[i]           # time since entering current epi_state

risk_perception[i]      # dynamic behavioral variable
fatigue[i]              # dynamic behavioral variable
```

Listing 6.4: Minimal agent state structure (Python-like pseudocode).

Next, for the third step, we need to define the environment and contact structure. The "environment" is where agents meet. In the epidemiological context, this is synonymous with the *contact process*. The main design decision is how contacts are generated, and that choice must match the research question and available data. A common hierarchy of contact representations is:

- **Well-mixed contacts:** Each agent draws contacts from the whole population. Use only as a baseline or if the question is not about heterogeneity (and you still want ABS for other reasons, e.g., individual interventions).
- **Structured mixing via contact matrices:** Agents are grouped (e.g., by age), and contact rates follow a matrix C_{ab}. Use when you have matrix estimates and need heterogeneity by group but not explicit networks.

- **Synthetic networks:** Edges represent repeated contacts. Use when clustering, degree heterogeneity, or superspreading driven by network structure is central.
- **Layered settings:** The most common pragmatic design: generate contacts in multiple layers, each with its own rules. This is ideal when interventions target specific settings (school closure, workplace distancing).

If your intervention acts at a setting level (close schools) you almost always need layered mixing. If your main output is age-specific incidence, but interventions are broad, a matrix-based approach can be sufficient and faster. In this context, it is good practice to pre-compute and visualize the implied mixing before adding disease dynamics with diagnostics such as:

- A heatmap of the realized age-by-age contact matrix (compare to target).
- Degree distribution of the contact network (mean, variance, tail behavior).
- Household size distribution and within-household contact density.

These figures often reveal implementation bugs that would otherwise be misattributed to *epidemic randomness*.

The forth step includes specifying the disease natural history and transmission, in which one defines how infection progresses and how infection occurs. Separate it into two parts: natural history and transmission. First, choose a state machine (e.g., SEIR, SEAIHRD) with dwell-time distributions. The key modeling decision is whether to use **fixed-duration** states (simple but unrealistic), **geometric/exponential** dwell times (memoryless, easy in discrete time), or **gamma/lognormal** dwell times (more realistic). If you care about timing-sensitive interventions (test-to-isolate delays), realistic dwell-time distributions matter. Second, transmission is typically modeled as a per-contact probability (Bernoulli) or as a hazard accumulating over contacts. In layered models, you may specify layer-specific intensities:

$$\Pr(\text{infection on contact}) = 1 - \exp(-\beta_\ell \cdot s_i \cdot \iota_j),$$

where β_ℓ is the layer intensity, s_i susceptibility modifier, and ι_j infectiousness modifier. Use additive hazards if multiple contacts in a day accumulate risk; use Bernoulli per contact if contacts are explicitly enumerated. An implementation of this approach takes the following format:

```
def infect_probability(beta_layer, susc_i=1.0, infec_j=1.0, duration=1.0):
    # hazard-based (stable for many contacts)
    hazard = beta_layer * susc_i * infec_j * duration
    return 1.0 - np.exp(-hazard)

# Example: household contact vs community contact
p_hh  = infect_probability(beta_layer=0.08, susc_i=1.2, infec_j=1.0)
p_com = infect_probability(beta_layer=0.01, susc_i=1.2, infec_j=1.0)
```

Listing 6.5: Layered transmission with susceptibility and infectiousness modifiers.

A common multi-option decision is whether to calibrate β to match R_0, the early growth rate, or secondary attack rates (SAR). Practically, prefer R_0 or growth calibration when early incidence data are reliable and mixing is approximately stable, and prefer SAR calibration (household/work) when you explicitly model settings and have setting-specific evidence.

Table 6.4 Intervention rule mapping (examples).

Intervention	Model lever	Implementation example
Masking	Transmission intensity	Multiply β_ℓ by $(1-\eta)$ in selected layers; optionally age/setting-specific adherence.
Isolation after test	Contact reduction	Set workplace/community contacts to zero for d days; keep household contacts.
School closure	Environment structure	Remove school layer edges or set school contact rates to near zero.
Vaccination	Susceptibility / severity	Reduce s_i and/or progression to severe states; allow waning over time.

The fifth step is about defining the behavioral and intervention rules. Interventions in ABS are best framed as *rule changes* in contact generation, transmission, or progression (e.g., vaccination reduces susceptibility; isolation reduces contacts; masking reduces β). The important design choice is whether behavior is: exogenous or endogenous. In exogenous behavior, compliance trajectories are imposed (e.g., 60% masking after day 30). Use when behavioral evidence is weak or behavior is not central to the question. On the other hand, in Endogenous (adaptive) behavior, agents respond to epidemic conditions (risk perception) and may fatigue. Use only if the research question requires feedback (e.g., why interventions lose effectiveness over time). Table 6.4 provides several examples of such interventions and their conceptual implementation in ABS.

After the model is well-defined, one moves to the six step of model initialization. Initialization defines the starting joint distribution of agent attributes and states. The choice here strongly affects early dynamics, so it should be treated as a modeling decision, not a technicality. Starting with the population synthesis. One may use: (i) **microdata-driven synthesis** (most realistic, data heavy), (ii) **marginal matching** (iterative proportional fitting) to match distributions of age/household size/occupation, or (iii) **simple generative rules** (fast, but risk misrepresenting correlations). Choose the simplest method that matches the correlations relevant to transmission (e.g., household size correlates with age composition). Next, the seeding infection also has several options, including: (i) seed a fixed number of infections uniformly, (ii) seed in specific subpopulations (e.g., travelers, certain neighborhoods), or (iii) back-calculate seeds to match observed early incidence. If you plan to compare interventions starting at a particular date, seeding should be calibrated so that the simulated epidemic state at that date is comparable across scenarios. Moreover, as started before, one should consider the randomness in the system – record random seeds and initialization draws. For credibility, report whether your results are conditioned on a fixed synthetic population or averaged over multiple population realizations (both are legitimate; the latter incorporates demographic uncertainty but costs more).

Finally, as the seventh step, run the simulation and collect outputs. Before running scenarios, define your output contract: which variables are logged, at what temporal resolution, and at what aggregation level (individual, household, neighborhood, city). Avoid logging everything-it slows simulations and encourages post-hoc fishing. Instead, log:

- Epidemiological time series (incidence, prevalence, hospitalizations).
- Stratified outcomes (by age, SES proxy, location) aligned with the estimand.
- Mechanism diagnostics (contacts by layer, isolation counts, effective reproduction estimates).

Since ABS is stochastic, a single run is not evidence. Decide whether you will report (i) medians and uncertainty intervals over many runs, (ii) probability of exceeding thresholds (e.g., hospital capacity), or (iii) distributions of outcomes. Choose based on decision context: threshold decisions benefit from exceedance probabilities; scientific explanation benefits from distributions and mechanism diagnostics.

A common multi-option decision is the number of replications. If outcomes are stable and the model is fast, use many runs (hundreds) and report uncertainty. However, if expensive, use fewer runs but prioritize variance reduction (common random numbers across scenarios) and targeted sensitivity analysis. See the following example:

```
seeds = [101, 202, 303, 404, 505]
for seed in seeds:
    pop = synthesize_population(seed=seed)  # fixed across scenarios
    for scenario in ["baseline", "masking", "school_closure"]:
        rng = np.random.default_rng(seed)   # common random numbers
        outputs = run_simulation(pop, scenario=scenario, rng=rng)
        save(outputs, seed=seed, scenario=scenario)
```

Listing 6.6: Scenario execution with common random numbers (sketch).

6.6 Sensitivity analysis and robustness checking

Designing an epidemiological ABS is only the first part of the modeling task. Once the model has been specified, initialized, and executed, the next methodological question is whether its conclusions are stable under the uncertainties that inevitably remain in the system description, the parameterization, the stochastic realization, and the structural assumptions built into the simulator. In this sense, sensitivity analysis is not an auxiliary exercise added after the main results have been produced, but a central component of model-based reasoning. Its role is to determine how variation in the conclusions of interest can be attributed to variation in the uncertain ingredients of the model, and thereby to distinguish robust mechanisms from fragile artifacts of specification. In the language of uncertainty quantification, if an epidemiological ABS is written abstractly as a mapping $Y = \mathcal{M}(X, U)$, where $X = (X_1, \ldots, X_d)$ denotes the vector of uncertain model inputs and U denotes the stochastic degrees of freedom of the simulator (random seeds, stochastic event realizations, randomized schedules, or synthetic-population draws), then any scalar quantity of interest $Q = h(Y)$ should be regarded not as a single number but as a random variable induced jointly by uncertainty in X and stochasticity in U. A mathematically useful decomposition is therefore $\mu(X) = \mathbb{E}[Q \mid X]$ and $\sigma^2(X) = \text{Var}(Q \mid X)$, so that $\mu(X)$ captures the systematic response of the model to the uncertain inputs and $\sigma^2(X)$ captures the intrinsic within-setting variability of the stochastic simulator. By the law of total variance, $\text{Var}(Q) = \text{Var}_X\big(\mu(X)\big) + \mathbb{E}_X\big[\sigma^2(X)\big]$, which shows that output variability has at least two conceptually distinct origins: i) uncertainty about the model inputs, and ii) irreducible randomness conditional on those inputs. This distinction is especially important in epidemiological ABS because a model may appear unstable simply because too few replications were used, or may appear stable only because the uncertainty in key assumptions was never explored in the first place [384,385].

Sensitivity analysis, in this setting, should be understood broadly. In the narrowest classical sense, it studies how changes in numerical inputs affect an output of interest. In agent-based epidemiological modeling, however, the relevant "inputs" often include not only numerical parameters such as transmissibility, latent duration, or compliance rates, but also discrete design choices such as contact topology, scheduler type, intervention trigger rules, and the representation of behavioral adaptation. This broader interpretation is essential because ABS are not merely parameterized equations; they are computational worlds whose results may depend on both parametric and non-parametric modeling choices. Accordingly, the goal of this section is twofold. First, we formalize why sensitivity analysis is indispensable for interpreting epidemiological ABS as scientific or policy-relevant objects. Second, in the subsequent subsections, we move from this general motivation to concrete methods, beginning with local sensitivity analysis, continuing to global sensitivity analysis, and then turning to practical reporting and workflow design. Throughout, the main principle is that a simulation conclusion should be trusted only to the extent that it survives systematic perturbation of the assumptions on which it depends [386,387].

6.6.1 Why sensitivity analysis is essential

Sensitivity analysis is essential because an epidemiological ABS is almost always many assumptions deep. Even a comparatively simple epidemic simulator requires the modeler to specify disease-natural-history parameters, contact-generation rules, initialization conditions, intervention adherence, behavioral response, and often a large number of implementation-level decisions whose effects are not visible in a single baseline run. A good fit to one historical trajectory does not, by itself, establish that the model has identified the correct mechanism, nor does it guarantee that the ranking of policy scenarios is stable under plausible perturbations of the uncertain inputs. Formally, let A and B denote two interventions, and let $\mu_A(X)$ and $\mu_B(X)$ be the corresponding conditional expected values of a quantity of interest, such as cumulative infections, peak hospital occupancy, or deaths. The policy conclusion that B outperforms A is then encoded in the sign of $\Delta(X) = \mu_A(X) - \mu_B(X)$. If $\Delta(X)$ is positive only on a narrow subset of the plausible input space and negative elsewhere, then the apparent superiority of B is fragile rather than robust. Conversely, if $\Delta(X)$ remains positive over a broad and epidemiologically defensible region of X, the policy ranking is much more credible. Sensitivity analysis is therefore the mechanism by which the modeler replaces an anecdotal result at a baseline point with a statement about behavior over a neighborhood – or, in global settings, over an entire probability distribution – of admissible assumptions. This is precisely why rigorous sensitivity analysis is repeatedly emphasized in the modeling literature, and also why one-factor-at-a-time perturbations are often inadequate: they can miss nonlinearities and interactions that dominate system behavior away from the baseline [386,388].

For epidemiological ABS models, the importance of sensitivity analysis is even greater because the model usually contains sources of variation that do not arise in simpler compartmental systems. First, stochastic simulation implies that the same parameter vector may produce a distribution of outcomes rather than a deterministic trajectory, so conclusions must be assessed against both between-setting and within-setting variability. Second, ABS often include structural choices that are not naturally represented on a single continuous axis, such as the scheduler, the contact network generator, the treatment of superspreading, or the rule by which agents update behavior in response to risk. Third, the policy questions posed to such models are often threshold-sensitive: whether ICU demand exceeds capacity, whether an intervention keeps the effective reproduction number below one, or whether tracing delays push the system beyond a control boundary. In such settings, a small perturbation in an uncertain input can induce a large qualitative shift in outcomes, making robustness analysis indispensable rather than optional. Empirical work in epidemic modeling illustrates this clearly. Uncertainty quantification and sensitivity analysis applied to COVID-19 individual-based transmission models have shown that intervention uptake, tracing efficiency, and transmission reduction assumptions can dominate policy conclusions, while broader global sensitivity studies of epidemic models show that intervention timing and quarantine-related parameters are often among the most influential drivers and can interact strongly with one another [385,388]. More generally, recent methodological work on ABMs has stressed that a proper analysis must examine the full set of "moving parts" of the simulator, including non-parametric elements, because otherwise one risks declaring a model robust only because the most consequential assumptions were never varied at all [387]. In this sense, sensitivity analysis is not merely a quality-control step appended to the end of an epidemiological simulation study; it is part of the argument that the model's proposed mechanism is scientifically interpretable and that its scenario comparisons are decision-relevant.

6.6.2 Local sensitivity analysis

Local sensitivity analysis is the natural first step once a baseline calibration or nominal parameter setting has been identified. Its purpose is not to characterize the full uncertainty structure of the model over the entire admissible input domain, but rather to determine how the outputs respond in the neighborhood of a reference point x^0. In an epidemiological ABS, this reference point is usually the calibrated or policy-relevant setting, and local analysis asks whether the main conclusions persist under small perturbations of transmissibility, latent duration, compliance, testing delay, or other uncertain quantities. Mathematically, if Q denotes a scalar quantity of interest, such as cumulative infections by day T, peak hospital occupancy, or the probability of exceeding capacity, and if the simulator is stochastic, then the appropriate object for local analysis is not a single realization but the conditional mean response $\mu(x) = \mathbb{E}[Q \mid X = x]$. For a continuous input x_i, the classical local sensitivity coefficient is the partial derivative $D_i(x^0) = \left.\frac{\partial \mu(x)}{\partial x_i}\right|_{x=x^0}$, which measures the first-order rate of change of the expected output with respect to the i-th parameter at the baseline. Because raw derivatives depend on units, it is often more informative to report the dimensionless local elasticity $E_i(x^0) = \frac{x_i^0}{\mu(x^0)} D_i(x^0)$, which has the usual comparative-statics interpretation that, near x^0, a one-percent increase in x_i produces approximately an $E_i(x^0)$-percent change in the expected output. This differential viewpoint is the classical meaning of local sensitivity, and it is especially useful when the modeler wants a directional answer near a trusted baseline rather than a full global ranking over the entire input space [384].

For an ABS, however, $\mu(x)$ is rarely available in closed form, so the derivative must be estimated numerically. The standard approach is finite differencing [389]. Formally, let e_i denote the i-th coordinate vector and let $h_i > 0$ be a small perturbation. The forward-difference estimator is $\widehat{D}_i^{(f)}(x^0) = \frac{\widehat{\mu}(x^0+h_ie_i)-\widehat{\mu}(x^0)}{h_i}$, while the central-difference estimator is $\widehat{D}_i^{(c)}(x^0) = \frac{\widehat{\mu}(x^0+h_ie_i)-\widehat{\mu}(x^0-h_ie_i)}{2h_i}$. Here $\widehat{\mu}(x)$ is a Monte Carlo estimate of $\mu(x)$, typically obtained from R independent replications, defined as: $\widehat{\mu}(x) = \frac{1}{R}\sum_{r=1}^{R} Q(x, U_r)$. When the simulator is sufficiently smooth in the perturbed parameter, the central estimator is generally preferable because its truncation error is of higher order than that of the forward estimator. In practice, the step size h_i must be chosen with care. If h_i is too large, the estimator ceases to reflect local behavior and instead measures a finite change over a region where curvature or threshold effects may dominate. If h_i is too small, then Monte Carlo noise and floating-point cancellation can overwhelm the signal. The appropriate perturbation is therefore problem-dependent and should be interpreted relative to the local scale of the parameter and to the stochastic variability of the output. For epidemiological ABS, this is particularly important near threshold regimes, for example when a small change in transmission reduction moves the system across the control boundary $R_{\text{eff}} = 1$, because the response surface may be sharply curved even in a seemingly narrow neighborhood. In such cases, a derivative estimate should be complemented by a short local response curve rather than taken as a complete description of sensitivity [388,390].

For stochastic ABS, local sensitivity estimation is improved substantially by coupling the perturbed runs through common random numbers. Suppose that for each replication r, one evaluates the model at $x^0 + h_i e_i$ and $x^0 - h_i e_i$ using the same underlying random seed or, more precisely, the same aligned stream of random draws. If the resulting paired outputs are denoted Q_r^+ and Q_r^-, then the central-difference estimator may be written $\widehat{D}_i^{(c)}(x^0) = \frac{1}{2h_i}\left(\frac{1}{R}\sum_{r=1}^{R}(Q_r^+ - Q_r^-)\right)$. Its variance is governed by

$$\operatorname{Var}\left(\widehat{D}_i^{(c)}\right) = \frac{1}{4h_i^2 R}\operatorname{Var}(Q^+ - Q^-) = \frac{1}{4h_i^2 R}\left(\operatorname{Var}(Q^+) + \operatorname{Var}(Q^-) - 2\operatorname{Cov}(Q^+, Q^-)\right).$$

When the coupling induces positive covariance between the paired outputs, the variance of the difference is reduced, often substantially. Intuitively, both perturbed systems are then exposed to the same "epidemic luck": the same random importations, the same stochastic realizations of contacts, and the same recovery draws insofar as these can be aligned. What remains in the paired difference is more nearly the effect of the parameter perturbation itself rather than the noise of independent stochastic histories. This makes common random numbers especially valuable in local sensitivity analysis, where the signal being measured is often small by construction and can otherwise be obscured by simulation variance [391].

A further complication is that not all relevant ABS inputs are differentiable. Some are discrete by design, such as the scheduler type, the choice between a household-layer contact process and a well-mixed community process, or the activation of a test-and-trace module. Others induce non-smooth responses because a perturbation changes the order of events, the activation of a policy trigger, or the combinatorial structure of the contact network. In such cases, derivative-based coefficients are not the right object. A more appropriate local summary is a finite-change sensitivity measure. If δ_i is a finite perturbation of input x_i, one may define the normalized finite-change effect

$$F_i(x^0; \delta_i) = \frac{\mu(x^0 + \delta_i e_i) - \mu(x^0)}{\mu(x^0)} \Big/ \frac{\delta_i}{x_i^0},$$

provided $x_i^0 \neq 0$ and $\mu(x^0) \neq 0$. This quantity reduces to the local elasticity in the limit $\delta_i \to 0$ when differentiability holds, but remains meaningful when the response is only piecewise smooth or when the perturbation of interest is intrinsically discrete. In ABMs this is often the more faithful local quantity, because many practical policy questions are themselves finite-change questions: what happens if quarantine compliance increases from 60% to 75%, if test delay falls from three days to one, or if school contacts are reduced by half? In those cases, the analyst is not approximating an infinitesimal derivative; the finite change is the estimand of interest, and the local analysis should be framed accordingly [387,392].

The following minimal Python sketch illustrates a practical local-sensitivity workflow for a stochastic epidemiological ABS. It estimates the local elasticity of a scalar output using central finite differences and common random numbers. The same structure can be used with cumulative infections, peak prevalence, hospital occupancy, or any other scalar metric extracted from a simulation run. The appropriate interpretation of such a calculation is deliberately local. If the estimated elasticity of cumulative infections with respect to `beta_community` is, say, 0.8, this means that near the baseline calibration, a one-percent increase in that parameter is associated with an approximate 0.8-percent increase in the expected outcome. It does not mean that the same proportional effect will hold over large perturbations, under different intervention regimes, or after structural assumptions such as the contact network or scheduler are changed. That limitation is not a weakness of the method but its defining scope. Local sensitivity analysis is most valuable as a diagnostic instrument: it tests whether the baseline behaves as expected, identifies parameters that warrant tighter calibration, reveals hidden thresholds in the neighborhood of the fitted model, and provides the first indication of whether a policy comparison is stable or fragile. Once these local diagnostics are in place, one can then move to the broader question of how sensitivity behaves over the full uncertainty space, which is the purpose of global sensitivity analysis in the next subsection.

```
import numpy as np

def estimate_mean_output(run_model, params, seeds):
    vals = []
    for seed in seeds:
        out = run_model(params=params, seed=seed)
        vals.append(out["cum_infections_T"])   # scalar quantity of interest
    vals = np.asarray(vals, dtype=float)
    return vals.mean(), vals.std(ddof=1) / np.sqrt(len(vals)), vals

def local_elasticity(run_model, base_params, par_name, rel_step=0.05, seeds=None):

    x0 = float(base_params[par_name])
    h  = rel_step * x0

    p_minus = dict(base_params)
    p_plus  = dict(base_params)
    p_minus[par_name] = x0 - h
    p_plus[par_name]  = x0 + h

    # Common random numbers: same seeds in baseline, minus, and plus runs
    mu0, se0, y0 = estimate_mean_output(run_model, base_params, seeds)
    mu_minus, se_minus, y_minus = estimate_mean_output(run_model, p_minus, seeds)
    mu_plus,  se_plus,  y_plus  = estimate_mean_output(run_model, p_plus, seeds)

    # Central finite-difference derivative and elasticity
    dmu_dx = (mu_plus - mu_minus) / (2.0 * h)
    elasticity = (x0 / mu0) * dmu_dx

    # Paired-difference standard error under common random numbers
    paired_diff = (y_plus - y_minus) / (2.0 * h)
    dmu_se = paired_diff.std(ddof=1) / np.sqrt(len(seeds))
```

```
    elas_se = abs(x0 / mu0) * dmu_se

    return {
        "parameter": par_name,
        "baseline_value": x0,
        "baseline_mean": mu0,
        "derivative": dmu_dx,
        "derivative_se": dmu_se,
        "elasticity": elasticity,
        "elasticity_se": elas_se,
        "mu_minus": mu_minus,
        "mu_plus": mu_plus
    }
```

Listing 6.7: Central-difference local sensitivity with common random numbers.

6.6.3 Global sensitivity analysis

While local sensitivity analysis interrogates the behavior of the simulator in a neighborhood of a baseline point, global sensitivity analysis treats the uncertain input vector $X = (X_1, \ldots, X_d)$ as a random object defined over its full admissible domain and asks how variation in the model output is distributed across the inputs over that domain as a whole. This distinction is fundamental in epidemiological ABS because such models are often nonlinear, interaction-rich, and threshold-sensitive, so conclusions derived from a local derivative at one calibration point may fail to describe the dominant drivers elsewhere in the parameter space. If $Q = f(X)$ denotes a scalar quantity of interest, where f should now be interpreted as the conditional mean response of the stochastic simulator or as a replication-aggregated summary thereof, then global sensitivity analysis studies how the distribution of Q changes when the components of X vary jointly according to their prescribed uncertainty distributions. The mathematical foundation of the most widely used variance-based framework is the functional ANOVA decomposition, under which f can be written as

$$f(X) = f_0 + \sum_{i=1}^{d} f_i(X_i) + \sum_{1 \le i < j \le d} f_{ij}(X_i, X_j) + \cdots + f_{1\cdots d}(X_1, \ldots, X_d),$$

with orthogonality conditions chosen so that the terms have zero mean with respect to their own arguments. This induces the variance decomposition

$$\mathrm{Var}(Q) = \sum_{i=1}^{d} V_i + \sum_{1 \le i < j \le d} V_{ij} + \cdots + V_{1\cdots d},$$

where $V_i = \mathrm{Var}\big(\mathbb{E}[Q \mid X_i]\big)$ is the first-order contribution of input X_i, V_{ij} is the pure interaction contribution of the pair (X_i, X_j), and so forth. The corresponding Sobol' first-order sensitivity index is $S_i = V_i / \mathrm{Var}(Q)$, which measures the fraction of the output variance explained by varying X_i alone, averaged over all values of the remaining inputs. When interactions are present, first-order indices are insufficient, and one therefore introduces the total-effect index $S_{T_i} = 1 - \mathrm{Var}\big(\mathbb{E}[Q \mid X_{\sim i}]\big) / \mathrm{Var}(Q)$, where $X_{\sim i}$ denotes all inputs except X_i. This quantity captures the full influence of X_i, including all interactions involving X_i, and satisfies $S_{T_i} \geq S_i$. In epidemiological terms, S_i tells us how much of the uncertainty in, say, peak ICU demand is attributable to uncertainty in a single parameter when averaged over everything else, while $S_{T_i} - S_i$ quantifies how strongly that parameter acts through interaction with the rest of the model, which is often substantial in transmission models with layered contacts, intervention timing, and behavioral feedback [393,394].

Variance-based indices are attractive because they admit a direct interpretation in terms of output variability, but they can be computationally expensive for large ABS. In a model with many uncertain inputs, or in settings where a single run is already costly, it is often useful to begin with a screening method whose purpose is not to estimate a full variance decomposition but to distinguish negligible factors from influential ones. The classical method of Morris does precisely this by replacing the local derivative with a randomized finite increment over the input hypercube. After rescaling inputs to [0, 1], the elementary effect of factor X_i at a sampled point x is defined as

$$EE_i(x) = \frac{f(x_1, \ldots, x_i + \Delta, \ldots, x_d) - f(x)}{\Delta},$$

for a prescribed grid jump Δ. Repeating this construction over multiple random trajectories yields a sample of elementary effects for each factor. The original Morris summaries are the mean μ_i, which reflects the overall directional effect, and the standard deviation σ_i, which reflects nonlinearity and interaction. In practice, because positive and negative effects can cancel, the more stable quantity is the mean absolute elementary effect, $\mu_i^* = \frac{1}{r}\sum_{k=1}^{r}\left|EE_i^{(k)}\right|$, where r is the number of trajectories. Factors with small μ_i^* are typically negligible over the chosen uncertainty domain, while large μ_i^* combined with large σ_i suggests either strong interaction, strong curvature, or both. This makes Morris screening especially well suited to epidemiological ABS, where the analyst may begin with dozens of uncertain quantities – layer-specific transmission rates, dwell-time parameters, vaccine protection modifiers, adherence parameters, seeding assumptions, and intervention triggers – but only a small subset may dominate outcomes. A practical workflow is therefore to use Morris as a first-stage dimensionality reduction procedure, retaining the influential factors for a second-stage variance-based analysis on a reduced space [395,396].

Once the influential inputs have been identified, one may estimate Sobol' first-order and total-effect indices using Monte Carlo or quasi-Monte Carlo designs. A widely used construction is based on two independent $N \times d$ sample matrices, conventionally denoted A and B, from the joint input distribution. For each input X_i, one constructs a mixed matrix $A_B^{(i)}$ by taking all columns from A except the i-th, which is taken from B. These matrices allow consistent estimation of the first-order and total-effect indices from a common set of model evaluations, and efficient sampling schemes have been proposed to reduce the computational cost of obtaining both sets of indices simultaneously. If $f(A^{(n)})$, $f(B^{(n)})$, and $f(A_B^{(i,n)})$ denote the outputs at the n-th sampled row of each design, then one common estimator family takes the schematic form

$$\widehat{S}_i \propto \frac{1}{N}\sum_{n=1}^{N} f(B^{(n)})\Big(f(A_B^{(i,n)}) - f(A^{(n)})\Big), \qquad \widehat{S}_{T_i} \propto \frac{1}{N}\sum_{n=1}^{N}\big(f(A^{(n)}) - f(A_B^{(i,n)})\big)^2,$$

with normalization by the sample variance of the pooled outputs. The exact estimator variant is less important here than the conceptual point: global sensitivity is inferred by repeatedly recombining input coordinates in a way that isolates the marginal role of one factor against the distribution of all the others. For stochastic ABS, the function f should be replaced by a replication-averaged response $\widehat{\mu}(x)$, otherwise, the estimated variance decomposition will conflate input uncertainty with simulator noise. This is not a technical nicety but a substantive modeling issue. If, for example, the analyst uses only a single stochastic realization at each Sobol design point, then a large portion of the measured output variance may reflect epidemic randomness rather than parametric uncertainty, and the resulting sensitivity ranking may be unstable. In practice, one either fixes the internal seed structure to study parametric sensitivity conditional on stochasticity, or averages multiple replications per design point to approximate the conditional mean response and then applies the variance decomposition to that smoother object [397,398].

Another important design decision concerns how the input space is sampled before the sensitivity indices are computed. Although Sobol' estimators are often implemented with Monte Carlo or quasi-random low-discrepancy sequences, other global workflows use Latin hypercube sampling to obtain broad stratified coverage of high-dimensional input spaces with fewer points than naive simple random sampling would require. This can be particularly useful when the goal is uncertainty propagation together with sensitivity ranking, or when the analyst wishes to fit an emulator on top of the ABS outputs. The central idea of Latin hypercube sampling is that each marginal input distribution is partitioned into N equiprobable intervals and each interval is sampled exactly once, with the coordinates then randomly paired across dimensions. For many estimands this yields better space-filling behavior than ordinary Monte Carlo and is therefore appealing in expensive epidemiological simulations where each design point may itself require many stochastic replications. In practical work, the sampling design, screening method, and index estimator should be viewed as parts of one coherent pipeline rather than as isolated technical choices. The method of Morris is appropriate when the number of candidate inputs is large and the first goal is to eliminate negligibly influential factors; Sobol' total-effect indices are appropriate when interaction matters and the analyst needs an interpretable variance decomposition; Latin hypercube designs are appropriate when one seeks efficient coverage for uncertainty propagation or surrogate modeling; and software libraries such as `SALib` make these designs operational in ordinary Python workflows [399,400].

The following Python code illustrates a compact two-stage workflow in which a Morris screen is first used to reduce the dimensionality of the uncertain input space, after which a Sobol analysis is carried out on the retained factors. The code assumes that each evaluation of the ABS returns a replication-averaged scalar summary, such as peak infectious prevalence or cumulative hospital admissions.

```
import numpy as np
from SALib.sample import morris as morris_sample
from SALib.analyze import morris as morris_analyze
from SALib.sample import sobol as sobol_sample
from SALib.analyze import sobol as sobol_analyze

def evaluate_mean_output(run_model, x, seeds):
    vals = []
    for seed in seeds:
        out = run_model(params=x, seed=seed)
```

```
        vals.append(out["peak_prevalence"])
    return float(np.mean(vals))

# Stage 1: Morris screening
problem_full = {
    "num_vars": 6,
    "names": [
        "beta_household",
        "beta_workplace",
        "latent_days",
        "infectious_days",
        "test_delay",
        "quarantine_compliance"
    ],
    "bounds": [
        [0.04, 0.12],
        [0.005, 0.03],
        [1.5, 4.0],
        [3.0, 8.0],
        [1.0, 5.0],
        [0.4, 1.0]
    ]
}

X_morris = morris_sample.sample(problem_full, N=40, num_levels=4)
Y_morris = np.array([
    evaluate_mean_output(my_epidemic_simulator,
                         dict(zip(problem_full["names"], row)),
                         seeds=[11, 22, 33, 44])
    for row in X_morris
])

Si_morris = morris_analyze.analyze(
    problem_full, X_morris, Y_morris, print_to_console=False
)

# Keep influential factors by thresholding mu_star
keep = [name for name, mu_star in zip(problem_full["names"], Si_morris["mu_star"])
        if mu_star > 0.05]

# Stage 2: Sobol analysis on reduced parameter set
reduced_bounds = {
    "beta_household": [0.04, 0.12],
    "beta_workplace": [0.005, 0.03],
    "latent_days": [1.5, 4.0],
    "infectious_days": [3.0, 8.0],
    "test_delay": [1.0, 5.0],
    "quarantine_compliance": [0.4, 1.0]
```

```
}

problem_reduced = {
    "num_vars": len(keep),
    "names": keep,
    "bounds": [reduced_bounds[k] for k in keep]
}

X_sobol = sobol_sample.sample(problem_reduced, N=512, calc_second_order=False)
Y_sobol = np.array([
    evaluate_mean_output(my_epidemic_simulator,
                         dict(zip(problem_reduced["names"], row)),
                         seeds=[11, 22, 33, 44, 55, 66])
    for row in X_sobol
])

Si_sobol = sobol_analyze.analyze(
    problem_reduced, Y_sobol, calc_second_order=False, print_to_console=False
)

for name, s1, st in zip(problem_reduced["names"], Si_sobol["S1"], Si_sobol["ST"]):
    print(name, "S1=", round(s1, 3), "ST=", round(st, 3))
```

Listing 6.8: Two-stage global sensitivity workflow with Morris screening and Sobol indices.

The interpretation of the resulting indices should remain tied to the scientific question. If, for example, the total-effect index of `test_delay` is large for the outcome `peak_prevalence`, this means that uncertainty in testing delay is an important driver of uncertainty in epidemic peak size over the chosen input domain, not that testing delay is necessarily the largest causal lever under every policy design. Likewise, if the first-order index of `beta_workplace` is small but the total-effect index is large, the correct conclusion is not that workplace transmission is unimportant, but that its effect is expressed mainly through interaction with other quantities such as isolation compliance or community transmission. Global sensitivity analysis is thus best understood as a map from uncertainty assumptions to output variability. Its value lies not merely in producing a ranked list of inputs, but in revealing whether the epidemiological conclusions drawn from the ABS are driven by isolated mechanisms, by high-order interaction, or by a small number of uncertain assumptions whose calibration and reporting therefore deserve special scrutiny.

6.6.4 A practical workflow

In practice, sensitivity analysis for an epidemiological ABS should be organized as a workflow rather than as an isolated computational routine. The reason is that the appropriate method depends not only on the simulator itself, but also on the purpose of the analysis, the nature of the uncertain inputs, the stochastic structure of the model, and the decision context in which the outputs will be interpreted. A scientifically useful workflow therefore begins before any sensitivity index is computed. One must first identify the quantities of interest, specify which model ingredients are to be treated as uncertain, and decide whether the analysis is intended to support explanation, calibration, prioritization of data collection, or comparison of interventions under uncertainty. In the general notation of this chapter, let

$$Q_j = h_j(Y), \qquad j = 1, \dots, m,$$

denote the outputs of interest extracted from the simulator, where $Y = \mathcal{M}(X, U)$, X is the vector of uncertain inputs, and U denotes the stochastic variables internal to the simulation. The first practical principle is that sensitivity should be conducted on outputs that are decision-relevant and epidemiologically interpretable. For example, if the scientific question concerns hospital overflow, then the principal output should not be only cumulative infections but also a threshold-sensitive statistic such as

$$Q_{\text{cap}} = \mathbf{1}\left\{\max_{t \le T} H_t > C\right\},$$

where H_t is hospital occupancy and C is system capacity. Likewise, if the goal is policy comparison, the object of interest is often a contrast rather than a raw outcome, for example,

$$\Delta(X) = \mu_A(X) - \mu_B(X),$$

where $\mu_A(X)$ and $\mu_B(X)$ are the conditional expected outcomes under scenarios A and B. This framing matters because sensitivity analysis should attach to the quantity that carries the substantive claim of the study, not merely to whatever output happens to be easiest to record. Methodological reviews emphasize that a sensible workflow begins precisely with this alignment between model purpose, uncertainty framing, and output definition, because otherwise one risks applying sophisticated sensitivity methods to quantities that are not actually central to the decision problem at hand [401,402].

Once the outputs have been fixed, the next step is to define the uncertain-input space in a way that is both mathematically explicit and epidemiologically defensible. This requires more than assigning arbitrary ranges to parameters. Each uncertain quantity should be accompanied by a justification of its support, distributional form, and relation to other quantities. If $X = (X_1, \dots, X_d)$, then a complete specification consists of a joint distribution $p(x)$, or at least a structured uncertainty set $\mathcal{X}$, together with any dependence assumptions judged necessary. In many epidemiological ABS applications, one separates inputs into operational variables, whose values correspond to intervention or design choices, and stochastic or aleatory variables, which represent internal randomness or exogenous variability. This distinction is not merely terminological. If it is ignored, then the analyst may inadvertently interpret simulator noise as substantive sensitivity, or may confound uncertainty about policy levers with uncertainty about random epidemic realization. Accordingly, a practical workflow should decide explicitly whether the purpose is to study sensitivity of the conditional mean response $\mu(x) = \mathbb{E}[Q \mid X = x]$, sensitivity of the full distribution of $Q \mid X = x$, or robustness of a decision rule under simultaneous operational and stochastic uncertainty. Recent work has stressed that such distinctions are central to an operationally meaningful sensitivity analysis, especially when the model is stochastic and used for policy support rather than for purely deterministic mechanism exploration [403,404].

After the uncertainty space has been defined, the analysis should proceed in stages rather than jumping immediately to the most computationally expensive method. A pragmatic strategy is to begin with a pilot design whose purpose is diagnostic. At this stage, the analyst checks whether the simulator runs reliably over the proposed uncertainty domain, whether the chosen outputs are numerically stable, and whether replication noise is large enough to require averaging at each design point. If, for a fixed input setting x, one computes R stochastic replicates $Q(x, U_1), \ldots, Q(x, U_R)$, then the natural estimators are

$$\widehat{\mu}(x) = \frac{1}{R}\sum_{r=1}^{R} Q(x, U_r), \qquad \widehat{\sigma}^2(x) = \frac{1}{R-1}\sum_{r=1}^{R}\left(Q(x, U_r) - \widehat{\mu}(x)\right)^2,$$

with Monte Carlo standard error

$$\text{MCSE}\left(\widehat{\mu}(x)\right) = \frac{\widehat{\sigma}(x)}{\sqrt{R}}.$$

This quantity should not be treated as an afterthought. If the MCSE is of the same order as the change induced by perturbing the input factors, then the sensitivity analysis is under-resolved and no stable ranking can be trusted. A practical stopping rule is therefore to increase R until the MCSE falls below a pre-specified tolerance ε, either in absolute terms or relative to the baseline output, that is, until

$$\frac{\text{MCSE}(\widehat{\mu}(x))}{|\widehat{\mu}(x)|} \le \varepsilon.$$

Only after this pilot stage is it sensible to decide whether the remaining task is essentially local, essentially global, or hybrid. Introductory overviews of sensitivity assessment stress that the analyst should choose the method only after clarifying the aim and the computational budget, because different methods answer different questions, and the temptation to treat them as interchangeable is a common source of poor practice [401,402].

The next phase is methodological escalation. If the purpose is baseline diagnosis, code checking, or ranking parameters near a calibrated setting, then local or extended OFAT methods are appropriate. If the purpose is to identify dominant uncertainty drivers over a broad domain, then global screening or variance-based methods are more appropriate. In high-dimensional epidemiological ABS, a staged pipeline is often the most efficient: first apply a screening method to eliminate negligibly influential factors, then apply a more expensive global method on the reduced set, and finally return to local perturbation analysis around policy-relevant settings for interpretation. In mathematical terms, one moves from a coarse map of influence over the entire uncertainty set to a fine-grained map of response near specific operating points. This staged logic is especially valuable when the simulator includes many policy levers and natural-history parameters at once, because it avoids spending most of the computational budget estimating high-precision indices for factors whose effect is practically zero. More broadly, current methodological discussions emphasize that sensitivity analysis should be embedded in a structured inferential process in which method choice, computational burden, and decision purpose are considered jointly rather than sequentially or ad hoc [401,404].

A complete workflow should end not with parameter rankings alone but with a statement about robustness of conclusions. Suppose that a decision-maker is comparing two interventions, A and B, and the relevant loss is $L_A(X, U)$ under policy A and $L_B(X, U)$ under policy B. Then the central robustness quantity is not merely the average difference

$$\mathbb{E}[L_A - L_B],$$

but also the probability that one policy outperforms the other over the uncertainty space,

$$\Pr\left(L_A(X, U) < L_B(X, U)\right),$$

or, in threshold terms,

$$\Pr(Q_A \le q^\star) \quad \text{versus} \quad \Pr(Q_B \le q^\star),$$

for some decision threshold $q^\star$. Reporting such quantities forces the analyst to translate sensitivity results into a form that is meaningful for policy choice. It also helps prevent a common misuse of sensitivity analysis, namely the presentation of an ordered list of parameters without any explanation of how that ordering affects the substantive conclusion of the study. In policy-facing settings, this broader inferential chain should itself be scrutinized, since the way uncertainty is framed, which scenarios are compared, and how thresholds are chosen all shape the message conveyed by the model. This is closely related to the wider idea of sensitivity auditing, in which one examines not only the numerical propagation of uncertainty but also the assumptions and framing choices that condition the interpretation of the model in a policy process [405].

The following Python code outlines an end-to-end workflow that puts these principles into an executable structure. It separates the pilot phase from the main design, averages over stochastic replications at each design point, and returns both the sensitivity object and a policy-robustness summary. In an actual epidemiological study, the functions `sample_inputs`, `screening_analysis`, and `global_analysis` would be implemented with the particular design and sensitivity method chosen by the analyst, but the logic of the workflow remains the same.

```python
import numpy as np

def run_replicated_mean(run_model, params, seeds, outcome_key):
    vals = []
    for seed in seeds:
        out = run_model(params=params, seed=seed)
        vals.append(out[outcome_key])
    vals = np.asarray(vals, dtype=float)
    return {
        "mean": vals.mean(),
        "var": vals.var(ddof=1) if len(vals) > 1 else 0.0,
        "mcse": vals.std(ddof=1) / np.sqrt(len(vals)) if len(vals) > 1 else 0.0,
        "values": vals
    }

def pilot_replication_check(run_model, x0, outcome_key, seed_batches, rel_tol=0.02):
    for seeds in seed_batches:
        stats = run_replicated_mean(run_model, x0, seeds, outcome_key)
        denom = max(abs(stats["mean"]), 1e-12)
        if stats["mcse"] / denom <= rel_tol:
            return {"accepted_seeds": seeds, "pilot_stats": stats}
    return {"accepted_seeds": seed_batches[-1], "pilot_stats": stats}

def sensitivity_workflow(run_model, baseline_params, sample_inputs,
                         screening_analysis, global_analysis,
                         policyA, policyB, outcome_key="cum_infections_T"):

    # 1. Pilot stage: choose enough replications
```

```
seed_batches = [
    [11, 22, 33, 44],
    [11, 22, 33, 44, 55, 66, 77, 88],
    [11, 22, 33, 44, 55, 66, 77, 88, 99, 111, 122, 133]
]
pilot = pilot_replication_check(run_model, baseline_params, outcome_key, seed_batches)
seeds = pilot["accepted_seeds"]

# 2. Generate experimental design
X = sample_inputs()

# 3. Evaluate replication averaged response
Y = []
for x in X:
    stats = run_replicated_mean(run_model, x, seeds, outcome_key)
    Y.append(stats["mean"])
Y = np.asarray(Y, dtype=float)

# 4. Optional screening
influential = screening_analysis(X, Y)

# 5. Global analysis on retained factors
G = global_analysis(X, Y, influential)

# 6. Decision robustness comparison
XA = [policyA(x) for x in X]
XB = [policyB(x) for x in X]

YA = np.array([run_replicated_mean(run_model, x, seeds, outcome_key)["mean"] for x in XA])
YB = np.array([run_replicated_mean(run_model, x, seeds, outcome_key)["mean"] for x in XB])

robustness = {
    "mean_difference_A_minus_B": float(np.mean(YA - YB)),
    "prob_A_better_than_B": float(np.mean(YA < YB))
}

return {
    "pilot": pilot,
    "screening": influential,
    "global_results": G,
    "robustness": robustness
}
```

Listing 6.9: Practical workflow for sensitivity analysis in a stochastic epidemiological ABS.

The main value of such a workflow is not only organizational but epistemic. It makes explicit which quantities are being protected from Monte Carlo noise, which uncertainties are being propagated, which parameters are screened out, and how the final policy claim is linked back to the sensitivity design.

In this sense, a good sensitivity-analysis workflow for an epidemiological ABS should be read as a traceable argument: the estimand is identified first, the uncertainty structure is made explicit, replication error is controlled, the method is matched to the inferential objective, and the final conclusions are reported in terms of robustness rather than in terms of a single nominal run. That is the point at which sensitivity analysis ceases to be a technical add-on and becomes part of the scientific credibility of the model itself [401,404].

6.7 Ethics, privacy, and policy risks

Epidemiological ABS are often, also in this book, introduced as technical instruments for representing transmission dynamics and evaluating intervention scenarios, yet they are never ethically neutral in any strong sense because the modeler must decide, at every stage of construction, which harms count, which benefits matter, which populations are visible, and which trade-offs are treated as acceptable. A choice to optimize cumulative infections, peak hospital occupancy, deaths, continuity of schooling, or continuity of economic activity is not merely a computational design decision; it is already a normative statement about what the model takes to be socially salient. The same is true of decisions about whether to encode unequal access to care, unequal capacity to comply with restrictions, or privacy loss as part of the modeled world, rather than as an external issue left outside the boundary of analysis. In that respect, an ABS does not merely represent an epidemic, but also a theory of what the epidemic means for a society, because it determines whose risks are measured, how those risks are aggregated, and which forms of burden disappear when outcomes are compressed into a small number of headline indicators. This ethical dimension becomes especially important in public-health settings, where simulation outputs are frequently mobilized in policy-facing contexts and can shape both governmental deliberation and public understanding. Accordingly, ethics should not be appended to epidemiological ABS only after calibration or validation, but treated as part of model specification itself, particularly when the simulation is used to support decisions that redistribute liberty, exposure, surveillance, or access to scarce protection and treatment resources [406,407].

In this context, privacy concerns arise well before any result is published, because realism at the agent level is often purchased by combining census marginals, administrative registers, mobility traces, household structure, clinical records, or other fine-grained data sources into a synthetic population whose statistical texture is intended to resemble a real community. The fact that the resulting agents are synthetic does not by itself remove disclosure risk, since re-identification can proceed through linkage on rare attribute combinations, unusual trajectories, or overly fine geographic detail, especially when models preserve small-area structure, rare subgroups, or richly cross-tabulated behavioral histories. For this reason, privacy in ABS is better understood as a managed continuum rather than a binary label: a model can preserve substantial analytical utility while still leaking more individual distinctiveness than necessary for the scientific question at hand. This is particularly important when modelers rely on synthetic-data pipelines, because such pipelines reduce some risks but do not magically transform all downstream products into ethically unproblematic artifacts. A prudent epidemiological practice is therefore to use the least detailed representation that still supports the estimand of interest, avoid releasing outputs at a granularity that invites linkage to external records, and to document clearly how source data were transformed, perturbed, synthesized, and evaluated. Otherwise, the pursuit of realism may create a false sense of methodological sophistication while quietly importing privacy risks through detail that is not actually needed for inference or decision support [408,409].

A further ethical challenge is that public-health interventions are rarely uniform in either their social cost or their epidemiological benefit, even when they are represented in a model as population-level levers. School closure, targeted vaccination, workplace distancing, mobility restrictions, and test–isolate policies all operate through contact structures that are themselves distributed unequally across class, occupation, household composition, neighborhood, and race or ethnicity, so a policy that appears effective on average may still protect already advantaged groups while shifting burdens onto those who are less able to reduce contacts, work remotely, or absorb interruptions in care and income. In this setting, an ABS that reports only total infections averted or aggregate reductions in transmission may conceal precisely the distributional effects that matter most for ethical evaluation. This is not merely a theoretical concern. Computational work on COVID-19 has shown that disadvantaged groups can face higher infection risk because they are less able to reduce mobility and because the locations they continue to visit are often more crowded and therefore riskier, while broader methodological commentary has argued that socioeconomic vulnerability remains underrepresented in infectious-disease modeling despite being central to both transmission and health inequity. The implication is that fairness cannot be read from a single epidemic curve. Instead, epidemiological ABS should treat subgroup outcomes as first-class results and should explicitly represent those structural constraints that determine whether an intervention is only efficient in aggregate or genuinely equitable in practice [410,411].

In addition, policy risk emerges when model outputs are granted more authority than the model can justify. In epidemic response, this commonly occurs when conditional scenario analysis is communicated as forecast, when a single calibrated trajectory is reported as though it were a stable expectation, or when large structural and behavioral uncertainties are compressed into a precise-looking number that travels more easily into media coverage and policy briefs than the caveats that should accompany it. Mechanistic epidemic models are often most useful not because they literally reveal the future, but because they clarify the implications of assumptions, expose thresholds, and allow decision-makers to compare counterfactuals under controlled conditions; their misuse begins when those conditional statements are forgotten and the model is treated as an oracle rather than as a device for disciplined reasoning. Responsible policy use therefore requires a clear distinction between exploration and prediction, explicit reporting of the time horizon over which the model is credible, and a willingness to show how conclusions change when behavioral, epidemiological, or structural assumptions are perturbed. This caution does not reduce the value of simulation. On the contrary, simulation is often indispensable for policy evaluation because it can examine unintended consequences, compare intervention designs that would be infeasible to test experimentally, and refine proposals before implementation. But that value depends on epistemic modesty: the model should organize judgment, not replace it, and it should support deliberation without being mistaken for a neutral machine that settles contested policy questions by itself [412,413].

Deepen your knowledge

Exercise 1: From research question to model scope (ABS or not?)

Task

For each question below, decide whether an ABS is *necessary* or whether a simpler model (e.g., compartmental / branching process) is likely superior. For the ones where ABS is justified, write: (i) the estimand, (ii) the minimum mechanisms required, and (iii) three outputs you must log.

1. Q1: "Estimate the final attack rate of an influenza-like pathogen in a population of one million over one season."
2. Q2: "Quantify how household size and age-dependent susceptibility interact to shape transmission, and whether targeted vaccination of multi-generational households reduces severe outcomes."
3. Q3: "Evaluate whether a policy that closes workplaces only when local hospital occupancy exceeds a threshold prevents capacity overflow."

Solution (worked)

- **Q1:** ABS is typically *not necessary*. If mixing is assumed homogeneous and the target is final size, a compartmental model gives transparency and speed. An ABS could be used for didactic purposes, but it is not the minimal tool.
- **Q2:** ABS is *well-justified*. The mechanism is explicitly about correlated heterogeneity (household structure × age) and a targeted policy operating on that structure. Minimum mechanisms: household clustering, age-stratified susceptibility and severity, vaccination targeting rules.
- **Q3:** ABS is *often justified*, but not always. If "local" is meaningful (spatial heterogeneity) and the intervention triggers endogenously based on realized occupancy, ABS or a metapopulation model is appropriate. Minimum mechanisms: spatial units (or neighborhoods), hospital occupancy state, policy trigger, and feedback to contact rates.

For Q2, a concrete "question-to-output map" could be:

- **Estimand:** difference in severe cases (or hospitalizations) over 120 days under targeted vaccination vs. baseline.
- **Minimum mechanisms:** household contact layer; age-dependent susceptibility; progression to severe disease; vaccination that reduces susceptibility and/or severity; targeting rule based on household composition.
- **Outputs:** daily incidence by age; cumulative hospitalizations; infections and severe cases stratified by household type (e.g., multi-generational vs. not).

Exercise 2: Design the agent schema (state variables with justification)

Task

You are modeling SARS-like transmission in a city with layered contacts (household, school/work, community). The policy question is: "Which is more effective: (A) school masking, or (B) household-focused vaccination?"

Design a minimal set of **agent attributes** and **dynamic states**. For each variable, state *why it is needed* (what causal pathway to outputs it enables). Put your answer in Table 6.5.

Table 6.5 Agent schema design template. Fill "Needed because..." with a one-sentence causal justification.

Variable	Type	Needed because...
age	static	
household_id	static	
setting_id (school/work)	static	
epi_state (S,E,I,R)	dynamic	
t_in_state	dynamic	
vaccinated	dynamic/static	
mask_adherence	dynamic/static	
severity_risk_mult	static	

Solution (worked)

A minimal schema that answers the policy comparison:

- `age`: drives susceptibility and severity; defines school-aged group.
- `household_id`: enables household transmission and household-targeted vaccination.
- `setting_id`: enables school/work contacts and school masking intervention.
- `epi_state` and `t_in_state`: govern natural history timing and infectiousness.
- `vaccinated`: modifies susceptibility and/or probability of severe outcomes; allows targeted allocation.
- `mask_adherence`: modifies transmission intensity in school layer; can be scenario-specific or individual-specific.
- `severity_risk_mult`: captures comorbidity/age-based severity without modeling full clinical detail.

Note what is *not* needed: detailed mobility trajectories, explicit friendships, or dozens of clinical states, unless the outputs require them.

Exercise 3: Build a two-layer contact structure (households + community)

Task

Create a toy population of $N = 500$ agents partitioned into households with sizes drawn from a small distribution (e.g., 1–5). Each day:

1. household contacts are fully connected within each household,
2. community contacts are random: each agent makes `k_community` contacts uniformly at random.

Return a daily contact list `contacts[i]` containing the IDs contacted by agent i that day (undirected contacts).

Solution (code)

```
import numpy as np

def make_households(N, rng, hh_sizes=(1,2,3,4,5), probs=(0.25,0.30,0.25,0.15,0.05)):
```

```
    """Assign agents to households; returns household_id array and a list of households (lists
        of agent ids)."""
    household_id = -np.ones(N, dtype=int)
    households = []
    i = 0
    h = 0
    while i < N:
        size = rng.choice(hh_sizes, p=probs)
        size = min(size, N - i)
        members = list(range(i, i + size))
        for m in members:
            household_id[m] = h
        households.append(members)
        i += size
        h += 1
    return household_id, households

def daily_contacts(N, households, k_community, rng):
    """Return list-of-lists contacts[i] with undirected contacts."""
    contacts = [set() for _ in range(N)]

    # Household: fully connected
    for members in households:
        for a in members:
            for b in members:
                if a != b:
                    contacts[a].add(b)

    # Community: random contacts (undirected)
    for a in range(N):
        # sample without replacement excluding self
        candidates = np.setdiff1d(np.arange(N), np.array([a]), assume_unique=False)
        k = min(k_community, len(candidates))
        chosen = rng.choice(candidates, size=k, replace=False)
        for b in chosen:
            contacts[a].add(int(b))
            contacts[int(b)].add(a)

    return [sorted(list(s)) for s in contacts]

# Example usage
rng = np.random.default_rng(123)
N = 500
household_id, households = make_households(N, rng)
contacts_today = daily_contacts(N, households, k_community=8, rng=rng)
```

Listing 6.10: Toy population synthesis and daily contacts for household + random community layers.

Exercise 4: Implement SEIR with realistic dwell times (Erlang stages)

Task

In discrete-time ABS, using a single-step geometric dwell time can distort timing. Implement an SEIR where:

- incubation period is approximately Gamma-like by using k_E exposed substages,
- infectious period is approximately Gamma-like by using k_I infectious substages.

Agents progress $E_1 \to E_2 \to \cdots \to E_{k_E} \to I_1 \to \cdots \to I_{k_I} \to R$ with per-day progression probabilities chosen to match desired mean durations.

Solution (code + explanation)

If each substage advances with probability p per day, the expected time to traverse k stages is approximately k/p (in discrete time, this is a negative binomial-like dwell). Choose $p_E = k_E/\mu_E$ and $p_I = k_I/\mu_I$, clipped to [0, 1].

```
import numpy as np

# Encode states as integers for speed:
# 0=S, 1..kE = E1..EkE, kE+1..kE+kI = I1..IkI, kE+kI+1 = R
def step_progression(state, rng, kE, kI, muE, muI):
    pE = min(1.0, kE / muE)
    pI = min(1.0, kI / muI)

    R_state = kE + kI + 1

    # Progress exposed stages
    for e in range(1, kE + 1):
        idx = np.where(state == e)[0]
        if len(idx) == 0:
            continue
        advance = rng.random(len(idx)) < pE
        # move E_j -> E_(j+1), and E_kE -> I_1
        state[idx[advance]] = (e + 1) if (e < kE) else (kE + 1)

    # Progress infectious stages
    for i_stage in range(kE + 1, kE + kI + 1):
        idx = np.where(state == i_stage)[0]
        if len(idx) == 0:
            continue
        advance = rng.random(len(idx)) < pI
        # move I_j -> I_(j+1), and I_kI -> R
        state[idx[advance]] = (i_stage + 1) if (i_stage < kE + kI) else R_state

    return state

# Example
```

```
rng = np.random.default_rng(1)
N = 1000
kE, kI = 3, 4
muE, muI = 4.0, 6.0
state = np.zeros(N, dtype=int)
state[:10] = 1  # seed 10 exposed in E1
state = step_progression(state, rng, kE, kI, muE, muI)
```

Listing 6.11: Erlang-stage SEIR state update for one day.

Sanity check (recommended)

Simulate many isolated agents starting at E_1 with no transmission and record time to reach I_1; the mean should be close to μ_E and the distribution should be less memoryless than geometric. Plot histograms of incubation/infectious durations.

Exercise 5: Calibrate transmission to a target early growth rate (simulation-based)

Task

In your toy ABS (from Exercises 3–4), calibrate a scalar transmission multiplier β such that the early epidemic grows with approximately a target exponential growth rate r (e.g., $r = 0.15$ per day over days 5–15). Use a simple search procedure.

Solution (code sketch)

Below is a minimal binary search: run short simulations at candidate β, estimate r from log(incidence + 1) over a window, and adjust.

```
import numpy as np

def simulate_short(N, households, k_community, beta, days, rng, kE=2, kI=3, muE=3.0, muI=5.0,
     seed_I=10):
    # States: 0=S, 1..kE E stages, kE+1..kE+kI I stages, kE+kI+1=R
    R_state = kE + kI + 1
    state = np.zeros(N, dtype=int)
    # seed infectious in I1
    initial = rng.choice(np.arange(N), size=seed_I, replace=False)
    state[initial] = kE + 1

    incidence = np.zeros(days, dtype=float)

    for t in range(days):
        contacts = daily_contacts(N, households, k_community, rng)

        # force of infection: infect S via contacts with infectious stages
        infectious = (state >= (kE + 1)) & (state <= (kE + kI))
        newly_exposed = []

        for i in range(N):
```

```
            if state[i] != 0:
                continue
            # if any infectious contact, infection occurs with probability 1-exp(-beta *
                n_inf_contacts)
            n_inf = sum(1 for j in contacts[i] if infectious[j])
            if n_inf > 0:
                p = 1.0 - np.exp(-beta * n_inf)
                if rng.random() < p:
                    newly_exposed.append(i)

        if newly_exposed:
            state[np.array(newly_exposed, dtype=int)] = 1  # E
        incidence[t] = len(newly_exposed)

        # disease progression
        state = step_progression(state, rng, kE, kI, muE, muI)

        # stop if extinct
        if incidence[t] == 0 and not infectious.any():
            break

    return incidence

def estimate_growth_rate(incidence, t0, t1):
    y = np.log(incidence[t0:t1] + 1.0)
    x = np.arange(t0, t1, dtype=float)
    # simple least squares slope
    x_mean = x.mean()
    y_mean = y.mean()
    denom = ((x - x_mean)**2).sum()
    if denom == 0:
        return 0.0
    slope = ((x - x_mean) * (y - y_mean)).sum() / denom
    return float(slope)

def calibrate_beta(target_r, N, households, k_community, rng, beta_lo=1e-4, beta_hi=0.2, iters
    =12):
    best = None
    for _ in range(iters):
        beta_mid = 0.5 * (beta_lo + beta_hi)
        inc = simulate_short(N, households, k_community, beta_mid, days=25, rng=rng)
        r_hat = estimate_growth_rate(inc, t0=5, t1=15)
        best = (beta_mid, r_hat, inc)

        if r_hat < target_r:
            beta_lo = beta_mid
        else:
            beta_hi = beta_mid
```

```
    return best  # (beta, r_hat, incidence)

# Example usage
rng = np.random.default_rng(7)
N = 500
household_id, households = make_households(N, rng)
beta_hat, r_hat, inc = calibrate_beta(target_r=0.15, N=N, households=households, k_community
    =6, rng=rng)
```

Listing 6.12: Binary search calibration of beta to match early growth rate.

CHAPTER

Data-driven models

7

Epidemiological modeling has traditionally been dominated by mechanistic approaches, since such models offer interpretability and causal narratives [414,415]. Yet, their fidelity is bounded by the correctness of their assumptions and by the extent to which the chosen structure captures reality. In contrast, *data-driven models* prioritize learning predictive structure directly from observations [213,416]. Rather than beginning with compartments, agents, or transmission equations, they begin with data streams and infer relationships that best explain or forecast the observed dynamics [417,418]. The rise of data-driven epidemiology reflects several converging developments: i) the increasing availability of high-frequency surveillance data; ii) the growing role of heterogeneous digital traces (e.g., mobility, web activity, and textual reports); iii) advances in machine learning and probabilistic forecasting; and iv) the operational demand for short-horizon predictions that can be updated rapidly as new data arrive. These trends have enabled models that are flexible, scalable, and often highly accurate in the regimes for which they are trained [419–421]. At the same time, data-driven approaches pose distinctive challenges, including sensitivity to non-stationarity, vulnerability to reporting artifacts, limited interpretability, difficulty in extrapolating across policy regimes, and the need for careful evaluation protocols that reflect real-time deployment constraints [422,423].

This chapter is organized around a progression of modeling capacity and data modality. We begin with time-series-based methods, which learn temporal dependencies directly from epidemiological observations. We then extend to machine learning models that incorporate covariates and nonlinear predictors, followed by deep learning architectures that learn representations from raw sequences and multimodal inputs, including spatio-temporal dependencies. The goal of this chapter is twofold. First, it provides a conceptual and algorithmic map of data-driven approaches that are genuinely learned from data, clarifying their assumptions and typical failure modes. Second, it equips the reader with practical guidance on how to evaluate and deploy such models in epidemiological contexts, with particular emphasis on uncertainty, validation under distribution shift, and the separation between predictive utility and causal interpretation.

Time-series-based data-driven models treat epidemiological signals as sequences and learn predictive structure directly from their temporal regularities [424–426]. The core object is an observed series $\{y_t\}_{t=1}^{T}$ where T is the event horizon (i.e., the farthest point of interest in future) – often a daily count: cases, admissions, deaths, positivity, wastewater concentration, optionally accompanied by exogenous covariates $\mathbf{x}_t \in \mathbb{R}^d$ such as mobility, testing volume, weather, and policy indicators. The goal is to predict the future without imposing an explicit transmission mechanism. Mathematically, the problem takes the form:

$$\hat{y}_{t+h} = f_\theta(y_{1:t}, \mathbf{x}_{1:t}), \qquad h \geq 1, \tag{7.1}$$

Epidemic Modeling. https://doi.org/10.1016/B978-0-44-355525-1.00013-8

or, in the probabilistic setting, to predict a distribution

$$\hat{F}_{t,h}(y) = \mathbb{P}(Y_{t+h} \leq y \mid y_{1:t}, \mathbf{x}_{1:t}) . \tag{7.2}$$

A practical complication is that epidemiological time series are rarely stationary due to interventions, behavioral responses, immunity, and reporting pipelines that shift the data-generating process [427,428]. Furthermore, the data are discrete, noisy, frequently overdispersed, and shaped by systematic artifacts [429,430]. These issues motivate three recurring design choices: i) stabilize the scale; ii) choose an observation distribution suitable for counts; and (iii) explicitly represent seasonality, latent trends, and structural breaks. Below, we discuss three types of models, in increasing computational complexity: auto-regressive, machine learning, and deep learning models.

To make the discussion of data-driven models concrete, this chapter uses a single running dataset that will reappear across methods, evaluation protocols, and failure-mode demonstrations. Namely, we work with a real-world daily surveillance signal derived from the Johns Hopkins University COVID-19 global time series, which was originally developed to support the real-time dashboard described by Dong et al. [431]. The purpose is not to claim this dataset is "clean" or "ideal", but rather the exact opposite. Its weekday oscillations, occasional backfills, and non-stationary phases provide a realistic substrate for illustrating how forecasting methods behave when the data-generating process is shaped by reporting operations as well as by epidemiology. In order to illustrate the data, Fig. 7.1 shows the raw daily incidence alongside a 7-day rolling mean, which is a convenient visualization of the latent trend under strong weekly reporting patterns. In a complementary manner, Fig. 7.2 summarizes the same series by weekday, revealing how large the operational periodicity can be. Throughout the rest of the chapter, we will treat the extracted daily incidence as the target series y_t, optionally paired with covariates derived from the same date index, so that different models can be compared on the same inputs and backtesting protocol.

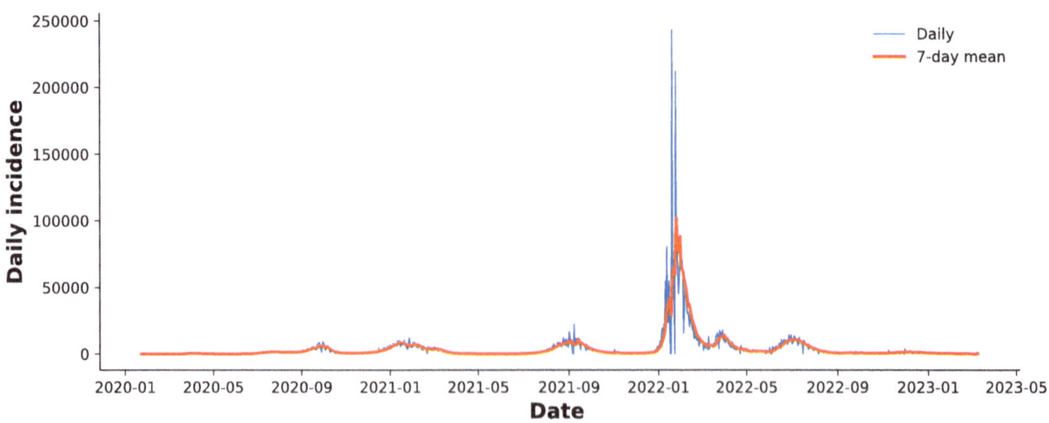

FIGURE 7.1

Daily incidence derived from the JHU global confirmed-case time series, with a 7-day rolling mean to highlight low-frequency dynamics under reporting noise.

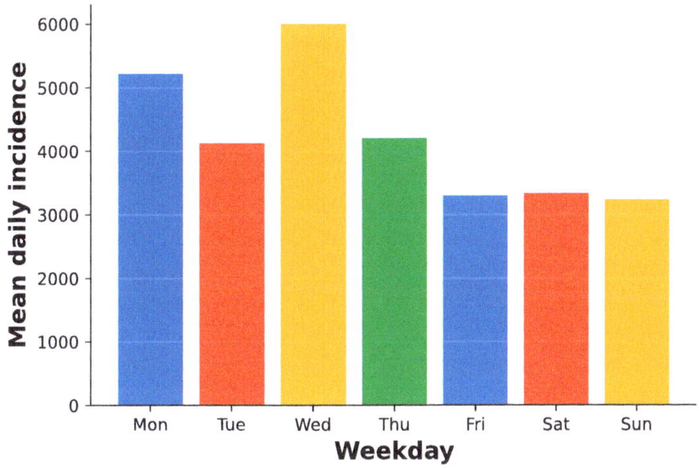

FIGURE 7.2

Weekday aggregation of the same daily incidence series. A strong day-of-week structure typically reflects reporting operations rather than changes in transmission.

7.1 Auto-regressive models

Many classical and modern time-series forecasting methods for epidemiological surveillance can be formulated as an observation model coupled to a learnable conditional mean process. This formulation is convenient because it separates a stochastic model for the data type from a flexible predictor for the conditional expectation, thereby enabling principled likelihood-based estimation within the generalized linear model (GLM) paradigm [432,433]. Formally, let $\{Y_t\}_{t=1}^{T}$ be a discrete-time epidemiological process, typically observed as daily counts. Define the filtration $\mathcal{F}_{t-1}$ as the σ-algebra (the set of sets that can be measured) generated by all information available up to time $t-1$, commonly including past observations and covariates $\mathcal{F}_{t-1} = \sigma\big(Y_{1:t-1}, \mathbf{X}_{1:t-1}\big)$. The central modeling object is the conditional mean $\mu_t = \mathbb{E}[Y_t \mid \mathcal{F}_{t-1}]$, which is mapped to a linear predictor η_t through a link function $g(\cdot)$, such that:

$$Y_t \mid \mu_t \sim \mathcal{D}(\mu_t, \psi), \text{ such that } g(\mu_t) = \eta_t = b + \sum_{\ell=1}^{p} \phi_\ell \, s(y_{t-\ell}) + \mathbf{x}_t^\top \beta + \gamma^\top \mathbf{w}_t, \tag{7.3}$$

where $\mathcal{D}$ denotes an observation distribution appropriate for the measurement scale, ψ denotes nuisance parameters (e.g., dispersion), and $\theta = (b, \phi_{1:p}, \beta, \gamma, \psi)$ is the parameter set to be learned from data. The term $s(\cdot)$ denotes an optional variance-stabilizing transform. In count surveillance applications, it is frequently advantageous to use $s(y) = \log(y + c)$ with a small constant $c > 0$ in order to reduce the influence of extreme observations and to facilitate approximately linear dependence on the transformed scale [434]. The vector $\mathbf{w}_t$ encodes deterministic seasonal indicators such as day-of-week effects; for daily counts, these effects are often substantial due to operational reporting patterns and health-system workflows [435].

Simply put, Eq. (7.3) says that the observed value at time t, denoted by Y_t, is modeled as a random quantity whose average depends on all information available from previous times. The most important part is $\mu_t = \mathbb{E}[Y_t \mid \mathcal{F}_{t-1}]$, which represents the expected value of Y_t given the past. The model then uses a link function $g(\cdot)$ to connect this expected value to a linear predictor η_t. This predictor is made up of several components: a baseline term b, the effect of previous observations $y_{t-\ell}$ through the coefficients ϕ_ℓ, the contribution of observed covariates $\mathbf{x}_t$ through β, and the effect of additional features $\mathbf{w}_t$ through γ. The transformation $s(\cdot)$ is sometimes applied to past observations, for example using a logarithm, to reduce the influence of very large values and make the relationship more stable. In simple terms, the equation says that the expected count at time t is predicted from past counts and other relevant information, while still allowing for random variation in the actual observed data. The specification in the equation should be understood as a strictly data-driven time-series construction where temporal dependence enters through lagged observations ($y_{t-\ell}$) and seasonal indicators ($\mathbf{w}_t$), while the model does not impose a mechanistic transmission structure.

The likelihood template in Eq. (7.3) is compatible with several observation models and latent mean parameterizations. In epidemiological surveillance, the most consequential modeling choices are typically not the autoregressive lag order itself, but the measurement scale, the treatment of overdispersion, and the representation of repeating reporting patterns and abrupt regime shifts.

7.1.1 Observation models for surveillance counts

Daily (or any other step in time, Δt) incidence and burden indicators are commonly represented as nonnegative integers, but their empirical variance is rarely equal to the mean. A Poisson observation model,

$$Y_t \mid \mathcal{F}_{t-1} \sim \text{Poisson}(\mu_t), \tag{7.4}$$

implies $\text{Var}(Y_t \mid \mathcal{F}_{t-1}) = \mu_t$ and is often too restrictive during waves, when clustering, super-spreading, and heterogeneous ascertainment produce heavy-tailed fluctuations. Recall that the Poisson distribution is a probability distribution used to model the number of times an event happens in a fixed interval of time or space, assuming the events occur independently and at a constant average rate. It is commonly used for count data, such as the number of disease cases reported in one day. A standard alternative is the negative binomial model,

$$Y_t \mid \mathcal{F}_{t-1} \sim \text{NegBin}(\mu_t, \kappa), \qquad \text{Var}(Y_t \mid \mathcal{F}_{t-1}) = \mu_t + \frac{\mu_t^2}{\kappa}, \tag{7.5}$$

where $\kappa > 0$ is a dispersion (size) parameter. Recall that the negative binomial distribution is a probability distribution used to model count data, especially when the observed values show substantial variability. It is suitable for non-negative integer outcomes and includes a parameter that controls how dispersed the counts are around their mean. This observation model appears frequently in applied epidemic forecasting and in statistical surveillance frameworks for infectious disease counts [436,437].

In this context, the negative binomial model is not merely a "more variable Poisson", but rather it provides a concrete interpretation of latent heterogeneity that is pervasive in surveillance. One widely spread construction is a Poisson–Gamma mixture: conditional on a latent multiplicative intensity Λ_t, counts are Poisson, while Λ_t varies across time because of unobserved contact-rate variation, changes in ascertainment, localized outbreaks, and operational factors. Concretely,

$$Y_t \mid \Lambda_t \sim \text{Poisson}(\Lambda_t), \qquad \Lambda_t \mid \mathcal{F}_{t-1} \sim \text{Gamma}(\kappa, \kappa/\mu_t), \tag{7.6}$$

implies the marginal negative binomial model in Eq. (7.5) with mean μ_t and quadratic mean–variance scaling [438,439]. This mixed view is useful in epidemiological time series because it clarifies a recurring modeling dilemma: variability in Y_t reflects not only process noise in incidence, but also the instability of observation and reporting mechanisms. In many datasets, the latter dominates at short horizons, and negative binomial dispersion is effectively a compact way to absorb unmodeled heterogeneity without requiring a full reporting-process model.

A distinct but equally important issue is the scale on which temporal dependence is modeled. Additive dynamics on the raw count scale tend to overweight large peaks, whereas multiplicative dynamics on a log scale better match the empirically observed proportional growth behavior early in waves. This motivates variance-stabilizing transforms such as $s(y) = \log(y + c)$ within Eq. (7.3), or the use of a log link $g(\mu_t) = \log(\mu_t)$ so that the linear predictor η_t models multiplicative changes directly. The choice is pragmatic rather than philosophical: the same model structure can behave very differently depending on whether lags are included on the raw scale, the log scale, or through a link function [440].

In practice, the observation model $\mathcal{D}$ should be viewed as part of the *forecast distribution* rather than merely a fitting convenience. When forecasts are evaluated as prediction intervals or quantiles, underestimating variability typically produces overly narrow intervals and systematically poor calibration; this failure mode is often more damaging for decision support than moderate bias in point forecasts [437]. From this perspective, the Poisson model can be understood as an optimistic uncertainty model, while negative binomial provides a first-order correction that frequently improves probabilistic calibration even when the conditional mean specification is unchanged.

The Poisson and negative binomial families do not cover every empirically relevant pattern. Surveillance data can exhibit underdispersion in stabilized or preprocessed series, for example, after aggressive smoothing, deduplication, or when counts represent constrained capacity processes (such as test quotas) rather than free incidence. In such cases, flexible two-parameter count distributions are useful because they can represent both over- and underdispersion while retaining a GLM-like estimation structure. The Conway–Maxwell–Poisson (COM-Poisson) distribution is a prominent example, introducing a dispersion parameter that interpolates between geometric-like overdispersion and Bernoulli-like underdispersion [441]. Although COM-Poisson is less common in operational epidemic forecasting than negative binomial, it is conceptually valuable because it emphasizes that dispersion mismatch can occur in both directions, and that mismatched dispersion is fundamentally a measurement-model problem.

Zero counts deserve special attention because they arise from multiple mechanisms that are not equivalent from a modeling standpoint. True absence of incidence, structural impossibility of observation, and reporting gaps can all produce zeros, but they imply different forecast behavior. Zero-inflated models explicitly treat zeros as arising from a mixture of a structural-zero state and a regular count state. For a zero-inflated Poisson (ZIP),

$$Y_t \mid \mathcal{F}_{t-1} \sim \begin{cases} 0, & \text{with probability } \pi_t, \\ \text{Poisson}(\mu_t), & \text{with probability } 1 - \pi_t, \end{cases} \tag{7.7}$$

with π_t typically linked to covariates via a logit model [442]. Hurdle (two-part) models represent a closely related alternative in which the zero vs. positive event is modeled separately from the positive counts; this separation can better match situations where "any cases at all" is governed by a different process than "how many cases conditional on being positive" [443]. In epidemic surveillance, mixture models can be helpful for rare outcomes, for small-area series, or for early-wave data, but they are also easy to misuse: a ZIP fit may capture reporting gaps rather than genuine structural zeros, and then extrapolate those gaps into the forecast distribution. This is one reason that evaluation on rolling-origin backtests, rather than in-sample fit, is indispensable when selecting between negative binomial and zero-inflated/hurdle alternatives.

A further extension addresses time-varying multiplicative random effects that induce overdispersion and correlation beyond the negative binomial quadratic mean–variance relationship. Poisson-lognormal models treat the Poisson intensity as lognormally distributed,

$$Y_t \mid Z_t \sim \text{Poisson}(\exp(Z_t)), \qquad Z_t \mid \mathcal{F}_{t-1} \sim \mathcal{N}(m_t, v_t), \tag{7.8}$$

which can be interpreted as adding a Gaussian random effect on the log-intensity scale. This construction supports richer dependence structures and is particularly natural when the mean process is modeled on a log scale [444]. While Poisson-lognormal models are more computationally demanding than negative binomial, they connect cleanly to state-space representations and to modern Bayesian formulations in which latent log-incidence evolves smoothly over time.

When the correct distributional form is uncertain, quasi-likelihood provides a useful compromise: inference is driven by the specified mean and a chosen mean–variance relationship, without requiring a fully specified likelihood. Wedderburn's quasi-likelihood framework formalizes this idea and underlies common "quasi-Poisson" and sandwich-robust approaches [445]. In epidemic time series, quasi-likelihood can be especially attractive for exploratory modeling and for sensitivity analysis, because it makes explicit what is being assumed about the mean–variance relationship while avoiding overcommitment to a particular parametric tail behavior.

Table 7.1 summarizes practical observation-model choices that recur in time-series-based epidemic forecasting. The intent is not to prescribe a single "correct" model, but to provide a menu that is aligned with recognizable data pathologies: overdispersion, underdispersion, excess zeros, and latent multiplicative variability.

Fig. 7.3 presents two panels, where the left panel is a rolling-window mean-variance plot: each point corresponds to a fixed-length window (e.g., 21 days) of the daily count series, where the window's empirical mean is plotted on the horizontal axis and the corresponding empirical variance on the vertical axis. If a Poisson observation model were adequate, these points would concentrate around the red reference line Var=Mean, because Poisson counts have variance equal to the mean by construction. Instead, the cloud typically lies substantially above that line during large parts of the epidemic trajectory, especially at moderate-to-high means. This systematic upward deviation is the empirical signature of overdispersion: fluctuations are larger than Poisson noise would allow. The negative binomial curves overlaid in the same panel illustrate how an negative binomial observation model accommodates this behavior by allowing variance to grow faster than the mean. The curvature becomes more pronounced as the dispersion parameter decreases, reflecting heavier tails and greater heterogeneity. Visually, the fact that many points align more closely with one of the negative binomial curves than with the Poisson line is an immediate motivation for replacing Poisson with negative

Table 7.1 Observation models for surveillance counts and the data pathologies they address.

Model	Defining feature	When it is typically useful
Poisson	$\mathrm{Var}(Y_t \mid \mathcal{F}_{t-1}) = \mu_t$	High counts with mild noise, or as a baseline for diagnosing dispersion mismatch [438].
Negative binomial	$\mathrm{Var}(Y_t \mid \mathcal{F}_{t-1}) = \mu_t + \mu_t^2/\kappa$	Overdispersed daily counts, heavy-tailed fluctuations, unstable ascertainment; improves probabilistic calibration in practice [436,437,439].
Quasi-likelihood / quasi-Poisson	Specified mean–variance without full likelihood	Sensitivity analysis and robust inference when the likelihood tails are uncertain [445].
Poisson-lognormal	Lognormal random effect on intensity	Latent multiplicative variability and richer dependence structures on the log scale [444].
COM-Poisson	Dispersion flexible in both directions	Underdispersion from preprocessing/capacity constraints, or mixed dispersion patterns [441].
ZIP / hurdle	Separate mechanism for zeros	Excess zeros in sparse series; early/late epidemic periods; small areas, with caution about reporting gaps [442,443].

binomial when the goal is not only point prediction but also calibrated predictive uncertainty. In addition, the right panel shows that dispersion is not only "greater than Poisson", it is often time-varying. Here, we plot the rolling overdispersion index (variance divided by mean) across time; values near 1 would be consistent with Poisson-like variability, while values above 1 indicate overdispersion. The series frequently oscillates and can exhibit pronounced peaks, which typically coincide with epidemic waves and periods where reporting pipelines are stressed or revised. This behavior emphasizes the section's claim that daily surveillance variability reflects both epidemiological process noise and instability of measurement and reporting mechanisms. In particular, a model that assumes Poisson variability (or a fixed low-dispersion negative binomial) will tend to produce prediction intervals that are too narrow precisely in the periods where decisions are most consequential. Conversely, an negative binomial model acts as a compact correction that widens the forecast distribution in response to the empirically observed mean–variance scaling, often improving interval calibration even when the conditional mean model is unchanged. When considered together, the two panels paint a picture of why observation models should be treated as part of the forecasting object rather than a fitting convenience, since they determine whether probabilistic forecasts reflect the real variability of surveillance data and whether uncertainty is realistically communicated during rapidly changing epidemic phases.

7.1.2 Seasonality and reporting artifacts as deterministic structure

Epidemiological surveillance time series frequently contain a strong deterministic structure that is generated by the measurement process rather than by infection dynamics [446,447]. The most representive example is the weekly reporting cycle, often called a day-of-week or weekend effect, where observed incidence dips during weekends and rebounds early in the week as laboratories and administrative backlogs clear. The result is a pronounced 7-day oscillation that can dominate short-horizon forecast error if it is not modeled explicitly. Empirically, such weekly oscillations have been documented across jurisdictions during COVID-19 and can be detected via spectral peaks near frequency 1/7 even when the underlying epidemic trend changes slowly [448]. In syndromic

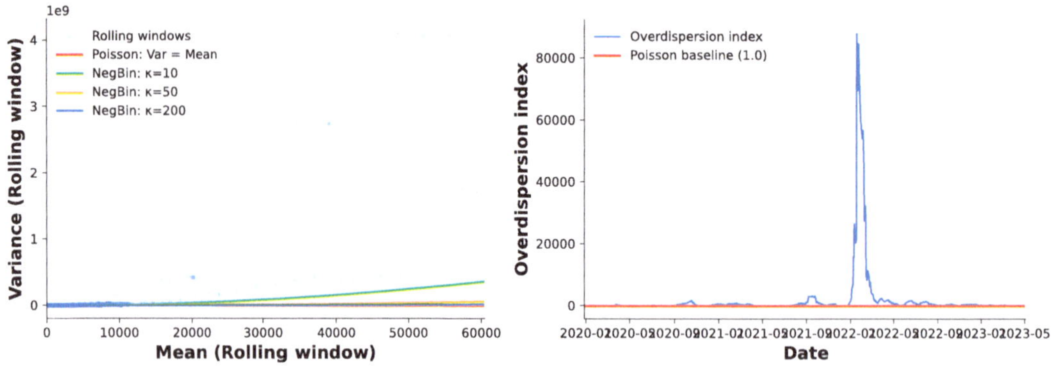

FIGURE 7.3

Observation-model diagnostics for daily surveillance counts computed in rolling windows. Left: empirical rolling variance versus rolling mean, with the Poisson constraint (Var=Mean) and several negative binomial mean–variance curves for different dispersion values κ. Right: the rolling overdispersion index (variance divided by mean) over time, with the Poisson baseline at 1.

surveillance, a similar calendar-induced structure is sufficiently ubiquitous that operational pipelines often include explicit correction for both weekday and public holiday effects [449]. Namely, if deterministic reporting patterns are omitted, the model reallocates that variance into latent trend or transmission-like components, producing biased local growth estimates and miscalibrated predictive intervals.

In the GLM formalization of Eq. (7.3), deterministic structure enters through the calendar feature vector $\mathbf{w}_t$. For daily data, the simplest specification uses weekday indicators. Let $d(t) \in \{1, \ldots, 7\}$ denote the weekday of time index t, and define $w_{t,j} = \mathbb{I}\{d(t) = j\}$ as one-hot regressors. With a log link, the predictor becomes

$$\log(\mu_t) = b + \sum_{\ell=1}^{p} \phi_\ell \, s(y_{t-\ell}) + \mathbf{x}_t^\top \beta + \sum_{j=1}^{6} \gamma_j \, \mathbb{I}\{d(t) = j\} + \delta \, \mathbb{I}\{t \in \mathcal{H}\}, \tag{7.9}$$

where one weekday is dropped as a baseline (or equivalently, a sum-to-zero constraint is imposed) and $\mathcal{H}$ denotes a set of holidays or exceptional reporting days. The holiday term is often essential because holiday closures create effects that are larger in magnitude than ordinary weekends and may differ by data stream (cases, admissions, deaths) [449]. This encoding is deliberately non-mechanistic: it asserts only that the observation process is periodic in the calendar, not that transmission is periodic.

Weekday dummies are flexible but can be parameter-inefficient in richer settings, especially when the model also carries many lags, covariates, and interactions. Harmonic encodings replace discrete indicators by a small Fourier basis, yielding smooth periodic effects with fewer parameters:

$$w_{t,k}^{(\sin)} = \sin\left(\frac{2\pi k t}{s}\right), \qquad w_{t,k}^{(\cos)} = \cos\left(\frac{2\pi k t}{s}\right), \qquad k = 1, \ldots, K, \tag{7.10}$$

with $s = 7$ for weekly structure. In regression terms, Fourier features act as a low-rank approximation of an arbitrary periodic function. They are particularly helpful when the seasonal period is long (e.g., annual seasonality in daily data with $s \approx 365$), where indicator-based seasonal effects would be infeasible. Modern forecasting practice often combines Fourier regressors with ARIMA-like error structures (dynamic regression), or more generally within state-space formulations that allow seasonal coefficients to evolve over time [450,451]. This is useful in surveillance because reporting systems change: laboratories alter batching, public health agencies revise pipelines, and the amplitude or phase of the weekday effect can drift. Allowing slow temporal variation in seasonal coefficients can be interpreted as a controlled relaxation of strict periodicity while retaining the essential deterministic scaffold [450].

Deterministic structure can also be handled through decomposition-based preconditioning, in which the series is separated into trend, seasonal, and remainder components before modeling the remainder. STL (Seasonal-Trend decomposition using Loess) is a widely used procedure:

$$y_t = T_t + S_t + R_t, \tag{7.11}$$

where T_t is a smooth trend, S_t is a seasonal component (e.g., weekly), and R_t is a residual process. STL is robust to outliers and allows local changes in the seasonal pattern through its smoothing windows [452]. In biosurveillance, such preconditioning is often motivated by downstream tasks (detection and alerting) that operate on residuals; subtracting estimated trend and weekday effects can stabilize control charts and reduce false alarms [453]. For forecasting, decomposition is best understood as a modeling choice about how to allocate capacity: deterministic filters absorb the predictable calendar structure, so that the remaining statistical model focuses on the less predictable components, such as short-term trend accelerations and noisy stochastic fluctuations.

Table 7.2 summarizes common deterministic structures used in daily surveillance forecasting. The entries should be read as composable building blocks: weekday indicators can coexist with Fourier terms for longer seasonalities, while decomposition can be combined with probabilistic count models by applying it on a transformed scale (e.g., $\log(y_t + c)$) and modeling the remainder with a distributional forecast.

Listing 7.1 shows a practical way to turn a daily count series into a small, explicit set of regressors and then fit a negative binomial GLM, which is a common baseline when counts are overdispersed. The function first builds a design matrix from three blocks: a lagged log-count (to capture short-term persistence), weekday indicators (to absorb reporting artifacts such as weekend dips), and weekly Fourier terms (a smooth periodic alternative that can represent a stable weekly cycle with few parameters). After dropping the first row because the lag feature is undefined there, the code fits an negative binomial-GLM in `statsmodels`, which yields mean predictions on the original count scale and is typically more robust than a Poisson GLM when variance exceeds the mean. For forecasting, it then constructs the same feature blocks for future time indices and calls `res.predict` to obtain an h-step mean forecast; the one simplification to notice is that the lag feature is "plugged in" using the last observed value for all future steps, so for multi-step horizons one would usually iterate recursively (feeding predicted means back into the lag) if the lag term is meant to carry genuine autoregressive signal. Finally, the weekday construction in the example uses `t % 7` only as a placeholder; in real surveillance data, weekday must be computed from actual calendar dates to avoid phase errors in the learned reporting pattern.

Table 7.2 Deterministic structure choices for daily surveillance time series and their typical roles.

Method	Form in the model	Practical notes
Weekday indicators	Add $\sum_j \gamma_j \mathbb{I}\{d(t) = j\}$ in η_t (Eq. (7.9))	Captures arbitrary day-specific biases; coefficients are easy to interpret and update; can drift over time if re-estimated on rolling windows [449].
Fourier terms	Add $\sum_{k=1}^{K} a_k \sin(2\pi kt/s) + b_k \cos(2\pi kt/s)$ (Eq. (7.10))	Parameter-efficient for long seasonalities (e.g., annual effects in daily series); supports multiple seasonalities; often paired with ARIMA errors or state-space dynamics [450,451].
Holiday indicators	Add $\delta \mathbb{I}\{t \in \mathcal{H}\}$	Needed when closures induce shifts larger than weekends; holidays can have asymmetric pre/-post effects and may require separate indicators by stream [449].
Decomposition (STL)	$y_t = T_t + S_t + R_t$ and model R_t (Eq. (7.11))	Robust separation of trend and seasonality; must be implemented causally for real-time forecasting; useful for residual-based monitoring [452, 453].
Nowcasting delay model	Explicit model of event time and reporting delay	Targets latent incidence and corrects right truncation; weekday-dependent delay distributions are often important in practice [454,455].

```
import numpy as np
import pandas as pd
import statsmodels.api as sm

def fourier_terms(t, period, K):
    t = np.asarray(t)
    cols = []
    for k in range(1, K+1):
        cols.append(np.sin(2*np.pi*k*t/period))
        cols.append(np.cos(2*np.pi*k*t/period))
    return np.column_stack(cols)

def fit_predict_nb_glm(y_train, h, K_week=3, add_weekday=True, c=0.5):
    y = np.asarray(y_train, dtype=float)
    n = len(y)
    t = np.arange(n)

    X_parts = []

    # optional: lag-1 on log scale to capture short term dependence
    lag1 = np.r_[np.nan, np.log(y[:-1] + c)]
    X_parts.append(lag1.reshape(-1, 1))

    # weekday indicators assume t=0 is a fixed weekday; in real data use actual calendar
```

```
    if add_weekday:
        dow = (t % 7).astype(int)
        D = pd.get_dummies(dow, drop_first=True).to_numpy()  # 6 columns
        X_parts.append(D)

    # Fourier terms for weekly periodicity (s=7)
    Fw = fourier_terms(t, period=7, K=K_week)
    X_parts.append(Fw)

    X = np.column_stack(X_parts)
    X = X[1:, :]            # drop first row due to lag
    y_fit = y[1:]

    X = sm.add_constant(X, has_constant='add')
    model = sm.GLM(y_fit, X, family=sm.families.NegativeBinomial())
    res = model.fit()

    # build forecast design matrix for t=n,...,n+h-1 using last observed y for lag1
    t_f = np.arange(n, n+h)
    Xf_parts = []
    lag1_f = np.full(h, np.log(y[-1] + c))   # simple plug-in; can be iterated recursively
    Xf_parts.append(lag1_f.reshape(-1, 1))

    if add_weekday:
        dow_f = (t_f % 7).astype(int)
        Df = pd.get_dummies(dow_f).reindex(columns=range(7), fill_value=0)
        Df = Df.drop(columns=[0]).to_numpy()  # match drop_first=True baseline
        Xf_parts.append(Df)

    Fw_f = fourier_terms(t_f, period=7, K=K_week)
    Xf_parts.append(Fw_f)

    Xf = np.column_stack(Xf_parts)
    Xf = sm.add_constant(Xf, has_constant='add')
    mu = res.predict(Xf)     # mean forecast; for full predictive distribution use res.scale/
        alpha
    return mu
```

Listing 7.1: Weekday and Fourier regressors with a negative binomial GLM for daily counts.

7.1.3 ARIMA and SARIMA as learned linear dependence

Autoregressive integrated moving-average (ARIMA) models represent one of the most widely used families of time-series predictors, and remain relevant in epidemiology as strong baselines and as components inside larger ensembles [456–459]. In its non-seasonal form, an ARIMA(p, d, q) model is:

$$\phi(B)\,(1 - B)^d\, y_t = c + \theta(B)\,\varepsilon_t, \tag{7.12}$$

where $\varepsilon_t \sim \mathcal{N}(0, \sigma^2)$, B is the backshift operator, $\phi(B) = 1 - \phi_1 B - \cdots - \phi_p B^p$ and $\theta(B) = 1 + \theta_1 B + \cdots + \theta_q B^q$. The differencing order d handles low-frequency non-stationarity, while p and q capture short-memory dependence. A seasonal extension with period s yields SARIMA$(p, d, q) \times (P, D, Q)_s$:

$$\Phi(B^s)\,\phi(B)\,(1 - B)^d (1 - B^s)^D\, y_t = c + \Theta(B^s)\,\theta(B)\,\varepsilon_t. \tag{7.13}$$

For daily surveillance counts, $s = 7$ is the dominant periodicity because of reporting structure, and SARIMA acts as a data-driven linear filter that learns predictable weekly distortions without invoking mechanistic drivers [460].

A useful way to read Eq. (7.13) is that differencing is a learned detrending operator and ARMA is a learned linear dependence model on the differenced series. Define the differenced process

$$y_t^\star = (1 - B)^d (1 - B^s)^D y_t, \tag{7.14}$$

so that the SARIMA equation becomes

$$\Phi(B^s)\,\phi(B)\,y_t^\star = c + \Theta(B^s)\,\theta(B)\,\varepsilon_t, \tag{7.15}$$

which is an ARMA model on $y_t^\star$. The stationarity and invertibility of the resulting ARMA component are governed by the roots of $\Phi(B^s)\phi(B)$ and $\Theta(B^s)\theta(B)$, respectively, which is why diagnostics and order selection are central to stable forecasting [459].

Despite the Gaussian innovation assumption in Eq. (7.12), ARIMA is commonly applied to transformed surveillance counts. One often fits SARIMA on a stabilized series $z_t = \log(y_t + c)$ and then back-transforms forecasts. This breaks the exact count likelihood but can produce accurate short-horizon predictions when counts are moderately large and reporting artifacts dominate fine-scale behavior [440]. Back-transform bias should then be handled explicitly by targeting forecast quantiles directly.

Model building in ARIMA practice is frequently framed in the Box–Jenkins cycle of identification, estimation, and diagnostics [459]. Identification uses the empirical autocorrelation function and partial autocorrelation function to suggest plausible orders and differencing, while diagnostics test whether fitted residuals resemble white noise. A standard diagnostic is the Ljung–Box test, which evaluates residual autocorrelation up to a chosen lag m, defined as follows:

$$Q(m) = n(n + 2) \sum_{k=1}^{m} \frac{\hat{\rho}_k^2}{n - k}, \tag{7.16}$$

where $\hat{\rho}_k$ is the residual autocorrelation at lag k and n is the residual length [461]. In infectious disease series, "good" residual whiteness is rarely perfect because of regime shifts and exogenous shocks; nevertheless, residual autocorrelation is a practical marker for unmodeled periodic artifacts or insufficient differencing [460,462].

Automatic order selection replaces manual identification with information criteria optimization, often using AICc to correct small-sample bias [463]. A widely used procedure is the stepwise selection strategy popularized in forecasting software, which searches over (p, d, q, P, D, Q) guided by AICc and unit-root heuristics [464]. In epidemiological deployment, automatic selection is attractive because models are refit frequently and must adapt to changing conditions, but it can also be brittle under

Table 7.3 Practical ARIMA-family algorithms used in epidemiological forecasting and surveillance. Notably, "selection" describes common operational fitting strategies rather than theoretical requirements.

Algorithm	Typical target series	Selection and fitting	Notable failure modes
SARIMA on y_t	Large-count, stable reporting	Manual Box–Jenkins or constrained grid search [459]	Overconfident intervals under overdispersion; spurious seasonality when reporting changes [460].
SARIMA on $\log(y_t + c)$	Daily counts with multiplicative growth	AIC/AICc selection; bias-corrected back-transform [463]	Poor tail behavior when counts are small; bias if back-transform is naive [440].
ARIMAX	Counts plus exogenous signals	Joint MLE with covariates; rolling validation [465]	Covariate shift; unavailable future covariates at forecast time.
ARIMA-STL	Strong periodic artifacts with drifting amplitude	Causal STL then ARIMA on remainder [469]	Leakage if STL uses future data; instability when seasonality changes abruptly.
Intervention ARIMA	Policy shocks, holidays, disruptions	Transfer-function and intervention terms [466]	Misspecified intervention shape; confounding with unmeasured concurrent changes.

abrupt regime changes, where short training windows and constrained order spaces can outperform unconstrained search.

Dynamic regression (often called ARIMAX) incorporates exogenous covariates:

$$\phi(B)\,(1-B)^d\, y_t = c + \beta^\top \mathbf{x}_t + \theta(B)\,\varepsilon_t, \tag{7.17}$$

and is useful when mobility, testing volume, weather, or digital traces explain predictable variation beyond autoregressive memory. Empirically, ARIMAX often improves nowcasting and short-horizon forecasting when exogenous signals are timely and stable, as shown for infectious disease surveillance augmented by search query data [465]. The main caveat is that exogenous covariates are themselves subject to distribution shift, so they can improve in-sample fit while harming future performance unless validated under realistic protocols with strict temporal separation [437].

Interventions and structural breaks can be handled within ARIMA via transfer functions and intervention analysis, which treat a policy or disruption as a deterministic input filtered through a dynamic response model [466]. This is particularly relevant for epidemiological time series because interventions can change both behavior and observation pipelines. In health policy evaluation, ARIMA-based interrupted time-series analysis is often used as an alternative to segmented regression when residual autocorrelation and seasonality are prominent [462]. From a data-driven forecasting perspective, the key point is that intervention terms help prevent the noise model from incorrectly attributing a deterministic break to persistent autoregressive structure, which would otherwise propagate as spurious forecasts. (See Table 7.3.)

ARIMA remains a practical epidemiological baseline across a range of diseases and settings. Classic surveillance discussions emphasize its utility for detecting and forecasting deviations in reportable disease series [460]. Empirical applications include seasonal ARIMA forecasting of hand, foot and mouth disease [467] and broader comparisons of ARIMA to alternative time-series methods in national surveillance data [468]. In near-term influenza-like illness forecasting, ARIMA is also used as a component within decomposition-based pipelines such as ARIMA-STL, which separates trend and seasonality before fitting a residual model [469].

Listing 7.2 is a small "try many options and keep the best one" routine for choosing a seasonal ARIMA model. It assumes you already have a training series `z_train` (often a stabilized version of the counts, such as $log(y + c)$) and it tests many reasonable combinations of ARIMA settings: how much the series should be differenced to remove trends, how many past days to look back for dependence, and whether to include a weekly seasonal pattern (with period 7 by default). For each candidate model, the code fits it only on the training window (so it does not peek into the future), then scores how good the model is using a criterion that rewards good fit but also penalizes unnecessary complexity. This score is the AICc: it starts from AIC (a standard measure of fit-vs-complexity) and adds an extra correction that is especially important when the training set is not very large, because complicated models can otherwise look falsely attractive. As the loops run, the function keeps track of the lowest AICc seen so far and stores the corresponding fitted model and its settings. At the end, it returns the best-scoring model, which you can then use for forecasting on the held-out period.

```python
import numpy as np
import statsmodels.api as sm

def aicc(aic, n, k):
    # AICc = AIC + 2k(k+1)/(n-k-1)
    return aic + (2.0*k*(k+1)) / max(n - k - 1, 1)

def fit_best_sarima(z_train, seasonal_period=7,
                    p_range=range(0,3), d_range=range(0,2), q_range=range(0,3),
                    P_range=range(0,2), D_range=range(0,2), Q_range=range(0,2)):
    best = None
    n = len(z_train)

    for p in p_range:
        for d in d_range:
            for q in q_range:
                for P in P_range:
                    for D in D_range:
                        for Q in Q_range:
                            try:
                                mod = sm.tsa.statespace.SARIMAX(
                                    z_train,
                                    order=(p,d,q),
                                    seasonal_order=(P,D,Q,seasonal_period),
                                    trend="c",
                                    enforce_stationarity=True,
                                    enforce_invertibility=True,
                                )
                                res = mod.fit(disp=False)
                                k = res.params.size
                                score = aicc(res.aic, n=n, k=k)
                                if (best is None) or (score < best["aicc"]):
```

```
                        best = {"res": res, "order": (p,d,q),
                                "seasonal_order": (P,D,Q,seasonal_period),
                                "aicc": score}
                except Exception:
                    pass
    return best
```

Listing 7.2: Seasonal ARIMA grid search with AICc selection using a causal training window.

Moreover, Listing 7.3 shows how to fit a seasonal ARIMA model that can also use extra information beyond the past values of the target series. The inputs are a training series `z_train` (often a transformed version of daily counts) and a matching table of covariates `X_train` (for example mobility, testing volume, weather, or policy indicators). The function builds an ARIMAX model by passing `X_train` as `exog`, which tells the model to learn how these external signals relate to the outcome while still modeling the remaining time dependence with ARIMA-style structure and a weekly seasonal component. After fitting, the code runs a quick "sanity check" on what the model failed to explain: it looks at the leftover errors (residuals) and applies the Ljung–Box test at a few lags (7, 14, and 21 days). This test helps answer a simple question: do the residuals still show a repeating pattern or autocorrelation that the model should have captured? If the test suggests strong remaining structure, it is a warning sign that the ARIMAX specification (orders, seasonality, covariates, or preprocessing) is missing something important, and the forecasts may be systematically biased or miscalibrated.

```
import numpy as np
import statsmodels.api as sm
from statsmodels.stats.diagnostic import acorr_ljungbox

def fit_arimax(z_train, X_train, order=(1,1,1), seasonal_order=(1,0,1,7)):
    mod = sm.tsa.statespace.SARIMAX(
        z_train,
        exog=X_train,
        order=order,
        seasonal_order=seasonal_order,
        trend="c",
        enforce_stationarity=True,
        enforce_invertibility=True,
    )
    res = mod.fit(disp=False)

    # Ljung Box on standardized one step ahead forecast errors
    resid = res.filter_results.standardized_forecasts_error[0]
    resid = resid[np.isfinite(resid)]
    lb = acorr_ljungbox(resid, lags=[7, 14, 21], return_df=True)
    return res, lb
```

Listing 7.3: ARIMAX with exogenous covariates and Ljung–Box residual check.

Probabilistic evaluation is essential because ARIMA-family models typically output Gaussian predictive distributions on the modeled scale, which are then transformed or discretized. In epidemic forecasting, forecasters often communicate predictive quantiles or central prediction intervals rather than full densities, and proper scoring rules for interval and quantile forecasts are therefore operationally important [437]. The weighted interval score (WIS) aggregates interval widths and penalties for miscoverage across multiple central intervals and approximates the continuous ranked probability score when many levels are used. When ARIMA is used as a baseline, computing WIS under rolling-origin evaluation helps distinguish models that look accurate in point forecasts from those that provide useful uncertainty. Listing 7.4 implements a simple way to score probabilistic forecasts when a model reports uncertainty as prediction intervals rather than as a full probability distribution. The first function, `interval_score`, takes the true observed value `y` and one central prediction interval $[lo, hi]$ (a 90% interval for instance) and assigns a penalty. The penalty is small when the interval is narrow and contains the truth, because that means the forecast was both confident and correct. The penalty grows if the interval is wide (reflecting less informative uncertainty) and grows even more if the true value falls outside the interval, with a stronger punishment for higher-confidence intervals (since missing the truth with a 90% interval is worse than missing it with a 50% interval). The second function, `wis`, repeats this calculation for several intervals at different confidence levels (e.g., 50%, 80%, 90%) and then averages the resulting penalties into a single score per time point. In practice, this gives a one-number summary of forecast quality that balances sharpness (not being overly wide) with calibration (actually covering what happened), which is exactly what we want when evaluating epidemic forecasts meant for decision-making.

```
import numpy as np

def interval_score(y, lo, hi, alpha):
    """
    Central (1-alpha) interval score:
    (hi-lo) + 2/alpha * (lo-y) * 1{y<lo} + 2/alpha * (y-hi) * 1{y>hi}
    """
    y = np.asarray(y)
    lo = np.asarray(lo)
    hi = np.asarray(hi)
    score = (hi - lo)
    score += (2.0/alpha) * np.maximum(lo - y, 0.0)
    score += (2.0/alpha) * np.maximum(y - hi, 0.0)
    return score

def wis(y, intervals):
    """
    intervals: list of tuples (alpha, lo, hi) for central (1-alpha) intervals
    Returns WIS per observation as average over intervals, including median term if desired.
    """
    scores = []
    for alpha, lo, hi in intervals:
        scores.append(interval_score(y, lo, hi, alpha))
    return np.mean(np.column_stack(scores), axis=1)
```

Listing 7.4: Weighted interval score (WIS) for a set of central prediction intervals.

Listing 7.5 shows a complete "practice loop" for evaluating a SARIMA forecasting approach in the way it would be used in real time. The helper function `sarima_fit_predict` takes the data available up to a given day (`y_train`), transforms counts to a log scale (to reduce the effect of big peaks and make changes more proportional), automatically chooses a good seasonal ARIMA model using the earlier grid-search routine, and then produces an h-day forecast. It returns two things: a point forecast (`yhat`) converted back to the original count scale, and a set of prediction intervals also converted back to counts so they can be interpreted and scored on the natural scale. The second function, `rolling_sarima_eval`, repeats this process many times in a rolling-origin backtest: it starts with a minimum training length (`min_train`), refits the model at each new time point using only the history up to that point, predicts the next h days, and compares those predictions to what actually happened. It then summarizes performance with two complementary metrics: MAE for point accuracy and WIS for uncertainty quality. The result is a single average MAE and WIS that reflect how the method would behave if it were re-run day after day during an outbreak, including the effects of changing regimes, reporting noise, and the fact that the model must be re-estimated using only information that would have been available at the time.

```
import numpy as np

def sarima_fit_predict(y_train, h, c=0.5):
    z = np.log(np.asarray(y_train, dtype=float) + c)
    best = fit_best_sarima(z, seasonal_period=7)
    res = best["res"]

    fc = res.get_forecast(steps=h)
    m = fc.predicted_mean
    v = fc.var_pred_mean

    # bias-corrected mean on count scale (approx)
    yhat = np.exp(m + 0.5*v) - c

    # build central intervals on z scale, back-transform to y scale
    levels = [0.5, 0.2, 0.1]   # alpha values -> 50%, 80%, 90% intervals
    intervals = []
    for alpha in levels:
        qlo = fc.conf_int(alpha=alpha).iloc[:, 0].to_numpy()
        qhi = fc.conf_int(alpha=alpha).iloc[:, 1].to_numpy()
        lo = np.maximum(np.exp(qlo) - c, 0.0)
        hi = np.maximum(np.exp(qhi) - c, 0.0)
        intervals.append((alpha, lo, hi))
    return yhat, intervals

def rolling_sarima_eval(y, dates, h=7, min_train=90):
    y = np.asarray(y, dtype=float)
    wis_vals = []
    mae_vals = []
    for t in range(min_train, len(y)-h):
        train = y[:t]
```

```
        true = y[t:t+h]
        yhat, intervals = sarima_fit_predict(train, h=h)
        mae_vals.append(np.mean(np.abs(true - yhat)))
        # compute WIS for each horizon point and average
        wis_h = np.mean(wis(true, intervals))
        wis_vals.append(wis_h)
    return float(np.mean(mae_vals)), float(np.mean(wis_vals))
```

Listing 7.5: Rolling-origin evaluation for SARIMA on log-counts with back-transform and WIS.

Continuing with the running example, Fig. 7.4 illustrates a typical real-time backtesting setup for classical linear time-series models: both ARIMA and SARIMA are fitted on the first 80% of the series (left of the vertical split), and then asked to forecast the remaining 20% (right of the split). In the training region, both models track the two epidemic waves reasonably well on average, with SARIMA showing slightly smoother alignment because it explicitly accounts for weekly periodic structure. After the split, the observed incidence (black) drops rapidly, indicating a regime change relative to the local dynamics near the end of the training period. The ARIMA forecast (blue, dotted) extrapolates an upward trajectory and diverges strongly, while the SARIMA forecast (orange, dotted) produces a more plausible continuation with a persistent weekly oscillation but still substantially overestimates the decline. The contrast highlights two points from the section: seasonality-aware structure can improve short-horizon realism (SARIMA vs. ARIMA), yet both models can fail under non-stationarity and structural breaks because they primarily extrapolate patterns encoded in recent history rather than anticipate abrupt shifts in transmission or reporting.

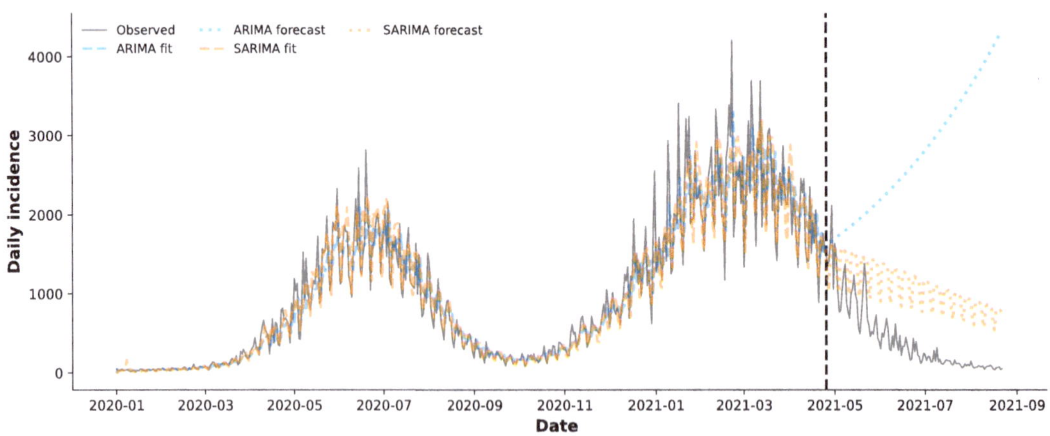

FIGURE 7.4

ARIMA vs. SARIMA on an 80/20 split of a daily incidence series. The vertical dashed line marks the transition from the training period (left) to the evaluation period (right). Model fits on the training segment are shown as dashed lines; forecasts on the held-out segment are shown as dotted lines. Observations are shown in black, ARIMA in blue, and SARIMA in orange.

7.1.4 Component and state-space formulations

A large fraction of time-series forecasting methods can be written as *state-space models*, in which the observed surveillance signal is treated as a noisy measurement of a lower-dimensional latent process that evolves over time [470]. This representation is especially useful in epidemiological surveillance because it provides a concrete separation between slowly varying structure and high-frequency perturbations. It also supports online updating and principled predictive uncertainty through the filtering distribution. Formally, in its linear-Gaussian form, the general state-space model is written as

$$Y_t = \mathbf{Z}_t \boldsymbol{\alpha}_t + \varepsilon_t, \qquad \varepsilon_t \sim \mathcal{N}(0, \mathbf{H}_t), \tag{7.18}$$

$$\boldsymbol{\alpha}_{t+1} = \mathbf{T}_t \boldsymbol{\alpha}_t + \mathbf{R}_t \boldsymbol{\eta}_t, \qquad \boldsymbol{\eta}_t \sim \mathcal{N}(0, \mathbf{Q}_t), \tag{7.19}$$

where $\boldsymbol{\alpha}_t$ is the latent state (trend/seasonal/regression components), $\mathbf{Z}_t$ maps the state to the observation, and $\mathbf{T}_t$ governs temporal evolution. This formulation is flexible enough to encode classical ARIMA models, structural trend-seasonal models, and dynamic regression, often with the same estimation machinery. The computational workhorse is the Kalman filter, which yields the one-step-ahead predictive distribution $p(Y_t \mid \mathcal{F}_{t-1})$ and the filtered state $p(\boldsymbol{\alpha}_t \mid \mathcal{F}_t)$ in $O(m^3)$ per step for an m-dimensional state which is typically small in surveillance settings [471,472].

Writing $\mathbf{a}_{t|t-1} = \mathbb{E}[\boldsymbol{\alpha}_t \mid \mathcal{F}_{t-1}]$ and $\mathbf{P}_{t|t-1} = \mathrm{Var}(\boldsymbol{\alpha}_t \mid \mathcal{F}_{t-1})$, the Kalman filter recursion takes the form

$$\text{Predict:} \quad \mathbf{a}_{t|t-1} = \mathbf{T}_{t-1} \mathbf{a}_{t-1|t-1}, \tag{7.20}$$

$$\mathbf{P}_{t|t-1} = \mathbf{T}_{t-1} \mathbf{P}_{t-1|t-1} \mathbf{T}_{t-1}^{\top} + \mathbf{R}_{t-1} \mathbf{Q}_{t-1} \mathbf{R}_{t-1}^{\top}, \tag{7.21}$$

$$\text{Innovations:} \quad v_t = y_t - \mathbf{Z}_t \mathbf{a}_{t|t-1}, \tag{7.22}$$

$$F_t = \mathbf{Z}_t \mathbf{P}_{t|t-1} \mathbf{Z}_t^{\top} + \mathbf{H}_t, \tag{7.23}$$

$$\text{Update:} \quad \mathbf{K}_t = \mathbf{P}_{t|t-1} \mathbf{Z}_t^{\top} F_t^{-1}, \tag{7.24}$$

$$\mathbf{a}_{t|t} = \mathbf{a}_{t|t-1} + \mathbf{K}_t v_t, \tag{7.25}$$

$$\mathbf{P}_{t|t} = \mathbf{P}_{t|t-1} - \mathbf{K}_t F_t \mathbf{K}_t^{\top}, \tag{7.26}$$

where v_t and F_t define the Gaussian predictive likelihood contribution at time t and enable efficient maximum-likelihood estimation of unknown variances and regression parameters via numerical optimization. Importantly, when y_t is unavailable (or flagged as unreliable), the update step is skipped and only the prediction step is applied, which is operationally attractive for irregular reporting pipelines and backfill patterns common in near-real-time surveillance.

A basic structural model is the local-level (random-walk) model,

$$Y_t = \mu_t + \epsilon_t, \qquad \epsilon_t \sim \mathcal{N}(0, \sigma_\epsilon^2), \tag{7.27}$$

$$\mu_t = \mu_{t-1} + \eta_t, \qquad \eta_t \sim \mathcal{N}(0, \sigma_\eta^2), \tag{7.28}$$

which treats the latent trend as a stochastic baseline with smoothness controlled by the signal-to-noise ratio $\sigma_\eta^2/\sigma_\epsilon^2$. In surveillance terms, μ_t often captures the slowly moving "true" burden signal after averaging out weekday-driven administrative oscillations. A slightly richer structural trend is the local linear trend model,

$$Y_t = \mu_t + \epsilon_t, \tag{7.29}$$
$$\mu_t = \mu_{t-1} + b_{t-1} + \eta_t, \tag{7.30}$$
$$b_t = b_{t-1} + \zeta_t, \tag{7.31}$$

where b_t is a latent slope that can drift over time, which is helpful when waves exhibit multi-week accelerations and decelerations that are not well-approximated by a simple random walk in level [470,473]. Weekly reporting artifacts can be introduced either as deterministic regressors (weekday indicators or Fourier terms) or as stochastic seasonal states. The stochastic seasonal construction is often preferable when the amplitude of the weekday effect changes over time (for example, during laboratory capacity saturation), because it allows the seasonal component to evolve rather than forcing a fixed pattern.

Dynamic regression fits naturally in the state-space form by augmenting the state with regression coefficients that are allowed to vary slowly:

$$Y_t = \mu_t + \mathbf{x}_t^\top \boldsymbol{\beta}_t + \epsilon_t, \tag{7.32}$$
$$\boldsymbol{\beta}_{t+1} = \boldsymbol{\beta}_t + \boldsymbol{\omega}_t, \tag{7.33}$$

with $\boldsymbol{\omega}_t$ controlling the degree of time-variation [474]. This is a pragmatic compromise between purely autoregressive dependence and fully mechanistic drivers: covariates such as testing volume, mobility, and reporting intensity can enter the conditional mean, while the model retains a memory of latent baseline dynamics through μ_t. In epidemiological deployment, this flexibility is both a strength and a failure mode: slowly drifting $\boldsymbol{\beta}_t$ can absorb distribution shift and thus stabilize short-horizon forecasts, but it can also overfit transient relationships that break under policy or behavioral changes unless evaluation explicitly mimics real-time operation.

Exponential smoothing (ETS) models can also be seen as innovations state-space models, which clarifies why they provide fast online updates and coherent predictive distributions despite their heuristic origins [475]. In this view, the familiar smoothing recursions correspond to a particular choice of $\mathbf{T}_t$, $\mathbf{Z}_t$, $\mathbf{Q}_t$, $\mathbf{H}_t$ and an innovations structure, enabling likelihood-based selection and simulation-based prediction intervals in a unified framework [476]. For surveillance, ETS-style models often serve as strong baselines when the objective is short-horizon operational forecasting and the dominant structure is trend plus stable weekly seasonality. The Gaussian assumption in Eq. (7.18) is imperfect for daily counts, but the state-space idea extends to non-Gaussian observation models by replacing the observation equation with an exponential-family likelihood, for example,

$$Y_t \mid \boldsymbol{\alpha}_t \sim \text{Poisson}(\lambda_t), \qquad \log \lambda_t = \mathbf{Z}_t \boldsymbol{\alpha}_t, \tag{7.34}$$

or a negative-binomial analog when overdispersion is pronounced. In such settings, exact Kalman recursions no longer apply, and inference typically proceeds via approximation (Laplace/extended Kalman methods) or simulation-based filtering.

Table 7.4 summarizes common component/state-space formulations used as data-driven baselines and as scaffolding for richer models in epidemiological forecasting.

Following the running example, Fig. 7.5 illustrates two practical reasons state-space models are attractive for surveillance forecasting. In the top panel, the observed daily counts (blue) are treated

Table 7.4 Common component/state-space formulations for surveillance forecasting.

Model family	Latent structure	Observation model	Typical surveillance role
Local level (LL)	Level μ_t as random walk	Gaussian on transformed scale	Robust trend extraction; short-horizon baseline when reporting noise dominates
Local linear trend (LLT)	Level μ_t and slope b_t drifting	Gaussian on transformed scale	Captures accelerating/decelerating waves without explicit mechanistic assumptions
Structural seasonal	Trend + stochastic seasonality (weekly)	Gaussian on transformed scale	Adapts weekday effects whose amplitude changes over time
Dynamic regression (DLM)	Trend + time-varying $\boldsymbol{\beta}_t$	Gaussian on transformed scale	Incorporates covariates while allowing relationships to drift under shift
ETS (innovations SSM)	Level/trend/seasonal states with innovations	Gaussian innovations SSM	Fast automatic forecasting and simulation intervals; strong operational baseline
Poisson/negative binomial state-space	Latent trend in log-rate $\log \lambda_t$	Poisson/negative binomial counts	Count-respecting forecasts via approximate or particle filtering

as a noisy measurement of an underlying latent signal, and the filtered trend estimates (orange for local level, green for local linear trend) track the broad epidemic waves while smoothing some of the day-to-day irregularity. The shaded band around the local-level estimate summarizes uncertainty about the latent signal, making explicit that confidence is higher when the data are stable and lower when fluctuations are large. The bottom panel demonstrates an operational advantage that is common in real reporting pipelines: when observations are missing or unreliable, the filter can simply skip the update step and continue propagating its best estimate forward. During the highlighted gap, the model's uncertainty expands sharply because it is effectively "flying blind," and once data collection resumes, the uncertainty rapidly tightens again as new observations re-anchor the latent state. This behavior matches the subsection's emphasis that state-space formulations separate slow structure from high-frequency perturbations and provide a natural, online notion of predictive uncertainty that responds to data availability.

7.1.5 Reporting delay and nowcasting as a pre-forecast layer

A defining friction in operational epidemic forecasting is that the most recent data points are often not "real" observations of incidence, but provisional aggregates that will be revised upward as delayed reports arrive. This discrepancy is not merely noise: it is a systematic right-truncation mechanism that produces an artificial downturn at the end of the series, precisely where short-horizon forecasting is most sensitive. Nowcasting addresses this problem by estimating the *occurred-but-not-yet-reported* incidence for recent event times using a statistical model of the reporting delay process [477,478]. It is helpful to distinguish event time from report time. Let t index event time, such as symptom onset date, specimen collection date, diagnosis date, or hospitalization date, depending on the surveillance definition. Let T denote the current real time at which the analyst is producing situational awareness products. The quantity of interest is the complete event-time incidence such that Y_t = number of events occurring at event time t, but at real time T one typically observes only

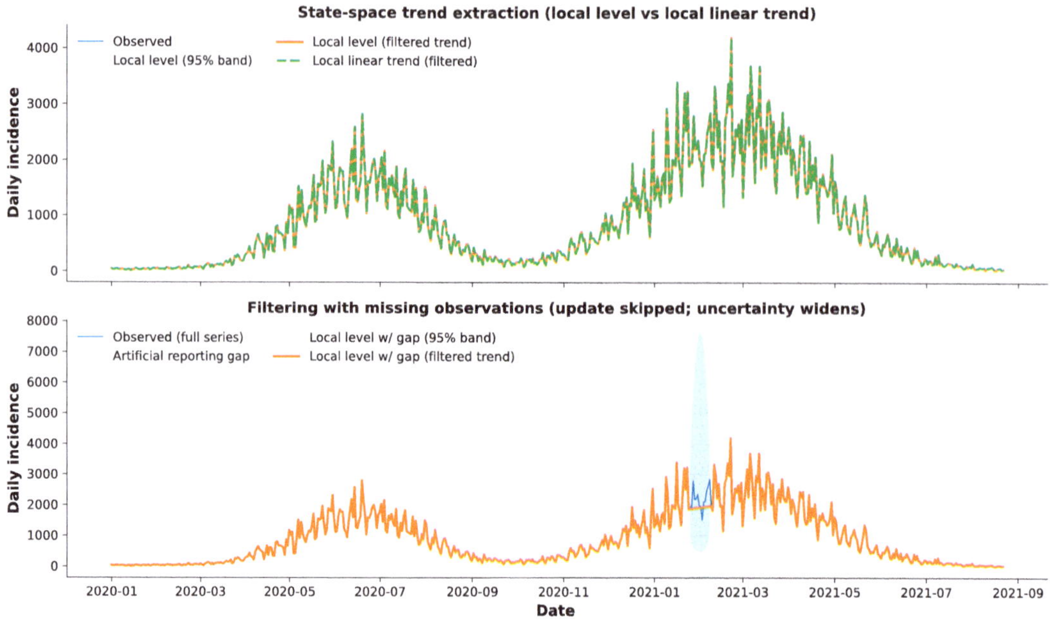

FIGURE 7.5

State-space behavior on the running daily-incidence example. **Top:** local-level and local-linear-trend state-space models produce a smoothed latent signal (with a 95% uncertainty band for the local-level model) overlaid on the observed counts. **Bottom:** the same local-level filter is applied after inserting an artificial reporting gap; during the gap, the filter skips measurement updates and relies on the state evolution, leading to visibly wider uncertainty that contracts once observations resume.

the partial count of events that have already been reported by T. A convenient representation uses the reporting-delay triangle, which tabulates the number of events that occurred at event time t and were reported with delay d days such that $N_{t,d}$ = number of events with event time t reported at time $t+d$. This, the total event-time incidence is $Y_t = \sum_{d=0}^{\infty} N_{t,d}$, while the information available at analysis time T is the right-truncated sum takes the form $Y_{t|T}^{\text{obs}} = \sum_{d=0}^{T-t} N_{t,d}$. All cells with $t+d > T$ are unobserved at time T, and the missing mass

$$U_{t|T} = Y_t - Y_{t|T}^{\text{obs}} = \sum_{d=T-t+1}^{\infty} N_{t,d},$$

is precisely what nowcasting aims to infer. The output of nowcasting is therefore a posterior or predictive distribution for Y_t for recent t close to T, $p\big(Y_t \mid \{N_{s,d} : s+d \le T\}\big)$, together with derived quantities such as a de-lagged series $\tilde{y}_t = \mathbb{E}[Y_t \mid \cdot]$ or predictive intervals for Y_t that reflect both incidence uncertainty and reporting uncertainty [479,480].

Nowcasting can be used as a pre-forecast layer in two distinct ways. In a pipeline approach, one first produces $\tilde{y}_t$ for recent t to remove the end-of-series truncation artifact, and then applies a standard forecasting model to the corrected series. In a joint approach, one fits a single model that couples the delay mechanism to temporal dependence in incidence and produces forecasts of future incidence while simultaneously imputing the recent unreported past [478,479]. The pipeline view is often easier to integrate into existing forecasting systems because it treats nowcasting as a modular data consolidation step, while the joint view is statistically cleaner because it propagates nowcast uncertainty forward into forecast uncertainty. A minimal nowcast begins by modeling the delay distribution. Let $p_{t,d} = \mathbb{P}(D = d \mid \text{event time } t)$ be the probability that an event at time t is reported after d days. Define the cumulative reporting fraction by time T for events at t as $F_t(T-t) = \sum_{d=0}^{T-t} p_{t,d}$. If $p_{t,d}$ were known and $F_t(T-t) > 0$, a natural completion-factor estimator is $\widehat{Y}_t^{\text{CF}} = Y_{t|T}^{\text{obs}}/F_t(T-t)$. This captures the basic logic of delay correction and is useful as a diagnostic baseline, but it highlights the two central statistical difficulties. First, $p_{t,d}$ is unknown and must be estimated from partially observed triangles in which the tail of the delay distribution is itself censored for recent t [477,481]. Second, $p_{t,d}$ typically varies with t due to operational changes, laboratory backlogs, altered testing practices, or changes in case definitions, so borrowing strength from the past introduces a bias–variance trade-off that must be managed explicitly [478,480].

Bayesian nowcasting models address these issues by treating the triangle $\{N_{t,d}\}$ as the primary data object and placing probabilistic structure on both the delay mechanism and the underlying epidemic curve. A common building block assumes that, conditional on the total event-time incidence Y_t and delay probabilities $\mathbf{p}_t = (p_{t,0}, \ldots, p_{t,D_{\max}})$ for a suitably chosen $D_{\max}$, the delay-specific counts follow a multinomial allocation, $(N_{t,0}, \ldots, N_{t,D_{\max}}) \mid Y_t, \mathbf{p}_t \sim \text{Multinomial}(Y_t, \mathbf{p}_t)$. At analysis time T, only the first $T-t+1$ components are observed. The truncation can be handled by re-normalizing the observed part of $\mathbf{p}_t$ and treating $Y_{t|T}^{\text{obs}}$ as a binomial thinning of Y_t with success probability $F_t(T-t)$. This perspective connects nowcasting to classical occurred-but-not-reported (OBNR) estimation and clarifies why uncertainty grows quickly near the present: when $F_t(T-t)$ is small, even modest uncertainty in $p_{t,d}$ induces large multiplicative uncertainty in Y_t [477,481].

The remaining ingredient is a model for the evolution of Y_t over event time. A purely data-driven choice is to assume a smooth latent epidemic curve on the log scale, often via a random walk or a penalized spline $\log \lambda_t = \log \lambda_{t-1} + \xi_t$, $\xi_t \sim \mathcal{N}(0, \sigma_\xi^2)$, with an observation model $Y_t \mid \lambda_t \sim \text{NegBin}(\lambda_t, \kappa)$ to accommodate overdispersion in incidence [477,478]. This is not a mechanistic transmission model; it is a temporal regularization prior that prevents the nowcast from fluctuating wildly due to sparse recent reporting, while still allowing waves and turning points if supported by the data. Höhle and an der Heiden combine a negative binomial count model with flexible delay modeling and explicit treatment of truncation, emphasizing proper scoring rule evaluation during an outbreak and highlighting that the delay distribution itself can change due to interventions or operational shifts [477]. This point is not academic: in practice, changes in reporting pipelines can create false apparent declines, and nowcasting is precisely an attempt to correct that artifact.

NobBS (Nowcasting by Bayesian Smoothing) operationalizes these ideas in a generalizable framework that couples a time-smoothed epidemic curve with a delay distribution learned from a moving historical window [478]. The moving window is an explicit knob that trades adaptability to changing delays against statistical stability when data are sparse. In an evaluation using New York City COVID-19 case line lists, Greene et al. show that shorter windows and negative binomial observation models improved accuracy and interval coverage relative to longer windows and Poisson assumptions, and that day-of-week effects can matter because diagnosis and reporting processes are themselves weekly [480]. A related hierarchical Bayesian nowcast for Bavaria incorporates weekday effects and time-varying delay distributions, and demonstrates retrospective validation using synthetic and historical snapshots while producing daily operational outputs for public health use [479]. These applied evaluations reinforce a practical lesson: for horizons of a few days, the dominant error source is often not the forecasting model but the consolidation model, meaning that improving nowcasting can yield larger gains than switching from one forecasting architecture to another. This motivation has been quantified directly in forecasting contexts. Ma et al. analyze reporting delays in COVID-19 time series and propose a framework to detect and remove such delays, reporting improved forecast accuracy after delay correction [482]. Although their emphasis differs from line-list nowcasting, the conceptual conclusion aligns with the pre-forecast framing: correcting the measurement process can shift the entire forecast evaluation regime by changing what the model sees near the present.

From a technical perspective, the implementation proceeds in two steps. First, Listing 7.6 converts the raw line-list data into a reporting triangle. The function `build_triangle` takes vectors of event dates and report dates, computes the reporting delay for each case as $d = r - e$, and stores the number of cases with event day t and delay d in the triangle entry $N_{t,d}$. In this matrix, each row corresponds to an event day and each column corresponds to a reporting delay, so the triangle provides a compact summary of how cases are gradually reported over time. Delays outside the chosen maximum $D_{\max}$ are ignored, which keeps the representation focused on the delay range of practical interest. Second, Listing 7.7 uses this triangle to produce a simple completion-factor nowcast. The function `empirical_delay_pmf` estimates the delay distribution from older, nearly complete rows of the triangle, using a trailing window of past data and excluding the most recent rows that may still be incomplete. This produces an empirical probability mass function p_d for the reporting delay. The function `completion_factor_nowcast` then uses these probabilities to inflate the currently observed partial counts. For each event day t, it sums the delays that are observable by the current analysis time, computes the cumulative reporting probability up to that delay, and divides the observed count by this fraction. In other words, if only a proportion of cases is expected to have been reported so far, the method rescales the partial total to estimate the final completed count. This makes the truncation adjustment explicit and easy to interpret, even though it remains a simple baseline method.

```
import numpy as np

def build_triangle(event_day, report_day, T=None, Dmax=21):
    event_day = np.asarray(event_day, dtype=int)
    report_day = np.asarray(report_day, dtype=int)
    if T is None:
        T = int(report_day.max())

    t_min = int(event_day.min())
    t_max = int(event_day.max())
    n_t = t_max - t_min + 1

    tri = np.zeros((n_t, Dmax + 1), dtype=int)
    for e, r in zip(event_day, report_day):
        d = r - e
        if 0 <= d <= Dmax:
            tri[e - t_min, d] += 1

```

```
    return tri, t_min, T
```

Listing 7.6: Constructing a reporting triangle from event and report dates.

```
import numpy as np

def empirical_delay_pmf(tri, now_row, window=28, Dmax=21, min_complete_delay=14):
    start = max(0, now_row - window - min_complete_delay)
    end = max(0, now_row - min_complete_delay)
    hist = tri[start:end]  # older rows
    counts = hist.sum(axis=0).astype(float)
    pmf = counts / max(1.0, counts.sum())
    pmf = pmf[:Dmax+1]
    pmf /= pmf.sum()
    return pmf

def completion_factor_nowcast(tri, now_row, pmf):
    Dmax = len(pmf) - 1
    Yhat = np.full(now_row + 1, np.nan, dtype=float)

    for t in range(now_row + 1):
        max_obs_d = min(Dmax, now_row - t)    # how many delays are observable by 'now'
        obs = tri[t, :max_obs_d+1].sum()
        F = pmf[:max_obs_d+1].sum()
        Yhat[t] = obs / max(F, 1e-12)

    return Yhat
```

Listing 7.7: Completion-factor nowcast with an empirical delay distribution.

Evaluation of nowcasting should mimic deployment. The relevant question is not "how close is the nowcast to the final curve when using finalized data," but "how close would the nowcast have been on each day, given only what was known then." This requires retaining snapshots or reconstructing historical triangles from revision logs. Listing 7.8 sketches a snapshot-based evaluation loop, analogous to rolling-origin forecasting but with the target being the finalized counts for recent event days rather than future incidence. This evaluation protocol is essential for choosing moving-window lengths, deciding whether day-of-week adjustments are necessary, and quantifying predictive interval calibration [477,480].

```
import numpy as np

def evaluate_nowcast(triangle_snapshots, nowcast_fn, horizon=14):
    trues = []
    preds = []
    for tri, now_row, final_Y in triangle_snapshots:
        # final_Y is the finalized event-time incidence for rows 0..now_row
        yhat = nowcast_fn(tri, now_row)

```

```
        # score only the last 'horizon' event days (most affected by truncation)
        start = max(0, now_row - horizon + 1)
        trues.append(final_Y[start:now_row+1])
        preds.append(yhat[start:now_row+1])

    return trues, preds
```

Listing 7.8: Snapshot-based evaluation loop for nowcasting under reporting delay.

Seen as a pre-forecast layer, nowcasting changes the effective boundary condition of forecasting. Without delay correction, the most recent observed points represent a biased measurement of incidence, and many forecasting models interpret the truncation-driven downturn as a true decline. With delay correction, the end of the series becomes a posterior distribution that is typically wider but centered closer to the eventual truth. This trade-off is desirable: it replaces systematic bias with quantifiable uncertainty. As a result, the forecast task becomes more stable, and differences between competing forecasting architectures may shrink relative to the gains from improved consolidation modeling [480,482].

Following the running example, Fig. 7.6 visualizes the core motivation for nowcasting as a pre-forecast layer. Near the present, the "observed at time T" event-time counts are systematically too small because many events have occurred but have not yet been reported, creating an artificial downturn exactly where short-horizon forecasting is most sensitive. The nowcast curve uses a delay-based completion adjustment to lift the tail upward toward what the finalized curve will eventually become. The key message is that the discrepancy at the end is not random noise but a structured measurement artifact; correcting it changes the boundary condition that forecasting models see, replacing a biased recent history with an uncertainty-aware estimate that is typically closer to the eventual truth.

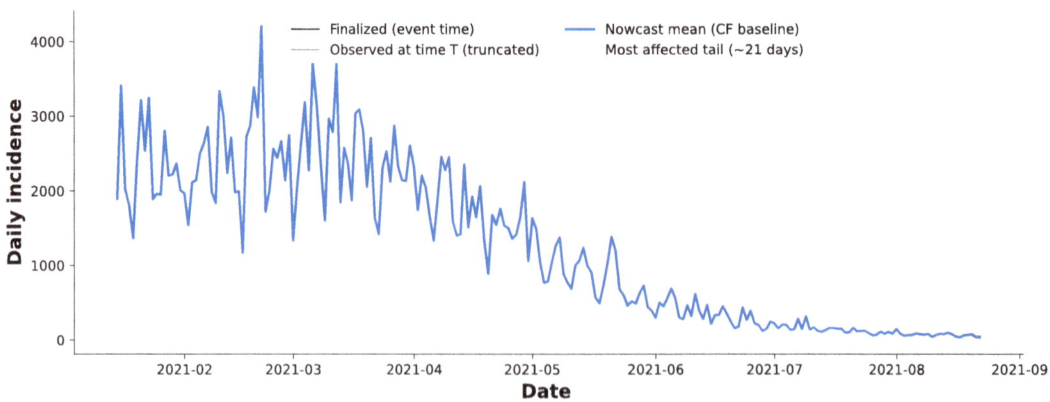

FIGURE 7.6

Right-truncation at the end of the series and its correction via nowcasting (completion-factor baseline). The shaded region highlights the most delay-affected tail (here, roughly the last 21 days).

Fig. 7.7 shows the data object that nowcasting models are built around: the reporting-delay triangle, where each cell represents how many events with a given event date were reported after a given delay. The diagonal boundary is the operational constraint imposed by time: at analysis time T, only cells with $t + d \leq T$ are observable, while the region beyond the boundary is missing by design. The concentration of mass at small delays reflects rapid reporting for many events, while the visible tail at larger delays reflects backfill and late reporting. This figure makes the right-truncation mechanism concrete and clarifies why the newest event times are most uncertain: they have the largest unobserved region and therefore the largest "missing mass" that must be inferred.

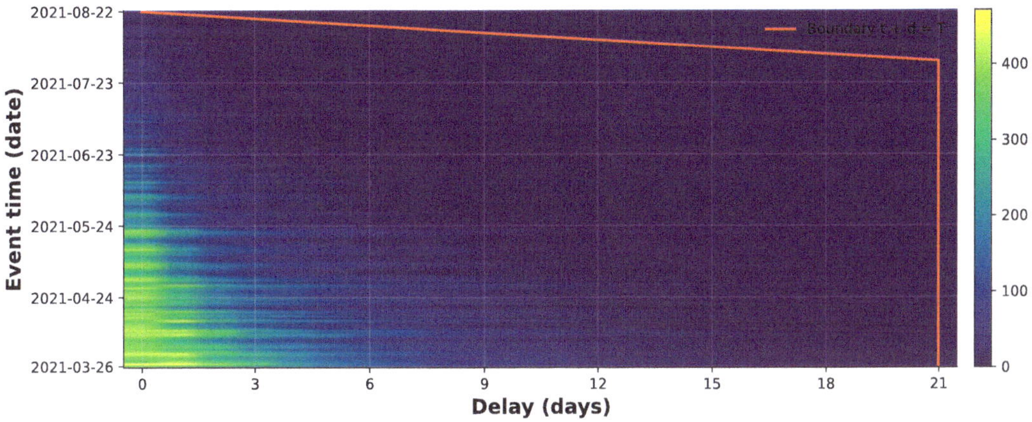

FIGURE 7.7

Reporting-delay triangle as a heatmap of counts by event time and reporting delay. The diagonal boundary marks the truncation line $t + d = T$ separating observed cells (below/left) from unobserved cells (above/right) at analysis time T.

Fig. 7.8 compresses the delay process into an intuitive one-dimensional view. The cumulative reporting fraction increases with the age of the event time because more late reports have had time to arrive; conversely, the completion factor (the multiplicative adjustment needed to "complete" the partial count) is largest for the most recent days and rapidly approaches 1 as event times become older. This emphasizes a central point from the subsection: nowcast uncertainty and instability are inherently largest near the present, because small changes in the estimated delay distribution translate into large multiplicative changes in the completed incidence. It also explains why improving delay modeling can have outsized impact on short-horizon forecasting: the forecast boundary is determined by the most recent few days, where completion factors are most extreme.

Fig. 7.9 illustrates how nowcasting performance should be assessed under real-time constraints and why design choices in the delay model matter most for the newest event times. The horizontal axis is "age since event," so the leftmost points correspond to the most recent days, which are the most truncated and therefore the hardest to estimate. Errors are largest at age 0–2 days and decay as age increases because more of the eventual reports have already arrived. The two curves compare different window lengths for estimating delays; the difference is most pronounced near the present, reflecting the bias–variance trade-off discussed in the text. The figure reinforces that the most consequential nowcasting decisions are those that govern the final few days of the series, since this is where both the truncation artifact is strongest and where downstream forecasts are most sensitive.

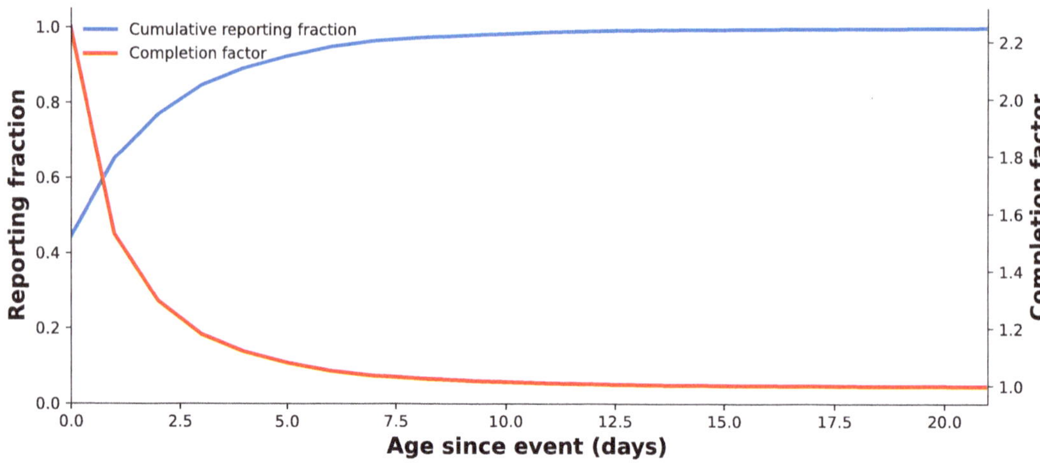

FIGURE 7.8

Empirical delay behavior summarized by the cumulative reporting fraction and the implied completion factor as a function of event-time age.

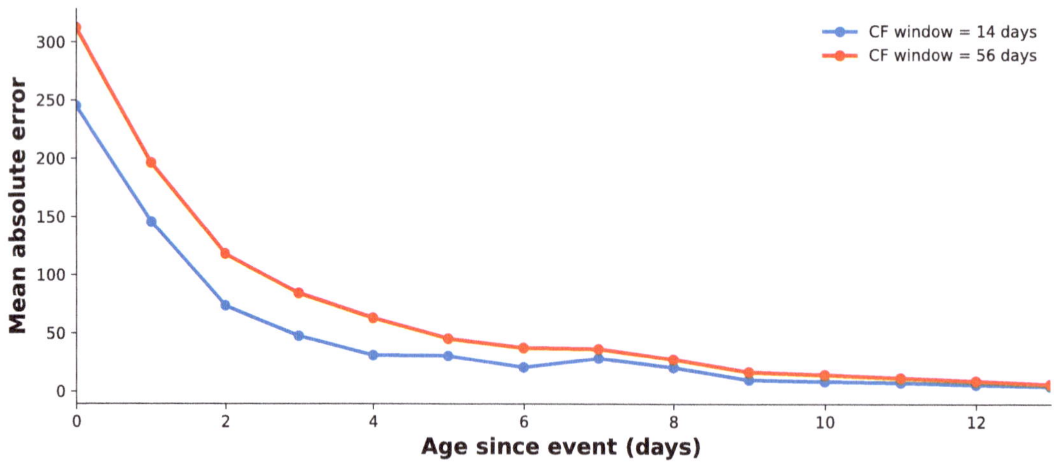

FIGURE 7.9

Snapshot-style evaluation of nowcasting: mean absolute error versus event-time age for two historical windows used to estimate the delay distribution.

7.2 Machine learning approaches

The previous section focused on "classical" time-series modeling, where stochastic structure is made explicit through parametric assumptions. Machine learning (ML) offers a complementary perspective – instead of specifying a small parametric family and estimating its parameters, we specify a function class that is flexible enough to approximate complex relationships, and we learn a particular function by optimizing predictive performance on observed data [483–486]. We represent the observed dataset as $\mathcal{D} = \{(\mathbf{x}_i, y_i)\}_{i=1}^{n}$, where $\mathbf{x}_i \in \mathbb{R}^p$ denotes a vector of predictors (also called *features* or *covariates*) and y_i is the target variable (also called the *label* or *outcome*). In epidemiology, y_i may be a continuous quantity (e.g., weekly incidence rate), a count (e.g., admissions), a category (e.g., outbreak vs. no outbreak), or a structured object (e.g., a future trajectory over multiple horizons). The goal of supervised learning is to learn a function $f(\cdot)$ such that $f(\mathbf{x})$ predicts y accurately on new, unseen data. This formalization is similar to the previous one as the underline temporal prediction problem is identical. Two elements are central. First, a *loss function* $\ell(y, f(\mathbf{x}))$ that quantifies predictive error (e.g., squared error for regression, log-loss for probabilistic classification). Second, a *learning procedure* that selects f from a hypothesis class $\mathcal{F}$ by minimizing an empirical objective, often of the form

$$\hat{f} \in \arg\min_{f \in \mathcal{F}} \frac{1}{n} \sum_{i=1}^{n} \ell(y_i, f(\mathbf{x}_i)) + \lambda\, \Omega(f),$$

where $\Omega(f)$ is a regularization term and $\lambda \geq 0$ controls the trade-off between data fit and model complexity.

The advantage of ML-based models is the representation richness in terms of the functions one can use for approximation. Despite its clear advantages, this richness pose some challnages. For example, a model can fit the training set extremely well yet fail to predict future observations-a phenomenon known as "overfitting" [487]. In ML, the primary objective is "generalization": performance on new data drawn from the same (or sufficiently similar) process. This is typically assessed by splitting data into training and test sets, or by cross-validation. For epidemiological time series, evaluation must respect temporal ordering: random shuffles can leak information from the future into the past. Common protocols include rolling-origin evaluation, blocked cross-validation, and horizon-specific scoring (e.g., evaluating 1-week-ahead and 4-week-ahead forecasts separately). Theoretical perspectives that formalize why complexity control and proper validation matter are a cornerstone of modern ML [488].

Many ML methods used in infectious-disease forecasting can be understood as a principled extension of the classical linear-model pipeline: one begins by rewriting prediction at time t as a supervised learning problem, where each observation is an input-output pair $(\mathbf{x}_t, y_t)$ and the feature vector $\mathbf{x}_t$ is constructed from lagged outcomes $\{y_{t-1}, \ldots, y_{t-k}\}$ together with exogenous signals, thereby converting temporal dependence into a regression design matrix [489]. For epidemiological outcomes that are not well modeled by Gaussian errors, the same construction naturally leads to GLMs with linear predictor $\eta_t = \beta_0 + \mathbf{x}_t^\top \beta$ and link $g(\mathbb{E}[Y_t \mid \mathbf{x}_t]) = \eta_t$, so that much of ML can be viewed as operating "around" the linear predictor while changing how features and complexity are handled [490]. The first major extension is representational: rather than committing to a small, hand-specified set of predictors, one expands $\mathbf{x}_t$ to a possibly high-dimensional feature map $\phi(\mathbf{x}_t)$ that includes transformations, interactions, and domain-motivated summaries, and then uses systematic feature selection to control redundancy and improve generalization [491]. The second major extension is statistical regularization: instead of unpenalized estimation, ML typically solves a penalized empirical-risk problem

$$\min_{\beta} \sum_{t} \ell\big(y_t, \beta^\top \phi(\mathbf{x}_t)\big) + \lambda\, \mathcal{P}(\beta),$$

where the penalty $\mathcal{P}$ constrains effective model complexity, stabilizing estimation under collinearity and noisy, high-dimensional covariates; the classical ℓ_2 penalty (ridge regression) shrinks coefficients continuously to reduce variance [492], whereas the ℓ_1 penalty (lasso) additionally induces sparsity, performing variable selection as part of fitting. Practical ML for linear/GLM backbones is then often about navigating the regularization path and tuning λ efficiently and reproducibly, which is supported by modern coordinate-descent algorithms for convex penalties (including elastic-net mixtures) [493]. Finally, as the function class is enriched beyond "linear in engineered features" (e.g., by allowing adaptive partitioning or learned representations), ML becomes most distinctive: nonlinear learners can capture interactions and heterogeneous effects without requiring them to be enumerated a priori, while still adhering to the same predictive-modeling workflow of preprocessing, resampling, hyperparameter tuning, and out-of-sample evaluation [494].

In what follows, we introduce a small set of widely used machine-learning model families for data-driven epidemiological analysis and forecasting: (i) regularized regression (as a disciplined extension of linear time-series models), (ii) tree ensembles (random forests and gradient-boosted trees), (iii) kernel methods (support vector regression), and (iv) Bayesian nonparametrics (Gaussian processes).

Regularized regression (LASSO). A direct bridge from classical linear modeling to machine learning is to retain a linear predictor while expanding the feature space and controlling complexity through penalization. With features $\mathbf{x}_t$ (e.g., lags of incidence and external signals), the LASSO estimator solves

$$\hat{\boldsymbol{\beta}} = \arg\min_{\boldsymbol{\beta}} \left\{ \frac{1}{n} \sum_{t=1}^{n} (y_t - \boldsymbol{\beta}^{\top}\mathbf{x}_t)^2 + \lambda \|\boldsymbol{\beta}\|_1 \right\},$$

where the ℓ_1 penalty induces sparsity and acts as automatic feature selection, which is especially useful when candidate predictors (search queries, mobility indicators, reporting artefacts) are numerous and collinear [495]. In epidemiological nowcasting, Yang et al. proposed ARGO (AutoRegression with Google search data), which couples autoregressive terms with a dynamically re-estimated sparse regression over search-query features to track U.S. influenza-like illness in real time, reporting improved accuracy relative to earlier search-based systems (including Google Flu Trends) while remaining statistically grounded and scalable [496].

Random forests are nonparametric ensembles of decision trees built via bootstrap resampling and randomized feature selection at each split, which reduces correlation among trees and stabilizes predictions by averaging [497]. Because trees naturally capture nonlinearities, threshold effects, and interactions (e.g., climate-by-season or surveillance-by-region effects) without manual specification [498–501], random forests are well-suited to heterogeneous epidemiological covariates. As an applied example, Zhao et al. compared random forests to artificial neural networks for forecasting dengue burden at national and sub-national scales in Colombia, using surveillance data and covariates; they evaluated performance across 1–12 week horizons and reported mean absolute errors ranging roughly from 9.32 to 24.56 (depending on horizon/setting), with particularly strong short-horizon performance [502].

Gradient-boosted decision trees (XGBoost). Boosting constructs an additive predictor $f(\mathbf{x}) = \sum_{m=1}^{M} \eta h_m(\mathbf{x})$ by fitting weak learners h_m sequentially to reduce a specified loss, thereby performing gradient-like optimization in function space; modern implementations add regularization, shrinkage, and efficient tree growth to improve generalization [503]. For epidemiological forecasting where many weak signals jointly predict short-term dynamics, this stagewise refinement can yield strong accuracy on tabular "feature stacks". Chang et al. used XGBoost with internet search data to predict short-term counts of COVID-19 positive tests, and in their evaluation, they reported a 3-day-ahead RMSE of 803.85 and MAPE of 9.96%, outperforming alternatives including random forests and ARIMA/ARIMAX baselines [504].

Support vector regression (SVR) extends the maximum-margin principle to regression by fitting a function that is both flat and accurate within an ε-insensitive tube, typically written (in primal form) as minimizing $\frac{1}{2}\|w\|^2 + C\sum_i(\xi_i + \xi_i^*)$ subject to ε-tube constraints, with nonlinear structure introduced through kernels $K(\mathbf{x}, \mathbf{x}')$ [505]. SVR can be effective when sample sizes are moderate and nonlinear relationships are expected (e.g., weather–vector–incidence links), but it requires careful kernel and hyperparameter selection and does not naturally yield calibrated predictive uncertainty. In an illustrative dengue application, Tanawi et al. modeled dengue incidents in DKI Jakarta using SVR with weather variables and lagged incidence (selected via cross-correlation), comparing linear and radial basis kernels and reporting that the linear-kernel SVR performed better under RMSE/MAE evaluation for this setting [506].

Gaussian process (GP) regression. Gaussian processes place a prior directly over functions, $f \sim \mathcal{GP}(m(\cdot), k(\cdot,\cdot))$, which yields closed-form posterior predictions under Gaussian observation models and naturally provides predictive uncertainty via the posterior variance [507]. Kernel choice encodes assumptions about smoothness, periodicity, and long-range dependence, while the Bayesian formulation supports principled uncertainty quantification—a key desideratum for decision-making under epidemic risk. Johnson et al. developed a heteroskedastic GP approach for dengue forecasting in the context of a U.S. government-sponsored forecasting challenge, emphasizing flexible nonlinear "trajectory matching" across seasons; they reported that their GP-based methods were among the top performers across multiple targets/locales (e.g., peak timing and magnitude), often outperforming a more classical autoregressive GLM comparator under the contest scoring criteria [508].

Table 7.5 provides a practitioner-oriented synthesis of the model families introduced above, aligning each method with the core trade-offs that typically govern epidemiological use: (i) *structural flexibility* (the capacity to represent nonlinearities, interactions, spatial coupling, or long-range temporal dependence), (ii) *data and operational requirements* (amount of historical data, feature engineering burden, computational cost, and retraining speed), and (iii) *decision relevance* (interpretability and uncertainty quantification).

From a coding perspective, below we outline the three main parts of using ML for forecasting: data preparation, model training, and model testing. Listing 7.9 presents a preparation of a supervised learning design matrix from epidemiological time series. The goal is to rewrite forecasting as supervised learning. Each time point t becomes a row with predictors $\mathbf{x}_t$ consisting of k lagged outcomes (an autoregressive backbone) and optional exogenous covariates (e.g., mobility indices, testing intensity, policy indicators). The label is defined as a *future* outcome y_{t+h} via the horizon shift. This construction ensures that when training on early dates and testing on later dates, the model uses only information available at the forecast origin, thereby avoiding leakage from the future into the past. The function returns (X, y) plus the aligned dates to support time-respecting splits and evaluation.

```
import numpy as np
import pandas as pd

```

Table 7.5 Summary of machine-learning model families for data-driven epidemiological forecasting.

Model family	Advantages in epidemiology	Limitations / risks	Use when...
Regularized regression (LASSO)	Interpretable; strong baseline; handles many correlated predictors via sparsity; fast to fit and update	Primarily linear in features; limited extrapolation to complex nonlinear dynamics; sensitive to feature leakage	You need transparency, rapid updates (nowcasting), and a robust first-line model with many candidate signals
Random forests	Captures nonlinearities/interactions; robust to scaling/outliers; good with mixed tabular features	Weaker for extrapolation; variable importance can be misleading under correlation; limited probabilistic outputs	You have moderate data with heterogeneous covariates and want reliable accuracy with minimal tuning
Gradient-boosted trees (XGBoost)	Often state-of-the-art on tabular prediction; captures complex interactions; flexible loss functions	Hyperparameter sensitivity; risk of overfitting without careful validation; less interpretable than linear models	You prioritize predictive accuracy for short-term forecasting with engineered features and structured covariates
Support vector regression	Effective for moderate sample sizes; kernel trick for nonlinear relationships; well-studied optimization	Scaling can be costly; kernel/hyperparameter selection critical; uncertainty not intrinsic	You have limited-to-moderate data and expect smooth nonlinear effects (e.g., climate–incidence), with careful tuning
Gaussian processes	Principled uncertainty; flexible nonlinear regression; kernels encode seasonality/smoothness assumptions	Computational cost for large n; kernel choice can dominate; may underperform with abrupt regime shifts	Uncertainty quantification is essential and data are modest (or can be approximated), especially for risk-aware decisions

```
def make_supervised_frame(df, date_col, y_col, exog_cols=None, lags=14, horizon=7):
    """

    Convert an epidemic time series into supervised learning rows:
      X_t = [y_{t-1},...,y_{t-lags}, exog_t],   target = y_{t+horizon}
    """

    exog_cols = exog_cols or []
    df = df.sort_values(date_col).reset_index(drop=True).copy()

    # Lag features from the outcome (autoregressive "backbone")
    for k in range(1, lags + 1):
        df[f"{y_col}_lag{k}"] = df[y_col].shift(k)

    # Forecast target: shift outcome backward so each row predicts y_{t+h}
    df["target"] = df[y_col].shift(-horizon)

    feature_cols = [f"{y_col}_lag{k}" for k in range(1, lags + 1)] + list(exog_cols)
    out = df[[date_col] + feature_cols + ["target"]].dropna()

    X = out[feature_cols].to_numpy(dtype=float)
    y = out["target"].to_numpy(dtype=float)
    dates = out[date_col].to_numpy()
```

```
    return X, y, dates, feature_cols
```

Listing 7.9: Preparing a supervised learning design matrix from epidemiological time series.

Listing 7.10 shows the training of a regularized ML model with a time-respecting split. This training routine enforces chronological separation between training and testing by splitting the supervised rows according to time order. The chosen estimator is an elastic-net regression, which minimizes a squared-error objective with a convex combination of ℓ_1 and ℓ_2 penalties, thereby stabilizing estimates under multicollinearity and performing automatic feature selection when the ℓ_1 component is strong. Standardization is included in a pipeline to ensure the penalty treats coefficients on comparable scales. Hyperparameters (α, `l1_ratio`) are selected by cross-validation *inside* the training segment, so that tuning does not "peek" at the held-out future observations.

```
from sklearn.pipeline import Pipeline
from sklearn.preprocessing import StandardScaler
from sklearn.linear_model import ElasticNetCV

def train_elastic_net(X, y, dates, train_frac=0.8, random_state=0):
    """
    Train on the earliest train_frac of observations; keep the most recent for testing.
    ElasticNetCV selects (alpha, l1_ratio) by internal cross-validation.
    """
    n = len(y)
    cutoff = int(np.floor(train_frac * n))

    # Time-respecting split: earlier rows for training, later rows for testing
    X_train, y_train = X[:cutoff], y[:cutoff]
    X_test,  y_test  = X[cutoff:], y[cutoff:]
    test_dates = dates[cutoff:]

    model = Pipeline([
        ("scaler", StandardScaler(with_mean=True, with_std=True)),
        ("enet", ElasticNetCV(
            l1_ratio=[0.1, 0.5, 0.9, 1.0],        # 1.0 corresponds to LASSO
            alphas=np.logspace(-4, 1, 30),
            cv=5,
            random_state=random_state
        ))
    ])

    model.fit(X_train, y_train)
    return model, (X_test, y_test, test_dates)
```

Listing 7.10: Training a regularized ML model with a time-respecting split.

Finally, Listing 7.11 presents a testing and reporting standard forecasting error metrics. The held-out test segment corresponds to the most recent dates (i.e., the operationally relevant regime). The model is applied once to the test design matrix to generate point forecasts $\hat{y}_{t+h}$. We summarize accuracy using

MAE (scale-sensitive and robust), RMSE (penalizes large errors), and MAPE (scale-normalized, with a small denominator safeguard). In practice, one may repeat this evaluation in a rolling-origin scheme (walk-forward testing) to obtain horizon-specific performance profiles and to reduce dependence on a single train/test split.

```
from sklearn.metrics import mean_absolute_error, mean_squared_error

def evaluate_forecast(model, X_test, y_test):
    yhat = model.predict(X_test)

    mae  = mean_absolute_error(y_test, yhat)
    rmse = mean_squared_error(y_test, yhat, squared=False)
    mape = np.mean(np.abs((y_test - yhat) / np.maximum(np.abs(y_test), 1e-8))) * 100.0

    return {"MAE": mae, "RMSE": rmse, "MAPE(%)": mape}, yhat
```

Listing 7.11: Testing and reporting standard forecasting error metrics.

Table 7.6 summarizes a rolling-origin backtest on the running example incidence series, comparing five non-deep-learning model families (Elastic Net, random forest, gradient-boosted trees, SVR, and a Gaussian process on $\log(1 + y)$) for horizons of 1, 7, and 14 days ahead. For each horizon and model, we report the mean and standard deviation of three common accuracy metrics (MAE, RMSE, and MAPE) across the evaluation folds. Reporting mean $\pm$ standard deviation is important in epidemic time series because different folds can fall into qualitatively different regimes (baseline, rising wave, peak, decline), leading to substantial performance variability even when the modeling pipeline is fixed. Several patterns are apparent. First, *errors increase with the forecast horizon* for all models, which is expected: from 1-day ahead to 14-days ahead, both MAE and RMSE rise markedly, reflecting the difficulty of maintaining accuracy through turning points and across changing growth rates. Second, the *regularized linear baseline (Elastic Net) achieves the best absolute accuracy* in this dataset for every horizon, with the lowest mean MAE and RMSE at 1, 7, and 14 days ahead. Concretely, Elastic Net attains RMSE of 276.0 at $h=1$ and 405.2 at $h=14$, while the tree ensembles are consistently higher (e.g., XGBoost RMSE 297.9 at $h=1$ and 478.5 at $h=14$). This suggests that, for this running example (which is driven largely by autoregressive structure plus systematic calendar/reporting effects), most of the predictive signal is already captured by a linear combination of lagged features with shrinkage, and the additional nonlinear flexibility of tree ensembles does not provide a net gain under the chosen feature set and validation scheme. Third, the *tree ensembles (RF and XGBoost) remain competitive* at short horizons but degrade more at longer horizons, consistent with the idea that highly flexible learners can fit local patterns well yet struggle when dynamics shift (e.g., near peaks or regime changes) unless the feature set explicitly encodes those transitions. Their performance is still reasonable in absolute terms, which is why they are often strong defaults once richer exogenous covariates are introduced (mobility, climate, interventions), where nonlinearities and interactions become more salient than in a purely univariate setting. Fourth, the *Gaussian-process model on* $\log(1 + y)$ *achieves the lowest MAPE* across horizons (e.g., 33.7% at $h=1$), but it performs substantially worse in MAE/RMSE and exhibits very large fold-to-fold variability (notably in MAE and RMSE). This combination typically indicates that the model can track *relative* changes reasonably well (hence lower MAPE), but has difficulty matching the *absolute* scale during high-incidence periods and/or is sensitive to kernel specification

Table 7.6 Rolling-origin backtest summary for the running example. Values are mean ± standard deviation across folds.

Horizon (days)	Model	MAE	RMSE	MAPE (%)
1	ElasticNet	229.4 ± 144.5	276.0 ± 184.7	48.7 ± 51.6
1	RF	250.1 ± 185.6	296.5 ± 213.4	39.4 ± 23.3
1	XGBoost	252.4 ± 193.8	297.9 ± 220.6	34.9 ± 15.9
1	SVR	457.3 ± 514.4	515.5 ± 555.5	80.5 ± 102.5
1	GP(log1p)	451.2 ± 693.3	506.5 ± 723.9	33.7 ± 19.0
7	ElasticNet	256.2 ± 173.7	319.2 ± 231.1	47.0 ± 41.7
7	RF	321.2 ± 223.9	377.3 ± 256.6	58.5 ± 43.7
7	XGBoost	321.2 ± 242.7	374.7 ± 268.9	54.0 ± 36.8
7	SVR	506.3 ± 510.0	565.3 ± 557.5	97.6 ± 122.4
7	GP(log1p)	498.2 ± 680.2	562.9 ± 713.0	42.0 ± 17.8
14	ElasticNet	335.9 ± 210.1	405.2 ± 267.5	64.1 ± 56.0
14	RF	403.4 ± 272.4	472.6 ± 301.7	91.8 ± 87.8
14	XGBoost	408.7 ± 288.1	478.5 ± 313.3	87.3 ± 78.2
14	SVR	584.6 ± 505.6	644.6 ± 556.7	121.6 ± 151.9
14	GP(log1p)	554.5 ± 655.1	632.4 ± 689.9	61.7 ± 35.2

and the back-transformation from the log scale. In practice, this is a reminder that GPs are often most valuable when one needs *probabilistic* forecasts and calibrated uncertainty bands; achieving strong point-forecast accuracy may require more careful kernel design and likelihood choices. Finally, the *SVR performs worst overall* in this configuration, with large mean errors and high variability across folds (e.g., MAPE 80.5% ± 102.5% at $h=1$), which is consistent with SVR being sensitive to feature scaling and hyperparameter choices (kernel width, C, ϵ), especially in nonstationary epidemic series. This does not imply that SVR is unsuitable in general, but rather that it typically benefits from more deliberate tuning and, often, from problem-specific feature construction.

To complement aggregate accuracy tables, we visualize performance in ways that expose how and when models succeed or fail. Fig. 7.10 summarizes error as the forecast horizon increases, Fig. 7.11 conditions errors on a simple epidemic "phase" proxy derived from recent growth, and Fig. 7.12 shows that the identity of the best-performing model can change over time even within a single dataset.

Fig. 7.10 plots RMSE versus horizon for multiple model families. The dominant pattern is monotone degradation: RMSE increases steadily from short horizons to longer horizons, reflecting the growing difficulty of anticipating turning points and regime shifts as the forecast window expands. A second key observation is that relative model ordering is not constant across horizons: while some models are competitive at short horizons, their advantage can erode as h grows, suggesting that model selection should be framed as a *horizon-specific* decision rather than a single global choice. The practical lesson is that epidemiological forecasting pipelines should report accuracy across a range of horizons (not only 1-week ahead), and should expect that feature sets and hyperparameters that work for $h = 1$ may not transfer to $h = 14$ without adjustment.

Fig. 7.11 asks a different question: *what kinds of epidemic situations produce large errors?* The x-axis is a 7-day log-growth proxy g_7, so negative values correspond roughly to declining incidence and positive values to rising incidence; points are also labeled into baseline/rising/declining/peak regimes.

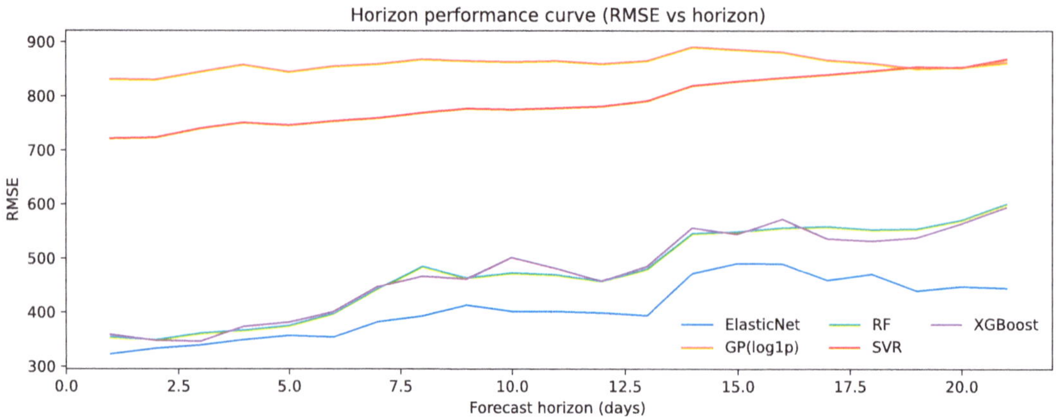

FIGURE 7.10

Horizon performance curve: RMSE as a function of forecast horizon (days) for multiple non-deep-learning ML models on the running example.

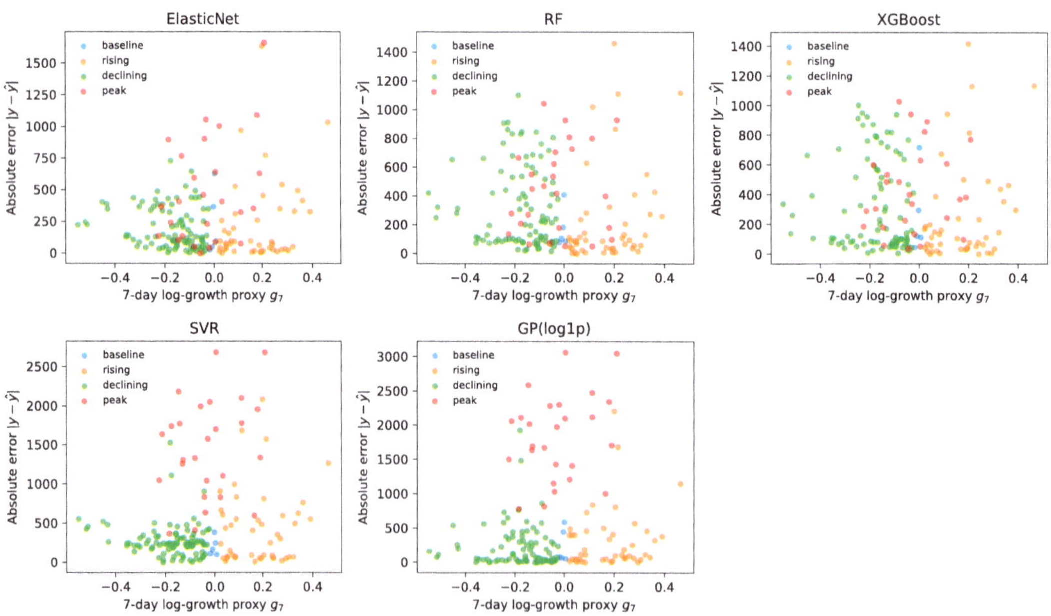

FIGURE 7.11

Error versus epidemic phase proxy: absolute error $|y - \hat{y}|$ plotted against a 7-day log-growth proxy g_7 (colored by baseline/rising/declining/peak), for horizon $h = 7$ days. Panels correspond to different model families.

Across panels, errors are typically smallest near baseline and moderate growth, and the largest errors tend to concentrate in peak and transition periods. This is a core epidemiological insight: operational decisions often depend most on peaks and turning points, precisely where purely lag-driven prediction is hardest. The important modeling takeaway is therefore not only "which method has lower average RMSE," but also "which method is more robust during peaks/rising phases," since those periods may dominate public-health utility. As a result, authors and practitioners should consider reporting stratified performance by phase, or designing loss functions and features (e.g., growth-rate features, peak indicators) that explicitly target these high-stakes regimes.

Fig. 7.12 highlights *nonstationary comparative performance*: rather than a single model winning uniformly, the rolling win-share reveals time windows where different models dominate. In other words, the "best" model is not a fixed property of the dataset; it can change across waves as the data-generating process shifts (e.g., reporting practices, behavioral responses, interventions, or variant-driven dynamics). The main lesson is methodological: evaluation should be time-respecting and should examine stability over time, because a model that performs well on average can still fail systematically during specific windows. Practically, this motivates (i) ensembling or model-switching strategies, (ii) frequent retraining with drift monitoring, and (iii) diagnostic plots that track performance over time rather than only aggregate statistics.

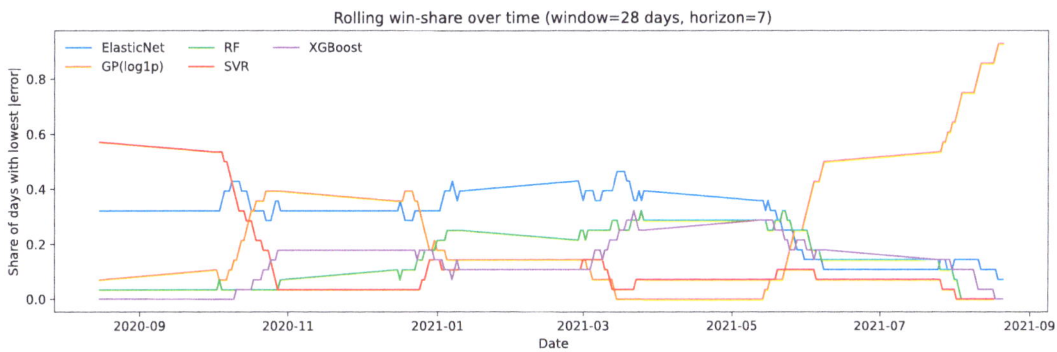

FIGURE 7.12

Rolling win-share over time (horizon $h = 7$ days): for each day, the "winner" is the model with lowest absolute error; the plot shows the rolling (28-day) fraction of days each model wins.

7.3 Deep learning models

Deep learning (DL) extends ML by replacing hand-crafted feature representations with representation learning, where predictors are expressed as compositions of many nonlinear layers that transform raw inputs (e.g., lagged incidence, covariates, and calendars) into increasingly abstract features [509,510]. Concretely, a feed-forward network can be written as $f_{\boldsymbol{\theta}}(\mathbf{x}) = (f_L \circ f_{L-1} \circ \cdots \circ f_1)(\mathbf{x})$, where $\boldsymbol{\theta}$ collects all weights and biases, and each $f_\ell(\cdot)$ is typically an affine map followed by a nonlinearity. This composition allows DL models to approximate complex, high-order interactions and regime changes that are difficult to encode explicitly in traditional regression or kernel methods, which is particularly relevant for epidemic surveillance data, where reporting artifacts, behavioral feedback, and policy shifts induce nonstationarity and heterogeneous effects [511,512]. DL models are trained by minimizing an empirical risk $\sum_t \mathcal{L}(y_t, f_\theta(\mathbf{x}_t))$ using stochastic optimization and automatic differentiation. In practice, modern training relies on adaptive gradient methods (notably Adam [513]) to stabilize updates across parameters with different scales, while regularization techniques and architectural normalization reduce overfitting and improve conditioning, which becomes essential when the model capacity is high relative to the effective sample size (often the case in outbreak data for small regions or short epidemic histories) [514–516].

For epidemic time series, the central modeling challenge is to encode temporal dependence across multiple scales (days to seasons) while remaining robust to noise and revisions. Recurrent neural networks (RNNs), especially long short-term memory (LSTM) architectures, introduce gating mechanisms that mitigate vanishing gradients and support long-range dependency learning in sequential data; temporal convolutional networks (TCNs) provide an alternative based on causal, dilated convolutions that yield long effective memory with parallelizable computation; and attention-based Transformers replace recurrence with learned relevance weights between time points, enabling flexible modeling of delayed and context-dependent effects [517–519]. To this end, a key advantage of DL in epidemiology is that forecasts can be formulated probabilistically rather than as point predictions. This is typically done by letting the network output parameters of a predictive distribution, allowing uncertainty to be propagated to decision-making. Deep autoregressive approaches such as DeepAR combine sequential encoders with likelihood-based training for calibrated predictive distributions; Temporal Fusion Transformers extend this to multi-horizon forecasting with explicit handling of time-varying and static covariates and interpretability mechanisms; and architectures such as N-BEATS emphasize strong performance for univariate forecasting via residual stacks that can also be structured for interpretable decompositions. For uncertainty quantification beyond a single fitted network, deep ensembles provide a simple, scalable approach that often improves calibration and robustness under distribution shift — a common feature of epidemic dynamics across waves [520–523].

In applied infectious-disease forecasting, DL models are most effective when they are embedded in a careful data pipeline that addresses missingness, backfill, changing reporting definitions, and the integration of heterogeneous signals (mobility, testing, syndromic surveillance, and exogenous indicators). A systematic review of ML/DL for infectious-disease prediction highlights both the breadth of tasks (temporal incidence, spatial spread, and risk) and recurring operational gaps such as limited uncertainty reporting and incomplete treatment of missing data. Representative deployed and task-focused examples include EpiDeep for influenza-like illness forecasting via embeddings of historical seasons, DeepCOVID for operational real-time COVID-19 forecasting with principled uncertainty propagation and explainability modules, and spatio-temporal graph neural approaches that encode geographic connectivity (e.g., mobility networks) to improve county-level COVID-19 case prediction [524–527].

Sequential deep models

Sequential deep models are designed for settings where observations arrive in time order and forecasts must condition on a history of past outcomes and covariates. In epidemiological surveillance, this structure is natural: incidence and burden time series exhibit delayed effects (e.g., interventions, behavior, school terms), multi-scale temporal dependence (week-to-week and seasonal), and regime changes across epidemic waves. The common principle is to learn a nonlinear mapping that consumes a sequence $(\mathbf{x}_1, \ldots, \mathbf{x}_t)$ and produces either a one-step or multi-step forecast, often by optimizing a loss over time indices and backpropagating gradients through the sequence computation (backpropagation-through-time).

A standard recurrent neural network (RNN) summarizes the past through a hidden state $\mathbf{h}_t$ updated recursively,

$$\mathbf{h}_t = \phi(\mathbf{W}_x \mathbf{x}_t + \mathbf{W}_h \mathbf{h}_{t-1} + \mathbf{b}), \qquad \hat{y}_{t+h} = g(\mathbf{W}_y \mathbf{h}_t),$$

where $\phi(\cdot)$ is typically a bounded nonlinearity (e.g., tanh) and $g(\cdot)$ maps to the appropriate output domain (e.g., nonnegative incidence). This "shared update rule" across time steps is the core inductive bias: the same parameters explain early and late epidemic phases, while the state $\mathbf{h}_t$ carries forward learned summaries of recent history [528,529]. A known limitation of plain RNNs is that, as the sequence length grows, gradients propagated backward can vanish or explode due to repeated multiplication by Jacobians, making long-term dependency learning difficult without careful stabilization [530]. Practical implementations therefore often include mechanisms such as gradient clipping to control exploding gradients during training [531].

Gated recurrent architectures modify the RNN update to explicitly control information flow, improving learning over longer temporal spans and enabling adaptive memory lengths. Long short-term memory (LSTM) networks introduce a separate cell state $\mathbf{c}_t$ and multiplicative gates (input, forget, output) that regulate writing new information, retaining prior information, and exposing internal memory to the output,

$$\mathbf{i}_t = \sigma(\cdot),\ \mathbf{f}_t = \sigma(\cdot),\ \mathbf{o}_t = \sigma(\cdot),\ \tilde{\mathbf{c}}_t = \tanh(\cdot),\ \mathbf{c}_t = \mathbf{f}_t \odot \mathbf{c}_{t-1} + \mathbf{i}_t \odot \tilde{\mathbf{c}}_t,\ \mathbf{h}_t = \mathbf{o}_t \odot \tanh(\mathbf{c}_t),$$

which stabilizes gradient flow via the (partially) additive evolution of $\mathbf{c}_t$ [517]. Gated recurrent units (GRUs) provide a simpler gating design that merges the cell and hidden state and uses update/reset gates to interpolate between carrying forward past state and overwriting it with new evidence [532]. For epidemiological series, gating is particularly useful because the relevant memory length is context-dependent: near peaks and turning points, recent growth signals and short lags may dominate, whereas in quieter phases, longer baseline and seasonal context may be more informative; gated updates allow the model to learn these shifts from data instead of fixing them by hand.

Temporal convolutional networks (TCNs) replace recurrence with causal convolutions over time, producing a representation at time t from a finite window of past inputs using convolutional filters that are shared across time. Long-range dependence can be achieved efficiently via *dilated* causal convolutions, which expand the receptive field exponentially with depth while keeping computation parallelizable across time steps, and are commonly combined with residual blocks for stable optimization [533]. Compared with recurrent models, TCN-style architectures often train faster (due to parallelism) and can be easier to optimize on long sequences, while still capturing multi-scale temporal patterns by design [518].

These model families have been used directly in epidemiological forecasting and nowcasting tasks. Chimmula and Zhang fit an LSTM-based model to COVID-19 case trajectories in Canada (using publicly reported data) to generate short-term forecasts and analyze outbreak trends [534]. Aiken et al. evaluate a GRU for influenza prediction at multiple spatial resolutions (state- and city-level in the United States) and report that the GRU improves predictive accuracy over baseline machine-learning approaches at longer horizons, while additional Google Trends features did not further improve the GRU in their experiments [535]. For convolutional sequential modeling, Lee et al. study temporal-convolution-based architectures (including TCNs) for influenza-like illness forecasting in the United

States and find that convolutional approaches are feasible and can be comparable to, and at times outperform, plain RNN baselines in multi-horizon settings [536].

Attention-based models

Attention-based deep models replace the strictly local, fixed-order dependence assumptions of classical autoregressive time-series models with a learnable mechanism that dynamically selects which past (and auxiliary) signals matter for predicting the present or future. The central idea is to compute, for each target time index, a content-dependent weighted combination of representations from many other time points, so that long-range dependencies, regime shifts, and context-specific interactions can be expressed without manually specifying lag structures. This flexibility is particularly attractive in epidemiology, where relevant temporal context may depend on changing interventions, behavior, or pathogen evolution, and where multiple surveillance streams can become informative at different times.

The technical foundation is the attention mechanism, originally introduced as a differentiable alignment operator that learns to focus on the most relevant parts of an input when producing each output, thereby mitigating the information bottleneck of fixed-length summaries in sequence models [537]. In modern sequence modeling, this becomes *self-attention*, where queries, keys, and values are all derived from the same input sequence. Given a sequence of hidden vectors arranged as a matrix $X \in \mathbb{R}^{T \times d}$, one forms $Q = XW^Q$, $K = XW^K$, and $V = XW^V$, and computes attention outputs via

$$\mathrm{Attn}(Q, K, V) = \mathrm{softmax}\left(\frac{QK^\top}{\sqrt{d_k}}\right) V,$$

so each time point aggregates information from all other time points, with weights determined by similarity in representation space. Because temporal order matters, attention models require explicit encoding of position or time; relative position representations provide a principled way to parameterize how interactions change with temporal displacement, improving generalization when dependencies depend on distance rather than absolute index [538].

Transformer-style architectures build deep stacks of self-attention and feed-forward blocks, but a practical issue for epidemiological sequences is that the naive attention matrix scales as $\mathcal{O}(T^2)$ in time and memory, which becomes burdensome for long daily series, multi-location panels, or fine-grained mobility streams. Several time-series Transformers therefore modify the attention computation to improve efficiency and inductive bias. Informer, for example, introduces a sparse approximation (ProbSparse attention) aimed at long-horizon forecasting, reducing computation while retaining the ability to capture long-range structure in extended sequences [539]. Autoformer takes a complementary route: it integrates series decomposition directly into the model and replaces pointwise attention with an Auto-Correlation mechanism designed to better exploit seasonality and trend structure for long-term forecasting [540]. In epidemiological practice, these design choices correspond to common data realities: long seasonal baselines (influenza), multi-wave dynamics (COVID-19), and dependence patterns that combine smooth trends with recurring periodicities.

A further complication in surveillance is that sampling can be irregular (e.g., intermittent testing campaigns, weekly administrative reporting, backfills) and observations may be missing or revised. Attention-based models typically handle missingness with masking (so attention weights ignore absent values) and by adding missingness indicators as features, but more specialized methods explicitly couple self-attention with imputation. SAITS is an example that uses self-attention with structured masking to learn imputations and forecasts jointly, which is valuable when downstream predictions are sensitive to systematic gaps or reporting artifacts [541]. For irregular timestamps, one can embed continuous time directly rather than relying on equally spaced positional indices; Time2Vec provides a learnable time representation that can be used as an input embedding so that models can condition on real-valued time gaps and calendar effects more flexibly than with fixed positional encodings [542].

In epidemiology, attention-based sequence models have been used both as forecasting engines and as multi-source fusion models that combine heterogeneous signals. For seasonal influenza, Li et al. propose a Transformer-based approach with a source-selection module that measures curve similarity to fuse multiple data sources and capture spatial dependence, and they evaluate performance against autoregressive and RNN baselines on USA and Japan datasets [543]. For localized COVID-19 forecasting, Wu et al. present a retrospective county-level study using a transformer neural network with multi-source data fusion that incorporates historical cases, deaths, and social-media sentiment, reporting strong multi-week predictive accuracy in their evaluation setting [544]. A complementary example is the influenza-like illness (ILI) forecasting study by Wu et al., which demonstrates that Transformer-based forecasting can be competitive with then-current state-of-the-art approaches on ILI time series, illustrating how attention mechanisms can extract predictive structure from epidemiological surveillance sequences [545].

Spatio-temporal deep learning

Many epidemiological surveillance tasks are inherently *multi-series*: we observe coupled incidence trajectories across regions, age groups, or facilities, where transmission induces *spatial dependence* (who affects whom) and interventions, behavior, and seasonality induce *temporal structure* (how the past affects the future). Spatio-temporal deep learning formalizes this setting by representing each unit (e.g., a county) as a node with time-varying features (e.g., recent incidence, testing volume, covariates), and defining edges that encode proximity or connectivity (e.g., adjacency, commuting, mobility flows). A central building block is the graph neural network (GNN), which updates node states via message passing; in the graph convolutional network (GCN) formulation, one layer can be written as $\mathbf{H}^{(\ell+1)} = \sigma(\tilde{\mathbf{D}}^{-1/2}\tilde{\mathbf{A}}\tilde{\mathbf{D}}^{-1/2}\mathbf{H}^{(\ell)}\mathbf{W}^{(\ell)})$, where $\tilde{\mathbf{A}} = \mathbf{A} + \mathbf{I}$ adds self-loops and $\tilde{\mathbf{D}}$ is its degree matrix, so each region aggregates normalized information from its neighbors to learn spatially-smoothed representations suited for prediction [546]. When connectivity is heterogeneous or context-dependent, attention-based graph layers replace fixed normalization by learned edge weights; graph attention networks compute coefficients α_{ij} from node features and use $\sum_{j\in\mathcal{N}(i)} \alpha_{ij}\mathbf{W}\mathbf{h}_j$ to focus propagation on the most relevant connections (e.g., highly connected commuting corridors or demographically similar regions) [547]. To capture temporal dynamics, spatio-temporal models combine graph propagation with sequence modeling in (at least) two common ways: (i) graph-aware recurrent updates that diffuse information along directed edges (useful when flows are asymmetric) and then roll representations forward through time, as in diffusion convolutional recurrent neural networks that couple diffusion-based graph operators with an encoder–decoder recurrent architecture [548]; and (ii) spatio-temporal convolutional blocks that alternate temporal convolutions (for local-to-medium-range lags) with graph convolutions (for instantaneous spatial coupling), yielding efficient fully-convolutional forecasting architectures such as STGCN [549]. A complementary theme is *learned spatial embeddings*: instead of hand-coding region similarity, models assign each region an embedding vector (either trained end-to-end as a lookup table or pre-trained from a connectivity graph) and let the predictor discover latent spatial structure; random-walk embedding methods such as node2vec generate such vectors by biased walks that preserve both homophily and structural equivalence, which can then be appended as static covariates or used to parameterize spatial effects [550]. Finally, mobility-derived signals naturally enter as *time-indexed* adjacency matrices $\mathbf{A}_t$ (or edge-feature tensors) so that the model learns not only who is connected, but how connectivity changes under policy, season, or behavioral shifts; this is particularly valuable around turning points, where information can propagate from regions that experience trend changes earlier to those that lag behind. Concrete epidemiological deployments illustrate these design choices: Cola-GNN models influenza-like illness (ILI) as a long-horizon forecasting problem over a location graph, combining dynamic message passing with cross-location attention to capture spatial dependence and improve long-term predictions on real ILI datasets from the United States and Japan while providing interpretable cross-location influences [551]. EpiGNN forecasts regional epidemics by encoding transmission risk and learning a region-aware graph that blends geographic dependency, temporal context, and optional external resources; across multiple real-world influenza and COVID-19 datasets, it reports improved accuracy, including a stated reduction in RMSE relative to strong baselines (reported as a 9.48% improvement in RMSE) [552]. HiSTGNN explicitly leverages sequences of dynamic mobility graphs (weekly flow networks between U.S. counties) and introduces a hierarchical spatial encoder (county- and region-level propagation) coupled to a Transformer-style temporal model; on COVID-19 case/death forecasting it demonstrates stronger predictive power than prior baselines while operating on large, evolving mobility graphs [553]. Notably, graph learning is not always necessary: PISID integrates a spatial embedding matrix (region-specific latent factors) with an SIR-based dynamical module, aiming to stabilize learning and retain interpretability; in multi-region COVID-19 forecasting it reports stable, superior performance with a compact parameterization (on the order of 27 K parameters) and fast training [554]. Evidence for the added value of mobility edges is provided by contextual evaluations such as an epiDAMIK extended abstract that compares a heterogeneous spatio-temporal GNN using mobility networks against a matched model without spatial information during Germany's first Omicron wave; mobility-conditioned graphs yield statistically significant improvements on several forecasting metrics and, importantly, improve earlier in time around change points [555]. Beyond single-resolution graphs, multiscale approaches such as temporal multiresolution GNNs learn hierarchical graph coarsenings and use attention to select informative resolutions, reporting competitive results on temporal epidemic datasets including COVID-19 and chickenpox [556].

A practical guide to designing, using, and testing deep learning forecasts for incidence

Deep learning forecasting for epidemiological incidence can be treated as a disciplined pipeline rather than an ad hoc modeling exercise. The core design choice is to specify (i) the *prediction target* (e.g., daily incidence y_{t+h} for one or multiple horizons h), (ii) the *information set* available at forecast time t (past incidence and covariates only, with explicit treatment of reporting delays and backfill), and (iii) the *evaluation protocol* that mirrors operational use (rolling-origin testing rather than random splits). Once these are fixed, the remaining steps become modular: input representation, architecture selection, loss/likelihood, training procedure, and diagnostics [557,558].

The first step is to cast the incidence series into supervised learning samples. Choose a lookback window L (e.g., 28–56 days) and construct input sequences $\mathbf{X}_t = [\mathbf{x}_{t-L+1}, \ldots, \mathbf{x}_t]$, where $\mathbf{x}_t$ may include y_t, calendar indicators (day-of-week, holidays), rolling statistics (e.g., MA_7), and exogenous signals if available. Define a horizon set $\mathcal{H}$ (e.g., $\{1, 7, 14\}$) and a target vector $\mathbf{y}_t = [y_{t+h}]_{h\in\mathcal{H}}$. Two target formulations are common: *direct multi-horizon* prediction (one model outputs all horizons simultaneously) and *iterated* prediction (one-step model rolled forward recursively). Direct multi-horizon training often reduces error accumulation and is compatible with modern sequence models that output a vector of future values [521,522]. In epidemiological settings, it is also often advisable to model a stabilized transform such as $\log(1 + y_t)$ to reduce heteroscedasticity and to ensure nonnegativity after inversion, while remembering that back-transformation changes the implied error structure and should be accounted for in interpretation [559].

Architecturally, a sensible baseline for a single-series task is a sequence encoder followed by a multi-horizon decoder. With an LSTM/GRU encoder, the final hidden state (or an attention-pooled summary) can be mapped to $\hat{\mathbf{y}}_t$ via a small multilayer perceptron; with a TCN, the final convolutional layer at time t serves the same role; with a Transformer, the output embedding at the final position (or a learned aggregation token) can be used for decoding. For most epidemic forecasting use-cases, the model should also incorporate a residual connection or skip path from simple baselines (e.g., persistence y_t or MA_7) so that the network focuses capacity on learning departures from naive dynamics, a strategy aligned with residual learning principles used in deep architectures [560]. If the objective is probabilistic forecasting rather than point prediction, one can train the network to output distribution parameters; for count data, this typically means Poisson or negative-binomial likelihoods, where the network outputs a mean (and optionally dispersion) and is trained by maximizing the corresponding log-likelihood [520]. When only prediction intervals are required, quantile regression losses can be used to estimate $\hat{Q}_\tau(y_{t+h} \mid \mathbf{X}_t)$ for several τ values, providing calibrated empirical intervals when combined with appropriate validation [561].

Training requires careful regularization and early stopping because epidemic time series are often short relative to model capacity, and the data-generating process can shift across waves. Standard practices include dropout, weight decay, and early stopping on a validation window that is strictly later in time than the training window (to avoid temporal leakage). Optimization is commonly performed with adaptive methods such as Adam, and gradient clipping is often beneficial for recurrent models to stabilize training when the sequence contains sharp changes [514,531]. In addition, it is good practice to standardize continuous covariates and, for incidence, to consider scaling by population (rates) if comparisons across regions or strata are intended; for a single-series running example, the emphasis is instead on transforming the marginal distribution (e.g., log1p) and encoding calendar effects explicitly.

Testing should follow a time-respecting protocol that reproduces the operational forecasting loop. A rolling-origin evaluation defines multiple forecast origins $t_1 < t_2 < \cdots$; for each origin t_i, fit (or update) the model on data up to t_i and score predictions on the subsequent horizon(s). This avoids the optimistic bias that arises from random shuffling and matches how real forecasts are produced under evolving dynamics [558]. Performance should be summarized across folds using metrics that reflect both absolute error (MAE/RMSE) and relative error (MAPE or symmetric MAPE), while noting that percentage metrics can behave poorly near zero incidence. If probabilistic forecasts are produced, calibration should be evaluated explicitly via empirical coverage of nominal intervals and proper scoring rules such as the continuous ranked probability score (CRPS) or the weighted interval score (WIS), which have become standard in epidemic forecast evaluation [562,563].

Finally, deep learning models should be subjected to epidemiology-specific diagnostics before being trusted. First, inspect residual structure across time and across epidemic phases (baseline, rising, peak, decline) to detect systematic failures near turning points. Second, perform ablations that remove calendar signals or recent lags to verify that improvements are not simply due to learning reporting artifacts. Third, check robustness under distribution shift by evaluating separately on different waves or seasons. Fourth, confirm that the deep model improves upon strong non-DL baselines (e.g., regularized linear models or boosted trees) under the same validation protocol; in many surveillance settings, this comparison is the deciding factor for deployment. These steps help ensure that the deep model is not only accurate on average, but also reliable during the regime changes that matter most for public-health decision-making.

7.4 Large language models

Large language models (LLMs) are generative models trained to predict the next token in a sequence [564]. Given tokens $w_{1:n}$, an autoregressive LLM defines

$$p_\theta(w_{1:n}) = \prod_{i=1}^{n} p_\theta(w_i \mid w_{<i}), \tag{7.35}$$

and can therefore map heterogeneous inputs (free text, tabular data serialized as text, and structured prompts) into summaries, labels, extracted entities, or forecasts. In epidemiological modeling context, the main role of LLMs is usually not to replace statistical or mechanistic models altogether, but to serve as a reasoning and representation layer for information that is abundant in text but difficult to encode numerically, such as outbreak bulletins, policy announcements, genomic notes, situation reports, and social-media narratives [565–567].

A convenient way to write this interface is

$$(\hat{\mathbf{z}}_t, \hat{y}_{t+h}) = g_\phi(y_{1:t}, \mathbf{x}_{1:t}, \mathcal{T}_t), \tag{7.36}$$

where $y_{1:t}$ is the observed surveillance series, $\mathbf{x}_{1:t}$ denotes standard numeric covariates, and $\mathcal{T}_t$ is a collection of texts available at forecast time t. The LLM may output structured intermediate variables $\hat{\mathbf{z}}_t$ (for example, a variant alert, a hospital-pressure indicator, or extracted case counts from a bulletin) that are then passed to a downstream forecaster, or it may directly produce the forecast itself after the numeric and textual context have been serialized into a prompt. In practice, the former is often easier to validate because extracted fields can be checked against schemas and domain rules before entering the forecasting pipeline.

Recent work illustrates several distinct uses. Consoli et al. used LLMs with in-context learning and an ensemble configuration to extract epidemic information from ProMED and WHO Disease Outbreak News for event-based surveillance [568]. Harris et al. evaluated LLMs on a broad set of public-health classification and extraction tasks and found encouraging performance together with substantial task-to-task variability, underscoring the need for task-specific validation before deployment [569]. Xie et al. used LLMs to identify self-reported COVID-19 infections from social-media data and reported that the resulting signal led official reports by several days, illustrating how LLMs can convert noisy text into a complementary surveillance stream [570]. Kwok et al. demonstrated a different role, namely LLM-assisted model development, where ChatGPT collaborated with a practitioner to co-design and debug a transmission model and estimate key epidemiological parameters [571]. At the forecasting end of the spectrum, Du et al. introduced PandemicLLM, a multimodal framework that combines policy text, genomic surveillance, spatial descriptors, and epidemiological time series for real-time COVID-19 forecasting across U.S. states [572], while Gong et al. proposed EpiLLM, which reformulates spatio-temporal epidemic forecasting as autoregressive next-token prediction with mobility-aware prompt learning [573]. These studies suggest that LLMs are most useful in epidemiology when the bottleneck is not merely nonlinear regression on a clean table, but the need to align heterogeneous evidence sources, convert narrative reports into structured variables, or fuse textual context with standard forecasting inputs. Their main limitations are equally important: outputs are prompt-sensitive, may hallucinate unsupported facts, and can drift when reporting language changes. For this reason, LLM-based epidemiological pipelines should prefer constrained outputs (for example, JSON schemas), retrieval from trusted sources, deterministic post-processing, and human validation for high-stakes decisions.

Practically, Listing 7.12 illustrates a deliberately cautious way to use an LLM in an epidemiological pipeline. Instead of asking the model to make a final forecast directly, the code assigns it a narrower and easier-to-validate task: reading a free-text outbreak bulletin and converting it into a structured JSON object with a predefined schema. This design is useful because outbreak reports are often written in natural language, with important details such as disease name, location, case counts, deaths, and reporting date embedded in prose rather than stored in clean columns. By forcing the model to return only a fixed set of fields, the user can check whether the output is complete, whether the values have the correct types, and whether the extracted information is plausible before it is used anywhere else in the modeling workflow. In other words, the listing demonstrates how an LLM can serve as a controlled information-extraction layer that transforms unstructured text into structured data while keeping the overall system easier to debug and validate. Furthermore, Listing 7.13 then extends the same idea one step further. Rather than extracting only factual fields such as cases or deaths, the LLM is asked to summarize more qualitative information from text into compact numerical indicators, for example whether a bulletin suggests policy tightening, increased hospital pressure, or elevated cross-border risk. These outputs can be interpreted as text-derived covariates: variables that do not come from the numerical surveillance series itself, but from narrative context that may still be epidemiologically relevant. The code then appends these LLM-generated features to standard lagged incidence variables and passes the resulting design matrix to a conventional forecasting model. This is an important pattern because it shows that LLMs need not replace established forecasting methods. Instead, they can complement them by translating textual evidence into machine-readable signals that a traditional forecaster can use alongside ordinary time-series inputs.

```
import json
import re
import torch
from transformers import AutoTokenizer, AutoModelForCausalLM

MODEL_NAME = "Qwen/Qwen2.5-7B-Instruct"

tokenizer = AutoTokenizer.from_pretrained(MODEL_NAME)
model = AutoModelForCausalLM.from_pretrained(
    MODEL_NAME,
    torch_dtype=torch.bfloat16,
    device_map="auto"
)

def llm_json(schema_description, document):
```

```
    messages = [
        {
            "role": "system",
            "content": (
                "You are an epidemiological information extraction model. "
                "Return only one valid JSON object and do not add commentary."
            ),
        },
        {
            "role": "user",
            "content": (
                "Extract structured information from the following document.\n\n"
                f"Schema:\n{schema_description}\n\n"
                f"Document:\n{document}"
            ),
        },
    ]

    prompt = tokenizer.apply_chat_template(
        messages,
        tokenize=False,
        add_generation_prompt=True
    )
    inputs = tokenizer(prompt, return_tensors="pt").to(model.device)

    outputs = model.generate(
        **inputs,
        max_new_tokens=256,
        do_sample=False
    )

    text = tokenizer.decode(outputs[0], skip_special_tokens=True)
    match = re.search(r"\{.*\}", text, flags=re.S)
    if match is None:
        raise ValueError("No JSON object found in model output.")
    return json.loads(match.group(0))

schema = """
{
  "disease": "string",
  "country": "string",
  "subnational_location": "string or null",
  "report_date": "YYYY-MM-DD or null",
  "suspected_cases": "integer or null",
  "confirmed_cases": "integer or null",
  "deaths": "integer or null",
  "hospital_pressure": "low | medium | high | unknown",
  "source": "string",
```

```
  "confidence": "low | medium | high"
}
"""

report = '''
WHO Disease Outbreak News, 2026-03-14. The Ministry of Health reported
23 confirmed cholera cases and 2 deaths in Coastal District. Hospital
admissions increased during the previous week and emergency water-treatment
measures were initiated.
'''

record = llm_json(schema, report)
print(record)
```

Listing 7.12: Schema-constrained extraction of structured outbreak records from free-text reports using an instruction-tuned LLM.

```
import pandas as pd
from sklearn.ensemble import HistGradientBoostingRegressor

def bulletin_to_features(bulletin_text):
    schema = """
    {
      "variant_alert": "0, 1, 2, or 3",
      "policy_tightening": "-1, 0, or 1",
      "hospital_pressure": "0, 1, 2, or 3",
      "cross_border_risk": "0, 1, 2, or 3"
    }
    """
    return llm_json(schema, bulletin_text)

def add_lags(df, y_col="cases", lags=(1, 7, 14)):
    df = df.copy()
    for lag in lags:
        df[f"{y_col}_lag{lag}"] = df[y_col].shift(lag)
    return df

# Example dataframe columns:
# date, cases, bulletin_text
df = add_lags(df, y_col="cases")

ctx = pd.json_normalize(df["bulletin_text"].apply(bulletin_to_features))
ctx = ctx.add_prefix("llm_")

design = pd.concat([df, ctx], axis=1).dropna()

feature_cols = [
    "cases_lag1", "cases_lag7", "cases_lag14",
```

```
    "llm_variant_alert",
    "llm_policy_tightening",
    "llm_hospital_pressure",
    "llm_cross_border_risk",
]

train = design.iloc[:-14]
test = design.iloc[-14:]

model = HistGradientBoostingRegressor(random_state=0)
model.fit(train[feature_cols], train["cases"])

pred = model.predict(test[feature_cols])

out = pd.DataFrame({
    "date": test["date"].to_numpy(),
    "y_true": test["cases"].to_numpy(),
    "y_pred": pred,
})
print(out)
```

Listing 7.13: Using LLM-derived textual context as exogenous features in a conventional short-horizon forecaster.

Deepen your knowledge

Exercise 1: From public-health question to data-driven model class (classical / ML / DL)

Task

For each question below, decide whether a classical time-series model, a feature-based ML model, or a deep sequential model is the most appropriate first choice. For each, write: (i) the forecast target (horizon and scale), (ii) the minimum inputs required at forecast time, and (iii) the primary failure mode you would watch for.

1. Q1: "Forecast national daily hospital admissions 7 days ahead using only the admissions series."
2. Q2: "Nowcast daily cases in the last 14 days under reporting delays using revision/backfill information and calendar effects."
3. Q3: "Forecast county-level cases 14 days ahead using mobility-derived connectivity and cross-county coupling."
4. Q4: "Forecast incidence 28 days ahead using multiple exogenous signals (testing, mobility, weather) with nonlinear interactions and changing relevance over time."

Solution (worked)

A practical decision rule is: use *classical* models when structure is mostly temporal and the input space is small; use *feature-based ML* when predictive information is spread across many engineered signals

Table 7.7 Leakage-safe feature design template for incidence forecasting.

Candidate feature	Available at t?	Justification / leakage-safe alternative
$y_{t-1}, \ldots, y_{t-28}$ (lags)		
$\mathrm{MA}_7(t)$, $\mathrm{MA}_{14}(t)$		
$\Delta_7(t) = \log(1 + \mathrm{MA}_7(t)) - \log(1 + \mathrm{MA}_7(t-7))$		
Day-of-week one-hot		
"Centered" moving average $\mathrm{MA}_7(t+3)$		
Peak indicator computed from full series		

and interactions; use *deep models* when you have enough data to learn representations (multi-series, long histories, rich covariates) or you need flexible long-range dependency handling.

- **Q1: Classical time-series (first choice).** Target: y_{t+7} on the original scale (or $\log(1+y)$ if variance grows with level). Inputs: past admissions only. Watch for: structural breaks and day-of-week reporting artifacts causing biased multi-day forecasts.
- **Q2: ML (first choice), possibly hybrid.** Target: event-time incidence for the most recent 14 days (nowcast triangle). Inputs: partial reports, delay indicators, calendar effects, and backfill features. Watch for: leakage (using finalized values) and overfitting to administrative revisions rather than transmission.
- **Q3: Deep spatio-temporal (first choice).** Target: county i cases $y^{(i)}_{t+14}$ (multi-output across counties). Inputs: per-county recent history plus a connectivity graph from mobility (possibly time-varying). Watch for: graph misspecification and instability under mobility regime changes (policy shifts).
- **Q4: Attention-based deep or strong ML baseline.** Target: y_{t+28} (multi-horizon recommended). Inputs: incidence history and exogenous signals; include time encoding and missingness indicators. Watch for: covariate shift (signals behave differently across waves) and brittle performance around turning points.

Exercise 2: Design a leakage-safe forecasting dataset (what is known at time t?)

Task

You have a daily incidence series y_t (running example) and want to forecast y_{t+h} with $h \in \{1, 7, 14\}$. Fill the table below for a candidate feature set. For each feature, indicate whether it is **available at forecast time** (yes/no) and give a one-sentence justification. If a feature is not safely available, propose a leakage-safe alternative. (See Table 7.7.)

Solution (worked)

All lags and trailing rolling statistics computed up to t are valid; anything that uses future values (even indirectly) is leakage.

- **Lags** ($y_{t-1}, \ldots, y_{t-28}$): *Yes*. They are observed by time t (subject to reporting delay caveats in real surveillance).
- **Trailing MAs** ($\mathrm{MA}_7(t)$, $\mathrm{MA}_{14}(t)$): *Yes* if computed as trailing windows ending at t.
- **Growth proxy** $\Delta_7(t)$: *Yes* if based on trailing means only; it summarizes phase without using future.

- **Day-of-week**: *Yes*. It is known deterministically from the date and captures reporting periodicity.
- **Centered moving average** $MA_7(t+3)$: *No*. It uses values after t. *Alternative:* trailing $MA_7(t)$.
- **Peak indicator from full series**: *No*. It requires future information to know whether a time is a peak. *Alternative:* a real-time phase proxy (e.g., $\Delta_7(t)$ thresholding) computed using only past data.

A useful discipline is to implement features in a function `make_features(df, as_of_date)` that explicitly truncates the data at the forecast origin before computing any transformation.

Exercise 3: A minimal rolling-origin backtest across classical, ML, and DL

Task

Implement a rolling-origin evaluation for the running example series to forecast $h = 7$ days ahead. Compare: (i) a classical ARIMA-family baseline, (ii) an ML baseline using lag/rolling/calendar features, and (iii) a deep sequential baseline (e.g., LSTM). Return a table with mean±std of RMSE and MAE across origins.

Solution (code)

```
import numpy as np
import pandas as pd
from pathlib import Path

# --- Classical baseline (ARIMA) ---
from statsmodels.tsa.statespace.sarimax import SARIMAX

# --- ML baseline ---
from sklearn.pipeline import Pipeline
from sklearn.preprocessing import StandardScaler
from sklearn.linear_model import ElasticNet

# --- Deep baseline (LSTM) ---
import torch
import torch.nn as nn


# ---------------------------------------------------------
# Load running example series
# ---------------------------------------------------------
def load_running_example(csv_path="example_daily_incidence.csv"):
    df = pd.read_csv(csv_path, parse_dates=["date"]).sort_values("date").reset_index(drop=True
        )
    df["y"] = df["y"].astype(float).clip(lower=0)
    return df

# ---------------------------------------------------------
# Supervised features for ML
# ---------------------------------------------------------
def make_ml_features(df, lags=28):
    d = df.copy()
```

```
    d["dow"] = d["date"].dt.dayofweek
    for k in range(1, lags + 1):
        d[f"y_lag{k}"] = d["y"].shift(k)
    d["ma7"] = d["y"].rolling(7).mean()
    d["ma14"] = d["y"].rolling(14).mean()
    dow_oh = pd.get_dummies(d["dow"], prefix="dow", drop_first=False)
    d = pd.concat([d, dow_oh], axis=1)
    feat_cols = [f"y_lag{k}" for k in range(1, lags + 1)] + ["ma7","ma14"] + list(dow_oh.
        columns)
    return d, feat_cols

# ----------------------------------------------------------
# Deep dataset (windowed)
# ----------------------------------------------------------
def make_windows(y, L):
    X, Y = [], []
    for t in range(L, len(y)):
        X.append(y[t-L:t])
        Y.append(y[t])
    return np.asarray(X), np.asarray(Y)

class SmallLSTM(nn.Module):
    def __init__(self, hidden=32):
        super().__init__()
        self.lstm = nn.LSTM(input_size=1, hidden_size=hidden, batch_first=True)
        self.head = nn.Linear(hidden, 1)

    def forward(self, x):
        # x: (B, L, 1)
        out, _ = self.lstm(x)
        z = out[:, -1, :]          # last step
        return self.head(z)        # (B, 1)

def fit_predict_lstm(y_train, L, y_context):
    # Train to predict next day on log1p scale, then roll to horizon with direct target below.
    device = "cpu"
    model = SmallLSTM(hidden=32).to(device)
    opt = torch.optim.Adam(model.parameters(), lr=1e-3)
    loss_fn = nn.MSELoss()

    Xtr, Ytr = make_windows(y_train, L)
    Xtr = torch.tensor(Xtr[:, :, None], dtype=torch.float32, device=device)
    Ytr = torch.tensor(Ytr[:, None], dtype=torch.float32, device=device)

    model.train()
    for _ in range(30):  # small epochs for template
        opt.zero_grad()
        pred = model(Xtr)
```

```
        loss = loss_fn(pred, Ytr)
        loss.backward()
        nn.utils.clip_grad_norm_(model.parameters(), 1.0)
        opt.step()

    # Predict from the last available window in y_context
    model.eval()
    x = torch.tensor(y_context[-L:][None, :, None], dtype=torch.float32, device=device)
    with torch.no_grad():
        yhat = model(x).cpu().numpy().ravel()[0]
    return yhat

# ---------------------------------------------
# Rolling-origin backtest
# ---------------------------------------------
def backtest(df, horizon=7, min_train=200, step=7, L=42):
    df = df.copy().reset_index(drop=True)
    y = np.log1p(df["y"].values)  # stabilize scale

    # ML feature frame (computed once, but we always slice up to origin to avoid leakage)
    feat_df, feat_cols = make_ml_features(df, lags=28)

    origins = list(range(min_train, len(df) - horizon, step))
    rows = []

    for t in origins:
        y_train = y[:t]                     # data available at origin
        y_true = y[t + horizon]             # evaluate at t+h (single point score)

        # --- Classical: SARIMAX (ARIMA-like) on y
        ar = SARIMAX(y_train, order=(2,1,2), trend="n",
                     enforce_stationarity=False, enforce_invertibility=False)
        ar_fit = ar.fit(disp=False)
        yhat_ar = ar_fit.forecast(steps=horizon)[-1]

        # --- ML: ElasticNet on engineered features ---
        train_block = feat_df.iloc[:t].dropna()
        Xtr = train_block[feat_cols].values
        Ytr = np.log1p(train_block["y"].shift(-horizon).dropna().values)
        Xtr = Xtr[:len(Ytr)]  # align
        ml = Pipeline([("scaler", StandardScaler()),
                       ("enet", ElasticNet(alpha=0.02, l1_ratio=0.9, max_iter=20000))])
        ml.fit(Xtr, Ytr)

        x_now = feat_df.iloc[t][feat_cols].values[None, :]
        yhat_ml = ml.predict(x_now)[0]

        # --- Deep: LSTM (direct horizon target) ---
```

```
        # simple direct target: train on pairs (window -> y_{t+h})
        # here: approximate by training next step and using horizon feature shifting in
            practice.
        # For a direct h setup, build windows and shift labels by +h.
        # Template keeps code short; see text for direct multi-horizon variants.
        y_context = y[:t+1]
        yhat_dl = fit_predict_lstm(y_train, L=L, y_context=y_context)

        # store errors on original scale (invert log1p)
        def inv(z): return np.expm1(z)
        yt = inv(y_true)
        rows.append({
            "origin_idx": t,
            "y_true": yt,
            "ARIMA": inv(yhat_ar),
            "ElasticNet": inv(yhat_ml),
            "LSTM": inv(yhat_dl),
        })

    out = pd.DataFrame(rows)
    for m in ["ARIMA","ElasticNet","LSTM"]:
        out[f"ae_{m}"] = np.abs(out["y_true"] - out[m])
        out[f"se_{m}"] = (out["y_true"] - out[m])**2

    summary = []
    for m in ["ARIMA","ElasticNet","LSTM"]:
        rmse = np.sqrt(out[f"se_{m}"].mean())
        mae = out[f"ae_{m}"].mean()
        summary.append({"model": m, "RMSE": rmse, "MAE": mae})

    return out, pd.DataFrame(summary)

# Example usage
df = load_running_example("example_daily_incidence.csv")
preds, summary = backtest(df, horizon=7, min_train=200, step=7, L=42)
print(summary)
```

Listing 7.14: Rolling-origin backtest template comparing classical, ML, and deep sequential models.

In a full implementation, the deep model should be trained *directly* for the target horizon (or for multiple horizons jointly) by shifting labels by $+h$ rather than relying on a next-step approximation, and the rolling-origin protocol should either refit or update models in a way that matches the intended deployment cadence. The key testing principle remains unchanged: all features and labels used for training at an origin must be computable from information available up to that origin, and performance must be aggregated across many origins to assess stability across epidemic phases.

CHAPTER

Parameter estimation and model calibration

8

Mathematical and data-driven epidemiological models are only as useful as their connection to reality [574–577]. To this end, parameter estimation and calibration are therefore central to making models operational, since they determine whether predictions are credible, whether uncertainty is properly quantified, and whether inferred epidemiological quantities can be interpreted and compared across settings [578]. In this chapter, we will cover several methods for model's parameter estimation and calibration. To start, let us first formally define what do we mean by parameter estimation for a model. Thus, let a mechanistic model be represented in state-space form. Formally, the latent state $x(t) \in \mathbb{R}^d$ evolves according to

$$\frac{dx(t)}{dt} = f\big(x(t), u(t), \theta\big), \tag{8.1}$$

where $x(t_0) = x_0$, $u(t)$ collects known inputs, and $\theta \in \Theta \subset \mathbb{R}^p$ denotes unknown parameters. Observations y_k at times t_k are linked to the latent state via an observation model,

$$y_k = g\big(x(t_k), \theta\big) + \varepsilon_k, \qquad \sim \text{(measurement / reporting noise)}, \tag{8.2}$$

where ε_k is assumed to be proportional to the rate between measurement and reporting noise. It is important to note this is an assumption, but it holds if the model (i.e., function g) is indeed captures the real dynamics one tries to model. This separation between disease dynamics and measurement is essential since epidemiological data are noisy, partially observed, and often biased by reporting processes and changes in surveillance [579].

From this point forward, we use the term "parameter estimation" to mean the statistical inference problem of learning θ (and sometimes x_0 or other latent quantities) assuming the model structure is fixed. This assumption is important as it implicatively holds that the underline dynamics are fixed over time and are not dependent on time explicitly. Technically, given data $y_{1:K} = \{y_k\}_{k-1}^{K}$, a likelihood-based formulation specifies a probability model $p_\theta(y_{1:K})$ induced by Eqs. (8.1)–(8.2), and seeks, for example, a maximum likelihood estimate

$$\hat{\theta}_{\text{MLE}} \in \arg\max_{\theta \in \Theta} \; p_\theta(y_{1:K}), \tag{8.3}$$

or, in a Bayesian formulation with prior $p(\theta)$, the posterior

$$p(\theta \mid y_{1:K}) \propto p_\theta(y_{1:K})\, p(\theta), \tag{8.4}$$

from which point estimates and uncertainty intervals follow [578]. For nonlinear, partially observed epidemic systems, evaluating or approximating $p_\theta(y_{1:K})$ can be nontrivial; modern approaches include

Epidemic Modeling. https://doi.org/10.1016/B978-0-44-355525-1.00014-X

sequential Monte Carlo and iterated filtering methods designed for such settings [579,580]. In data-driven models, the same principle applies: parameter estimation targets coefficients, states, and noise terms in time-series models, or weights in learned predictors, under an explicit or implicit statistical objective [575–577].

In a complementary manner, we use the term "model calibration" in a broader, practice-oriented sense, in which the process of selecting parameter values (and sometimes initial conditions, latent states, and discrepancy terms) so that model outputs reproduce salient features of observed epidemics for a specified purpose. Calibration often emphasizes predictive adequacy and operational performance rather than strict statistical identifiability under a fully specified likelihood. A common formulation chooses θ to minimize a loss (or negative log-likelihood) comparing observations and model outputs, formalized as follows:

$$\hat{\theta}_{\text{cal}} \in \arg\min_{\theta \in \Theta} \mathcal{L}\big(y_{1:K}, \hat{y}_{1:K}(\theta)\big) + \lambda\, \mathcal{R}(\theta), \tag{8.5}$$

where $\hat{y}_{1:K}(\theta)$ are model-predicted observables, $\mathcal{R}$ is a regularizer encoding plausibility or stability constraints, and $\lambda \geq 0$ controls the fit–complexity trade-off. Importantly, calibration may explicitly account for "model discrepancy". Namely, even a well-motivated model can be systematically wrong due to aggregation, unmodeled heterogeneity, or behavioral feedback. Bayesian calibration frameworks formalize this by augmenting the observation equation with a discrepancy term and propagating the resulting uncertainty [581,582]. This perspective is especially relevant in extended SIR models (where simplifying assumptions can be consequential) and in ML/time-series settings (where misspecification manifests as distribution shift and degraded out-of-sample performance) [576,577].

8.1 Model parameter fitting

Across mechanistic epidemic models and data-driven predictors, parameter fitting is commonly posed as numerical optimization in which, one choose parameters $\theta \in \Theta$ to minimize a discrepancy between observed data $y_{1:K}$ and model outputs $\hat{y}_{1:K}(\theta)$, optionally subject to constraints and regularization [582,583]. A generic formulation takes the form of Eq. (8.5). In epidemic settings, the map $\theta \mapsto \hat{y}_{1:K}(\theta)$ typically involves solving differential equations or running stochastic simulations, making the loss potentially nonconvex, noisy, and expensive to evaluate [584]. The subsections below summarize widely used fitting strategies.

8.1.1 Least mean squares and nonlinear least squares

A baseline approach fits parameters by minimizing a sum of squared residuals between observed and predicted quantities [585,586]:

$$\hat{\theta}_{\text{LS}} \in \arg\min_{\theta \in \Theta} \sum_{k=1}^{K} w_k \big(y_k - \hat{y}_k(\theta)\big)^2, \tag{8.6}$$

with optional weights $w_k > 0$. When measurement errors are approximately Gaussian with known (or proportional) variances, Eq. (8.6) corresponds to maximum likelihood under a normal noise model (with w_k reflecting inverse variances), and weighted least squares can partially correct heteroscedasticity commonly seen in incidence and hospitalization data [583,587]. In epidemic models, least squares is frequently applied to smoothed incidence curves, cumulative counts, or transformed data, with constraints enforcing epidemiological plausibility (e.g., nonnegativity, bounds on rates) [574].

In implementation, it is often helpful to expose the residual structure directly. Define the weighted residual vector

$$r(\theta) = \big(r_1(\theta), \ldots, r_K(\theta)\big)^\top, \qquad r_k(\theta) = \sqrt{w_k}\,\big(y_k - \hat{y}_k(\theta)\big), \tag{8.7}$$

so that minimizing Eq. (8.6) is equivalent to minimizing $\|r(\theta)\|_2^2$. Many solvers are built around Eq. (8.7): instead of requiring the scalar objective, they request the vector of residuals and internally compute the sum of squares (and, when possible, approximate curvature information). This framing also clarifies what it means to "use weights": the modeler is not changing the simulator, but changing which time points or data streams are emphasized by the fit.

A smooth fitting workflow begins by choosing what is being matched and how the model produces predictions. In mechanistic modeling, $\hat{y}_k(\theta)$ rarely equals a compartment value directly; it is typically a measurement mapping from latent states to observables, possibly including reporting fractions, delays, or aggregation (for example, converting infections to reported cases through a reporting rate and a delay distribution). Once this mapping is fixed, the next decision is scaling. Uniform weights $w_k \equiv 1$ tend to prioritize peak periods because residual magnitudes grow with the level of the signal. If early growth or low-count regimes are epidemiologically important, weights can be chosen to balance influence over time, and when multiple observables are fitted jointly (e.g., cases and hospitalizations), separate weighting and normalization are often needed so that one stream does not dominate the objective purely by its numerical scale.

Because $\hat{y}_k(\theta)$ is typically nonlinear in θ, solvers rely on iterative updates such as Gauss–Newton and Levenberg–Marquardt, which trade off speed and robustness and can exploit sensitivities (gradients/Jacobians) when available [587,588]. In practice, two refinements are particularly common. The first is to replace the pure squared loss with a robust alternative when data contain outliers or reporting anomalies; such losses preserve least-squares behavior near zero residuals but reduce the influence of large deviations [589]. The second is to stabilize the fit with regularization or prior information, which is especially important when multiple parameters have similar effects on the observed output and the optimization becomes weakly identified [582]. Both refinements can be interpreted as ways to encode "what errors should matter" and "what parameter values are plausible" within the same optimization framework.

Nonlinear least squares objectives are often nonconvex, so initialization matters. A practical approach is to start from a domain-informed guess (for example, using early exponential growth to seed a plausible transmission rate), then perform several restarts from randomized initial points within epidemiologically plausible bounds. The best solution by objective value is then inspected for plausibility and for residual structure. Persistent patterns in residuals, such as systematic underprediction near peaks or phase shifts in timing, often indicate model mismatch (missing mechanisms or mis-specified measurement mapping) rather than simply an optimizer failure.

Listing 8.1 shows a minimal and reusable implementation pattern. The key idea is that the residual function returns the weighted residual vector from Eq. (8.7), while the solver handles the iterative minimization and respects parameter bounds. This structure applies equally well to ODE-based models, stochastic simulators (with appropriate smoothing or replication), and even data-driven models when a differentiable forward mapping is available.

```
import numpy as np
from scipy.optimize import least_squares

def simulate_outputs(theta, t, known_inputs):
    """
    User provided forward model:
      - solves the mechanistic system (or runs a simulator)
      - maps latent states to observables
    Returns yhat with shape (K,).
    """
    # yhat = ...
    return yhat

def residuals(theta, t, y_obs, w, known_inputs):
    yhat = simulate_outputs(theta, t, known_inputs)
    return np.sqrt(w) * (y_obs - yhat)

result = least_squares(
    residuals, x0=theta0, bounds=(lb, ub),
    args=(t, y_obs, w, known_inputs),
    # robust alternative: loss="huber" or loss="soft_l1"
    loss="linear"
)

theta_hat = result.x
```

Listing 8.1: Weighted nonlinear least squares fitting via residuals and bounds.

Once parameters are fitted, it is useful to evaluate performance in a way that is easy to compare across models and datasets. Even if the fitting objective is squared error, reporting additional metrics on a held-out set helps reveal whether the model is merely matching training noise or genuinely capturing dynamics that generalize. Listing 8.2 computes three standard forecasting metrics from a predictor: MAE, RMSE, and MAPE. In a data-driven model, `model.predict` naturally returns forecasts. For a mechanistic model, the same interface can be achieved by wrapping the simulator (run with $\hat{\theta}$) in a small adapter that takes inputs describing the forecast window and returns $\hat{y}$ in the same format as the observed series.

```
from sklearn.metrics import mean_absolute_error, mean_squared_error

def evaluate_forecast(model, X_test, y_test):
    yhat = model.predict(X_test)

    mae  = mean_absolute_error(y_test, yhat)
    rmse = mean_squared_error(y_test, yhat, squared=False)
    mape = np.mean(np.abs((y_test - yhat) / np.maximum(np.abs(y_test), 1e-8))) * 100.0

```

```
    return {"MAE": mae, "RMSE": rmse, "MAPE(%)": mape}, yhat
```

Listing 8.2: Testing and reporting standard forecasting error metrics.

The three metrics emphasize different aspects of predictive error. MAE averages absolute deviations and is relatively stable under occasional spikes. RMSE penalizes larger errors more strongly and therefore tends to reflect peak misspecification, aligning closely with least-squares training objectives. MAPE expresses errors as percentages, which is appealing for communication, but can become unstable when counts are near zero; the denominator safeguard in Listing 8.2 prevents numerical issues, yet interpretation still benefits from pairing MAPE with absolute metrics. A smooth end-to-end process is therefore to fit θ on a training interval using (8.6), generate predictions on a held-out interval, report MAE/RMSE/MAPE, and examine residual plots over time to diagnose whether remaining errors are random noise or systematic structure that suggests revising the observation model, weights, or mechanistic assumptions.

8.1.2 Grid search (and coarse-to-fine sweeps)

Grid search discretizes the feasible set Θ into a Cartesian grid $\mathcal{G} \subset \Theta$, evaluates the objective (e.g., Eq. (8.3)) at all grid points, and returns the best value:

$$\hat{\theta}_{\text{grid}} \in \arg\min_{\theta \in \mathcal{G}} \mathcal{L}\big(y_{1:K}, \hat{y}_{1:K}(\theta)\big). \tag{8.8}$$

Its main advantages are conceptual simplicity, straightforward parallelization, and transparency (useful for sanity-checking identifiability and multimodality) [583]. In practice, grid search is most viable when the number of tuned parameters is small (often $p \leq 3$ or 4), or when domain knowledge narrows Θ substantially. However, grid search suffers from the "curse of dimensionality" [590], where the number of evaluations grows exponentially with p, quickly becoming infeasible for extended compartmental models with many rates, change points, or covariate effects [591].

A smooth way to think about grid search is as an explicit map of the loss landscape. Instead of trusting an optimizer to move through parameter space, we enumerate candidate parameter values, compute the model predictions for each candidate, and directly measure how well each candidate matches the data. This "brute-force" character makes grid search an excellent diagnostic tool, even when it is not the final fitting method. For example, if a two-parameter grid over (β, γ) reveals a long, flat valley of nearly identical loss values, that is an immediate sign that the data and the chosen observation mapping do not uniquely identify those parameters. Conversely, a sharp, isolated basin suggests the problem is locally well-posed and that a local optimizer initialized near the basin is likely to converge reliably.

One way to improve grid search is by constructing a useful grid with bounding and scaling. Bounds should be epidemiologically plausible and wide enough to include realistic alternatives, while scaling should reflect what constitutes a meaningful change in a parameter. Many epidemic parameters act multiplicatively (rates, contact modifiers, reporting fractions), so it is often more informative to create grids in transformed space, such as a log grid for positive rates. This avoids wasting resolution where changes would have negligible effect and improves the chances that a coarse grid can still locate the region containing the optimum.

The primary practical limitation is computational. Simply put, if each parameter has m grid points, then a full Cartesian grid requires m^p model evaluations. This exponential growth is the operational meaning of the curse of dimensionality [590]. For extended SIR models, even modest p can become infeasible if the forward model is expensive (e.g., fine time resolution, multiple coupled compartments, stochastic simulation). For this reason, a common compromise is coarse-to-fine searching. One starts with a coarse grid to identify a small neighborhood of promising candidates, then refines the grid only within that neighborhood, and finally hands off the best candidate to a faster local method such as least squares refinement. This workflow maintains the transparency of grid search while concentrating computation where it matters.

Listing 8.3 shows a minimal pattern for grid search over a small number of parameters. The code uses nested iteration over a grid, evaluates an objective, and returns the best parameter set together with a record of all evaluated candidates. Keeping the full record is often as valuable as the best-fit value: it enables contour plots, uncertainty heuristics based on near-optimal regions, and quick checks for multimodality.

```
import numpy as np
from itertools import product

def objective(theta, t, y_obs, known_inputs):
    """
    Compute loss for candidate theta.
    For least squares style objectives, this could return
    sum_k (y_k - yhat_k(theta))^2 or a weighted variant.
    """
    yhat = simulate_outputs(theta, t, known_inputs)
    return np.sum((y_obs - yhat)**2)

def grid_search(param_grids, t, y_obs, known_inputs):
    """
    param_grids: list of 1D arrays, one per parameter.
    Returns: best_theta, best_loss, all_results
    """
    best_theta, best_loss = None, np.inf
    all_results = []

    for values in product(*param_grids):
        theta = np.array(values, dtype=float)
        loss = objective(theta, t, y_obs, known_inputs)
        all_results.append((theta, loss))

        if loss < best_loss:
            best_theta, best_loss = theta, loss

    return best_theta, best_loss, all_results
```

Listing 8.3: Basic grid search over a small parameter set.

Coarse-to-fine sweeps can be implemented by reusing Listing 8.3 iteratively. After an initial run, the best candidate $\hat{\theta}$ defines a center, and one constructs a narrower grid around it with higher resolution. In two dimensions, this can be done by shrinking the range on each parameter; in higher dimensions, it is often more effective to refine only a subset of parameters at a time, holding the others fixed at their current best values. This sequential refinement is a practical way to control computation while still benefiting from the interpretability of grid-based exploration.

In applied epidemiological work, grid search is therefore best understood not only as a fitting method, but as a modeling instrument. It helps expose sensitivity, diagnose identifiability and multimodality, and provide robust starting points for subsequent optimization. When combined with coarse-to-fine refinement, it remains viable for small to moderate parameter subsets even in complex extended SIR formulations.

8.1.3 Bayesian search

When each model evaluation is expensive (ODE solves with many time points, agent-based simulations, stochastic likelihood approximations), *Bayesian optimization* provides a sample-efficient strategy for minimizing $\mathcal{L}(\theta)$ using a probabilistic surrogate model [592,593]. The key idea is to treat the objective as a black-box function that is costly to query. Instead of evaluating $\mathcal{L}(\theta)$ on a dense grid or moving via local gradients, Bayesian optimization builds an explicit statistical model of the objective from the evaluations already performed, and uses that model to decide where to sample next. This is especially attractive in epidemiology because one forward simulation may involve solving stiff dynamical systems, running multiple stochastic replicates, and applying an observation mapping, so the total optimization budget is usually limited.

A standard approach places a prior over $\mathcal{L}(\theta)$, most commonly via a Gaussian process (GP), and updates the posterior after each evaluated parameter vector θ_i. The surrogate model provides both a mean prediction $\mu(\theta)$ and an uncertainty estimate $\sigma(\theta)$, which together drive an *acquisition function* $a(\theta)$. The acquisition function formalizes the exploration–exploitation trade-off: it prefers parameters that either look promising under the current surrogate (low predicted loss) or are uncertain enough that evaluating them could improve the surrogate meaningfully. The next evaluation is chosen by maximizing the acquisition function,

$$\theta_{n+1} \in \arg\max_{\theta \in \Theta} a_n(\theta), \qquad \mathcal{D}_{n+1} = \mathcal{D}_n \cup \{(\theta_{n+1}, \mathcal{L}(\theta_{n+1}))\}, \tag{8.9}$$

and the loop repeats until a budget of evaluations is exhausted [593]. Expected improvement and upper confidence bounds are common acquisition choices because they are simple, effective, and provide intuitive knobs for controlling how aggressively the method explores uncertain regions [593]. In practice, this sequential design is what produces sample-efficiency: compared to uninformed search, the algorithm spends few evaluations in regions that the surrogate already believes are poor.

Using Bayesian optimization smoothly in an epidemiological workflow begins with defining a scalar objective that reflects the intended fitting target. This can be the same least-squares criterion used earlier, a deviance under a count model, or a forecasting error computed on a validation window. Because Bayesian optimization only sees the final scalar value, it is common to wrap all model-specific logic inside the objective: simulation, mapping to observables, and loss computation. It is also helpful to encode constraints directly through parameter bounds and transformations. Positive rates and scale parameters often benefit from log-parameterization so that the search explores multiplicative changes evenly, rather than allocating many samples to large absolute values that behave similarly.

Noisy objectives are the rule rather than the exception in simulation-based calibration. Stochastic simulators, subsampling, and Monte Carlo likelihood approximations introduce variance in $\mathcal{L}(\theta)$, and the surrogate needs to tolerate this. A practical approach is to either (i) evaluate multiple replicates at the same θ and average the loss, or (ii) tell the Bayesian optimizer that the objective is noisy so the surrogate does not overfit single evaluations. Another pragmatic detail is failure handling: some parameter combinations can cause numerical instability or unphysical trajectories. In these cases, the objective should return a large penalty rather than raising an exception, so that the optimizer learns to avoid problematic regions without terminating the run.

Listing 8.4 shows a concise implementation pattern for Bayesian optimization. The code assumes a user-defined `simulate_outputs` that returns model predictions, and defines an objective function that computes a scalar loss. The optimizer then runs a limited number of expensive evaluations, starting from a small initial design and progressively focusing on regions that look promising under the surrogate.

```
import numpy as np

from skopt import gp_minimize
from skopt.space import Real

def objective(theta, t, y_obs, known_inputs):
    """
    theta: list/array of parameters proposed by the optimizer.
    Returns a scalar loss value to minimize.
    """
    try:
        yhat = simulate_outputs(theta, t, known_inputs)
        loss = np.mean((y_obs - yhat)**2)   # MSE as an example
        return float(loss)
    except Exception:
        # Penalize parameter regions that lead to simulation failure
        return 1e12

# Define the search space: log-uniform is often useful for positive rates
space = [
    Real(1e-3, 5.0, prior="log-uniform", name="beta"),
    Real(1e-3, 2.0, prior="log-uniform", name="gamma"),
]

res = gp_minimize(
    func=lambda th: objective(th, t, y_obs, known_inputs),
    dimensions=space,
    n_calls=80,                # total expensive evaluations
    n_initial_points=15,       # initial space-filling evaluations
    acq_func="EI",             # or "LCB" / "PI"
    random_state=0
)
```

```
theta_hat = res.x
best_loss = res.fun
```

Listing 8.4: Bayesian optimization pattern for expensive epidemiological objectives.

The inputs in Listing 8.4 correspond directly to practical decisions. The list `space` encodes the feasible region Θ and the scaling of each parameter; log-uniform priors are a convenient way to explore rates over orders of magnitude. The `n_initial_points` setting determines how much the optimizer explores before it begins exploiting the surrogate; if the initial design is too small, the GP can become overconfident in a misleading region early on. The acquisition function controls exploration: expected improvement tends to focus on improving the current best while still sampling uncertain areas, whereas lower confidence bounds introduce an explicit exploration parameter (often exposed as κ in implementations) that can be increased when the objective is noisy or when the current surrogate seems overconfident.

A common refinement in epidemiological calibration is to use Bayesian optimization as a front-end that finds a good region of parameter space, then switch to a local method (such as nonlinear least squares) to polish the solution. This hybrid workflow is often effective because Bayesian optimization is good at global navigation under a limited budget, while local solvers are good at fine-grained convergence once near a basin. The main practical limitation is dimensionality: as the number of tuned parameters grows, surrogate modeling and acquisition maximization become harder, so it is usually beneficial to restrict Bayesian search to the most influential parameters (for example, a small set of transmission modifiers and reporting parameters) and keep the rest fixed or structured through a low-dimensional parameterization [592,594].

8.1.4 Genetic algorithms

Genetic algorithms (GAs) are population-based, stochastic global optimization methods inspired by evolutionary principles [595–597]. They are particularly useful in epidemiological parameter fitting when the loss surface is rugged, multimodal, or includes discontinuities introduced by thresholds, event-driven interventions, integer decisions, or simulation noise. Rather than maintaining a single iterate θ_n, a GA maintains a population of candidates $\{\theta^{(i)}\}_{i=1}^{M}$. Each generation evaluates the objective for all candidates, then uses a small set of simple operators to produce a new population that is biased toward better solutions while preserving diversity.

The process begins by defining an encoding for θ. For continuous parameters, the encoding is typically the parameter vector itself, with each coordinate constrained to a feasible interval. For mixed problems, the encoding may concatenate continuous components (e.g., β, γ) with discrete components (e.g., a change-point day or a policy on/off indicator). The population is initialized by sampling uniformly (or from epidemiologically informed priors) within bounds, often supplemented with a few hand-crafted plausible parameter sets to improve early progress. Each candidate is then scored by the objective $\mathcal{L}(\theta)$, which can be a least-squares mismatch, a deviance, or a validation metric computed on a forecasting horizon.

Selection is the mechanism that converts scores into reproductive advantage. In practice, selection is commonly implemented via tournament selection or rank-based selection: candidates with lower loss are more likely to be chosen as parents, but weaker candidates still have a chance, which helps prevent the algorithm from collapsing too early into a single region. Once parents are chosen, crossover creates offspring by combining parts of the parents. For continuous vectors, crossover can be implemented as arithmetic recombination, such as taking convex combinations of parent parameters, or by exchanging coordinate blocks. For discrete components, crossover naturally corresponds to swapping sub-structures, such as exchanging change-point locations or intervention schedules. Mutation then perturbs offspring parameters, injecting new variation. In continuous spaces, mutation is often additive Gaussian noise (possibly scaled per-parameter); in discrete spaces, it may be random resampling or a small integer shift. Together, crossover exploits good partial solutions by mixing them, while mutation explores new regions and helps recover from premature convergence.

Using GAs smoothly requires attention to constraints and numerical stability. Because offspring are created mechanically, it is common for crossover or mutation to propose parameters that violate feasibility (negative rates, inconsistent ordering of change points) or cause simulation failures. A practical implementation, therefore, includes a repair or projection step, which maps invalid offspring back into the feasible region, and uses penalization when the simulator fails. Penalization means that if the forward model cannot produce predictions, the objective returns a large loss, ensuring that selection drives the population away from the problematic region without terminating the run.

GAs are especially valuable as a front-end global search. Once a GA has found a high-quality region, it is often efficient to switch to a local optimizer for refinement. In extended SIR calibration, a common pattern is to run a GA for a fixed number of generations, take the best individual (or a small elite set), and then apply nonlinear least squares or a derivative-free local method starting from those points. This hybrid approach leverages the strength of GAs in global exploration and the strength of local solvers in rapid convergence within a basin.

Fig. 8.1 presents a schematic view of the genetic algorithm process.

Listing 8.5 shows a practical genetic algorithm (GA) pattern for fitting epidemiological model parameters in Python using the `DEAP` library [598]. In this example, the unknown parameter vector has two components, which may be interpreted as quantities such as a transmission rate β and a recovery rate γ. The line defining `BOUNDS` specifies the allowed interval for each parameter. These bounds are important because epidemiological parameters must usually remain within biologically meaningful ranges. The helper function `clip_to_bounds` is a repair step: after crossover or mutation, a candidate may fall outside the feasible range, and this function pushes each coordinate back into the allowed interval. This keeps the search numerically stable without discarding the candidate entirely. The function `fitness` defines what the GA is trying to optimize. For each candidate `individual`, the code converts the parameter list into a NumPy array θ, runs the forward model through `simulate_outputs`, and compares the simulated outputs $\hat{y}$ with the observations y_{obs}. Here the loss is the mean squared error,

$$\frac{1}{n}\sum_{i=1}^{n}(y_i - \hat{y}_i)^2,$$

so smaller values correspond to better fits. The fitness must be returned as a one-entry tuple because this is the format expected by `DEAP`. The `try/except` block is also important: if a parameter choice causes the simulator to fail, the code returns a very large penalty value rather than stopping the optimization. The two `creator.create(...)` lines define the data structures used by `DEAP`. `FitnessMin` indicates that the objective should be minimized, and `Individual` defines each candidate as a list with an attached fitness value. The `toolbox` then collects the main ingredients of the

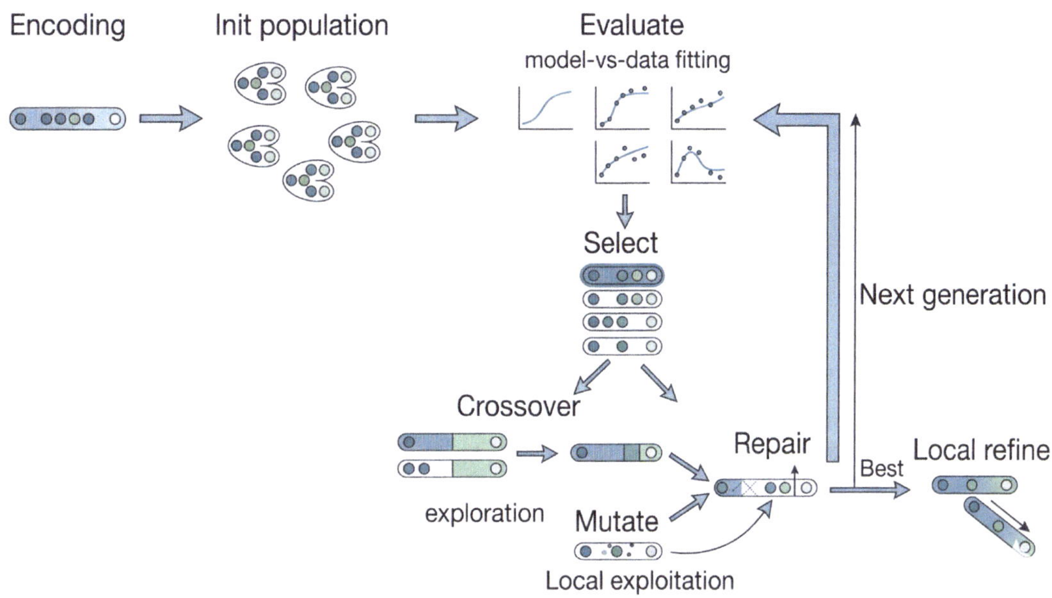

FIGURE 8.1

A schematic view of the genetic algorithm process.

GA. It registers how to sample each parameter uniformly within its bounds, how to assemble one candidate individual, and how to repeat this process to generate an initial population. The registered operators determine how the population evolves. The crossover operator `cxBlend` mixes two parent vectors to create offspring with intermediate values, which is natural for continuous parameters. The mutation operator `mutGaussian` adds random Gaussian perturbations to some coordinates, allowing the search to explore new regions. The selection operator `selTournament` introduces reproductive pressure: better candidates are more likely to survive, but weaker ones still have some chance, which helps preserve diversity. The line registering `evaluate` attaches the fixed data inputs (t, y_{obs}, `known_inputs`), so that later the code only needs to call `toolbox.evaluate(ind)`. The function `run_ga` executes the optimization. It first sets the random seed for reproducibility, creates the initial population, repairs each candidate if needed, and evaluates its loss. The main loop then runs for `ngen` generations. In each generation, `algorithms.varAnd` creates offspring by applying crossover and mutation, the offspring are repaired and evaluated, and tournament selection chooses which candidates form the next generation. This repeated cycle of variation, repair, evaluation, and selection is the core of the GA. After the final generation, the code extracts the best individual in the population and returns both its parameter vector $\hat{\theta}$ and its loss value. Conceptually, the example illustrates why GAs are useful for epidemiological calibration: the forward model is treated as a black box, no derivatives are required, and simulator failures can be handled through penalization. The same pattern also extends naturally to mixed discrete–continuous problems by changing the representation of individuals and using mutation and crossover operators that respect the variable types.

```
import random
import numpy as np
from deap import base, creator, tools, algorithms

# Example: two positive rates with bounds
BOUNDS = [(1e-3, 5.0), (1e-3, 2.0)]   # (beta_min, beta_max), (gamma_min, gamma_max)

def clip_to_bounds(individual):
    for j, (lo, hi) in enumerate(BOUNDS):
        if individual[j] < lo: individual[j] = lo
        if individual[j] > hi: individual[j] = hi
    return individual

def fitness(individual, t, y_obs, known_inputs):
    theta = np.array(individual, dtype=float)
    try:
        yhat = simulate_outputs(theta, t, known_inputs)
        loss = np.mean((y_obs - yhat)**2)   # MSE example
        return (loss,)                      # DEAP expects a tuple
    except Exception:
        return (1e12,)                      # penalty for invalid/singular simulations

creator.create("FitnessMin", base.Fitness, weights=(-1.0,))
creator.create("Individual", list, fitness=creator.FitnessMin)

toolbox = base.Toolbox()
# Initialization: uniform within bounds
for j, (lo, hi) in enumerate(BOUNDS):
    toolbox.register(f"attr_{j}", random.uniform, lo, hi)

toolbox.register("individual", tools.initCycle, creator.Individual,
                 (toolbox.attr_0, toolbox.attr_1), n=1)
toolbox.register("population", tools.initRepeat, list, toolbox.individual)

# Operators
toolbox.register("mate", tools.cxBlend, alpha=0.5)          # arithmetic crossover
toolbox.register("mutate", tools.mutGaussian, mu=0.0, sigma=0.2, indpb=0.5)
toolbox.register("select", tools.selTournament, tournsize=3)

# Fitness evaluation (injects data)
toolbox.register("evaluate", fitness, t=t, y_obs=y_obs, known_inputs=known_inputs)

def run_ga(seed=0, pop_size=80, ngen=60, cxpb=0.6, mutpb=0.3):
    random.seed(seed)
    pop = toolbox.population(n=pop_size)

    # Evaluate initial population
```

```
    for ind in pop:
        clip_to_bounds(ind)
        ind.fitness.values = toolbox.evaluate(ind)

    for gen in range(ngen):
        offspring = algorithms.varAnd(pop, toolbox, cxpb=cxpb, mutpb=mutpb)

        for ind in offspring:
            clip_to_bounds(ind)
            ind.fitness.values = toolbox.evaluate(ind)

        pop = toolbox.select(offspring, k=len(pop))

    best = tools.selBest(pop, k=1)[0]
    return np.array(best, dtype=float), best.fitness.values[0]

theta_hat, best_loss = run_ga()
```

Listing 8.5: Genetic algorithm pattern for epidemiological parameter fitting.

The main knobs in Listing 8.5 correspond to practical trade-offs. The population size and number of generations determine the evaluation budget $M \times G$. Larger populations increase diversity and reduce the risk of premature convergence, but they consume more evaluations per generation. The crossover probability `cxpb` controls how aggressively the GA recombines solutions, while the mutation probability `mutpb`, together with mutation scale `sigma`, controls exploration. If mutation is too weak, the population can collapse into a suboptimal region; if it is too strong, the search becomes nearly random. Tournament selection size controls selection pressure: larger tournaments favor rapid improvement but can reduce diversity too quickly. A robust workflow is to begin with conservative selection pressure and moderate mutation, then adjust based on whether the population becomes homogeneous early or fails to improve over generations.

Because GAs produce a full population rather than a single fit, they naturally support diagnostic use. Inspecting the spread of near-best individuals provides a rough view of practical non-identifiability: if many substantially different parameter vectors produce similar objective values, the fitted parameters should not be over-interpreted, and downstream uncertainty quantification becomes essential. In this way, genetic algorithms serve not only as optimizers but also as exploratory tools that reveal structure in the parameter landscape before committing to a single calibrated model [599].

8.1.5 Gradient-based optimization

If gradients $\nabla_\theta \mathcal{L}(\theta)$ are available or can be reliably approximated, gradient-based methods can be highly efficient. A gradient provides local directional information about how to change parameters to reduce the loss, so each evaluation can be used more productively than in purely exploratory search. This is especially valuable when the number of tuned parameters p is moderate-to-large, as occurs when transmission is parameterized with covariates, splines, or time-varying modifiers, or when the fitted model includes multiple coupled data streams and nuisance parameters.

In mechanistic epidemic models, gradients can be produced in several ways. A simple option is finite differences, which perturbs each parameter and re-runs the forward model to approximate partial derivatives. This is easy to implement but can be expensive and sensitive to step-size choice, particularly in stiff systems. More structured approaches compute sensitivities alongside the state dynamics, either by augmenting the ODE system with sensitivity equations or by using adjoint methods that scale well when p is large. In data-driven models and differentiable simulators, gradients are typically obtained through automatic differentiation, which makes first-order methods natural and often very fast.

Once gradients are available, the simplest algorithm is gradient descent [600], which updates parameters by stepping in the negative gradient direction. In practice, pure gradient descent is rarely the most efficient choice for smooth deterministic objectives, because step sizes must be tuned carefully and curvature can vary widely across parameters. Quasi-Newton methods address this by incorporating curvature information without explicitly forming a Hessian. A commonly used representative is (L-)BFGS, which constructs an approximation H_n of the inverse Hessian from successive gradients and iterates. A generic iteration has the form

$$\theta_{n+1} = \theta_n - \alpha_n H_n \nabla_\theta \mathcal{L}(\theta_n), \tag{8.10}$$

where α_n is chosen by a line search or schedule. Intuitively, H_n rescales the gradient so that the algorithm takes smaller steps along directions that are steep or poorly conditioned, and larger steps along directions that are flat, yielding faster and more stable convergence than naïve gradient descent.

A smooth workflow for using gradient-based optimization begins with careful parameterization and constraints. Epidemiological rates must remain positive and often behave better on a log scale; transforming parameters so that the optimizer works in an unconstrained space can prevent invalid proposals and reduce numerical instability. When hard bounds are required, modern solvers support box constraints directly, but even then, transformations can improve conditioning. The next step is defining the objective in a way that is differentiable with respect to θ. Least-squares and Gaussian negative log-likelihoods are straightforward; objectives that include discontinuous logic (for example, hard thresholds for policy triggers or piecewise definitions without smoothing) can lead to unreliable gradients and should be smoothed or handled with alternative methods if possible.

Initialization plays a larger role than is sometimes appreciated. Gradient-based methods are local: they move downhill from the starting point and therefore converge to a nearby basin of attraction. In extended SIR models, multiple basins are common, and ill-conditioning arises when different parameter combinations produce similar observed outputs. A practical solution is to combine gradient-based refinement with a global or semi-global initialization strategy, such as coarse grid search, Bayesian search, or evolutionary methods, and then use a quasi-Newton solver to rapidly converge once a good region has been found. Even within a single basin, scaling matters; standardizing parameters (or working with dimensionless combinations) can significantly improve convergence behavior because it makes gradients comparable across coordinates.

Listing 8.6 shows a minimal pattern for gradient-based fitting using a bounded L-BFGS-B solver. The example uses finite differences for gradients to keep the interface simple, but the same structure applies when gradients are computed by sensitivities, adjoints, or automatic differentiation. The objective is written as a scalar loss that internally runs the forward model and computes mismatch to the observed data.

```
import numpy as np
from scipy.optimize import minimize

def objective(theta, t, y_obs, known_inputs):
    yhat = simulate_outputs(theta, t, known_inputs)
    return np.mean((y_obs - yhat)**2)  # example: MSE

# Bounds enforce feasibility directly (e.g., positive rates)
bounds = [(1e-6, 10.0), (1e-6, 5.0)]  # example bounds for (beta, gamma)

res = minimize(
    fun=lambda th: objective(th, t, y_obs, known_inputs),
    x0=theta0,
    method="L-BFGS-B",
    bounds=bounds,
    options={"maxiter": 500}
)

theta_hat = res.x
best_loss = res.fun
```

Listing 8.6: Gradient-based fitting with a quasi-Newton method (L-BFGS-B).

The main practical checks after running Listing 8.6 are straightforward. Convergence diagnostics from the solver should be inspected, but the more informative tests are external: whether the fitted parameters are epidemiologically plausible, whether residuals show systematic structure over time, and whether performance generalizes to a held-out period or alternative observable. When these checks fail, the issue is often not the optimizer but the modeling choices that define $\hat{y}(\theta)$ and $\mathcal{L}$: missing delay structure, mismatched observation mapping, overly flexible time variation, or insufficient constraints. In that sense, gradient-based optimization is best viewed as a powerful engine within a broader modeling loop, where parameterization, constraints, and objective design determine whether the optimization problem is well-conditioned enough for gradients to be trustworthy and informative.

8.1.6 Derivative-free local search

When gradients are unavailable, unreliable, or the objective is non-smooth, derivative-free local optimizers provide a practical way to improve parameters using only function evaluations. This situation arises frequently in epidemiological modeling: the forward model may include discontinuities (piecewise interventions, threshold-based behavior), stochastic components (agent-based simulations), or numerical solvers whose output is not smoothly differentiable with respect to parameters. Even when gradients exist in principle, finite-difference approximations can become unstable if the model is stiff or if small parameter perturbations produce qualitatively different trajectories. In these cases, derivative-free methods are attractive because they require only the ability to compute $\mathcal{L}(\theta)$ for proposed θ, making them easy to integrate with black-box simulators.

A widely used method for low-dimensional continuous problems is the Nelder–Mead simplex algorithm [601]. The method maintains a simplex consisting of $p+1$ vertices in p-dimensional parameter space. At each iteration, it ranks simplex vertices by their objective values and attempts to replace the worst vertex with a better point using geometric transformations of the simplex. Reflection proposes a point across the face opposite the worst vertex; if reflection yields a substantial improvement, expansion pushes further in the same direction; if reflection fails, contraction searches closer to the current best region; and if contraction also fails, shrink pulls the entire simplex toward the current best vertex. These simple operations produce a local search behavior that adapts step sizes automatically based on whether improvements are found, often making Nelder–Mead effective as a refinement tool even when the loss is moderately irregular.

In practical epidemiological fitting, Nelder–Mead is most useful when paired with a sensible starting point, because it is a local method and can converge to different minima depending on initialization. A common workflow is to first obtain a candidate $\theta^{(0)}$ from a global strategy such as a coarse grid, Bayesian search, or a population-based method, and then run Nelder–Mead to reduce the loss without requiring any sensitivities. Because simplex methods do not natively enforce bounds, feasibility is typically maintained by re-parameterization (e.g., optimizing over $\log \beta$ to ensure positivity), by adding penalties for infeasible parameters, or by projecting proposed points back into a feasible region. These simple mechanisms are often sufficient for compartmental models where bounds and positivity constraints are central.

Pattern search and related direct search methods generalize the same principle while providing stronger theoretical guarantees under mild conditions and more natural handling of constraints [602]. Instead of moving a simplex, pattern search evaluates the objective on a structured set of poll directions around the current point and accepts a move when an improvement is found, adapting the step size based on success or failure. This makes pattern search particularly convenient when bounds are important, since proposed points can be generated directly within the feasible region. In practice, direct search is a good choice when the objective is expensive and one wants controlled, monotone improvement, or when Nelder–Mead's geometry becomes unstable due to scaling issues or irregular loss surfaces.

Listing 8.7 illustrates a minimal Nelder–Mead refinement step. The objective wraps the epidemiological forward model and returns a scalar loss. When positivity is required, the example demonstrates a log-parameterization so that the optimizer operates in an unconstrained space while the simulator receives valid parameters.

```
import numpy as np
from scipy.optimize import minimize

def objective_log_params(eta, t, y_obs, known_inputs):
    # Enforce positivity through exponentiation
    theta = np.exp(eta)
    yhat = simulate_outputs(theta, t, known_inputs)
    return np.mean((y_obs - yhat)**2)

# theta0 is an initial guess in original scale (positive)
eta0 = np.log(np.maximum(theta0, 1e-12))

res = minimize(
    fun=lambda et: objective_log_params(et, t, y_obs, known_inputs),
    x0=eta0,
    method="Nelder-Mead",
```

```
    options={"maxiter": 2000, "xatol": 1e-6, "fatol": 1e-8}
)

theta_hat = np.exp(res.x)
best_loss = res.fun
```

Listing 8.7: Nelder–Mead refinement with log-parameterization for positivity.

The same wrapper style can be used for pattern search, with the main difference being the optimizer implementation. In all cases, the most important practical decisions are the choice of initialization, the handling of feasibility, and the stopping criteria. Because derivative-free methods can spend many evaluations making small local improvements, they are most effective when used deliberately as a refinement stage, when the modeler expects that a good basin has already been located and wants to reduce loss further without investing in sensitivity calculations.

8.1.7 Swarm-based methods

Particle swarm optimization (PSO) is a population-based heuristic in which candidate solutions (particles) move through parameter space, guided by their own best historical position and the swarm's best-known position [603]. It is well suited to epidemiological objectives that behave like black boxes: expensive forward simulations, noisy losses, and limited differentiability. Unlike genetic algorithms, PSO does not require designing crossover or mutation operators or choosing an explicit encoding beyond the parameter vector itself. Each particle simply represents a candidate parameter set $\theta^{(i)}$, and the algorithm repeatedly updates the swarm by combining memory of good solutions with randomized exploration.

A convenient way to view PSO is as a compromise between greedy local refinement and global search. Each particle carries three pieces of information: its current position $\theta_n^{(i)}$, its current velocity $v_n^{(i)}$, and its personal best position $p_n^{(i)}$, meaning the best solution it has seen so far. The swarm as a whole maintains a global best position g_n, which is the best solution found by any particle. At each iteration, every particle proposes a new velocity that blends inertia with two attraction terms: a cognitive pull toward its own best and a social pull toward the global best. A common update has the form

$$v_{n+1}^{(i)} = \omega\, v_n^{(i)} + c_1 r_{1,n}^{(i)} \left(p_n^{(i)} - \theta_n^{(i)}\right) + c_2 r_{2,n}^{(i)} \left(g_n - \theta_n^{(i)}\right), \tag{8.11}$$

$$\theta_{n+1}^{(i)} = \theta_n^{(i)} + v_{n+1}^{(i)}, \tag{8.12}$$

where ω is an inertia weight, c_1 and c_2 are acceleration coefficients, and $r_{1,n}$, $r_{2,n}$ are vectors of i.i.d. uniform random numbers on $[0, 1]$. The random factors keep particles from collapsing deterministically onto the same trajectory, while the two pulls ensure that the swarm keeps exploiting regions that have already produced good fits. The algorithm evaluates $\mathcal{L}(\theta_{n+1}^{(i)})$ for each particle, updates personal bests and the global best, and repeats.

Using PSO smoothly for epidemiological parameter fitting usually begins with careful constraint handling. Compartmental model parameters are often bounded and frequently require positivity. A simple practical approach is to enforce bounds after each position update by clipping or reflecting particles back into the feasible region. When parameters span orders of magnitude, performing PSO in a transformed space (such as log space for positive rates) can improve stability because it makes step sizes comparable across parameters. Objective noise is also common; PSO is typically robust to moderate noise because it aggregates evidence across many particles, but if the simulator variance is large, it can help to average the loss over a small number of replicates when updating the personal and global bests.

PSO is naturally parallelizable because the most expensive operation is evaluating $\mathcal{L}(\theta)$ for many particles each iteration, and these evaluations are independent. This is an important practical advantage in simulation-heavy calibration: one can trade more cores for fewer wall-clock hours without changing the algorithm. The main cost is the number of objective evaluations, roughly (number of particles) × (number of iterations). For this reason, PSO is often used to locate a good region of parameter space and then followed by a local refinement method to converge more precisely.

Listing 8.8 provides a minimal PSO loop for bounded continuous parameters. The implementation keeps the algorithm explicit: initialization, objective evaluation, personal/global best updates, and the velocity-position recursion. The same structure extends to mixed discrete–continuous problems by modifying the proposal step for discrete coordinates, although PSO is most commonly used for continuous parameters.

```
import numpy as np

def clip(theta, lb, ub):
    return np.minimum(np.maximum(theta, lb), ub)

def pso(theta_lb, theta_ub, t, y_obs, known_inputs,
        n_particles=40, n_iter=200, w=0.7, c1=1.4, c2=1.4, seed=0):
    rng = np.random.default_rng(seed)

    lb = np.array(theta_lb, dtype=float)
    ub = np.array(theta_ub, dtype=float)
    dim = lb.size

    # Initialize positions uniformly and velocities to small random values
    X = rng.uniform(lb, ub, size=(n_particles, dim))
    V = rng.normal(0.0, 0.1, size=(n_particles, dim))

    # Evaluate initial swarm
    L = np.array([objective(x, t, y_obs, known_inputs) for x in X])

    P = X.copy()             # personal best positions
    LP = L.copy()            # personal best losses
    g_idx = np.argmin(LP)
    G = P[g_idx].copy()      # global best position
    LG = LP[g_idx]           # global best loss

    for _ in range(n_iter):
        r1 = rng.uniform(0.0, 1.0, size=(n_particles, dim))
        r2 = rng.uniform(0.0, 1.0, size=(n_particles, dim))

        # Velocity and position updates
```

```
        V = w * V + c1 * r1 * (P - X) + c2 * r2 * (G - X)
        X = clip(X + V, lb, ub)

        # Evaluate swarm
        L = np.array([objective(x, t, y_obs, known_inputs) for x in X])

        # Update personal bests
        improved = L < LP
        P[improved] = X[improved]
        LP[improved] = L[improved]

        # Update global best
        g_idx = np.argmin(LP)
        if LP[g_idx] < LG:
            G = P[g_idx].copy()
            LG = LP[g_idx]

    return G, LG
```

Listing 8.8: A minimal particle swarm optimization loop for bounded parameters.

The parameters in Listing 8.8 control the balance between exploration and exploitation. A larger inertia w tends to keep particles moving and can help exploration early, while smaller w yields tighter convergence late in the run. The coefficients c_1 and c_2 determine whether particles rely more on their own history or on the swarm's collective best; in noisy objectives, slightly stronger social influence can help focus the swarm, while stronger cognitive influence can preserve diversity and reduce premature collapse. As with other global methods, PSO is most effective when it is used as part of a workflow: it finds a strong candidate solution under limited modeling assumptions, and that candidate can then be refined and validated using the local methods and out-of-sample evaluation criteria introduced elsewhere in this chapter [599].

8.1.8 Simulation-based inference via discrepancy matching

In many realistic epidemic simulators, the likelihood $p_\theta(y_{1:K})$ is unavailable or prohibitively expensive. This is common when the forward model is stochastic and high-dimensional, as in branching-process models, agent-based simulations, network epidemics, or compartmental models with complex reporting and intervention logic. In such settings, it may be easy to simulate synthetic epidemic trajectories y^{sim} for a proposed parameter vector θ, yet hard to write down (or evaluate) the probability of observing the real data under that θ. Approximate Bayesian computation (ABC) addresses this by replacing likelihood evaluation with a simulation-and-comparison procedure [604,605]. The core idea is to accept parameters that generate simulated data "close enough" to the observed data, where closeness is defined through summary statistics and a discrepancy measure.

ABC begins by choosing a prior $p(\theta)$ over feasible parameter values, then deciding how to compare simulated and observed epidemics. The comparison is typically done through a vector of summaries $s(\cdot)$, such as early growth rate, peak timing, peak height, cumulative attack rate, or age-specific incidence patterns. A distance function $d(\cdot, \cdot)$ measures mismatch between summaries computed on simulated and observed data. Given a tolerance $\varepsilon > 0$, a basic ABC rejection scheme samples $\theta \sim p(\theta)$, simulates pseudo-data $y^{\text{sim}} \sim p_\theta(\cdot)$, computes $s(y^{\text{sim}})$, and accepts θ if

$$d\Big(s(y^{\text{sim}}), s(y)\Big) \leq \varepsilon, \tag{8.13}$$

producing an approximate posterior distribution over θ [604]. As ε decreases, accepted simulations must match the data more closely, so the approximation becomes sharper; at the same time, acceptance becomes rarer, which increases computational cost. In practice, the quality of ABC hinges on the choice of summaries: if $s(\cdot)$ preserves the information in the data that matters for θ, then ABC can yield meaningful posterior uncertainty; if summaries discard key information, the resulting posterior may be overly diffuse or biased even with very small ε.

From a model-fitting perspective, ABC can be read as discrepancy matching under uncertainty. Instead of producing a single best-fit parameter set, it returns a distribution of plausible parameters whose simulations resemble observed epidemic features. This is particularly useful in epidemiology because different parameter combinations can generate similar observed curves, and the spread of accepted θ values naturally reflects this practical non-identifiability. ABC is also well aligned with scenario analysis: one can propagate the accepted parameter uncertainty through the simulator to obtain predictive envelopes for outcomes under alternative interventions, while keeping the simulator faithful to realistic dynamics.

The simplest rejection sampler is rarely the most efficient way to use ABC because it wastes many proposals that are far from the data. Practical variants improve efficiency by exploring parameter space more systematically [605,606]. ABC-MCMC embeds the accept/reject rule inside a Markov chain, so the algorithm proposes new θ values near previously accepted ones, concentrating effort on promising regions. ABC-SMC (sequential Monte Carlo) organizes the computation into stages with decreasing tolerances $\varepsilon_1 > \varepsilon_2 > \cdots > \varepsilon_L$. Early stages accept coarse matches and identify broad regions of plausibility; later stages focus on tighter matches by reweighting and perturbing particles from the previous stage. This staged refinement is often a smoother way to control the computational trade-off between accuracy and acceptance rate.

A practical ABC workflow therefore starts with engineering the discrepancy. The summaries should be chosen to reflect what the model is intended to capture and what the data can reliably inform. When the model is used for near-term forecasting, summaries might emphasize recent growth, turning points, and short-horizon predictive errors. When the model is used for burden estimation, summaries may emphasize cumulative counts, peak burdens, and timing. Distances should also be scaled so that no single component dominates simply due to units; standardizing summaries or using a weighted distance is common. Because epidemic data often contain reporting noise and artifacts, ε is not only a computational knob but also an implicit statement of how much mismatch is considered acceptable given measurement limitations and model discrepancy.

Listing 8.9 shows a minimal ABC rejection sampler. The code makes the ingredients explicit: sampling from a prior, simulating pseudo-data, computing summaries, computing a distance, and accepting samples that fall below a tolerance. Although rejection ABC is computationally demanding, it is easy to implement and serves as a clear reference point before moving to ABC-MCMC or ABC-SMC.

```
import numpy as np

def summarize(y):
```

```
    # Example summaries: choose summaries appropriate to the application
    peak = np.max(y)
    t_peak = int(np.argmax(y))
    total = np.sum(y)
    return np.array([peak, t_peak, total], dtype=float)

def distance(s_sim, s_obs, w=None):
    if w is None:
        w = np.ones_like(s_obs)
    # Weighted Euclidean distance in summary space
    z = (s_sim - s_obs) * np.sqrt(w)
    return float(np.sqrt(np.sum(z**2)))

def abc_rejection(prior_sampler, simulator, y_obs, epsilon,
                  n_accept=500, max_tries=200000, seed=0):
    rng = np.random.default_rng(seed)

    s_obs = summarize(y_obs)
    accepted = []
    tries = 0

    while len(accepted) < n_accept and tries < max_tries:
        tries += 1
        theta = prior_sampler(rng)
        y_sim = simulator(theta, rng)
        d = distance(summarize(y_sim), s_obs)

        if d <= epsilon:
            accepted.append(theta)

    return np.array(accepted), tries
```

Listing 8.9: A minimal ABC rejection sampler for simulator-based fitting.

The output of Listing 8.9 is a set of accepted parameter samples and the number of proposals required to obtain them. As ε is reduced, the accepted sample set typically becomes more concentrated, but the number of proposals required grows. In practice, one often begins with a relatively large tolerance to verify that the simulator and summaries behave sensibly, then gradually tightens ε while monitoring acceptance rates and the stability of posterior summaries. This tuning process is not merely computational: it encourages explicit thinking about which epidemic features the model is expected to match and how measurement noise and model discrepancy should be reflected in the inference procedure [606].

8.1.9 Joint state–parameter inference in state-space models

When epidemic models are cast as partially observed Markov processes, inference typically involves two intertwined unknowns: the latent epidemic trajectory $x_{0:K}$ (the unobserved compartment counts or other hidden states over time) and the parameter vector θ that governs transmission, progression, and observation mechanisms. In many practical settings, the data do not directly reveal the latent states, and naive parameter fitting that ignores latent uncertainty can yield unstable or biased estimates, especially when reporting noise, delays, and partial observability play a central role. State-space modeling addresses this explicitly by specifying a transition model $p_\theta(x_k \mid x_{k-1})$ and an observation model $p_\theta(y_k \mid x_k)$, which together induce a joint posterior $p(\theta, x_{0:K} \mid y_{1:K})$.

The computational challenge is that the likelihood $p_\theta(y_{1:K})$ is rarely available in closed form for nonlinear and/or non-Gaussian epidemic dynamics, even when simulation of the transition and observation models is straightforward. Particle filtering (sequential Monte Carlo) provides a way to approximate filtering distributions over states by propagating and reweighting a set of particles through time, producing an estimate of the likelihood as a by-product. Particle Markov chain Monte Carlo (PMCMC) leverages this fact by embedding a particle filter within an MCMC scheme so that parameters can be updated using an unbiased likelihood estimate [607]. Conceptually, PMCMC replaces the exact likelihood evaluation in a Metropolis–Hastings ratio with a particle-filter-based estimate, yet still targets the correct posterior under appropriate conditions, which is what makes it an appealing bridge between mechanistic stochastic models and fully Bayesian inference.

A typical PMCMC workflow alternates between proposing new parameters and running a particle filter conditioned on those parameters. At iteration n, the algorithm proposes $\theta' \sim q(\theta' \mid \theta_n)$. A particle filter is then run under θ' to (i) approximate the distribution over latent trajectories consistent with the data and (ii) produce an unbiased estimate $\widehat{p}_{\theta'}(y_{1:K})$ of the likelihood. The proposal is accepted or rejected using the standard Metropolis–Hastings logic, with the likelihood replaced by its estimate. When accepted, the chain moves to $\theta_{n+1} = \theta'$; otherwise, it remains at $\theta_{n+1} = \theta_n$. In many implementations, the particle filter also yields a representative latent trajectory or a set of weighted trajectories that can be retained for posterior summaries of hidden epidemic states.

Using PMCMC smoothly requires several practical design choices. The first is selecting a particle filter appropriate for the model and observation process. For epidemic data, weight degeneracy can occur quickly if the observation model is sharp (for example, when reported cases tightly constrain latent infections), and resampling strategies are essential to keep particle diversity. The second is choosing the number of particles N. Too few particles yield a highly variable likelihood estimate, which can cause poor MCMC mixing because acceptance probabilities become erratic; too many particles increase computational cost without commensurate gains. The third is proposing parameters effectively. Because epidemic parameters can be correlated, proposal distributions often need to reflect scale and correlation structure, which is commonly handled by working in transformed coordinates (log rates, logit probabilities) and adapting proposal covariance during warm-up.

Computational intensity is the main trade-off. Each MCMC step requires running a full particle filter across all observation times, so the total cost scales with the product of chain length and the number of particles. This can become substantial for long time series, high-dimensional state representations, or weakly identified models where the chain mixes slowly. Nonetheless, PMCMC remains a cornerstone approach when uncertainty quantification is central, because it propagates uncertainty in latent states jointly with parameter uncertainty, rather than treating the latent trajectory as fixed or implicitly optimized. In applied epidemiological analysis, it is therefore often used in a targeted way: as a high-fidelity inference engine for a carefully chosen model and a restricted set of key parameters, or as a benchmark against which faster approximate fitting methods can be compared.

Listing 8.10 provides a minimal skeleton of a particle marginal Metropolis–Hastings scheme. The code is intentionally schematic: it highlights the moving parts that must be supplied for a concrete model, namely a parameter proposal mechanism, a particle filter that returns an unbiased likelihood estimate, and optional storage of latent-state summaries.

```
import numpy as np

def pmmh(theta0, n_steps, propose, log_prior, particle_filter, seed=0):
    """
    propose(theta, rng) -> (theta_proposed, log_q_forward, log_q_backward)
    particle_filter(theta, rng) -> (log_lik_hat, optional_state_summary)
    log_prior(theta) -> log p(theta)
    """
    rng = np.random.default_rng(seed)

    theta = np.array(theta0, dtype=float)
    logp = log_prior(theta)

    loglik_hat, state_sum = particle_filter(theta, rng)

    chain = [theta.copy()]
    loglik_chain = [loglik_hat]
    state_summaries = [state_sum]

    for _ in range(n_steps):
        theta_p, logq_fwd, logq_bwd = propose(theta, rng)

        logp_p = log_prior(theta_p)
        if np.isfinite(logp_p):
            loglik_hat_p, state_sum_p = particle_filter(theta_p, rng)

            log_alpha = (logp_p + loglik_hat_p + logq_bwd) - (logp + loglik_hat + logq_fwd)
            if np.log(rng.uniform()) < log_alpha:
                theta, logp = theta_p, logp_p
                loglik_hat, state_sum = loglik_hat_p, state_sum_p

        chain.append(theta.copy())
        loglik_chain.append(loglik_hat)
        state_summaries.append(state_sum)

    return np.array(chain), np.array(loglik_chain), state_summaries
```

Listing 8.10: Skeleton of particle marginal Metropolis–Hastings (PMMH), a common PMCMC variant.

The PMCMC loop in Listing 8.10 makes the logic transparent: every proposed parameter value requires a new particle-filter run, and the acceptance probability depends on both the prior and the estimated likelihood. In practice, the most critical ingredient is the particle filter itself. It must be implemented so that the likelihood estimate is unbiased and numerically stable, and it must be configured with enough particles that the log-likelihood estimate is not excessively noisy. When these conditions are met, PMCMC provides a principled route to posterior inference for stochastic epidemic state-space models, yielding both parameter uncertainty and latent-state uncertainty in a unified framework [607].

8.2 Model calibration

Model calibration addresses a different question than parameter estimation. Parameter estimation treats the model class as fixed and asks which parameter values are supported by the data under an assumed statistical data-generating mechanism. Calibration starts from the practical reality that models are approximations and asks how to tune the model so that it is fit for a specific purpose, such as forecasting hospital load, reproducing peak timing, or comparing intervention scenarios. In this sense, calibration is purpose-driven: the definition of a "good" parameter set depends on which outputs matter, on which time horizon is relevant, and on how errors should be penalized [578,582].

A useful formalization begins with separating three ingredients: the simulator, the observation mapping, and the calibration objective. Let the simulator (mechanistic or data-driven) produce predicted observables $\hat{y}_{1:K}(\theta)$ for a given θ, and let $y_{1:K}$ denote the corresponding measurements. Calibration chooses θ to minimize a loss that reflects the intended use:

$$\hat{\theta}_{\text{cal}} \in \arg\min_{\theta \in \Theta} \mathcal{L}_{\text{cal}}\big(y_{1:K}, \hat{y}_{1:K}(\theta)\big) + \lambda \mathcal{R}(\theta), \tag{8.14}$$

where $\mathcal{L}_{\text{cal}}$ is explicitly chosen by the modeler and may differ from a likelihood-based negative log-probability. The regularizer $\mathcal{R}$ encodes stability, smoothness, plausibility, or simplicity, and is often the most direct way to prevent calibration from "chasing noise" when the model is flexible or the data are limited. Eq. (8.14) intentionally mirrors generic fitting, but the interpretation is different: $\hat{\theta}_{\text{cal}}$ is the parameter set that best supports the task as defined by $\mathcal{L}_{\text{cal}}$, not necessarily the parameter set that is closest to a hypothetical true θ under a fully correct model.

Calibration becomes especially distinct from estimation when the model is known to be misspecified. A standard way to write this is to acknowledge discrepancy between the simulator and reality by adding an explicit discrepancy term,

$$y_k = \hat{y}_k(\theta) + \delta_k + \varepsilon_k, \tag{8.15}$$

where δ_k represents systematic mismatch (due to unmodeled heterogeneity, aggregation, behavioral feedback, or simplifications) and ε_k represents measurement noise. In calibration, δ_k is not an inconvenience to be ignored; it is part of the reason why a purpose-specific loss and validation protocol are necessary. Bayesian calibration frameworks treat δ as a structured stochastic process (often smooth in time) and propagate its uncertainty into predictions, which helps prevent overconfident inference even when a tuned simulator matches observed data closely [581,582].

A practical calibration design also separates fitting from evaluation. One typically selects a calibration window (or calibration dataset) and a validation window that reflects how the model will be used. For time-indexed epidemic data, validation is rarely an i.i.d. split; it is more naturally defined as a holdout period after the calibration period, or as a rolling-origin protocol that repeatedly calibrates on past data and scores predictions on the subsequent horizon. This makes the calibration objective operational: a model that matches history but fails to forecast the near future is not well calibrated for forecasting, even if it achieves a low in-sample loss. When probabilistic forecasts are produced, calibration should also be evaluated with proper scoring rules, since point-wise errors alone can reward overconfident predictions [562].

In multi-stream calibration, several observables are fitted jointly, such as cases, hospitalizations, and deaths. A common approach is to define $\mathcal{L}_{\text{cal}}$ as a weighted sum of per-stream losses, with weights chosen to reflect importance and scale:

$$\mathcal{L}_{\text{cal}}(y, \hat{y}) = \sum_{m=1}^{M} \alpha_m \, \mathcal{L}_{\text{cal}}^{(m)}\Big(y^{(m)}, \hat{y}^{(m)}(\theta)\Big), \tag{8.16}$$

where m indexes data streams and $\alpha_m > 0$ controls trade-offs. The role of α_m is not merely numerical; it encodes the operational preference of what the model must get right. A calibration aimed at ICU capacity planning may reasonably prioritize hospitalization fit even if case incidence fit degrades, whereas transmission inference may prioritize matching early growth patterns.

The calibration process itself is most reliable when it is treated as a loop rather than a one-shot optimization. One begins by choosing targets and defining a loss aligned with the application, then fits parameters using one of the optimization strategies described earlier. The fitted model is then stress-tested by out-of-sample evaluation, residual diagnostics, and plausibility checks on implied quantities (for mechanistic models, quantities like reproduction numbers or reporting fractions). When failures are systematic, the correction is often not "optimize harder" but revise one of the modeling ingredients: the observation mapping, the loss scaling, the regularization, or the model structure and discrepancy assumptions [578,581].

Listing 8.11 shows a minimal calibration workflow that makes the separation between calibration and validation explicit. The code calibrates by minimizing an in-sample loss on a calibration window, then reports forecasting-style metrics on a holdout window using the evaluation routine introduced earlier. The same pattern applies to mechanistic models by wrapping the simulator as a predictor with a `predict` method.

```
import numpy as np
from scipy.optimize import minimize

def calibrate(theta0, bounds, X_cal, y_cal, model_builder):
    """
    model_builder(theta) -> fitted model object with predict(X) method.
    Calibration objective is defined on the calibration window.
    """

    def cal_objective(theta):
        model = model_builder(theta)
        yhat = model.predict(X_cal)
        return float(np.mean((y_cal - yhat)**2))  # example calibration loss

    res = minimize(cal_objective, x0=theta0, method="L-BFGS-B", bounds=bounds)
    theta_hat = res.x
    return theta_hat, model_builder(theta_hat)

# After calibration: evaluate on holdout (validation) window
```

```
# metrics, yhat_val = evaluate_forecast(fitted_model, X_val, y_val)
```

Listing 8.11: A minimal calibration workflow with an explicit holdout window.

The rest of this section develops calibration in a structured way. We first discuss how to choose calibration objectives, weights, constraints, and regularization so that $\mathcal{L}_{\text{cal}}$ matches the modeling purpose. We then formalize validation protocols for time-indexed epidemic data and show how discrepancy-aware calibration prevents overconfident conclusions when a simulator is inevitably imperfect.

8.2.1 Calibration targets and intended use

Calibration is meaningful only relative to a purpose. Before choosing an optimizer, a loss function, or even a dataset window, one must specify the decision or scientific task the calibrated model will support. In epidemiology, the same mechanistic structure can be calibrated in substantially different ways depending on whether the objective is short-horizon forecasting, capacity planning, scenario comparison, or retrospective reconstruction. This purpose-first stance is what distinguishes calibration from generic parameter fitting: it determines which outputs matter, which time periods are emphasized, and what kinds of errors are acceptable.

We formalize this by defining a vector of target quantities (calibration targets) $z_{1:K}$, derived from the raw observations $y_{1:K}$ through a target-extraction operator ϕ. Examples include daily incidence, weekly totals, smoothed incidence, peak timing, peak magnitude, cumulative burden, or hospitalization occupancy. Writing $z_k = \phi_k(y_{1:K})$ allows calibration to operate on quantities that are closer to the decision-relevant signal than the raw data, while remaining explicit about the transformation. The model produces corresponding predicted targets $\hat{z}_k(\theta) = \phi_k(\hat{y}_{1:K}(\theta))$, and calibration seeks parameters that make $\hat{z}$ reproduce z under a purpose-specific loss:

$$\hat{\theta}_{\text{cal}} \in \arg\min_{\theta \in \Theta} \sum_{k=1}^{K} \rho_k \, \ell\big(z_k, \hat{z}_k(\theta)\big) \; + \; \lambda \, \mathcal{R}(\theta), \tag{8.17}$$

where $\ell(\cdot, \cdot)$ compares target values, $\rho_k \geq 0$ controls temporal emphasis, and $\mathcal{R}$ encodes regularization or plausibility. The weights ρ_k capture the intended use directly. For instance, a forecasting task naturally assigns higher weight to recent observations (or to errors on a validation horizon), while retrospective reconstruction may weight the entire historical window more uniformly.

The intended use determines not only which targets appear in z, but also their semantics. For near-term forecasting, calibration typically emphasizes short-horizon predictive accuracy and timing, because operational decisions are sensitive to what happens in the next days to weeks. This often motivates targets such as recent growth rates, turning-point behavior, and horizon-weighted losses that penalize errors near the forecast origin more strongly than errors far in the past. For capacity planning, calibration targets shift toward burden-relevant observables such as hospital admissions, occupancy, ICU utilization, or deaths, because these are the quantities that constrain resources. In such cases, it can be reasonable to prioritize matching peaks and upper tails over matching low-incidence periods, and to accept larger relative error in cases if the hospitalization pathway is well captured.

Scenario comparison and counterfactual evaluation impose a different requirement: calibration must preserve the causal structure by which interventions or covariates affect transmission and outcomes.

Targets may include not only marginal fits (how well the model matches observed curves) but also invariance properties (whether fitted intervention effects remain stable across time windows or regions). Calibration here commonly emphasizes internal consistency of inferred mechanisms, because scenario forecasts can be misleading if the model fits historical data by compensating errors across parameters in a way that breaks under changed conditions. Retrospective reconstruction, in turn, may prioritize long-run coherence and smoothness, using targets such as cumulative burden, wave timing, or seroprevalence-consistent attack rates, since the goal is often to infer latent quantities (true infections, effective reproduction number trajectories) rather than to optimize short-horizon forecasts.

Calibration targets also differ between mechanistic and data-driven models, even when the same raw data are used. Mechanistic models often calibrate to multiple linked observables simultaneously (e.g., cases and hospitalizations), because the model's value lies in its ability to translate latent infections into downstream outcomes. Data-driven models may instead calibrate directly to the forecasted series (e.g., incidence), with targets defined by the evaluation horizon and the forecasting metric. In both cases, clarity about intended use prevents a common failure mode: selecting a calibration objective that yields excellent in-sample fit on an easy-to-match signal but poor performance on the quantity that actually matters for decisions.

A practical way to make intended use operational is to explicitly define the calibration window, the validation window, and the horizon. Let $\mathcal{T}_{\text{cal}}$ denote time indices used to fit, and $\mathcal{T}_{\text{val}}$ denote indices used to score performance. In time-indexed epidemic data, $\mathcal{T}_{\text{val}}$ is typically chosen as a future period following $\mathcal{T}_{\text{cal}}$, possibly repeated in a rolling-origin scheme. Writing the objective as

$$\hat{\theta}_{\text{cal}} \in \arg\min_{\theta \in \Theta} \sum_{k \in \mathcal{T}_{\text{cal}}} \rho_k \, \ell\big(z_k, \hat{z}_k(\theta)\big) + \lambda \mathcal{R}(\theta), \qquad \text{score}(\hat{\theta}) = \sum_{k \in \mathcal{T}_{\text{val}}} \tilde{\rho}_k \, \tilde{\ell}\Big(z_k, \hat{z}_k(\hat{\theta})\Big), \tag{8.18}$$

makes explicit that calibration and evaluation can legitimately use different losses and different weights. The calibration loss encodes how parameters are chosen, while the validation score encodes whether the calibrated model is fit for the intended deployment.

By making calibration targets explicit—what is matched, on which time window, and with what priority—the remainder of the calibration section can be presented as a sequence of concrete design choices. Observation mapping determines how $\hat{y}$ is produced from the simulator; loss design determines ℓ and weights ρ_k; constraints and discrepancy assumptions determine what kinds of mismatches are tolerated; and validation protocols determine whether the resulting calibrated model can be trusted for the stated use.

8.2.2 Observation mapping and measurement process

Calibration connects a model to data through an observation mapping, and many calibration failures trace back to getting this mapping wrong. Mechanistic epidemic models evolve latent states $x(t)$ such as susceptible, exposed, infectious, and removed compartments, while surveillance systems report quantities such as notified cases, test positivity, hospital admissions, occupancy, and deaths. These are not direct readings of $x(t)$; they are derived, delayed, filtered, and biased measurements. Calibration must therefore specify how a simulated latent trajectory becomes a predicted observable, and which stochastic effects represent measurement noise versus systematic mismatch [574,582].

A convenient formalization distinguishes the simulator from the measurement process. Let $x(t;\theta)$ denote the latent state implied by parameters θ. The model produces an idealized measurement signal through a deterministic mapping h,

$$\mu_k(\theta) = h\big(x(t_k;\theta),\theta\big), \tag{8.19}$$

and the data are then generated by a stochastic observation model around $\mu_k(\theta)$,

$$y_k \sim p\big(\cdot \mid \mu_k(\theta), \psi\big), \tag{8.20}$$

where ψ collects noise or dispersion parameters. In calibration, $\mu_k(\theta)$ is the critical object: it embeds reporting fractions, delays, aggregation rules, censoring, and transformations. If h is misspecified, the optimizer may still find a parameter set that matches the observed curve, but the fit is often achieved by distorting epidemiologically meaningful parameters, leading to poor transfer to new time periods or new intervention regimes.

Delays are a central feature of observation mapping. For instance, infections occurring on day t are typically observed as reported cases after a random delay (incubation, testing, reporting), and many data streams are effectively convolutions of an underlying incidence signal with a delay distribution. Let $i_t(\theta)$ denote model-implied incidence (new infections) on day t, and let π_d be a discrete delay probability mass function over delays $d = 0, 1, \ldots, D$. A simple reported-cases mean can be written as

$$\mu_t^{\text{cases}}(\theta) = \rho_t \sum_{d=0}^{D} \pi_d \, i_{t-d}(\theta), \tag{8.21}$$

where $\rho_t \in (0, 1]$ is a reporting fraction that may vary over time. Even when ρ_t is constant, the convolution alone changes the shape and timing of observed peaks relative to the latent epidemic, and ignoring it typically forces calibration to compensate by shifting transmission or progression parameters. Similar constructions apply to hospital admissions and deaths, each with their own delay kernels. Occupancy data require additional care because they represent a stock rather than a flow; one often maps simulated admissions and discharges into an occupancy process, so that the observable is consistent with length-of-stay dynamics rather than a direct transformation of incidence [574].

Aggregation and censoring matter in the same way. Many public health systems report weekly totals, weekend effects create periodic artifacts in daily series, and late reporting leads to right-censoring in the most recent days. Calibration can accommodate this either by mapping model outputs into the same aggregated units (so μ_k matches what is reported), or by pre-processing the data and explicitly stating what information is being retained in the target z. If a model is calibrated on cumulative counts, one obtains a smooth objective but introduces strong temporal dependence in errors because cumulative series share most of their mass between adjacent time points. Calibrating on incidence (or on increments of cumulative counts) typically yields objectives that better reflect day-to-day predictive errors, at the cost of higher sensitivity to reporting noise. The right choice depends on the intended use, but the observation mapping should remain explicit so that these trade-offs are visible rather than accidental [578].

The stochastic component $p(\cdot \mid \mu_k(\theta), \psi)$ should reflect data characteristics. Count data often exhibit overdispersion relative to a Poisson model, and the same mean μ_k can be paired with different noise models depending on how variability is expected to scale with the mean. In calibration practice, this choice functions as an implicit weighting: heavier-tailed observation models reduce the penalty assigned to large deviations, while tighter models force the fit to track peaks and abrupt changes more closely. When multiple data streams are calibrated jointly, the observation mapping must also ensure that the streams are aligned in time and consistent in definitions, since a mismatch in one stream can force implausible trade-offs in the inferred parameters.

Listing 8.12 shows a minimal observation mapping for turning simulated incidence into an expected reported series via a delay convolution and a reporting fraction, with an optional weekly aggregation. This is not meant to prescribe a single "correct" mapping, but to make the structure concrete: the calibration objective should be computed against $\mu(\theta)$ that already incorporates the reporting process.

```
import numpy as np

def convolve_delay(signal, delay_pmf):
    """
    Discrete convolution for nonnegative delays.
    signal[t] corresponds to time t.
    delay_pmf[d] = P(delay = d), d=0..D, sum(delay_pmf)=1.
    Returns same-length array with causal convolution.
    """
    D = len(delay_pmf) - 1
    out = np.zeros_like(signal, dtype=float)
    for t in range(len(signal)):
        acc = 0.0
        for d in range(D + 1):
            if t - d >= 0:
                acc += delay_pmf[d] * signal[t - d]
        out[t] = acc
    return out

def observation_mean_from_incidence(incidence, delay_pmf, report_frac=1.0, weekly=False):
    """
    incidence: array I_t(theta)
    report_frac: scalar or array rho_t
    """
    mu = convolve_delay(incidence, delay_pmf)
    mu = mu * report_frac

    if weekly:
        # aggregate into consecutive 7-day sums
        n_weeks = len(mu) // 7
        mu = mu[: 7 * n_weeks].reshape(n_weeks, 7).sum(axis=1)

    return mu
```

Listing 8.12: Observation mapping: delay convolution, reporting fraction, and optional weekly aggregation.

In a full calibration pipeline, the simulator produces latent trajectories, the observation mapping produces $\mu(\theta)$ in the same units and timing as the data, and only then is a loss computed. Treating the observation mapping as a first-class component of the model has two practical benefits. It prevents compensatory parameter distortions that arise when the fit tries to explain reporting artifacts through transmission dynamics, and it makes validation more interpretable, since out-of-sample failures can be attributed to either the epidemic dynamics or the measurement assumptions rather than being blended into a single opaque mismatch [582].

8.2.3 Calibration objectives and loss design

Once calibration targets and an observation mapping are defined, calibration reduces to an explicit choice of objective. In parameter estimation, the objective is typically dictated by a probabilistic model through a likelihood. In calibration, the objective is chosen to reflect intended use, acceptable error types, and the practical limitations of data and model fidelity. This is not a cosmetic choice: different losses can lead to materially different calibrated parameters, especially when models are misspecified or data streams conflict. Treating the loss as part of the model specification makes these trade-offs explicit and easier to validate.

A general calibration objective can be written as

$$\hat{\theta}_{\text{cal}} \in \arg\min_{\theta \in \Theta} \sum_{k \in \mathcal{T}_{\text{cal}}} \rho_k \, \ell\big(z_k, \hat{z}_k(\theta)\big) + \lambda \, \mathcal{R}(\theta), \tag{8.22}$$

where z_k are the chosen calibration targets, $\hat{z}_k(\theta)$ are the corresponding model-predicted targets, ρ_k are temporal weights, and $\mathcal{R}$ is a regularizer that encodes stability or plausibility. The role of ρ_k is often operational: when recent fit matters most (e.g., for near-term forecasting), ρ_k can increase toward the end of the calibration window; when the goal is retrospective reconstruction, ρ_k is typically more uniform. The loss ℓ determines what kinds of errors are punished most severely, and therefore what the optimizer will prioritize matching.

Squared-error losses are a common default because they are simple and align naturally with least-squares solvers. When targets have a large dynamic range, squared error on the raw scale tends to overweight peaks, so a frequent alternative is to calibrate on a transformed scale. A typical choice is a variance-stabilizing transform such as $\log(1 + z_k)$, yielding

$$\ell_{\log\text{-}2}(z_k, \hat{z}_k) = \big(\log(1 + z_k) - \log(1 + \hat{z}_k)\big)^2, \tag{8.23}$$

which makes proportional errors more comparable across epidemic phases. This is often useful when matching both growth and peak periods is important, or when the data include long tails of low counts where absolute errors are small but relative errors matter for timing.

For count-like targets, losses based on deviance can be more appropriate than pure squared error because they scale naturally with the mean. A widely used example is the Poisson deviance (up to an additive constant independent of θ),

$$\ell_{\text{Pois}}(z_k, \hat{z}_k) = \hat{z}_k - z_k \log(\hat{z}_k), \qquad \hat{z}_k > 0, \tag{8.24}$$

which penalizes underprediction more strongly at high counts and reflects the intuition that variability increases with the level of the signal. In practice, $\hat{z}_k$ is lower-bounded by a small constant to avoid numerical issues. When overdispersion is expected, calibration can still use deviance-style losses while softening peak penalties through robustification or by designing the targets (weekly aggregates, smoothed series) so that the objective does not chase day-to-day reporting artifacts.

Robust losses are useful when the data contain outliers from reporting backlogs, redefinitions, or sudden changes in testing behavior. A standard robust alternative is the Huber loss, applied to residuals $e_k = z_k - \hat{z}_k(\theta)$,

$$\ell_{\text{Huber}}(e_k) = \begin{cases} \frac{1}{2}e_k^2, & |e_k| \leq \delta, \\ \delta\left(|e_k| - \frac{1}{2}\delta\right), & |e_k| > \delta, \end{cases} \tag{8.25}$$

which behaves like squared error near zero but grows only linearly for large deviations. In calibration, this often produces parameter sets that track the overall epidemic trajectory while not being overly pulled by a few anomalous points.

Multi-stream calibration requires extra care because different streams have different scales, delays, and operational importance. A convenient design is a weighted sum of per-stream losses,

$$\mathcal{L}_{\text{cal}}(\theta) = \sum_{m=1}^{M} \alpha_m \sum_{k \in \mathcal{T}_{\text{cal}}} \rho_k^{(m)} \ell^{(m)}\left(z_k^{(m)}, \hat{z}_k^{(m)}(\theta)\right), \tag{8.26}$$

where m indexes streams such as cases, admissions, occupancy, or deaths. The weights α_m should not be viewed as nuisance constants; they encode which outcomes the calibration must get right. A common practical approach is to scale each stream so that typical residual magnitudes are comparable (for example, by dividing by a reference scale such as the stream's median or peak), and then adjust α_m based on the intended use. If the calibration is aimed at hospital planning, it is defensible for hospital-related streams to dominate, even when that degrades case fit, because the calibrated model is being tuned for that operational target.

When the output of interest is probabilistic rather than point-valued, loss design should reward both accuracy and honest uncertainty. Proper scoring rules provide a principled way to do this, ensuring that calibrated predictive distributions are not incentivized to be overly narrow [562]. In practice, this means that if a calibrated model outputs a predictive distribution $F_{\theta,k}$ for z_k, the calibration objective can be built from a sum of proper scores $S(F_{\theta,k}, z_k)$ across time, rather than from pointwise errors. This becomes especially important when the model is used to quantify risk (e.g., the probability of exceeding ICU capacity), where miscalibrated uncertainty can be more damaging than a small bias in the mean.

Listing 8.13 shows a simple implementation pattern that makes loss choices explicit. The function computes a calibration loss from observed and predicted target series, supports temporal weights, and allows switching between squared error, log-squared error, a deviance-like loss, and Huber robustness. The same pattern can be applied per stream and then combined via (8.26).

```
import numpy as np

def huber(e, delta=1.0):
    a = np.abs(e)
    return np.where(a <= delta, 0.5 * e**2, delta * (a - 0.5 * delta))

def calibration_loss(z, zhat, kind="sse", rho=None, huber_delta=1.0):
    """
    z, zhat: arrays of targets and predicted targets, aligned in time
    rho: optional temporal weights (same length)
```

```python
    kind: "sse", "log_sse", "poisson", "huber"
    """
    z = np.asarray(z, dtype=float)
    zhat = np.asarray(zhat, dtype=float)

    if rho is None:
        rho = np.ones_like(z)
    rho = np.asarray(rho, dtype=float)

    eps = 1e-12
    if kind == "sse":
        per_t = (z - zhat)**2

    elif kind == "log_sse":
        per_t = (np.log1p(z) - np.log1p(np.maximum(zhat, 0.0)))**2

    elif kind == "poisson":
        # deviance-like (up to constant): zhat - z*log(zhat)
        per_t = np.maximum(zhat, eps) - z * np.log(np.maximum(zhat, eps))

    elif kind == "huber":
        per_t = huber(z - zhat, delta=huber_delta)

    else:
        raise ValueError(f"Unknown kind: {kind}")

    return float(np.sum(rho * per_t))
```

Listing 8.13: A configurable calibration loss for different target scales and robustness needs.

Loss design is best treated as iterative rather than fixed from the start. A smooth workflow is to begin with a simple objective that matches the primary target on an appropriate scale, calibrate, inspect residual structure and failure modes, then revise ℓ, ρ_k, and α_m to better reflect the intended use and the realities of measurement. This is one of the main practical ways to prevent calibration from becoming an exercise in curve-fitting: the loss encodes what the model is allowed to trade off, and validation later determines whether those trade-offs produce a model that is genuinely fit for purpose [578,582].

8.2.4 Constraints, regularization, and identifiability control

Calibration objectives can typically be driven to lower values by adding flexibility: more free parameters, more time variation, more degrees of freedom in the observation mapping. Without additional structure, this flexibility often produces parameter sets that fit historical data well yet behave unrealistically under perturbations, fail to generalize to new windows, or imply epidemiologically implausible mechanisms. Constraints and regularization are the primary tools for controlling this behavior. They define what parameter values are admissible, how much complexity is permitted, and which trade-offs are allowed when data do not uniquely identify the parameters.

Constraints encode hard scientific or operational knowledge. The most common are simple bound constraints, such as positivity of rates and probabilities, upper limits on transition rates, and plausible ranges for reporting fractions. These constraints can be enforced directly by bounded optimizers, or indirectly by re-parameterization. For example, optimizing over unconstrained $\eta \in \mathbb{R}$ and setting $\beta = \exp(\eta)$ guarantees positivity of a transmission rate, while $\rho = \text{logit}^{-1}(\eta)$ guarantees a reporting fraction in (0, 1). Transformations often improve numerical conditioning as well, because they make multiplicative changes in rates correspond to additive steps in the optimizer's working space.

Constraints also arise from structure in the model rather than simple bounds. In extended SIR models, several parameters must satisfy ordering or conservation properties, and multi-wave fits often require that some parameters are shared across time periods or regions. A simple but effective identifiability control is to fix a subset of weakly identifiable parameters to literature values or values obtained from external studies, and calibrate only the subset that the available data can realistically inform. This approach is not a concession; it is a statement about the information content of the calibration targets, and it can markedly improve stability and interpretability when the model is otherwise underdetermined.

Regularization complements constraints by imposing soft preferences rather than hard feasibility rules. In calibration, regularization appears as an additive term $\lambda\mathcal{R}(\theta)$ in the objective (8.22). The role of $\mathcal{R}$ is to prefer parameter values that are stable, smooth, or close to a baseline unless the data provide strong evidence otherwise. This is particularly important when the model includes time-varying parameters. A common example is a time-varying transmission modifier $\beta(t)$ represented by a vector of values $\beta_{1:K}$ or by coefficients b of a basis expansion. Without regularization, the calibration can set $\beta(t)$ to track every fluctuation in reported cases, effectively absorbing measurement artifacts into transmission. A smoothness penalty discourages rapid changes:

$$\mathcal{R}_{\text{smooth}}(\beta_{1:K}) = \sum_{k=2}^{K} (\beta_k - \beta_{k-1})^2, \tag{8.27}$$

and, more strongly, a second-difference penalty discourages changes in slope:

$$\mathcal{R}_{\text{curv}}(\beta_{1:K}) = \sum_{k=3}^{K} (\beta_k - 2\beta_{k-1} + \beta_{k-2})^2. \tag{8.28}$$

These penalties make the calibration problem better posed by restricting the effective complexity of $\beta(t)$, and they tend to improve forecast stability because the fitted transmission does not overreact to short-lived anomalies.

Shrinkage toward a baseline is another regularization strategy that is useful when parameters are expected to remain near nominal values. If θ_{ref} denotes a reference parameter vector (from literature or a previous fit), then a quadratic penalty

$$\mathcal{R}_{\text{shrink}}(\theta) = \|\theta - \theta_{\text{ref}}\|_2^2 \tag{8.29}$$

encourages conservative updates unless the calibration objective strongly favors deviation. This is particularly useful in rolling recalibration settings where models are re-tuned frequently: shrinkage can prevent dramatic parameter swings driven by short-term noise and helps maintain temporal coherence across successive calibration windows.

Regularization parameters such as λ are not arbitrary knobs; they control an explicit bias–variance trade-off. Small λ allows the model to chase the calibration loss aggressively, increasing the risk of overfitting and instability. Large λ forces stability and smoothness, potentially increasing bias if the model cannot adapt to real regime changes. In calibration practice, λ is typically selected by validation: one tries a range of values, calibrates on $\mathcal{T}_{\text{cal}}$, and scores on $\mathcal{T}_{\text{val}}$. The value that yields the best out-of-sample performance is preferred, because it best matches the intended deployment behavior. This makes regularization a calibration design choice rather than an ad hoc patch.

Another common identifiability control is to reduce dimensionality through structured parameterizations. Instead of estimating a free parameter per day for $\beta(t)$, one can represent $\log \beta(t)$ with a small number of spline coefficients or with piecewise-constant segments separated by change points. This preserves the ability to represent major regime shifts while limiting the number of degrees of freedom, which improves identifiability and reduces the tendency to absorb noise. In scenario modeling, such low-dimensional structure is often crucial: it encourages calibrated parameters to represent stable mechanisms rather than incidental fluctuations.

Listing 8.14 shows a simple pattern for adding smoothness regularization to a calibration loss when $\beta(t)$ is time-varying. The example assumes that the objective compares observed targets to model predictions and then adds a penalty on successive differences of the time-varying parameter vector. The same idea applies to spline coefficients, seasonal effects, or any parameter vector whose variation should be controlled.

```
import numpy as np

def smoothness_penalty(v):
    dv = np.diff(v)
    return float(np.sum(dv**2))

def calibration_objective(theta_static, beta_tv, z_obs, zhat_fn, lam=1.0):
    """
    theta_static: other parameters (fixed over time)
    beta_tv: time-varying transmission modifier (length K)
    zhat_fn: function producing predicted targets given parameters
    """
    zhat = zhat_fn(theta_static, beta_tv)
    data_loss = float(np.mean((z_obs - zhat)**2))
    reg = lam * smoothness_penalty(beta_tv)
    return data_loss + reg
```

Listing 8.14: Adding smoothness regularization for time-varying parameters in calibration.

Constraints and regularization should be treated as part of the calibration specification, not as afterthoughts. They define which solutions are admissible, which solutions are preferred, and how much the calibrated model is allowed to adapt to idiosyncrasies of the data. When chosen carefully and validated explicitly, they turn calibration from a fragile curve-fitting exercise into a controlled procedure that yields stable parameters and reliable downstream behavior, even when the data do not uniquely determine every aspect of the model.

8.2.5 Model discrepancy and misspecification-aware calibration

Even with a carefully chosen model structure, a realistic observation mapping, and a well-designed calibration loss, a calibrated epidemic model will almost never be "true." Aggregation across individuals, unmodeled heterogeneity, behavioral feedback, changing surveillance regimes, and simplifying assumptions about contact structure or immunity all introduce systematic differences between what the simulator can produce and what is observed. Treating these differences as if they were purely measurement noise encourages overconfident conclusions: the calibration may find parameters that match historical data extremely well, but the apparent fit is partly achieved by forcing parameters to compensate for structural errors. Misspecification-aware calibration makes this explicit by separating measurement noise from structural mismatch.

A common formalization adds a discrepancy term to the data-model relationship:

$$z_k = \hat{z}_k(\theta) + \delta_k + \varepsilon_k, \tag{8.30}$$

where z_k are the calibration targets, $\hat{z}_k(\theta)$ are the model-predicted targets, ε_k captures observation noise and reporting variability, and δ_k captures systematic discrepancy. In this view, calibration is not expected to drive residuals to zero; rather, calibration seeks parameter values θ that make $\hat{z}(\theta)$ match the data up to a discrepancy component whose structure is plausible. The distinction matters because ε_k is typically assumed to average out with more data, while δ_k represents persistent bias that can carry into forecasts and counterfactual scenarios if it is ignored.

In practice, discrepancy can be introduced in several ways, each corresponding to a different calibration philosophy. The simplest is implicit discrepancy, where one acknowledges misspecification by choosing robust losses, smoothing targets, and using validation-based regularization so that the calibrated model is not forced to explain every feature. This approach is often sufficient for short-horizon forecasting, where the goal is stable predictive performance rather than mechanistic interpretability. A more explicit approach models discrepancy as a structured function of time (or covariates), typically constrained to be smooth so that it represents persistent bias rather than noise. For example, one may treat $\delta_{1:K}$ as an unknown low-frequency process and calibrate θ jointly with δ, penalizing rough discrepancy trajectories:

$$(\hat{\theta}, \hat{\delta}) \in \arg\min_{\theta,\delta} \sum_{k \in \mathcal{T}_{\text{cal}}} \rho_k\, \ell\big(z_k, \hat{z}_k(\theta) + \delta_k\big) + \lambda_\theta \mathcal{R}(\theta) + \lambda_\delta \mathcal{R}_\delta(\delta), \tag{8.31}$$

where $\mathcal{R}_\delta$ can be a smoothness penalty, such as a first- or second-difference penalty. This formulation prevents discrepancy from trivially absorbing all mismatch: λ_δ controls how much discrepancy is allowed, and validation can be used to select λ_δ so that predictive performance improves rather than deteriorates.

The benefits of explicit discrepancy are clearest when the model is used for counterfactual scenarios. If mismatch is absorbed entirely into parameters, then the calibrated parameters may encode both true mechanisms and compensatory corrections. Under a counterfactual intervention, the compensatory component may no longer apply, producing biased projections. When discrepancy is modeled separately, the fitted parameters can remain closer to mechanistic meaning, while discrepancy captures the portion of mismatch that should not be extrapolated mechanistically. This does not eliminate uncertainty, but it makes the source of uncertainty more interpretable and helps avoid false precision.

Misspecification-aware calibration also changes how validation is interpreted. If a model fits the calibration window well but consistently fails on the validation window in a structured way (for example, consistently late peaks or systematic underprediction after policy changes), this is evidence of discrepancy that is not well captured by the current structure. The response is typically not to enlarge the optimizer budget, but to revise the model class, revise the observation mapping, or introduce a discrepancy term with an appropriate structure. This viewpoint encourages calibration to be treated as a modeling loop: systematic residual patterns are information about misspecification.

Listing 8.15 shows a minimal pattern for explicit discrepancy as a smooth additive correction. The example fits θ while also allowing a discrepancy vector δ that is penalized for roughness. The code is schematic, but it makes two operational points clear: discrepancy is a variable being estimated, and regularization is what prevents discrepancy from becoming a perfect but meaningless interpolator.

```
import numpy as np

def smooth_penalty(delta):
    d2 = np.diff(delta, n=2)
    return float(np.sum(d2**2))  # second difference penalty

def loss_with_discrepancy(z, zhat, delta):
    # example: squared error on corrected mean
    return float(np.mean((z - (zhat + delta))**2))

def objective(theta, delta, z_obs, zhat_fn, lam_theta=0.0, lam_delta=10.0):
    zhat = zhat_fn(theta)
    data = loss_with_discrepancy(z_obs, zhat, delta)
    reg_delta = lam_delta * smooth_penalty(delta)
    # reg_theta can be added analogously if needed
    return data + reg_delta
```

Listing 8.15: Penalized discrepancy calibration with a smooth additive correction term.

In many applied projects, the practical goal is not to identify a single "true" discrepancy process, but to avoid calibrated models that are overconfident and brittle. Misspecification-aware calibration supports this goal by preventing structural errors from being silently reallocated into parameters and by encouraging validation-driven choices about how much mismatch the model should be expected to explain. When the model is deployed for forecasting or scenario analysis, this explicit separation between mechanistic signal and discrepancy provides a clearer basis for communicating uncertainty and for deciding when a model.

Validation protocols for time-indexed epidemic data

Calibration without validation is indistinguishable from curve fitting. Because epidemic data are time-indexed and strongly non-stationary, validation must respect temporal structure: the model is typically used to predict what happens next, not to interpolate randomly withheld points. This makes standard i.i.d. train–test splits inappropriate in most calibration contexts. Instead, validation protocols are designed to mimic deployment: calibrate on information available up to a time point, then score predictions on a subsequent horizon. The protocol chosen should reflect the intended use, the typical update frequency, and the timescale over which conditions change.

The simplest protocol is a single holdout window. One selects a calibration window $\mathcal{T}_{\text{cal}}$ and a future validation window $\mathcal{T}_{\text{val}}$ that immediately follows it. The model is calibrated by minimizing a loss on $\mathcal{T}_{\text{cal}}$, then evaluated by computing forecast errors on $\mathcal{T}_{\text{val}}$. This approach is easy to implement and is often sufficient for a first assessment, but it can be brittle if the validation period happens to be atypical (for example, a sudden regime change or a reporting anomaly). It also provides only one realization of out-of-sample performance, which can be misleading in highly variable epidemic dynamics.

Rolling-origin validation is a smoother alternative that produces a more stable picture of generalization. One chooses a sequence of forecast origins $t^{(1)} < t^{(2)} < \cdots < t^{(J)}$. For each origin $t^{(j)}$, calibration uses data up to $t^{(j)}$ (often a fixed-length trailing window), and validation scores predictions on the next H time steps. The collection of J backtests yields a distribution of errors rather than a single score, which is more informative for deciding whether calibration choices (loss design, regularization strength, discrepancy assumptions) produce robust performance. It also mirrors operational practice when models are recalibrated repeatedly as new data arrive.

A closely related protocol is multi-wave backtesting, in which calibration and validation are repeated across distinct epidemic phases or waves. The purpose is not only to estimate average forecast error, but also to evaluate failure modes that occur during transitions: onset of rapid growth, turning points near peaks, and declines that may be shaped by interventions or behavior. In such settings, it is often informative to stratify validation scores by phase, since a model that performs well during stable phases may still fail systematically at turning points. This phase-aware evaluation can guide calibration targets and weights; for example, emphasizing near-peak timing in the loss may be justified if peak timing is the most operationally critical component of performance.

When models are intended to transfer across regions, validation can be designed to test spatial generalization. A common approach is leave-one-region-out validation: calibrate parameters on a subset of regions and evaluate on held-out regions, possibly with a limited set of region-specific nuisance parameters (such as initial conditions) refit for the held-out region. This design tests whether the calibrated mechanisms are portable or merely overfit to local artifacts. Similarly, when models are intended to transfer across time periods with changed surveillance, one can validate across policy regimes, treating a regime change as a natural stress test for observation mapping assumptions.

Validation requires that the scoring rule match the intended use. Point forecast evaluation can use absolute or squared errors, but when predictive uncertainty matters, validation should reward calibrated uncertainty rather than just accurate means. This is particularly important in hospital capacity planning and risk-based decision-making, where underestimating uncertainty can lead to systematically underestimated tail risk. When probabilistic forecasts are produced, proper scoring rules provide a principled validation target [562]. Even for point forecasts, it is often valuable to report multiple metrics that emphasize different error aspects, such as MAE for overall deviation, RMSE for peak sensitivity, and a percentage-based metric when relative errors are a key communication target.

A practical implementation of rolling-origin validation follows a consistent pattern: define origins, calibrate at each origin using the same calibration specification, generate forecasts for a fixed horizon, compute evaluation metrics, and aggregate results. Listing 8.16 shows a minimal template. The function `calibrate_at_origin` represents any calibration procedure (least squares, Bayesian search, PSO, etc.) that returns a fitted model, and evaluation is performed with the same metric routine used elsewhere in the chapter.

```
import numpy as np

def rolling_origin_backtest(origins, horizon, X, y, calibrate_at_origin, evaluate_fn):
    """
    origins: list of forecast origin indices
    horizon: number of steps to forecast after each origin
    X, y: full covariates and targets indexed by time
    calibrate_at_origin(j) -> fitted model using data up to origin j
    evaluate_fn(model, X_test, y_test) -> (metrics_dict, yhat)
    """
    all_metrics = []

    for j in origins:
        model = calibrate_at_origin(j)

        X_test = X[j : j + horizon]
        y_test = y[j : j + horizon]

        metrics, _ = evaluate_fn(model, X_test, y_test)
        metrics["origin"] = j
        all_metrics.append(metrics)

    return all_metrics
```

Listing 8.16: Rolling-origin validation template for epidemic forecasting calibration.

Two practical details determine whether validation results are informative. The first is alignment with deployment: the calibration window length, the update frequency of recalibration, and the forecast horizon should match how the model is intended to be used. The second is stability under data revision: epidemic data are often revised retrospectively, so validation ideally uses the data-as-known-at-the-time when evaluating historical forecast performance, or at least acknowledge that revised data may slightly distort backtests. When these details are handled carefully, validation becomes the mechanism that connects calibration design choices to operational reliability. It provides an empirical basis for selecting losses, weights, regularization strengths, and discrepancy structures, and it is the primary safeguard against calibrating models that look accurate in-sample but fail when used as decision-support tools.

8.2.6 Calibration diagnostics and stress testing

Validation scores summarize performance, but they rarely explain *why* a calibrated model succeeds or fails. Diagnostics and stress tests fill this gap by revealing whether errors are random, structured, or phase-dependent, and whether the calibration is stable under reasonable perturbations of the pipeline. In epidemic modeling, where both data and mechanisms can change abruptly, these checks are often the difference between a calibrated model that is merely well fit to one window and a calibrated model that is trustworthy as a reusable tool.

A first diagnostic is residual structure over time. Define residuals on the calibration targets as $e_k = z_k - \hat{z}_k(\hat{\theta})$. Plotting e_k versus time (or examining its autocorrelation) can reveal systematic lag, amplitude mismatch, and phase shifts. For instance, residuals that are predominantly positive during growth and predominantly negative during decline often indicate that the calibrated transmission trajectory is too smooth or that the delay mapping is misaligned. Residuals that spike on particular weekdays point to reporting seasonality that the observation mapping or target extraction did not handle. Such patterns are actionable: they usually suggest revisiting the observation mapping, the target definition, or the loss scaling rather than merely changing the optimizer.

A second diagnostic is decomposition by epidemic phase. Calibration and validation errors can be stratified into growth, peak, and decline periods, or by wave. This is useful because different mechanisms dominate in different phases, and many losses overweight peaks by construction. A model can achieve a low average error while still consistently missing turning points, which are often the most consequential for decision-making. Phase stratification can be implemented by defining phase labels using observed data (e.g., sign of a smoothed derivative, or proximity to local maxima) and computing metrics conditional on phase. The resulting error profile provides a concrete basis for adjusting temporal weights ρ_k or redesigning the calibration objective to emphasize the difficult but operationally important parts of the trajectory.

Stability diagnostics check whether calibration results are robust to choices that should not materially change the conclusion. One basic check is multi-start sensitivity: run the same calibration procedure from multiple initializations and compare both the fitted loss values and the resulting parameter vectors. If many distinct parameter sets achieve similar loss, the problem is practically non-identifiable under the chosen targets, and parameter interpretation should be treated cautiously. Another stability check is window sensitivity: recalibrate on slightly shifted calibration windows (e.g., dropping the first week or the last week) and evaluate how much $\hat{\theta}$ and forecasts change. Large swings can indicate that the model is compensating for mismatch, that regularization is too weak, or that the objective is too dependent on a small number of points.

Stress testing goes beyond perturbing initial conditions and instead challenges the model under conditions that mimic likely deployment shocks. A common stress test in epidemiology is to evaluate robustness to reporting artifacts by injecting or removing weekly seasonality, simulating backlogs (single-day spikes redistributed across prior days), or censoring the most recent days to reflect nowcasting uncertainty. Another stress test is structural: recalibrate after introducing a plausible regime change (e.g., a change in contact rate or reporting fraction) and examine whether the model can adapt smoothly without producing implausible parameter excursions. These tests help identify whether the calibration specification is overly brittle and whether the observation mapping and discrepancy assumptions are adequate.

Diagnostics should also include plausibility checks on implied epidemiological quantities. Even when calibration is purpose-driven rather than strictly inferential, mechanistic models produce interpretable derived quantities such as effective reproduction numbers, generation-interval-consistent growth rates, or implied reporting fractions. Checking whether these quantities fall within credible bounds is an important guardrail. A calibrated model that matches incidence perfectly but implies a reporting fraction greater than one, or an implausibly large day-to-day swing in transmission, is signaling that the calibration objective is being satisfied through unrealistic compensations.

A practical way to operationalize these checks is to treat them as part of the calibration report produced for each run. Listing 8.17 shows a minimal template that computes residual summaries, reports performance by phase (assuming phase labels are available), and captures multi-start variability by repeating calibration with different seeds. The example is intentionally generic: it applies whether the calibrated model is mechanistic or data-driven, as long as it can produce $\hat{z}$ aligned with z.

```
import numpy as np

def diagnostic_report(z, zhat, phase=None):
    e = z - zhat
    out = {}
    out["bias"] = float(np.mean(e))
    out["mae"]  = float(np.mean(np.abs(e)))
    out["rmse"] = float(np.sqrt(np.mean(e**2)))

    # residual autocorrelation at lag 1 (simple indicator of structure)
    if len(e) > 1:
        out["acf_lag1"] = float(np.corrcoef(e[:-1], e[1:])[0, 1])

    # phase conditional MAE
    if phase is not None:
        for p in np.unique(phase):
            idx = (phase == p)
            out[f"mae_phase_{p}"] = float(np.mean(np.abs(e[idx])))

    return out

def multi_start_calibration(calibrate_fn, n_starts=10, seed=0):
    rng = np.random.default_rng(seed)
    results = []
    for _ in range(n_starts):
        theta0 = calibrate_fn.sample_initial(rng)
        theta_hat, loss_hat = calibrate_fn.run(theta0)
        results.append((theta_hat, loss_hat))
    return results
```

Listing 8.17: Template for a simple calibration diagnostic report.

Interpreting diagnostics should lead to concrete calibration actions. Residual seasonality suggests revising target extraction (weekly aggregation) or adding a weekday component to the observation mapping. Phase-dependent peak errors suggest reweighting the objective or using a loss that emphasizes peak timing and magnitude. Multi-start non-identifiability suggests stronger regularization, reduced dimensionality, or shifting focus from parameter interpretation to predictive ensembles. Window instability suggests that the calibration window is too short for the model structure, that the objective is too sensitive to recent censoring, or that discrepancy needs to be modeled more explicitly.

The central goal of diagnostics and stress testing is not to achieve a perfect fit, but to understand the behavior of the calibrated model under the kinds of perturbations it will face in practice. By making these checks systematic, calibration becomes a controlled process: validation tells whether the model generalizes, diagnostics explain failure modes, and stress tests determine whether the model is robust enough for the intended use.

8.2.7 Multi-objective and multi-criteria calibration

Epidemiological calibration is rarely single-objective in practice. Even when a model is intended for a specific task, it is common to calibrate against multiple targets or multiple data streams that do not perfectly agree, such as cases, hospital admissions, occupancy, and deaths. These streams differ in scale, delay structure, and susceptibility to reporting artifacts, and a parameter vector that improves fit for one stream can degrade fit for another. Multi-objective calibration makes these trade-offs explicit and provides a structured way to choose θ when no single solution is best for all criteria.

A simple and widely used approach collapses multiple criteria into a single scalar objective via a weighted sum. If $\mathcal{L}^{(m)}(\theta)$ is a loss for stream or criterion m, the aggregated objective is

$$\mathcal{L}_{\text{sum}}(\theta) = \sum_{m=1}^{M} \alpha_m \, \mathcal{L}^{(m)}(\theta), \qquad \alpha_m > 0, \tag{8.32}$$

which can be optimized using any of the fitting methods described earlier. This design is attractive because it is easy to implement and yields a single calibrated parameter set. However, the choice of α_m is substantive: it encodes the operational priority assigned to each criterion. To make the choice meaningful, it is common to normalize each loss so that its typical magnitude is comparable across streams (for example, dividing by a reference scale such as the median or peak of the stream), and then interpret α_m as importance weights rather than as numerical scaling fixes.

In many settings, it is more natural to treat calibration as a true multi-objective problem rather than collapsing objectives immediately. One then considers the vector-valued loss

$$\mathbf{L}(\theta) = \left(\mathcal{L}^{(1)}(\theta), \ldots, \mathcal{L}^{(M)}(\theta)\right), \tag{8.33}$$

and seeks parameter vectors that are not dominated by any other candidate. A parameter vector θ_a is said to dominate θ_b if it is no worse on all criteria and strictly better on at least one. The set of non-dominated solutions forms a Pareto front. In calibration terms, the Pareto front summarizes the achievable trade-offs: moving along the front improves one criterion only by sacrificing another. This is often more informative than returning a single calibrated point, because it exposes whether trade-offs are mild (a wide region of near-equivalent solutions) or sharp (small improvements in one stream require large sacrifices in another).

Once a Pareto set is obtained, selecting a final calibrated model becomes a decision problem. A common strategy is to choose a knee point, where the trade-off curve bends sharply and further improvement in one objective requires a disproportionate sacrifice in others. Another strategy is to translate stakeholder preferences into a utility function after viewing the Pareto front, effectively choosing weights α_m informed by the observed trade-offs rather than by arbitrary prior selection. In capacity planning, for example, one might accept a visibly worse case fit if it yields meaningfully better hospitalization peak prediction, while in transmission inference, one might prioritize early-growth case dynamics and treat hospitalization fit as secondary.

Multi-criteria calibration also arises when objectives are qualitatively different rather than just different data streams. A typical example is combining a fit-to-data objective with a plausibility objective, such as penalizing unrealistic parameter values, overly rapid time variation, or implausible implied reproduction numbers. In such cases, the regularization term $\lambda\mathcal{R}(\theta)$ can be interpreted as an additional criterion that trades off against fit. Another example is combining in-sample fit with out-of-sample performance, where the calibration step may minimize in-sample loss while model selection among calibration specifications is performed by validation scores. Both are multi-criteria decisions, even when the inner optimizer uses a single scalar objective.

A practical workflow for multi-objective calibration often combines exploration with refinement. One can first generate a diverse set of candidate solutions using a population-based optimizer (GA or PSO) while tracking multiple losses per candidate. The non-dominated subset provides an empirical approximation to the Pareto front. One can then refine a small number of selected candidates using a local optimizer on a chosen scalarization (for example, the weighted sum corresponding to an agreed utility). This hybrid approach is convenient because it reuses the same simulation and loss evaluation code while making the trade-offs visible early, before committing to a single set of weights.

Listing 8.18 shows a minimal Pareto filtering utility. Given a set of candidate parameter vectors and their corresponding loss vectors, it returns the indices of non-dominated candidates. This can be used with any candidate generator: a grid sweep, a GA/PSO run, or repeated restarts of local optimization.

```
import numpy as np

def pareto_nondominated(losses):
    """
    losses: array of shape (N, M), each row is a loss vector to minimize.
    returns: boolean mask of length N indicating non dominated points.
    """
    losses = np.asarray(losses, dtype=float)
    N = losses.shape[0]
    is_nd = np.ones(N, dtype=bool)

    for i in range(N):
        if not is_nd[i]:
            continue
        # i is dominated if there exists j with losses[j] <= losses[i] componentwise
        # and strict inequality in at least one component
        dominated_by_any = np.any(
            np.all(losses <= losses[i], axis=1) & np.any(losses < losses[i], axis=1)
        )
        if dominated_by_any:
            is_nd[i] = False
            continue
        # if i is non dominated, it can dominate others
        dominates = np.all(losses[i] <= losses, axis=1) & np.any(losses[i] < losses, axis=1)
        is_nd[dominates] = False
        is_nd[i] = True

    return is_nd

# Example:
# losses = np.array([[L_cases, L_hosp], ...])  # minimize both
```

```
# nd_mask = pareto_nondominated(losses)
# pareto_candidates = candidates[nd_mask]
```

Listing 8.18: Pareto filtering for multi-objective calibration candidates.

Multi-objective calibration does not remove the need for clear intended use; it operationalizes it. The purpose determines which objectives belong in $\mathbf{L}(\theta)$, which parts of the Pareto front are acceptable, and which trade-offs are defensible. By making trade-offs explicit, multi-criteria calibration reduces the risk of inadvertently optimizing for the wrong quantity and provides a clearer basis for communicating why a particular calibrated model was selected for deployment.

8.2.8 Practical implementation patterns

A reliable calibration pipeline is built from a small number of reusable patterns. The first is separation of concerns: simulation, observation mapping, target extraction, loss computation, optimization, and evaluation should be implemented as distinct components that can be swapped without rewriting the entire workflow. This modularity matters because calibration is iterative; changes are often needed in the observation mapping or the loss, and it should be possible to make those changes without entangling the forward model.

A concrete implementation begins by defining a single function that evaluates the calibration objective for a proposed θ. This function should (i) run the simulator to produce latent trajectories, (ii) apply the observation mapping to obtain predicted observables, (iii) apply any target extraction steps used to define z, and (iv) compute the scalar loss. Centralizing this logic ensures that all optimizers see the same objective, whether the search is local (least squares), global (PSO/GA/SA), or surrogate-driven (Bayesian search). It also simplifies caching and debugging, as every calibration run becomes a sequence of objective evaluations.

The second pattern is global-to-local optimization. In many epidemic models, the loss surface is nonconvex and sensitive to initialization, so it is often efficient to spend a limited budget on a global or semi-global search to locate a good region, and then refine with a fast local solver. For example, a coarse grid or Bayesian search can identify plausible transmission and reporting parameters, and a nonlinear least squares solver can then polish the solution using sensitivities if available. This pattern tends to reduce failures due to poor initialization while still achieving precise convergence when the model is locally well behaved.

The third pattern is explicit management of constraints and numerical failures. Realistic simulators can fail for some parameters due to stiffness, invalid rates, or unstable regimes. Objective evaluation should therefore be defensive: it should enforce feasibility through parameter transforms or bounds, and it should return a large penalty when the simulation fails instead of crashing the optimization loop. This makes calibration robust to occasional invalid proposals, which is especially important for global methods that explore broadly.

The fourth pattern is explicit time splits and repeatable validation. Calibration should be run on a well-defined calibration window, and scoring should be computed on a well-defined validation horizon. Rolling-origin validation can be implemented as a loop that repeatedly calls the same calibration routine at different forecast origins. This encourages reproducibility and prevents over-tuning the pipeline to a single window. It also makes it natural to tune regularization strength and loss weights by selecting the configuration that performs best across backtests rather than on a single split.

The fifth pattern is consistent reporting. Every calibration run should produce a standard report containing fitted parameters, calibration loss, validation metrics (MAE/RMSE/MAPE for point forecasts and proper scores for probabilistic forecasts when available), residual diagnostics, and a small set of derived epidemiological quantities relevant to plausibility. Producing this report automatically makes calibration outcomes comparable across model variants, data streams, and time periods, and it reduces the temptation to judge calibration quality by in-sample fit alone.

Listing 8.19 shows a compact template that implements these patterns. The code defines a single objective function that wraps simulation and mapping, a calibration routine using any optimizer, and an evaluation step on a holdout window using the forecasting metrics routine introduced earlier. The template is intentionally generic; replacing the optimizer changes only one line, while the objective and evaluation remain unchanged.

```
import numpy as np
from scipy.optimize import minimize

def objective(theta, t_cal, y_cal, known_inputs, simulate_outputs, map_to_targets, loss_fn):
    """
    simulate_outputs(theta, t, known_inputs) -> predicted observables aligned with y
    map_to_targets(y) -> calibration targets z (may include smoothing/aggregation)
    loss_fn(z_obs, zhat) -> scalar loss
    """
    try:
        yhat = simulate_outputs(theta, t_cal, known_inputs)
        z_obs = map_to_targets(y_cal)
        z_hat = map_to_targets(yhat)
        return float(loss_fn(z_obs, z_hat))
    except Exception:
        return 1e12

def calibrate(theta0, bounds, t_cal, y_cal, known_inputs,
              simulate_outputs, map_to_targets, loss_fn):
    res = minimize(
        fun=lambda th: objective(th, t_cal, y_cal, known_inputs,
                                 simulate_outputs, map_to_targets, loss_fn),
        x0=theta0,
        method="L-BFGS-B",
        bounds=bounds,
        options={"maxiter": 500}
    )
    return res.x, res.fun

class WrappedModel:
    def __init__(self, theta, known_inputs, simulate_outputs, map_to_targets):
        self.theta = theta
        self.known_inputs = known_inputs
        self.simulate_outputs = simulate_outputs
        self.map_to_targets = map_to_targets
```

```
    def predict(self, t):
        yhat = self.simulate_outputs(self.theta, t, self.known_inputs)
        return self.map_to_targets(yhat)

# Usage pattern:
# theta_hat, _ = calibrate(theta0, bounds, t_cal, y_cal, known_inputs,
#                          simulate_outputs, map_to_targets, loss_fn)
# model = WrappedModel(theta_hat, known_inputs, simulate_outputs, map_to_targets)
# metrics, yhat_val = evaluate_forecast(model, t_val, map_to_targets(y_val))
```

Listing 8.19: Template for a modular calibration pipeline with holdout evaluation.

In a mature pipeline, this template is extended with small additions rather than structural rewrites. Caching can store simulator outputs for repeated evaluations in Bayesian search or population-based methods. Parallel execution can evaluate multiple particles or multiple initializations simultaneously. Logging can capture objective values and parameter proposals over time, enabling post-hoc debugging and sensitivity analysis. Most importantly, the calibration specification can be versioned: the definition of targets, observation mapping, loss, constraints, and validation protocol can be recorded alongside results so that model performance claims remain reproducible even as the code evolves.

Deepen your knowledge

Exercise 1: From calibration intent to a concrete calibration specification

Task

For each question below, propose a reasonable *first* calibration specification. For each one, write: (i) the calibration target(s) (horizon and scale), (ii) the key observation-mapping elements you must include, (iii) the calibration loss (including any temporal or multi-stream weights), (iv) the validation protocol you would use, and (v) the primary failure mode you would watch for.

1. Q1: "Calibrate a model for 14-day ICU occupancy forecasting using daily admissions and ICU occupancy."
2. Q2: "Calibrate for capacity planning at the peak: we care most about peak height and peak timing of hospital occupancy, less about the tail."
3. Q3: "Calibrate a model for scenario comparison of interventions (e.g., school closure vs. no closure) across multiple waves."
4. Q4: "Calibrate a multi-region model intended to transfer to a new region with limited local history."
5. Q5: "Calibrate a model with time-varying transmission $\beta(t)$ represented daily, but you want stable forecasts and interpretable changes."
6. Q6: "Calibrate an agent-based simulator when the likelihood is unavailable; your goal is plausible scenario exploration rather than point-accurate day-by-day fit."

Solution (worked)

Q1. A suitable target is ICU occupancy on the original scale over a short horizon, ideally multi-horizon ($h \in \{1, \ldots, 14\}$) rather than only $h = 14$, because near-term errors and drift both matter.

The observation mapping should explicitly connect admissions to occupancy through length-of-stay dynamics (stock/flow consistency), and account for reporting timing (weekly effects, censoring of the most recent days). A reasonable loss is a horizon-weighted squared error or robust squared error on occupancy, optionally combined with an admissions stream using a secondary weight; recency weights can increase toward the forecast origin. Validation should be rolling-origin backtests with the same 14-day horizon, because single holdouts are brittle under regime shifts. The main failure mode is a model that matches historical occupancy by compensating errors in admissions-to-occupancy mapping (length-of-stay or discharge assumptions), producing biased peaks when conditions change.

Q2. Here the target should be peak timing and peak magnitude of occupancy (and possibly the maximum over a window), not the full trajectory equally. The observation mapping must still preserve occupancy consistency, but the loss can be designed to emphasize peak behavior, for example, by giving higher weights to times near the observed peak or by using a loss on derived targets such as $\max_t \hat{z}_t(\theta)$ and $\arg\max_t \hat{z}_t(\theta)$. Validation should include multiple historical peaks (or multiple waves) if available, because peak-focused calibration can overfit a single event. The primary failure mode is brittle peak chasing: a specification that nails the historical peak but becomes unstable when the peak shifts due to a new intervention or reporting change.

Q3. Targets should include both the time series (to anchor realism) and mechanism-relevant summaries (to stabilize intervention effects), such as pre/post intervention growth rates, turning-point timing, or wave-level attack rates. The observation mapping should include intervention timing, any delays, and reporting changes; it should avoid letting intervention effects silently absorb observation artifacts. A good loss is multi-stream and phase-aware: it may weight growth and turning points more than quiet periods, and it should discourage implausible parameter compensation through regularization. Validation should be structured across regimes: fit on one set of waves or periods and score on another, or use rolling-origin backtests stratified by regime. The main failure mode is non-transportable compensation: the model fits history by trading off intervention parameters against unmodeled behavior, so counterfactual scenarios become unreliable.

Q4. The targets should separate shared mechanisms from region-specific nuisance components. A common choice is to calibrate shared parameters on a pool of regions (or a subset), while allowing region-specific initial conditions and, possibly, reporting factors. The observation mapping must be consistent across regions (definitions, aggregation, reporting delays) or explicitly allow region-level adjustments. The loss is naturally multi-region, either as a sum across regions with normalization so no single region dominates, or as a robust aggregate. Validation should include leave-one-region-out (or leave-a-group-out) evaluation to test transfer. The main failure mode is memorization of region-specific idiosyncrasies: calibration appears excellent in-sample but performs poorly when the model is deployed in a region whose reporting or contact patterns differ.

Q5. The target can remain the primary observable (e.g., cases or admissions), but the calibration specification must explicitly control the complexity of $\beta(t)$. The observation mapping should handle delays and reporting so that $\beta(t)$ is not forced to explain administrative artifacts. The loss is typically a data misfit term plus a smoothness penalty on $\beta(t)$ (first- or second-difference), and possibly shrinkage toward a baseline. Validation should tune the smoothness strength by rolling-origin forecasts: the chosen regularization is the one that improves out-of-sample performance and stabilizes inferred $\beta(t)$. The main failure mode is overfitting through time variation: $\beta(t)$ becomes a flexible curve that tracks noise and produces unstable forward projections.

Q6. Since day-by-day likelihood fitting is unavailable (and often undesirable for complex ABMs), targets should be discrepancy-based summaries aligned with the use case: peak timing, peak height, final size (attack rate), wave duration, age-structured burden, or other decision-relevant features. The observation mapping should reflect what the simulator actually outputs versus what the data represent (e.g., reported cases vs. infections) and should include any necessary aggregation or delay kernels. The loss is a distance in summary space, with explicit scaling/weights so that no single summary dominates; the tolerance or acceptance criterion should reflect measurement noise and model mismatch. Validation should focus on whether accepted/calibrated ensembles reproduce the same summaries on held-out windows or held-out regions, and whether the ensemble uncertainty is realistic for scenario exploration. The main failure mode is summary misspecification: choosing summaries that are easy to match but not sufficient for the questions of interest, producing plausible-looking fits that do not constrain the outcomes stakeholders actually care about.

A useful discipline is to write each specification as a short checklist: "targets → mapping → loss → constraints/regularization → validation." If any of these items is ambiguous, the calibration will usually be fragile in practice.

Exercise 2: Diagnose a calibration failure from residual patterns

Task

You calibrated a mechanistic model to daily reported cases y_t over a 90-day window using a squared-error loss on the original scale. You observe the following residual patterns $e_t = y_t - \hat{y}_t(\hat{\theta})$: (i) residuals are strongly positive during weekdays and negative on weekends, (ii) residuals are mostly positive during the growth phase and mostly negative during the decline phase, (iii) the last 7 days show large negative residuals and frequent data revisions.

For each pattern, answer: (a) the most likely *cause* (observation mapping, target definition, loss/weights, model dynamics, or data revision), (b) a concrete *fix* to the calibration specification (mapping/loss/regularization), and (c) a validation check that would confirm the fix helped without introducing leakage.

Solution (worked)

Weekday/weekend oscillations are a classic signature of reporting periodicity that the model is being forced to explain with transmission dynamics. The likely cause is an observation mapping that treats reported cases as a direct reflection of latent incidence, without a weekday reporting component or aggregation. A concrete fix is to calibrate on weekly totals or a trailing moving average target, or to add a day-of-week reporting multiplier in the observation mapping and regularize it toward 1. A leakage-safe validation is rolling-origin forecasting where each origin uses only data available up to that day; improvement should appear as reduced periodic residual autocorrelation without degrading forecast accuracy.

Growth-positive and decline-negative residuals often indicate a phase mismatch: the model's inferred transmission trajectory is too smooth, or the delay mapping is misaligned, causing the fitted curve to lag the data. The cause is frequently a misspecified delay kernel or an overly rigid $\beta(t)$ parameterization. A concrete fix is to include an explicit delay convolution in the observation mapping and/or allow time-varying transmission with a smoothness penalty tuned by validation. The corresponding validation check is phase-stratified forecast error (growth vs. decline) on repeated backtests; success means reduced phase-dependent bias and improved turning-point timing.

Large negative residuals in the last week combined with frequent revisions, typically reflect right-censoring and backfill: recent observations are incomplete at the time of forecast and later revised upward. The cause is not model dynamics but the measurement process. A concrete fix is to censor the last d days from calibration targets, incorporate a nowcasting layer that models reporting delay, or calibrate on a revised-as-of snapshot. The validation check is to backtest using data-as-known-at-the-time (or simulate censoring) so that the pipeline does not inadvertently learn from finalized values.

The general lesson is to treat persistent residual structure as information: if residuals align with calendar effects, phase, or revision patterns, the calibration should be modified at the level of targets and observation mapping rather than by changing optimizers.

Exercise 3: Choose calibration weights and regularization under competing goals

Task

You calibrate a model jointly to two streams: cases $z_t^{(1)}$ and hospital admissions $z_t^{(2)}$. Your stakeholders care primarily about hospital planning, but they also want cases to be "reasonable." You can choose: (i) stream weights α_1, α_2 in a weighted-sum loss, (ii) temporal weights ρ_t that can emphasize recent days, (iii) a smoothness penalty strength λ for a time-varying transmission modifier $\beta(t)$.

Answer the following:

1. Q1: Propose a principled way to scale the two stream losses so weights α_m are interpretable.
2. Q2: Give one reasonable choice of α_1, α_2 for hospital planning and justify it.
3. Q3: Describe how you would choose ρ_t if the model is updated weekly for 14-day forecasts.
4. Q4: Describe a validation procedure to pick λ and explain what failure mode too-small and too-large λ would create.
5. Q5: If improving hospital fit substantially worsens case fit, list two actions you would take before changing the mechanistic model structure.

Solution (worked)

A practical scaling step is to normalize each stream's residuals by a reference scale so that "one unit" of loss corresponds to a comparable relative error. For example, define per-stream losses on standardized residuals $e_t^{(m)}/s_m$, where s_m could be the median, interquartile range, or peak of the stream on the calibration window. After this normalization, α_m becomes a direct importance weight rather than a hidden unit-conversion factor.

For hospital planning, a reasonable first choice is $\alpha_2 \gg \alpha_1$, such as $\alpha_2 = 1$ and $\alpha_1 = 0.2$, with the understanding that scaling has already made losses comparable. This encodes that admissions must be accurate, while cases should not be wildly inconsistent. If the organization is risk-averse about capacity exceedance, one can further emphasize upper-tail errors by using an asymmetric or peak-sensitive loss on admissions.

If the model is updated weekly for 14-day forecasts, ρ_t should emphasize recent information because it governs the immediate forecast origin. A common design is an exponential recency weighting on the calibration window, or a piecewise scheme that weights the last 14–21 days more heavily than earlier days, while still retaining enough history to avoid overreacting to short-lived anomalies.

To pick λ, use rolling-origin backtests that mirror deployment: for each origin, calibrate with a candidate λ, forecast 14 days, and score hospital and case metrics. Choose λ that yields the best *validation* performance (and stability) on the hospital stream while keeping case errors within acceptable bounds. Too-small λ typically yields a highly variable $\beta(t)$ that tracks noise and produces unstable forecasts; too-large λ produces an over-smoothed $\beta(t)$ that lags regime changes and misses turning points.

If hospital fit improves at the expense of cases, two actions are often effective before changing core dynamics. First, revisit the observation mapping to ensure the case and hospital streams are aligned through a plausible delay and severity pathway; misalignment can force the model to choose incompatible parameter values. Second, revise the loss design to reflect the intended use more precisely, for example by calibrating cases on a transformed or aggregated target (weekly totals) while calibrating admissions on the original scale, reducing the extent to which case reporting artifacts dominate the joint fit.

CHAPTER

9 Integrating surveillance, mobility, and genomic data in pandemic models

Pandemic modeling has historically been anchored in clinical surveillance: case counts, hospital admissions, deaths, and (when available) line lists [608]. These streams remain indispensable, but they are also intrinsically filtered through measurement processes such as diagnostic capacity, reporting delays, policy changes, and shifting care-seeking behavior. A central theme of this chapter is that modern pandemic inference is less about choosing a single "best" dataset, and more about building coherent probabilistic mappings from multiple imperfect data streams to shared latent epidemic quantities [609]. Delay-aware reconstruction provides one of the basic building blocks for this integration by converting reported counts into estimates that are closer to the epidemic events that generated them [610]. Once epidemic timing is represented on a defensible scale, time-varying transmissibility can be estimated and compared across regions and periods using renewal-type approaches [149]. In parallel, mobility data has emerged as a practical proxy for the opportunity structure of transmission: where and when people mix, and how interventions reshape flows between places. During COVID-19, large-scale mobility analyses demonstrated how population movement could explain early spatial dissemination patterns and how movement restrictions altered subsequent spread [256]. At the same time, the use of mobile-phone-derived indicators raises governance questions that must be treated as first-class modeling considerations known to shape what can be inferred and what should be deployed operationally [611]. In addition, genomic data adds a different kind of resolution. Rather than measuring incidence directly, sequences encode ancestry and, through that, information about introductions, growth, and replacement dynamics. Real-time genomic epidemiology platforms helped make these insights actionable during recent outbreaks by tightly coupling sequence processing, phylogenetic reconstruction, and contextual metadata [612]. Increasingly, phylodynamic methods allow mechanistic epidemic models to be fitted to genetic data, enabling inference on quantities such as reproduction numbers and epidemic trajectories under explicit evolutionary and sampling assumptions [613]. Finally, the "surveillance stack" is expanding beyond clinical reporting. Wastewater measurements, for example, can provide a community-level signal that is partially decoupled from testing behavior; importantly, it can be transformed into transmission indicators comparable to case-based metrics [614]. To this end, this chapter is organized around the objective of turning multiple data streams into a consistent, uncertainty-aware picture of epidemic state and drivers. The goal is not merely to *add more data* for models, but to show how each stream constrains different parts of the same latent epidemic process and how careful integration can improve inference, forecasting, and intervention evaluation without overstating what the data can support.

Epidemic Modeling. https://doi.org/10.1016/B978-0-44-355525-1.00015-1

9.1 Surveillance as a measurement process

Surveillance streams are often spoken about as if they were direct measurements of epidemic activity ("cases are rising"). For modeling, however, they are best treated as *noisy, delayed, and biased projections* of a latent process that is closer to transmission events [615]. The core methodological move in this section is therefore to separate (i) a latent epidemic timeline (infections, symptom onsets, hospital admissions) from (ii) observation operators that map those events into the data products actually reported by health systems.

In this context, let x_t denote the (unobserved) number of new events occurring at discrete time t (e.g., new infections or new symptom onsets). Surveillance typically reports y_t, the number of events reported at time t, not the number that occurred at time t. If individual-level line lists are available, it is often useful to represent reporting explicitly via a reporting triangle where $n_{t,d}$ stands for the number of events that occurred at time t and were reported with delay d, for $d \in \{0, 1, 2, \ldots\}$. The observed count at report time r is then $y_r = \sum_{d \geq 0} n_{r-d,d}$. This representation makes clear why the most recent y_r are systematically incomplete (right-truncated) and why naive trend interpretation can be misleading.

Thus, a workhorse observation model expresses the expected reported counts as a convolution of latent events with a delay distribution, scaled by a time-varying ascertainment (or detection) fraction. A generic formulation can take the form:

$$y_t \mid \mu_t, \kappa \sim \text{NegBinom}(\mu_t, \kappa), \tag{9.1}$$

$$\mu_t = \rho_t \sum_{\tau \geq 0} x_{t-\tau}\, g_\tau, \tag{9.2}$$

where $\rho_t \in (0, 1]$ is the probability that an event is captured by the surveillance pipeline at time t (testing policy, care-seeking, coding practices), g_τ is a discrete delay kernel (e.g., onset-to-report or infection-to-report), and κ controls overdispersion. This template accommodates most surveillance indicators by changing the latent event definition and the kernel: e.g., case reports as a delayed, under-ascertained image of infections; hospital admissions as a delayed image of symptomatic incidence; deaths as a delayed image of severe outcomes. A systematic treatment of such surveillance likelihoods and their practical variants (Poisson/NegBinom, hierarchical stratification, and covariate-linked ρ_t) can be found in [616].

As discussed in previous chapter of this book, since recent occurrence times are under-reported, real-time modeling usually requires nowcasting, i.e., estimating the latent x_t (or the completed counts $\sum_d n_{t,d}$) from partially reported triangles. Here, Bayesian hierarchical approaches to nowcasting treat the delay distribution as an estimand and propagate its uncertainty into the nowcast, which is crucial when reporting processes shift mid-outbreak (weekends, lab backlogs, policy changes) [617]. A canonical formulation that jointly models the epidemic curve and reporting delays in real time is given in [618]. Hence, in many settings, additional signals arrive earlier in the clinical pathway. These can be used as "leading indicators" to improve nowcasting precision for late outcomes, effectively turning nowcasting into a small multi-stream fusion problem within the surveillance layer itself. For example, one study used correlated time series (cases and ICU admissions) to improve real-time inference of death incidence under reporting delays [619].

Serology offers a complementary lens on epidemic burden because it targets cumulative exposure rather than incident reports, and is therefore less mechanically coupled to short-term testing capacity, eligibility rules, and care-seeking behavior. In integrated models, this makes serosurveillance particularly valuable as a scale-setting constraint: it can anchor the total number of infections implied by a transmission model and, in doing so, help identify time-varying under-ascertainment in case-based surveillance. A practical way to express this idea is to treat reported cases (or admissions) as a noisy and incomplete observation of a latent epidemic process, while seroprevalence provides an additional observation that is (approximately) an integral transform of the same latent infections, offering leverage to infer the fraction observed over time. The value of this anchor is especially clear in global syntheses of SARS-CoV-2 seroprevalence, which consistently indicate large infection-to-case gaps and substantial heterogeneity driven by geography, timing, and sampling design [620].

In order to use serology coherently, the observation model must explicitly represent the laboratory and immunological pathways from infection to a positive assay. At minimum, this requires adjusting for imperfect sensitivity and specificity, because even small deviations can materially bias prevalence estimates when true prevalence is low or when comparing across assays. In an epidemiological model, this adjustment can be encoded by mapping the model-implied probability of true seropositivity $p_t^\star$ to an observed positivity probability $p_t = (1 - \mathrm{sp}) + (\mathrm{se} + \mathrm{sp} - 1)p_t^\star$, and then linking observed positives to p_t through an appropriate sampling likelihood (often binomial, sometimes extended to incorporate design effects or cluster sampling). Evidence accumulated early in COVID-19 underscores how strongly inference can depend on test performance and on the heterogeneity of assay validation contexts, motivating explicit propagation of uncertainty in se and sp rather than treating them as fixed constants [621].

Beyond test performance, the crucial epidemiological complication is that seropositivity is not a perfect cumulative counter as antibodies rise after infection with a delay (seroconversion), may wane below detection (seroreversion), and—depending on the pathogen and the assay—may respond differently to infection versus vaccination. Mechanistic integration therefore often treats seroprevalence as a convolution of past infections with a kernel that captures the distribution of time-to-seroconversion and subsequent decay, so that $p_t^\star$ depends on the recent history of incidence and not only its cumulative sum. When vaccination contributes to measured antibodies, models frequently add parallel components for infection-induced and vaccine-induced seropositivity (potentially distinguished using antigen targets such as nucleocapsid versus spike, where available), and then reconcile these components with survey metadata on vaccination coverage, timing, and assay type. Large representative studies quantifying post-infection antibody trajectories provide empirical guidance for these kernels and help calibrate plausible waning rates and heterogeneity by age and disease severity [622].

From a practical point of view, before coupling surveillance to mobility or genomics, it is useful to stress-test the surveillance layer on its own, because any mismatch between the latent epidemic clock and the dataset clock will otherwise propagate into the integrated system and masquerade as effects of movement or lineage dynamics. The first task is to define the model's event time with precision: decide whether the latent state x_t represents infections, symptom onsets, diagnoses, admissions, ICU entries, or deaths, and then document which of these the dataset most closely follows. This choice is important as infection-time models are natural for coupling to mobility (via contact opportunity) and to genomics (via lineage growth and introductions), but the data typically arrive on later clocks; the resulting time shift must be expressed explicitly rather than absorbed into ad hoc smoothing.

Once the event time is fixed, the model should specify a clear observation operator that maps latent events to observed counts. In practice, this means representing reporting delays, censoring, and right truncation (the systematic undercount near "now"), and allowing the delay distribution to vary when reporting practices change – for example, by weekday, policy, or system load. Hierarchical nowcasting formulations that jointly model the epidemic curve and time-varying delay mechanisms illustrate how treating delay dynamics as part of the likelihood can stabilize real-time inference and prevent spurious short-term oscillations from being interpreted as epidemiological signals [623].

Then, the next step is to give a justified and testable form for the ascertainment process ρ_t, rather than treating it as a residual fudge factor. Even if ρ_t is modeled flexibly (e.g., as a random walk or spline), it should be tied to plausible drivers such as testing volume, eligibility rules, or care-seeking proxies, and it should be placed downstream of the delay model so that the inferred time variation is not an artifact of truncation or backfill. A useful pattern is to reconstruct latent infections via deconvolution/renewal-style models with explicit delay distributions, then express observed cases or admissions as a thinning of that latent incidence, enabling coherent uncertainty propagation into downstream components like mobility effects or lineage-specific growth [624].

Finally, surveillance streams should be audited for overdispersion and for revisions. Overdispersion is the rule rather than the exception in operational reporting, so likelihoods should permit variance inflation beyond Poisson sampling (e.g., negative binomial components or mixture formulations), and model checking should target day-of-week artifacts, sudden definition changes, and reconciliation operations. Revisions are increasingly central in modern epidemic data pipelines: values may be backfilled or corrected repeatedly, so it is often necessary to version the data, quantify revision magnitudes as a function of *reporting age*, and ensure that retrospective validation uses only the information that would have been available in real time. Open data systems that expose revision histories alongside multiple indicators (cases, deaths, hospitalizations, surveys, claims, and mobility) make such audits feasible and also provide a natural substrate for cross-indicator coherence checks [625]. Where revisions are substantial, one can go further and model them explicitly, treating provisional observations as noisy early snapshots of a latent finalized process and forecasting the revision trajectory so that downstream inference can condition on likely finalized values rather than on raw preliminary counts [626]. With these components in place, surveillance becomes a modular likelihood layer that can be joined coherently with mobility-driven contact opportunity and genomic evidence on introductions and lineage growth.

9.2 Mobility data

Population movement shapes epidemics through two intertwined mechanisms. First, it imports infection between places (seeding new foci). Second, it modulates who mixes with whom by changing the time people spend away from their home environment. Modern mobility data streams make these mechanisms observable at unprecedented scale, but they also force the modeler to make explicit choices about spatial aggregation, temporal resolution, and which notion of "contact opportunity" is actually being measured [627–629].

A useful mental model is that mobility is simultaneously predictable in aggregate and heterogeneous at the individual level. Namely, most people exhibit repeated routines (home-work-home, weekday/seasonal cycles), while a minority contribute disproportionally to long-range connectivity and novel introductions. What makes this combination analytically powerful is that "predictable" and "heterogeneous" are not in tension once we distinguish *within-person regularity* from *between-person variability*. In large-scale trajectory data, the typical individual revisits a small set of locations with high frequency, so the next location is often well constrained, even if the characteristic distance traveled differs dramatically across users. This can be formalized by defining an individual's empirical visitation distribution $p_i(\ell)$ over discrete locations $\ell \in \{1, \ldots, L\}$ and the corresponding Shannon entropy [630]

$$S_i = -\sum_{\ell=1}^{L} p_i(\ell) \log p_i(\ell),$$

which measures how concentrated (and thus predictable) the trajectory is. One can translate S_i into an upper bound on the best-achievable next-location predictability Π_i, emphasizing that high predictability can coexist with large variation in total distance traveled. Empirically, this predictability is remarkably high and relatively stable across users, even when their travel ranges differ substantially [631].

At the same time, the distribution of displacements that generates aggregate spreading is heavy-tailed across multiple scales, indicating that rare long trips are far more common than a purely diffusive (Gaussian) picture would imply. A canonical finding is that the probability density of travel distances r (or jump lengths Δr) can be well approximated, over relevant ranges, by algebraic tails, often summarized as

$$P(\Delta r) \propto (\Delta r)^{-(1+\alpha)}$$

for $\Delta r \geq r_{\min}$, alongside broadly distributed waiting times between moves. In a continuous-time random walk view, such scaling can yield effective superdiffusion (superdiffusion is a class of diffusion process with a non-linear relationship between the mean squared displacement with jumps with a heavy-tail distribution) at intermediate horizons, while long waiting times and return-to-origin behavior temper spread over longer horizons. Analyses leveraging banknote circulation as a proxy for human movement provide a particularly vivid illustration of this phenomenon, showing that the *macroscopic* dispersal kernel relevant for spatial epidemic spread is shaped by a mixture of frequent short moves and rare but consequential long-distance steps [632]. For epidemic modeling, the practical implication is that coupling terms driven by mobility should not be tuned solely to mean flow; higher moments (tail weight) can govern invasion timing and the probability of early seeding events.

Between-person heterogeneity becomes especially visible when mobility is summarized by individual-scale metrics. A common choice is the radius of gyration,

$$r_{g,i} = \sqrt{\frac{1}{n_i} \sum_{k=1}^{n_i} \left\| \mathbf{r}_{i,k} - \bar{\mathbf{r}}_i \right\|^2},$$

where $\mathbf{r}_{i,k}$ are the visited coordinates and $\bar{\mathbf{r}}_i$ is the individual's center of mass. Across populations, $r_{g,i}$ is broadly distributed, reflecting that some individuals live and move within tight spatial "orbits" while others routinely traverse larger regions. This motivates mixture perspectives in which the population is not modeled as a single mobility type but as multiple behavioral classes. One empirically grounded dichotomy distinguishes "returners", who repeatedly circulate among a stable, small set of places, from "explorers", who continue to add new places over time and thereby contribute disproportionately to spatial connectivity and seeding potential [633]. In epidemic terms, explorers disproportionately contribute to the stochastic importation process that initiates outbreaks in new locations, while returners disproportionately shape sustained within-location transmission because their repeated contacts stabilize local mixing.

Daily mobility also exhibits low-dimensional structure. When individual trajectories are converted into networks of visited places and transitions, only a small number of recurring "motifs" accounts for the overwhelming majority of observed daily patterns. This finding matters because it suggests that high-resolution mobility can often be compressed without losing epidemiologically salient structure: rather than treating each person as an idiosyncratic random walker, one can model them as sampling from a limited library of daily graphs (with stable motif identity over months) and then simulate mixing over those graphs. Such motif-based compressions provide a mechanistic explanation for the coexistence of predictability and heterogeneity: many different individuals share the same topology of daily movement, but differ in spatial scale, timing, and the specific places that instantiate the motif [634]. For model design, this opens a path to parsimonious hybrid coupling layers in which mobility is represented as (i) a distribution over motifs (structure), (ii) a distribution over characteristic scales (geometry), and (iii) time-of-day or day-of-week schedules (temporal modulation).

The epidemiological consequences of this mobility structure are clearest when we write importations explicitly. Suppose location i has prevalence proportional to $I_i(t)/N_i(t)$ and a set of travelers from i to j at time t. If the per-traveler probability of initiating at least one successful transmission chain upon arrival is approximated by η (absorbing contact opportunity and early branching survival), then a simple first-order approximation for the expected number of successful importations from i to j is

$$\mathbb{E}\big[Z_{ij}(t)\big] \approx \eta\, M_{ij}(t)\, \frac{I_i(t)}{N_i(t)},$$

where $M_{ij}(t)$ is the flow. Under heavy-tailed individual mobility, $M_{ij}(t)$ itself is effectively a sum of highly unequal contributions, so the realized $Z_{ij}(t)$ is often dominated by a small subset of individuals and routes. This helps explain why large-scale spread can appear irregular in geographic space while remaining regular in the network geometry induced by traffic. By defining an "effective distance" based on passenger fluxes rather than Euclidean distance, network-driven epidemics can often be described by near-wavelike propagation, enabling robust prediction of relative arrival times even when geographic intuition fails [253]. Complementarily, explicitly simulating epidemic diffusion on airline networks demonstrates that heterogeneity in connectivity and flow volumes constrains both predictability and forecast uncertainty of global invasion patterns; importantly, it clarifies why early-stage forecasts are intrinsically variable even under fixed parameters, because the seeding process is governed by stochastic transmission and heterogeneous traffic [250].

A practical modeling takeaway is that mobility coupling layers should be built to preserve routine regularity and tail-driven connectivity simultaneously. One defensible strategy is to decompose the mobility operator into a high-volume recurrent component and a low-volume exploratory component, for example,

$$M_{ij}(t) = (1-\varepsilon)\, M_{ij}^{\text{routine}}(t) + \varepsilon\, M_{ij}^{\text{explore}}(t),$$

where M^{routine} captures stable commuting and repeated flows, and M^{explore} captures rare long-range links and novelty. The parameter ε can be learned from data or treated as a regime-dependent quantity (e.g., shrinking under movement restrictions while not vanishing due to essential travel). Reviews of mobility models emphasize that the key scientific problem is not merely fitting M_{ij}, but selecting the representation that preserves the mechanisms relevant to the modeling task at hand: invasion timing, local attack rates, intervention counterfactuals, or equity impacts [635].

Mobility enters pandemic models through multiple observation layers, and the central modeling move is to treat each layer as a noisy proxy for an underlying, latent movement-and-contact process. At one extreme are census and survey commuting tables: they are often well-curated, geographically complete, and stable enough to serve as a structural prior for who mixes with whom across administrative units, but they are typically infrequent and slow to reflect disruptions such as lockdowns, telework adoption, or fuel-price shocks. A pragmatic use is therefore to interpret commuting tables as a baseline mixing scaffold, then overlay faster data streams as multiplicative (or additive) perturbations around that scaffold, in the spirit of movement-coupled metapopulation formulations discussed in standard modeling references [150]. A concrete example is the use of detailed demographic and movement assumptions to generate realistic daily activity patterns for epidemic simulation; such constructions make explicit that "commuting" is not merely a flow between zones but a repeated, time-structured relocation that changes where contacts can occur across the day [636]. Even when the model is not agent-based, the same logic motivates converting commuting tables into a location-presence operator (who is where, when) rather than using raw flows as if they were contacts.

A second layer is transportation-system data: ticketing records, aviation itineraries, rail ridership, road sensors, and freight manifests. These datasets are often route-specific and institutionally fragmented, but they are invaluable for capturing long-range connectivity and importation pressure, especially early in an outbreak when introductions dominate local dynamics. Reviews of air-travel–mediated spread emphasize that international aviation can connect distant outbreaks on timescales far shorter than pathogen generation intervals, making it natural to represent air travel as an exogenous introduction process superimposed on local transmission [637]. More broadly, transport network syntheses highlight that sea and land transport (and the movement of vectors and goods) can matter as much as passengers, depending on pathogen ecology, motivating models that distinguish passenger-mediated seeding from other forms of connectivity [638]. In practice, transport datasets often under-observe informal travel, non-ticketed movements, and mode substitution (e.g., when flights drop and road travel rises), so they are best treated as informative but incomplete views of a multi-modal mobility system.

A third layer comprises mobile network call-detail records (CDR) and app-based GPS traces. These are high volume and temporally granular, but they are privacy-constrained and sample-biased in ways that directly affect inference if ignored. Empirical work has shown that unequal access to phones and unequal data generation across socioeconomic groups can induce structural distortions in inferred mobility networks (e.g., over-representing long-range ties), and that such distortions can materially change epidemic forecasts built on those networks [639]. Crucially, even before bias correction, one must define what a "trip" means under the collection process (tower handoffs, pings, dwell thresholds) and how aggregation is performed. Recent evidence demonstrates that different aggregation methodologies applied to the same underlying CDR can yield meaningfully different epidemic predictions, underscoring that the mobility pre-processing pipeline is not a neutral step but part of the statistical model [640]. A practical implication for integrated modeling is to propagate mobility-layer uncertainty explicitly, for example by treating OD estimates as random objects with scenario or posterior ensembles rather than as fixed inputs.

A fourth layer is platform-derived indices (e.g., category-specific visit changes), which are attractive because they are global, standardized, and easy to ingest—but they are typically not directly interpretable as flows. For instance, Google's Community Mobility Reports are expressed as percentage deviations from a baseline and incorporate privacy mechanisms (including differential privacy) and thresholding rules that can introduce both noise and missingness in sparse regions [641]. These indices can be extremely useful covariates, but they should be interpreted as proxies for activity intensity or time-in-category rather than as OD matrices. Moreover, multi-country analyses of mobility restrictions and transmission show that the mobility–transmission relationship is neither stationary nor universal, so treating a single index as a direct surrogate for contact rate is often an over-simplification [642].

The translation step of turning "whatever is available" into a model-ready object should therefore be made explicit and, ideally, parameterized. One common target object is a time-indexed origin-destination matrix $M(t)$, where $M_{ij}(t)$ is the expected number of trips (or the probability mass of visits) from origin i to destination j over a time window. When raw counts $\widehat{F}_{ij}(t)$ are available (from ticketing, CDR, or GPS-derived stops), a robust transformation is to form a row-stochastic presence operator

$$\pi_{ij}(t) = \frac{\widehat{F}_{ij}(t)}{\sum_k \widehat{F}_{ik}(t)}, \tag{9.3}$$

and then couple transmission through a mixing kernel, for example,

$$\lambda_i(t) = \beta(t) \sum_j \pi_{ij}(t) \frac{I_j(t)}{N_j(t)}, \tag{9.4}$$

where $\lambda_i(t)$ is the force of infection experienced by residents of i and $\beta(t)$ is allowed to vary with interventions and seasonality. This construction makes clear what mobility is doing: it reweights exposure toward the places where individuals are present, rather than assuming contacts occur only within home regions. In settings where only platform indices $z_c(t)$ are available (e.g., workplace, retail, residential), the model-ready object may instead be a scaled contact operator. A useful pattern is to decompose baseline contacts into settings c and scale each by a function of the corresponding index,

$$C(t) = \sum_c f_c\big(z_c(t)\big)\, C_0^{(c)}, \tag{9.5}$$

with $f_c(\cdot)$ estimated from jurisdictions where both mobility and contact surveys exist. Recent work comparing "naïve" scalings to regression-based transformations finds that borrowing such fitted mappings can improve how mobility data proxy setting-specific contact rates, particularly when detailed contact surveys are unavailable in the target setting [643]. Finally, when high-resolution GPS/POI traces exist, a dynamic bipartite network between neighborhoods and venues (with time-varying edge weights representing visit intensity) can be the most faithful intermediate object, because it preserves heterogeneity in where contacts plausibly occur and allows risk to be attached to venue attributes (e.g., dwell time, occupancy, ventilation) rather than only to geographic zones.

Across all layers, the modeling message is the same: mobility data are not interchangeable, and "mobility" itself is not a single variable. Each observation layer comes with a distinct clock, population coverage, and semantic meaning (flows, presence, activity intensity). Integrated pandemic models benefit when the mobility layer is treated as a modular, explicitly-defined measurement model that outputs a clearly specified mixing object (OD matrix, time-in-location matrix, bipartite visitation network, or contact-rate scaler) together with uncertainty and sensitivity analyses, rather than as a black-box covariate plugged into $\beta(t)$ without interpretation.

9.3 Genomics: from sequences to transmission inference

Genomic epidemiology adds a third observational axis to pandemic models, complementing time series surveillance (when did the epidemic grow?) and mobility (where could it have moved?) with directly heritable variation that records ancestry (how are infections related?). In practice, genomes help answer questions that are weakly identified from counts alone: how many independent introductions occurred, whether apparent resurgences reflect local persistence versus re-seeding, and whether a growth advantage is attributable to a specific lineage or to contemporaneous behavioral change. The operational value of this viewpoint are well summarized by the observation that outbreak reconstruction is no longer limited to case curves, but can be reframed as inference over a partially observed transmission genealogy augmented by metadata such as sampling time and location [644].

9.3.1 From reads to lineages: the measurement pipeline as model structure

A modeling chapter benefits from treating "genomic data" not as an abstract oracle but as an explicit measurement process, because every step between a patient sample and a lineage time series induces selection, distortion, and uncertainty that can propagate into downstream epidemic inference. The raw object is a set of reads produced by a particular laboratory workflow (e.g., amplicon-based enrichment versus metagenomic sequencing), which are then filtered, aligned to a reference genome, and collapsed into a consensus sequence together with associated metadata (collection date, location, sometimes age/risk group). These transformations are not merely technical; they define the implicit observation model under which genomes are available and comparable. For example, amplicon-based protocols are designed to recover viral genomes directly from clinical material and have become a standard operational approach for outbreak genomics, but they necessarily introduce primer- and coverage-related failure modes that are structured in genome coordinate space rather than random [645].

A useful formalization is to view the pipeline as a stochastic map from an unobserved "true" within-host viral population (or a dominant haplotype) to an observed consensus sequence with missing data and errors. Let G denote the true consensus (a string over $\{A, C, G, T\}^S$ for genome length S), and let $\widehat{G}$ be the called consensus after sequencing and processing. Site-by-site, one can represent observation as

$$\Pr(\widehat{G}_s = b \mid G_s = a, D_s) = \begin{cases} 1 - \varepsilon_s(D_s) & b = a, \\ \varepsilon_s(D_s)\, q_{a \to b,s} & b \neq a, \end{cases} \qquad \Pr(\widehat{G}_s = \mathtt{N} \mid G_s = a, D_s) = m_s(D_s),$$

where D_s is the read depth at site s, $\varepsilon_s(D_s)$ is an effective error probability, $q_{a \to b,s}$ is a substitution kernel induced by platform chemistry and processing decisions, and $m_s(D_s)$ is the probability that the site is masked as missing (reported as N). In practice, D_s itself is generated by the upstream wet-lab design and bioinformatic filters, so both $\varepsilon_s(\cdot)$ and $m_s(\cdot)$ inherit structured dependence on factors such as amplicon coverage, primer mismatches, and sample quality. Empirical audits of SARS-CoV-2 sequence datasets show that apparent recurrent mutations and spurious phylogenetic signal can arise from such systematic artifacts (including contamination and uneven coverage), motivating explicit quality control and, when feasible, explicit modeling of error and masking mechanisms rather than treating consensus genomes as error-free [646].

Selection enters even earlier, at the moment of sequencing success. A simple success model makes the dependency on viral load explicit: let C be a proxy for viral load such as PCR cycle threshold (Ct), and let S indicate whether sequencing yields a usable consensus. One can write

$$\Pr(S = 1 \mid C, \mathbf{w}) = \text{logit}^{-1}(\alpha + \beta C + \mathbf{w}^{\top}\gamma),$$

where $\mathbf{w}$ includes laboratory batch, extraction protocol, and sample type. Because C_t correlates with time since infection, symptoms, and potentially age or immune status, conditioning on $S = 1$ can induce preferential sampling in calendar time and across demographic strata even if the intent is "random" sequencing. In integrated models, this is epidemiologically consequential: the genomic dataset may over-represent early infections in a household outbreak, severe cases presenting to hospitals, or travelers screened at borders, and thus does not form a simple random sample of all infections.

The alignment stage similarly embodies model structure. Read mapping and multiple sequence alignment determine which positions are treated as comparable and how indels and low-complexity regions are handled. These choices matter because phylogenetic and lineage assignment algorithms assume that homologous sites are aligned and that the remaining differences are meaningful evolutionary signal. Widely used aligners such as minimap2 (for read-to-reference mapping) and MAFFT (for multiple sequence alignment) represent different statistical and algorithmic compromises between speed, sensitivity, and error modes, and their use should be acknowledged as part of the measurement pipeline that can influence downstream branch lengths, clustering, and inferred introductions [647,648]. A practical implication is that a modeling workflow that consumes genomic outputs (trees, lineages, or variant calls) should record the pipeline versioning and masking rules as carefully as it records the epidemic model code, because those choices define the effective likelihood of the genomic layer.

Metadata are an equally important component of measurement. Collection date is often substituted by report date or upload date when the former is missing, and location can be recorded at different granularities (patient residence, sampling site, reporting laboratory). Each substitution changes the implied time-location stamp that drives phylodynamic timing and phylogeographic transitions. In operational data sharing, these metadata travel with governance constraints and standardization practices, which directly influence what can be integrated and how quickly it can be done. For example, the GISAID sharing model enabled rapid global access to influenza and SARS-CoV-2 sequences while enforcing specific terms of use and attribution practices; from a modeling perspective, it also shaped the temporal and geographic completeness of genomic availability in ways that can be difficult to reconstruct retrospectively [649].

Downstream analyses often reduce sequences to discrete labels (lineages or clades) to obtain a low-dimensional representation that is robust to limited divergence in early epidemics and scalable for high-throughput surveillance. The key modeling point is that "lineage" is itself an inferred variable produced by a classification procedure that depends on the current state of the phylogeny, reference set composition, and decision thresholds. The Pango nomenclature formalizes this representation as a dynamically updated set of epidemiologically meaningful lineages, and lineage assignment software (such as Pangolin) operationalizes the mapping from genome to label, enabling routine lineage counts and frequency trajectories [650]. When integrating lineage counts into transmission models, it is therefore natural to treat lineage labels as observations with potential misclassification and missingness rather than as ground truth. One lightweight way to do this is to introduce a confusion matrix A with entries $A_{uv} = \Pr(\widehat{L} = u \mid L = v)$ and model observed lineage counts as multinomial draws over $\widehat{L}$, where A can absorb known challenges such as reduced resolution among closely related sublineages during periods of low divergence or when many sites are masked. This perspective makes the genomic layer compatible with the same "measurement-first" philosophy used for surveillance and mobility: what we call a lineage trajectory is a derived data product, and its integration is strongest when the derivation is represented explicitly enough to support uncertainty propagation, bias diagnosis, and principled sensitivity analysis.

9.3.2 Genomic sampling as an observation model

A minimal integration strategy is to treat sequences as a biased sample from the latent infection process, in direct analogy to the case ascertainment layer. The key modeling move is to separate the epidemic process (which generates infections and onward transmission) from the sampling process (which generates sequences). Let $I_{t,\ell}$ denote latent incident infections at time t in location ℓ. Suppose sequencing occurs with intensity $\eta_{t,\ell}$, producing $n_{t,\ell}$ genomes $n_{t,\ell} \sim \text{Poisson}(\eta_{t,\ell})$ such that $\eta_{t,\ell} = \kappa_\ell I_{t,\ell} \pi_{t,\ell}$, where κ_ℓ captures laboratory capacity and infrastructure, and $\pi_{t,\ell}$ is an (unknown and potentially time-varying) propensity to sequence conditional on infection. The decomposition makes explicit that the expected number of genomes is proportional not only to epidemic size but also to institutional and policy constraints.

In practice, $\pi_{t,\ell}$ is rarely constant. It may increase during outbreak investigations, decrease when case numbers overwhelm laboratory throughput, or spike when a "variant of concern" triggers targeted sequencing. A useful parameterization is

$$\text{logit}(\pi_{t,\ell}) = \alpha_\ell + \mathbf{x}_{t,\ell}^\top \boldsymbol{\beta} + \gamma \log I_{t,\ell}, \tag{9.6}$$

where $\mathbf{x}_{t,\ell}$ may include policy indicators (e.g., sentinel programs, travel screening), testing intensity, or laboratory backlog metrics. The inclusion of $\log I_{t,\ell}$ allows preferential scaling of sequencing with epidemic size, capturing the empirically common situation in which sequencing volume increases as case counts increase but saturates under capacity constraints.

The phenomenon of *preferential sampling* arises precisely when $\eta_{t,\ell}$ depends on the epidemic process in ways not accounted for by the likelihood. If sampling intensity increases during periods of rapid growth, then genealogies will tend to contain more tips in those periods, even absent any change in transmission. Phylodynamic estimators that treat sampling times as fixed or exogenous can then misattribute sampling-induced clustering of tips to underlying demographic changes. Theoretical and empirical analyses demonstrate that ignoring this coupling can bias inference of effective population size trajectories and reproduction numbers [651,652].

A principled correction is to jointly model the sampling process and the transmission process. One approach is to treat sampling times as realizations of an inhomogeneous Poisson process with a rate proportional to effective population size or incidence. If $N_e(t)$ denotes effective population size under a coalescent framework, one can write

$$\lambda_{\text{sample}}(t) = \psi(t)\, N_e(t), \tag{9.7}$$

with $\psi(t)$ parameterizing time-varying sampling intensity. Under this joint formulation, the posterior over epidemic trajectories conditions simultaneously on branching events and sampling times, restoring coherence between the genealogical and observation components [653]. In birth–death formulations, sampling is already embedded as a process with explicit rate $\psi(t)$; allowing $\psi(t)$ to vary or to depend on covariates yields similar corrections [654].

Beyond temporal dependence, spatial and demographic preferential sampling can induce additional distortions. For example, if hospitalized patients are sequenced at higher rates than mild cases, then lineages associated with severe disease may appear overrepresented, biasing naive estimates of growth advantage. Let g index a demographic stratum (e.g., age group or severity class) and define stratum-specific sampling propensities $\pi_{t,\ell,g}$. Then the observed lineage frequency among sequences,

$$\widehat{\pi}_{v,t,\ell} = \frac{\sum_g \pi_{t,\ell,g} I_{v,t,\ell,g}}{\sum_{v',g} \pi_{t,\ell,g} I_{v',t,\ell,g}}, \tag{9.8}$$

may differ systematically from the true population frequency

$$\pi_{v,t,\ell} = \frac{\sum_g I_{v,t,\ell,g}}{\sum_{v',g} I_{v',t,\ell,g}}.$$

This expression makes clear that differential sequencing by stratum induces compositional bias unless either (i) sequencing is uniform across strata or (ii) the model explicitly adjusts for $\pi_{t,\ell,g}$ using auxiliary data.

In integrated models, the genomic sampling layer plays a role closely analogous to time-varying ascertainment ρ_t in surveillance models. Just as reported cases represent a thinned and delayed view of infections, sequenced genomes represent a further thinned and structured subsample of those cases. Treating $n_{t,\ell}$ merely as "available genomes" discards information contained in sequencing volume dynamics. By contrast, modeling $n_{t,\ell}$ jointly with $I_{t,\ell}$ allows sequencing effort itself to become informative about epidemic size and growth, provided that capacity limits and policy shifts are encoded transparently. This perspective aligns genomic data with the same measurement-first philosophy applied to case counts and mobility: the scientific signal lies not only in the genealogical structure of sequences, but also in when, where, and how intensively sequencing occurred.

9.3.3 Lineage frequencies as a scalable genomic likelihood

Many policy-relevant genomic questions can be answered without reconstructing a full transmission tree by using lineage frequency dynamics as the main summary statistic. This approach treats genomic surveillance as repeated cross-sectional sampling of a categorical variable (lineage identity) from the infected population. Let $\pi_{v,t,\ell}$ denote the true proportion of infections due to lineage v at time t and location ℓ. Conditional on observing $n_{t,\ell}$ genomes, the lineage counts $m_{v,t,\ell}$ can be represented as

$$(m_{1,t,\ell}, \ldots, m_{V,t,\ell}) \sim \text{Multinomial}\left(n_{t,\ell}, \tilde{\pi}_{1,t,\ell}, \ldots, \tilde{\pi}_{V,t,\ell}\right), \tag{9.9}$$

such that $\tilde{\pi}_{v,t,\ell} \propto \pi_{v,t,\ell} b_{v,t,\ell}$, where $b_{v,t,\ell}$ captures lineage-specific sampling bias. This bias term is not merely theoretical: it can arise from targeted sequencing of suspected variants of concern, lineage-specific PCR dropout (e.g., S-gene target failure), or systematic differences in viral load or disease severity that influence sequencing success [655].

The key epidemiological quantity of interest is typically a lineage-specific transmission advantage. Under a renewal formulation of incidence, suppose lineage v generates infections $I_{v,t}$ according to

$$I_{v,t} = R_{v,t} \sum_{\tau=1}^{\infty} I_{v,t-\tau}\, w(\tau), \tag{9.10}$$

where $w(\tau)$ is the generation-time distribution. If two lineages v and u circulate in the same epidemiological environment (shared contact structure and susceptible pool), and if $R_{v,t} = R_t \exp(\delta_v)$ while $R_{u,t} = R_t$, then the ratio of incidences evolves approximately exponentially:

$$\frac{I_{v,t}}{I_{u,t}} \approx \frac{I_{v,0}}{I_{u,0}} \exp(r_v t), \qquad r_v \approx \frac{\delta_v}{T_g}, \tag{9.11}$$

where T_g is the mean generation time and r_v is the effective Malthusian growth-rate difference [656]. This exponential growth in the ratio translates directly into logistic growth of lineage frequency, motivating the logit-linear model

$$\operatorname{logit}(\pi_{v,t,\ell}) = \alpha_\ell + s_\ell t. \tag{9.12}$$

Here, s_ℓ is interpretable as a relative growth rate; with generation time T_g, a common approximation is that the multiplicative reproduction number advantage satisfies

$$\frac{R_v}{R_u} \approx \exp(s_\ell T_g). \tag{9.13}$$

This mapping clarifies why frequency-based models are attractive: they allow estimation of relative transmissibility without requiring full reconstruction of the transmission tree or absolute incidence levels.

The statistical structure of the multinomial likelihood also admits natural extensions. Overdispersion in lineage counts—arising from clustering of related samples in outbreaks or laboratory batch effects—can be accommodated by a Dirichlet–multinomial formulation:

$$(m_{1,t,\ell}, \ldots, m_{V,t,\ell}) \sim \text{DirichletMultinomial}\big(n_{t,\ell}, \boldsymbol{\theta}_{t,\ell}\big), \tag{9.14}$$

where $\boldsymbol{\theta}_{t,\ell}$ parameterizes extra-multinomial variance. Such extensions are particularly important in settings where sequencing is concentrated in localized outbreak investigations rather than in representative sampling.

An additional advantage of frequency-based likelihoods is scalability. Full Bayesian phylogenetic inference scales superlinearly with the number of sequences, whereas lineage counts can be aggregated by time and region, reducing millions of genomes to manageable contingency tables. This reduction is especially useful during periods of explosive growth in genomic surveillance, such as the emergence of SARS-CoV-2 Alpha and Delta variants, where frequency trajectories alone provided early, robust evidence of transmission advantage before detailed phylodynamic analyses were complete [657]. In such contexts, logistic growth fits to regional lineage frequencies were used to estimate selective advantage and inform rapid policy responses.

However, the simplicity of frequency models imposes assumptions that must be articulated. First, they implicitly assume that competing lineages share the same contact environment and generation-time distribution; violations (e.g., lineage-specific changes in infectious period) can bias inferred advantages [658]. Second, spatial aggregation can mask heterogeneous dynamics; a lineage may appear to grow logistically at national scale while actually reflecting sequential regional introductions. Hierarchical models that allow region-specific slopes s_ℓ and intercepts α_ℓ partially address this, and shrinkage priors can borrow strength across locations while preserving heterogeneity.

Conceptually, lineage-frequency models provide a clean interface between genomics and transmission models. Let total incidence at (t, ℓ) be $I_{t,\ell} = \sum_v I_{v,t,\ell}$. Then a coupled system can evolve total incidence through a renewal or compartmental model, while allocating incidence among lineages according to the logistic dynamics above. The genomic likelihood then informs s_ℓ and initial conditions for $I_{v,t,\ell}$, while surveillance data inform $I_{t,\ell}$; the two layers meet through the identity $I_{v,t,\ell} = \pi_{v,t,\ell} I_{t,\ell}$. This factorization separates questions of "how big is the epidemic?" from "which lineage is growing faster?", allowing each data stream to constrain the component it measures most directly.

9.4 Surveillance beyond clinical reporting

Clinical surveillance-cases, hospitalizations, and deaths captures individuals who intersect with the healthcare system. Secondary measurements, by contrast, observe population-level signals that arise indirectly from infection. These include wastewater viral concentrations, syndromic indicators, digital traces (e.g., symptom searches), over-the-counter medication sales, and occupational absenteeism. The modeling challenge is to connect these signals to latent infections through explicit mechanistic or statistical observation operators, thereby integrating them coherently with primary surveillance.

9.4.1 Wastewater-based epidemiology as a population-level shedding signal

Wastewater-based epidemiology (WBE) measures viral RNA concentrations in sewage, providing a pooled, community-level signal of infection that is partially decoupled from healthcare-seeking behavior and diagnostic testing policy [659,660]. For SARS-CoV-2, fecal shedding occurs in a substantial fraction of infected individuals, often beginning near symptom onset and sometimes preceding testing [661]. This temporal positioning makes wastewater an attractive early indicator of changes in transmission intensity.

A minimal mechanistic link between latent infections and observed wastewater concentration can be expressed as a convolution:

$$C_t = \sum_{\tau=0}^{\infty} I_{t-\tau}\, s(\tau)\, \phi_t + \epsilon_t, \tag{9.15}$$

where I_t is latent incident infections, $s(\tau)$ is the shedding profile (viral load per infected individual τ days since infection), ϕ_t captures flow and dilution effects in the sewer system, and ϵ_t represents measurement noise and laboratory variability. The shedding kernel $s(\tau)$ plays a role analogous to the delay distribution in case reporting, but with a biological rather than administrative origin.

In practice, measured concentrations are affected by environmental and infrastructural factors: precipitation, industrial discharge, temperature-dependent RNA degradation, and sampling design (grab versus composite samples). Flow normalization (e.g., dividing by total wastewater volume or by fecal strength indicators such as pepper mild mottle virus) attempts to isolate the infection-driven component of signal variation [614]. A convenient statistical formulation treats normalized concentration $\widetilde{C}_t$ as log-normally distributed:

$$\log \widetilde{C}_t \sim \mathcal{N}(\log \mu_t, \sigma^2), \qquad \mu_t = \sum_\tau I_{t-\tau}\, s(\tau), \tag{9.16}$$

thereby capturing multiplicative measurement noise.

Empirical work during COVID-19 demonstrated that wastewater trends often precede or smooth reported case trends, particularly when clinical testing availability changes [662]. From an integrated modeling perspective, wastewater can therefore reduce reliance on unstable ascertainment fractions. If case counts are modeled as

$$Y_t \sim \text{NegBinom}\left(\rho_t \sum_\tau I_{t-\tau}\, g(\tau), \kappa\right), \tag{9.17}$$

then simultaneous fitting of Y_t and C_t allows identification of ρ_t (case ascertainment) separately from I_t, provided the shedding and reporting kernels are sufficiently distinct.

Wastewater can also support lineage-specific inference. Amplicon sequencing of wastewater samples yields variant proportions aggregated at community scale. Let $C_{v,t}$ denote variant-specific RNA signal. Then, under a similar convolution,

$$C_{v,t} = \sum_\tau I_{v,t-\tau}\, s(\tau)\, \phi_t + \epsilon_{v,t}, \tag{9.18}$$

with $I_{v,t}$ lineage-specific incidence. Such data provide early evidence of emerging variants before clinical sequencing coverage is extensive, though interpretation must account for differential shedding and amplification bias [663].

9.4.2 Syndromic and behavioral indicators

Syndromic surveillance aggregates non-specific clinical or behavioral signals – emergency department chief complaints, telehealth consultations, influenza-like illness (ILI) reports, over-the-counter medication sales, school absenteeism, or even search-engine queries. These indicators often respond rapidly to symptom onset and can therefore provide leading information when diagnostic confirmation is delayed or testing capacity is constrained. During influenza seasons, for example, digital epidemiology approaches demonstrated that aggregated internet search activity for symptom-related terms could correlate with incidence and, in some contexts, anticipate seasonal peaks [664]. More broadly, participatory and digital reporting systems have shown that near-real-time symptom data can augment traditional surveillance, albeit with important caveats regarding representativeness and drift in user behavior [665].

From a modeling perspective, syndromic signals are indirect projections of symptomatic incidence rather than infection itself. Let I_t denote latent infections and let O_t denote latent symptom onsets. A natural intermediate representation is

$$O_t = \sum_{\tau=0}^{\infty} I_{t-\tau}\, f(\tau), \tag{9.19}$$

where $f(\tau)$ is the incubation period distribution. A parsimonious observation model for a syndromic signal S_t then writes

$$S_t = \beta_0 + \beta_1 \sum_\tau O_{t-\tau} h(\tau) + \xi_t, \tag{9.20}$$

where $h(\tau)$ captures the delay between symptom onset and the behavioral act being measured (e.g., seeking care, performing a search query, purchasing medication), and ξ_t represents background seasonality and stochastic noise. In many cases, $h(\tau)$ is sharply peaked near $\tau = 0$, reflecting near-contemporaneous reporting relative to symptom onset, which explains why syndromic indicators can precede confirmed case reports that depend on diagnostic and reporting delays.

Unlike wastewater, syndromic signals are typically not pathogen-specific. Influenza-like illness, for example, may reflect influenza, RSV, SARS-CoV-2, or other respiratory viruses circulating concurrently. A more general formulation therefore decomposes S_t as a mixture of multiple latent processes:

$$S_t = \beta_0 + \sum_{k=1}^{K} \beta_k O_t^{(k)} + B_t + \varepsilon_t, \tag{9.21}$$

where $O_t^{(k)}$ is onset incidence for pathogen k, B_t is a latent baseline capturing non-epidemic seasonal variation (e.g., allergies, background respiratory illness), and ε_t is measurement noise. Identifiability in such systems depends critically on external information—laboratory confirmation data, genomic surveillance, or prior knowledge about seasonal patterns.

State-space models provide a principled way to separate epidemic-driven deviations from baseline dynamics. One common strategy is to model the baseline B_t as a smooth seasonal component:

$$B_t = \mu + \gamma \sin\left(\frac{2\pi t}{52}\right) + \delta \cos\left(\frac{2\pi t}{52}\right) + u_t, \tag{9.22}$$

$$u_t = \phi u_{t-1} + \eta_t, \tag{9.23}$$

where the sinusoidal terms encode annual seasonality, and u_t captures residual autocorrelation. The epidemic signal is then inferred as departures from this baseline. Such decomposition has been widely applied in syndromic systems to distinguish emerging outbreaks from predictable seasonal fluctuations [666].

Behavioral indicators introduce additional complexities because they reflect perception and media effects as much as biological processes. Search query volume, for instance, can spike following news coverage independent of true incidence, leading to overestimation if media-driven feedback is not modeled. A simple extension incorporates an exogenous information variable M_t (e.g., media intensity):

$$S_t = \beta_0 + \beta_1 O_t + \beta_2 M_t + \xi_t, \tag{9.24}$$

allowing the model to partition variance between epidemiological and informational drivers. Empirical analyses following early enthusiasm for search-based influenza tracking highlighted how shifts in user behavior and media coverage could degrade predictive performance when not explicitly modeled [667].

Another important property is demographic and digital divide bias. Participatory surveillance platforms and search data over-represent certain age, socioeconomic, and geographic groups. Let w_g

denote the participation weight of demographic group g and $O_{t,g}$ its symptom incidence. Then

$$S_t \propto \sum_g w_g O_{t,g}, \tag{9.25}$$

so that time-varying participation weights can distort apparent trends. Integrated models can partially correct this by calibrating β_k or w_g against representative clinical data.

Despite these limitations, syndromic and behavioral indicators offer several advantages. They often have high temporal resolution (daily or sub-daily), low reporting lag, and broad geographic coverage. When combined with laboratory-confirmed data, they can improve short-term forecasts and enhance early outbreak detection sensitivity. In multi-stream state-space frameworks, syndromic signals contribute primarily to identifying the timing of growth and decline, while clinical surveillance anchors absolute scale and severity. Their errors are typically not perfectly correlated with those of diagnostic case counts, which enhances robustness in integrated inference.

When considering all above together, syndromic and behavioral indicators are best understood as rapid, indirect sensors of symptomatic burden. Their integration into pandemic models requires explicit modeling of symptom onset distributions, seasonal baselines, media and behavioral feedback, and demographic participation structure. When treated as structured observation layers rather than naive proxies, they provide valuable early information that complements both wastewater and genomic surveillance in a comprehensive multi-sensor pandemic monitoring system.

9.4.3 Pharmaceutical and absenteeism data

Over-the-counter (OTC) medication sales, school or workplace absenteeism, health insurance claims, and related administrative records provide additional indirect measurements of population health that are often available at fine temporal resolution (daily or weekly) and across wide geographic coverage. Unlike laboratory-confirmed cases, these signals capture behavioral responses to symptoms—purchasing antipyretics or cough suppressants, staying home from school, filing sick leave claims—rather than pathogen-specific diagnoses. Their strength lies in timeliness and breadth; their limitation lies in low etiological specificity.

In modeling terms, these indicators are best viewed as composite signals reflecting multiple latent disease processes and non-disease-related drivers. Let $I_t^{(k)}$ denote incident infections for pathogen k at time t, and let $O_t^{(k)}$ denote the corresponding symptomatic onsets (after convolution with incubation distribution $f_k(\tau)$). A parsimonious representation writes the observed administrative signal A_t as

$$A_t = \sum_{k=1}^{K} \gamma_k O_t^{(k)} + B_t + \nu_t, \tag{9.26}$$

where γ_k reflects the propensity of symptomatic cases of pathogen k to generate the observed behavior (e.g., purchasing medication or missing school), B_t is a baseline capturing secular and seasonal variation unrelated to infectious disease, and ν_t is stochastic noise. In settings where one pathogen dominates (e.g., a clear influenza season), the model may be reduced to a single dominant component; however, in multi-pathogen environments such as respiratory virus seasons with influenza, RSV, and SARS-CoV-2 co-circulating, identifiability of γ_k requires auxiliary laboratory data or informative priors.

OTC pharmaceutical sales have been shown to correlate with influenza-like illness and laboratory-confirmed influenza incidence, often leading traditional surveillance by several days [668]. The mechanism is intuitive: individuals frequently self-medicate before seeking clinical care. However, medication purchasing is influenced by pricing, advertising, stockpiling behavior, and weather, all of which must be considered in the baseline B_t. A useful decomposition for B_t is

$$B_t = \alpha + \delta_1 \sin\left(\frac{2\pi t}{52}\right) + \delta_2 \cos\left(\frac{2\pi t}{52}\right) + \mathbf{z}_t^\top \boldsymbol{\theta} + u_t, \tag{9.27}$$

where seasonal harmonics capture predictable annual cycles, $\mathbf{z}_t$ may include temperature or holiday indicators, and u_t is an autoregressive residual component.

School and workplace absenteeism data operate through a related but distinct pathway. Let $N_{t,g}$ denote the population at risk in group g (e.g., students in a district), and let $p_{t,g}$ be the probability of absence due to illness. Then absentee counts can be modeled as

$$A_{t,g} \sim \text{Binomial}\left(N_{t,g}, p_{t,g}\right), \qquad \text{logit}(p_{t,g}) = \alpha_g + \sum_k \gamma_{k,g} O_t^{(k)} + \eta_{t,g}. \tag{9.28}$$

Here, $\gamma_{k,g}$ may differ by age or occupational category, reflecting heterogeneous symptom severity or sick-leave policies. Empirical work has shown that school absenteeism can provide early warning of influenza outbreaks and strongly correlates with community transmission, particularly in pediatric-dominated epidemics [669].

Insurance claims and electronic health record (EHR) data provide a somewhat more specific signal because they encode diagnostic codes. However, they still reflect healthcare-seeking behavior and reimbursement practices. Claims data often have systematic reporting delays and batching effects, which can be modeled analogously to case-report delays:

$$C_t^{\text{claims}} = \sum_\tau O_{t-\tau}\, g_{\text{claims}}(\tau) + \epsilon_t, \tag{9.29}$$

where $g_{\text{claims}}(\tau)$ captures time from symptom onset to claim submission and adjudication. During the COVID-19 pandemic, insurance and outpatient visit data were incorporated into ensemble forecasting systems to improve short-term situational awareness, especially when testing capacity fluctuated [625].

A common challenge across pharmaceutical and absenteeism signals is behavioral confounding. For example, heightened media attention may increase precautionary absenteeism or medication purchases even without corresponding increases in infection. Let M_t denote a proxy for media intensity or public risk perception. Then a more complete model is

$$A_t = \sum_k \gamma_k O_t^{(k)} + \delta M_t + B_t + \nu_t. \tag{9.30}$$

This formulation helps distinguish epidemiological drivers from informational feedback loops, which can otherwise inflate early outbreak signals.

From an integration standpoint, pharmaceutical and absenteeism data function as high-frequency, low-specificity sensors that are complementary to laboratory-confirmed case counts and wastewater. When incorporated into a joint state-space model, they primarily constrain the timing of symptomatic waves, while clinical surveillance constrains scale and severity. Their value is greatest when laboratory testing is limited or when rapid detection of resurgence is critical. However, because they aggregate multiple disease processes and behavioral influences, their interpretation depends on careful baseline modeling, cross-validation with pathogen-specific data, and sensitivity analysis of the γ_k parameters.

9.4.4 Joint state-space integration

A coherent integration of secondary measurements employs a shared latent infection process I_t and multiple observation equations that map this latent trajectory into heterogeneous data streams. The essential modeling principle is that infections are unobserved, while cases, wastewater concentrations, syndromic indicators, pharmaceutical sales, and genomic counts are noisy and delayed projections of the same underlying process. By embedding these projections within a single state-space framework, one enforces internal consistency while allowing each sensor to contribute information according to its signal-to-noise characteristics.

Let I_t denote latent incident infections. A generic renewal-type process model writes

$$I_t = R_t \sum_{\tau=1}^{\infty} I_{t-\tau} w(\tau), \tag{9.31}$$

where $w(\tau)$ is the generation-time distribution and R_t is the time-varying effective reproduction number. Alternatively, in compartmental form, I_t may be derived from transitions between susceptible and infectious compartments. Regardless of parameterization, the latent state evolves according to epidemiological dynamics, while observation equations connect I_t to data streams:

$$Y_t \sim \text{NegBinom}\left(\rho_t \sum_{\tau} I_{t-\tau} g(\tau), \kappa\right), \tag{9.32}$$

$$\log \widetilde{C}_t \sim \mathcal{N}\left(\log \sum_{\tau} I_{t-\tau} s(\tau), \sigma^2\right), \tag{9.33}$$

$$S_t \sim \mathcal{N}\left(\beta_0 + \beta_1 \sum_{\tau} I_{t-\tau} h(\tau), \sigma_S^2\right). \tag{9.34}$$

Here, $g(\tau)$ is a reporting-delay kernel for cases, $s(\tau)$ is a shedding kernel for wastewater, and $h(\tau)$ is a symptom-to-behavior kernel for syndromic signals. Each observation equation has its own dispersion parameters and bias terms (e.g., ρ_t, β_0), reflecting distinct measurement processes.

In compact form, the system can be written as a nonlinear state-space model:

$$\textbf{State equation:} \quad \mathbf{x}_t = f(\mathbf{x}_{t-1}, \theta, \omega_t), \tag{9.35}$$

$$\textbf{Observation equation:} \quad \mathbf{y}_t = h(\mathbf{x}_t, \phi, \varepsilon_t), \tag{9.36}$$

where $\mathbf{x}_t$ contains latent epidemic states (e.g., I_t, R_t), $\mathbf{y}_t$ aggregates observed streams $(Y_t, \widetilde{C}_t, S_t, \ldots)$, θ and ϕ are process and observation parameters, and ω_t, ε_t are process and measurement noise terms. The convolutional structure in the observation equations introduces memory into $h(\cdot)$, linking current observations to past latent states.

Estimation proceeds by filtering and smoothing methods. In linear-Gaussian special cases, the Kalman filter provides closed-form recursive updates. For nonlinear or non-Gaussian settings—such as negative binomial case counts or log-normal wastewater concentrations—particle filters (sequential Monte Carlo) or ensemble Kalman filters can approximate the posterior distribution of latent states [670,671]. Let $\{\mathbf{x}_t^{(i)}, w_t^{(i)}\}_{i=1}^N$ denote particles and weights at time t. The recursive update involves:

1. Propagation: $\mathbf{x}_t^{(i)} \sim p(\mathbf{x}_t \mid \mathbf{x}_{t-1}^{(i)})$,
2. Weighting: $w_t^{(i)} \propto p(\mathbf{y}_t \mid \mathbf{x}_t^{(i)})$,
3. Resampling: to mitigate weight degeneracy.

Because each data stream contributes multiplicatively to the likelihood $p(\mathbf{y}_t \mid \mathbf{x}_t)$, discrepancies between streams naturally influence posterior weights rather than forcing arbitrary reconciliation.

A key advantage of joint state-space integration is identifiability through complementary kernels. Suppose case ascertainment ρ_t declines due to reduced testing, causing Y_t to underestimate infections. Wastewater $\widetilde{C}_t$, whose observation equation lacks ρ_t, may remain stable. Joint inference can then attribute divergence between Y_t and $\widetilde{C}_t$ to changes in ρ_t rather than to changes in I_t. Formally, if two observation equations share I_t but differ in multiplicative bias terms, the posterior distribution of I_t is constrained by their joint likelihood:

$$p(I_{1:T} \mid Y_{1:T}, C_{1:T}) \propto p(I_{1:T}) \prod_{t=1}^{T} p(Y_t \mid I_{1:T}) p(C_t \mid I_{1:T}). \tag{9.37}$$

The multiplicative structure ensures that agreement between streams sharpens inference, while disagreement widens posterior uncertainty and signals potential misspecification.

Another benefit is improved short-term forecasting. Because different streams have different lag structures, combining them effectively increases temporal resolution of inference. Wastewater may provide early signal of growth, syndromic indicators may reflect symptom onset dynamics, and case counts may stabilize estimates of magnitude. Empirical work during COVID-19 showed that multi-stream ensemble models outperformed single-stream models in predictive accuracy, particularly during regime shifts [672].

Conceptually, joint integration transforms secondary measurements from auxiliary validation tools into primary components of inference. Rather than comparing wastewater and cases qualitatively, the state-space framework quantifies their consistency and propagates uncertainty into derived quantities such as R_t or projected hospitalizations. Moreover, the structure is extensible: genomic lineage frequencies, mobility-derived contact modifiers, and hospitalization counts can be appended as additional observation equations, all linked through the same latent infection trajectory.

CHAPTER

Modeling and optimizing intervention policies

10

The previous chapters developed the mathematical foundations of epidemic dynamics, beginning with historical pandemics and progressing through compartmental, spatial, stochastic, and multi-strain models. These frameworks describe how infectious diseases propagate through populations under wide range of constraints. Yet understanding epidemic dynamics is only part of the challenge. The native follow-up question is how should we intervene?

Interventions such as vaccination, isolation, mobility restrictions, and behavioral mandates act as deliberate perturbations to epidemiological systems, but their consequences are mediated by nonlinear feedback, heterogeneous contact patterns, and time-varying behavior [673–675]. Because epidemic dynamics are nonlinear and often exhibit threshold behavior, small changes in intervention timing or intensity can produce disproportionately large differences in outcomes [676–678]. This nonlinearity makes intuition alone an unreliable guide for policy design and creates a strong case for formal modeling. Indeed, experience during COVID-19 illustrates both the power and the limitations of non-optimized decision-making, as empirical inference studies found that non-pharmaceutical interventions substantially altered epidemic trajectories when implemented decisively [128], yet the timing, duration, and intensity of interventions varied widely across jurisdictions, even among regions with similar epidemic signals and comparable resources. That heterogeneity reflects differences in values and constraints, but it also reflects the lack of a unified, operational optimization framework that could translate data into consistent, goal-directed policy recommendations in real time.

Interventions alter the parameters and structure of epidemic systems. Unlike passive forecasting, policy design requires normative judgment: one must define objectives, weigh competing costs, and choose actions under uncertainty. This transforms epidemic modeling from a descriptive enterprise into a prescriptive one [679,680]. Mathematically, intervention design may be framed as a controlled dynamical system problem as follows. Let $\mathbf{x}(t)$ denote the epidemiological state variables and $\mathbf{u}(t)$ denote policy controls. The epidemic evolves according to

$$\frac{d\mathbf{x}}{dt} = f(\mathbf{x}(t), \mathbf{u}(t), \theta), \tag{10.1}$$

where f is the pandemic dynamics and θ represents biological and behavioral parameters. The policymaker seeks to choose $\mathbf{u}(t)$ over a time horizon $[0, T]$ to optimize a specified objective functional

$$J(\mathbf{u}) = \int_0^T f(\mathbf{x}(t), \mathbf{u}(t))\, dt + \Phi(\mathbf{x}(T)), \tag{10.2}$$

subject to dynamical and feasibility constraints.

Epidemic Modeling. https://doi.org/10.1016/B978-0-44-355525-1.00016-3

This formulation connects epidemiology with optimal control theory, economic decision theory, and systems engineering. Early applications of control-theoretic methods to infectious diseases demonstrated that vaccination and treatment strategies could be derived from principled optimization criteria rather than heuristic reasoning [681]. Subsequent work has expanded these ideas to include time-varying policies, constrained healthcare capacity, and resource allocation problems [682]. The necessity of formal optimization arises from three structural properties of epidemic systems:

1. **Nonlinearity and Threshold Effects.** Transmission dynamics are governed by feedback processes that generate tipping points and exponential growth phases.
2. **Competing Objectives.** Health outcomes, economic costs, social disruption, and equity considerations often conflict.
3. **Uncertainty and Adaptation.** Parameter values, behavioral responses, and pathogen evolution are not known with certainty and may change over time.

These properties imply that intuitive or reactive policies may perform poorly. Instead, intervention design requires explicit mathematical formulation of objectives, constraints, and uncertainties.

This chapter develops a systematic framework for modeling and optimizing intervention policies. We begin by formalizing how interventions enter compartmental and spatial epidemic models. We then introduce optimal control theory and numerical solution techniques. Extensions incorporate economic coupling, vaccination prioritization, spatial targeting, evolutionary pressures, and robust decision-making under uncertainty.

10.1 Representing interventions in compartmental models

To optimize intervention policies, one must first represent them formally within epidemiological models. In compartmental systems, interventions act by changing the rates and pathways through which individuals move between health states, by reshaping the effective pattern of contacts, and by imposing feasibility constraints that reflect real-world limits on resources and compliance. The key modeling task is therefore to embed intervention mechanisms into the governing equations in a way that is both mechanistically interpretable and mathematically compatible with optimization. A convenient unifying view treats interventions as control inputs that modify either parameters, state transitions, or model structure.

Recall the classical SIR system,

$$\frac{dS}{dt} = -\beta SI, \quad \frac{dI}{dt} = \beta SI - \gamma I, \quad \frac{dR}{dt} = \gamma I. \tag{10.3}$$

Building on this formalization, we describe four common intervention policies.

10.1.1 Transmission-reducing intervention

A transmission-reducing intervention, such as distancing or masking, is naturally represented by replacing β with a time-dependent effective rate

$$\beta(t) = \beta_0(1 - u(t)), \tag{10.4}$$

where $0 \leq u(t) \leq 1$ encodes intervention intensity and β_0 stands for the infection rate at $t = 0$. This parameter modulation approach is mathematically simple, allows continuous adjustment of policy, and fits directly into optimal control formulations in which $u(t)$ is the decision variable [683].

For instance, [684] let a planner choose a time-varying lockdown intensity $L(t)$ (the fraction of susceptible and infected individuals placed under lockdown) to balance epidemic harms against economic output losses for data collected during the COVID-19 pandemic. They integrate the policy directly into the transmission term by assuming that when a fraction $L(t)$ is locked down, only a fraction $(1 - \theta L(t))$ of agents can effectively transmit (and analogously be involved in infectious contacts), where $\theta \in (0, 1]$ measures lockdown effectiveness. This yields the modified SIR infection flow

$$\dot{S}(t) = -\beta\, S(t)\big(1 - \theta L(t)\big)\, I(t)\big(1 - \theta L(t)\big), \qquad \dot{I}(t) = \beta\, S(t)\big(1 - \theta L(t)\big)\, I(t)\big(1 - \theta L(t)\big) - \gamma I(t),$$

so the intervention is equivalent to a time-dependent effective rate

$$\beta(t) = \beta_0\big(1 - \theta L(t)\big)^2 \quad \Longleftrightarrow \quad u(t) = 1 - \big(1 - \theta L(t)\big)^2.$$

Similarly, [685] explicitly frame social distancing, quarantines, and lockdowns used during COVID-19 as time-dependent controls on transmission by letting policymakers vary a time-dependent reproduction number $\sigma(t)$ during an intervention window. Writing the SIR dynamics in normalized variables $(x(t), y(t))$ (susceptible and infected fractions), they model the controlled epidemic as

$$x'(t) = -\gamma\, \sigma(t)\, x(t)\, y(t), \qquad y'(t) = \gamma\, \sigma(t)\, x(t)\, y(t) - \gamma\, y(t),$$

so the intervention enters the SIR system through

$$\beta(t) = \gamma\, \sigma(t) \quad \Longleftrightarrow \quad u(t) = 1 - \frac{\beta(t)}{\beta_0} = 1 - \frac{\sigma(t)}{\sigma_0}.$$

They then optimize over admissible $\sigma(t)$ (interpreted as policy stringency), subject to feasibility limits on how long the strictest quarantine can be maintained, and (for a linear intervention-cost assumption) show the optimal policy is bang-bang: a single interval of strict quarantine turned on once and then turned off.

10.1.2 Structured mixing control

Any interventions are more naturally expressed through structured mixing. If individuals are partitioned into groups (by age, risk, or activity), the force of infection for group i can be written as

$$\lambda_i(t) = \sum_j \beta_{ij}(t) I_j(t), \tag{10.5}$$

where $\beta_{ij}(t)$ captures the effective contact rate between groups i and j. In such settings, interventions like school closures, workplace restrictions, or shielding of vulnerable groups correspond to targeted reductions in selected entries of the contact matrix. To this purpose, let C_0 denote baseline contacts, and let $C^{(k)}$ encode the contact contribution of sector k; then a policy vector $\mathbf{u}(t)$ can be represented as

$$C(t) = C_0 - \sum_k u_k(t)\, C^{(k)}, \tag{10.6}$$

so that the model expresses both selective and time-varying suppression of contacts while preserving interpretability. This matrix-based view is particularly useful because it connects directly to structured next-generation formulations and provides a principled way to represent heterogeneous interventions within the same dynamical system [686].

As an example, [687] study the early COVID-19 outbreak in Wuhan using an age-structured transmission model with location-specific (home, school, workplace, other/community) contact matrices, so that the force of infection for age group i depends on weighted mixing with all other groups through a contact matrix $C(t)$ (their model writes the force of infection in terms of β and C_{ij}, separating clinical and subclinical infectiousness). Their non-pharmasutical interventionss are implemented by modifying the relevant components of the mixing structure: school closures set school contacts to (near) zero, workplace distancing reduces the effective workplace component (e.g., "intense control" assumes schools closed and only about 10% of the workforce mixing at work), and additional physical distancing reduces community/other-location contacts; reopening is represented as a time-varying relaxation with a phased return to work (e.g., 25% → 50% → 100% workforce) and eventual school resumption. In the notation of SIR models, their intervention can be written as a setting-decomposed contact matrix

$$C(t) = C^{\text{home}} + w_{\text{work}}(t)\, C^{\text{work}} + w_{\text{school}}(t)\, C^{\text{school}} + w_{\text{other}}(t)\, C^{\text{other}},$$

so that (up to a per-contact transmission probability q) one may identify $\beta_{ij}(t) = q\, C_{ij}(t)$, and policies like school closure/workplace restriction correspond to targeted reductions of selected blocks/entries through $w_{\text{school}}(t)$ and $w_{\text{work}}(t)$. Alike, [688] propose a "fine-grained" intervention representation by augmenting social contact matrices with time-use data, enabling policies to act on specific activities (e.g., work, schooling, transport, social visits, hospitality, shopping, parks) rather than only on an aggregate transmission parameter. In a closely related UK SAGE technical note, they operationalize this by mapping time-use categories to POLYMOD contact locations and then assuming that reductions in time spent in an activity are proportional to reductions in contacts in the corresponding location/activity; scenario-specific assumptions (e.g., strong reductions in social visits, work participation, and near-closure of schools) are converted into location/activity weights that scale each component of the contact matrix [689]. In structured-mixing notation this corresponds to

$$C(t) = \sum_k w_k(t)\, C^{(k)} \qquad \Longrightarrow \qquad \beta_{ij}(t) = q \sum_k w_k(t)\, C_{ij}^{(k)},$$

where k indexes activities/locations (home, school, work, leisure, transport, etc.). For example, in their SAGE scenarios, the home weight is held at $w_{\text{home}}(t) = 1$, while other settings receive reduced weights (e.g., in one "current restrictions" scenario, school ≈ 0.03 and work ≈ 0.33, with additional reductions in leisure/other places), directly encoding targeted non-pharmaceutical interventions, such as school closure, workplace reductions, and limits on social visits, into selected entries of $C(t)$ rather than into a single scalar $\beta(t)$ [689].

10.1.3 Vaccination

Vaccination differs from purely contact-reducing interventions because it directly moves individuals between compartments. In an SIR model with vaccination rate $v(t)$, one can write

$$\frac{dS}{dt} = -\beta S I - v(t) S, \tag{10.7}$$

$$\frac{dR}{dt} = \gamma I + v(t)S, \tag{10.8}$$

which treats vaccination as a controlled flow from susceptibility to immunity. If immunity is partial, delayed, or waning, additional compartments and transitions can be introduced without violating mass balance, and the same conceptual structure remains: the intervention is a control-dependent transition that reshapes susceptibility and therefore future transmission potential.

As vaccination implementation can vary significant, below are several implementations. [690] formulate an optimal control problem for COVID-19 using a SIR model with exposed individuals and a split infectious class (symptomatic I_s and asymptomatic I_a), plus an isolation compartment Q and removed R (the isolation state is defined in the following subsection). Vaccination is integrated exactly as a controlled transition from susceptibility to removal:

$$\frac{dS(t)}{dt} = -\frac{\theta}{T_{\text{inf}}}\frac{I_s(t)+\alpha I_a(t)}{N(t)}S(t) - v(t)S(t), \qquad \frac{dR(t)}{dt} = \frac{I_s(t)+I_a(t)}{T_{\text{inf}}} + \frac{Q(t)}{T_{\text{ser}}} + v(t)S(t),$$

so that the decision variable $v(t) \in [0, v_U]$ directly scales the mass-action vaccination flow $v(t)S(t)$. Technically, they also encode realistic vaccine logistics through (i) an isoperimetric (total-supply) constraint $\int_{t_I}^{t_F} v(t)S(t)\,dt = V_T$ (implemented via an auxiliary state $W'(t) = v(t)S(t)$ with terminal constraint $W(t_F) = V_T$), and/or (ii) a mixed state-control daily-capacity constraint $v(t)S(t) \le V_0$ [690]. Their objective functional penalizes infectious prevalence (both I_s and I_a) together with intervention costs (vaccination effort and testing/isolation effort), producing time-varying vaccination schedules that are explicitly constrained by supply and administration limits.

[691] study measles using an SEIR model with demographic growth and calibrate/interpret it against U.S. measles dynamics, including the pre- and post-vaccine era [692]. Vaccination is modeled as a direct depletion of susceptibles via a (possibly time-dependent) vaccination coverage/control ℓ:

$$\begin{aligned}
\frac{dS}{dt} &= rS(1-bS) - \beta SI - \ell S \\
\frac{dE}{dt} &= \beta SI - (d+a)E \qquad , \\
\frac{dI}{dt} &= aE \quad (d+\delta)I
\end{aligned} \tag{10.9}$$

so ℓS is exactly the controlled flow leaving susceptibility (in the simplest SIR-with-vaccination form, it would appear as $-v(t)S$ in dS/dt and $+v(t)S$ into the removed class). They then pose an optimal control problem to find an optimal vaccination strategy $\ell^*(t)$, deriving the usual adjoint system and a projection-type characterization of the optimal control (bounded between 0 and an upper limit), and numerically demonstrate front-loaded vaccination (high early $\ell^*(t)$ followed by gradual relaxation) as a cost-effective way to suppress outbreaks.

[693] argue that many real vaccination campaigns are better represented by discrete-time injections rather than continuous rates, and propose a hybrid SIR-type model with a temporary-vaccinated compartment V to represent short-term or imperfect immunity. In their continuous-time controlled

formulation, vaccination appears as a classical controlled transition

$$\begin{aligned}\frac{dS(t)}{dt} &= \Pi(t) - \beta S(t)I(t) - a\,\theta(t)\,S(t) - \mu S(t)\\ \frac{dV(t)}{dt} &= a\,\theta(t)\,S(t) - b\beta V(t)I(t) - \mu V(t)\end{aligned}, \tag{10.10}$$

so the flow $a\theta(t)S(t)$ moves individuals from S into V, while vaccinated individuals retain residual susceptibility via the $b\beta VI$ term (with $b \in [0, 1]$). The key technical step is then to replace continuous control by *impulses* at specified times t_i: for $t \neq t_i$, the system evolves without the vaccination term, and at each $t = t_i$ the state experiences a jump

$$\begin{aligned}S(t_i^+) &= S(t_i) - a\,\theta(t_i)\,S(t_i)\\ V(t_i^+) &= V(t_i) + a\,\theta(t_i)\,S(t_i)\end{aligned}, \tag{10.11}$$

with I and R unchanged at the impulse [693]. They then solve an optimal impulse control problem (minimizing cumulative infections plus quadratic vaccination costs over impulses), using an impulse version of Pontryagin's principle to compute optimal pulse magnitudes $\theta(t_i)$.

[694] analyze optimal vaccination when immunity is not absorbing, using a controlled SIRS model in which immunized individuals return to susceptibility at rate $\nu > 0$ (waning or incomplete immunization). Vaccination remains a controlled transition out of S, but now competes with the replenishment of susceptibility from R:

$$\begin{aligned}\frac{dS(t)}{dt} &= -\beta S(t)I(t) - u(t)S(t) + \nu R(t)\\ \frac{dI(t)}{dt} &= \beta S(t)I(t) - \gamma I(t)\\ \frac{dR(t)}{dt} &= \gamma I(t) + u(t)S(t) - \nu R(t)\end{aligned}, \tag{10.12}$$

with control bounds $u(t) \in [0, u_{\max}]$. Technically, this is the minimal mass-balance extension of the vaccinated-SIR flow picture: vaccination still moves mass $S \to R$, but waning creates an additional transition $R \to S$ that can sustain endemicity and alters long-run optimal policy. They formulate a discounted infinite-horizon planner problem with convex running costs (infection burden and vaccination effort), derive the associated HJB equation, and characterize optimal feedback vaccination policies that may require sustained nonzero effort to keep infections at (or near) zero when $\nu > 0$.

10.1.4 Infectious class control

Testing, tracing, and isolation are often represented by refining the infectious class to distinguish unobserved infectious individuals from detected and isolated individuals. A simple formulation splits I into a community infectious compartment I and a quarantined compartment Q:

$$\begin{aligned}\frac{dI}{dt} &= \beta SI - (\gamma + \delta(t))I\\ \frac{dQ}{dt} &= \delta(t)I - \gamma_Q Q\end{aligned}, \tag{10.13}$$

where $\delta(t)$ is a policy-controlled detection or isolation intensity and γ_Q is the recovery rate in quarantine. The epidemiological effect arises because Q is assumed to have reduced transmission, either implicitly by removing it from the force of infection or explicitly by assigning it a smaller transmission coefficient. Such refinements are common in intervention design studies because they create a natural representation of targeted control: instead of suppressing contacts for everyone, one attempts to identify and remove the most infectious contacts from the network of transmission [695,696].

For example, Aronna et al. propose a COVID-19 compartmental model that distinguishes (i) infectious individuals circulating in the community and (ii) confirmed positives placed into enforced quarantine Q [697]. Their intervention lever is a *time-dependent testing intensity* $\rho(t)$ which removes individuals from multiple pre-confirmation compartments and moves them into Q. In their normalized system with free vs. reduced-contact subpopulations, the quarantine inflow is

$$\dot{Q}(t) = \sigma(1-\alpha)\big(I_f(t) + I_r(t)\big) + \rho(t)\Big[\delta\big(E_f(t) + E_r(t)\big) + I_f(t) + I_r(t) + A_f(t) + A_r(t)\Big] - \gamma_2 Q(t) - \mu Q(t),$$

while the force of infection depends only on *unquarantined* infectious compartments (with an optional extension that allows a small leakage term εQ) [697]. In the reduced template

$$\dot{I} = \beta S I - (\gamma + \delta(t))I, \qquad \dot{Q} = \delta(t) I - \gamma_Q Q,$$

their construction corresponds to a policy-controlled detection/isolation intensity $\delta(t) \approx \rho(t)$ (up to eligibility weights and additional symptom-driven confirmation), with quarantined individuals removed from transmission (or assigned a much smaller effective transmission coefficient via ε).

Song et al. build a COVID-19 model with compartments S, E, A, I, Q, R (susceptible, exposed, asymptomatic infectious, symptomatic infectious, isolated, recovered) and incorporate case finding/isolation by splitting infectiousness into *community* (A, I) versus *isolated* (Q) [698]. Isolation is implemented as linear removal terms from A and I with rates f_t and f_q, but the inflow into Q is *delayed* to represent operational lags:

$$\frac{dA}{dt} = (1-p)\alpha E(t) - f_t A(t) - \gamma_1 A(t) - dA(t), \qquad \frac{dI}{dt} = p\alpha E(t) - \gamma_2 I(t) - f_q I(t) - dI(t),$$
$$\frac{dQ}{dt} = f_t A(t-\tau_1) + f_q I(t-\tau_2) - \gamma_3 Q(t) - dQ(t) - d_1 Q(t).$$

Transmission is driven by $(\sigma A(t) + I(t))$ and Q does not appear in the incidence term [698]. Mapping to the (I, Q) split, the policy-controlled detection/isolation intensities are $\delta_A(t) = f_t$ and $\delta_I(t) \equiv f_q$, and the model refines $\dot{Q} = \delta(t)I - \gamma_Q Q$ into a delay-differential form. In their control formulation, they further introduce bounded, time-varying controls $u_i(t) \in [0, 1]$ to optimize time-dependent isolation intensity under resource/cost trade-offs [698].

Sturniolo et al. develop an SEIR-TTI construction that duplicates each epidemiological compartment into *unconfined/undiagnosed* (X_U) and *diagnosed/isolated* (X_D) branches [699]. Testing is encoded as a direct transition from unconfined infectious to isolated infectious,

$$\Delta^T_{I_U \to I_D} = \theta I_U,$$

and (in their idealized baseline) I_D is removed from the infection term, i.e. only I_U contributes to incidence [699]. Contact tracing then induces additional transitions $X_U \to X_D$ (for $X \in \{S, E, I, R\}$), derived via a probabilistic argument that yields a *global tracing rate*

$$\mathcal{D}(CT) = \chi\, \theta\, \eta\, \Pr(CI),$$

and, crucially, a compartment-specific tracing isolation flow

$$\Delta^{CT}_{X_U \to X_D} = \eta\theta\chi\, \Pr(CI \mid X_U)\, X_U,$$

where η is tracing success (fraction of contacts identified) and χ is tracing speed [699]. In the reduced (I, Q) template, the effective detection intensity becomes

$$\delta(t) = \theta + \delta_{\text{trace}}(t), \qquad \delta_{\text{trace}}(t) \propto \eta\theta\chi\, \Pr(CI \mid I_U),$$

which is *state-dependent* through the conditional contact-history probability, and isolation acts by transferring mass from the transmitting class to a non-(or weakly-)transmitting class.

Heidecke et al. propose a delay-differential TTIQ model inspired by COVID-19 dynamics (Germany, second wave 2020), in which undetected infectious individuals U can be (i) detected by testing and move into confirmed isolation ($U \to I$) and (ii) generate traced infected contacts who enter quarantine either while latent ($E \to Q_E$) or infectious ($U \to Q_U$), with a tracing delay κ [700]. A key technical feature is that quarantined/isolated individuals retain infectiousness but with *reduced transmission coefficients*. Writing β_X for the transmission rate contributed by infectious compartment X, they set

$$\beta_Q = \rho_Q\, \beta_U, \qquad \beta_I = \rho_I\, \beta_U, \qquad 1 > \rho_Q > \rho_I,$$

so Q and I contribute less to the force of infection than undetected U [700]. Testing-driven detection is modeled through *state-dependent* per-capita rates $\eta_X(t)$ that decrease with prevalence due to finite lab capacity; they define $\eta_X(\cdot)$ via an explicit functional form (their Eq. (2)) and interpret it through the induced *total* detection flow in each compartment [700]. Contact tracing terms Tr_X are derived mechanistically and are proportional to successful detection of index cases at time $t - \kappa$, yielding delayed quarantine inflows [700]. In the simplified (I, Q) split, this corresponds to $\delta(t) \equiv \eta_U(t)$ for $U \to I$ plus additional delayed quarantine removals driven by tracing, with the important refinement that Q is *not* assumed non-infectious but instead transmits with $\beta_Q < \beta_U$.

Browne et al. formulate a mechanistic SEIR-type model for the West Africa Ebola outbreak that explicitly links transmissions from reported cases to the initiation of tracing and monitoring [701]. Infectious individuals are split into reported/hospitalized I_h and unreported I_u, and a fraction ϕ of contacts of reported cases are traced; traced infections enter traced-exposed E_c and then a monitored infectious class I_c which transmits at a reduced rate β_m and is removed faster (mean infectious period T_m) [701]. In a reduced special-case system, they write

$$S'(t) = -\beta S(t)\big(I_h(t) + I_u(t)\big) - \beta_m S(t) I_c(t),$$
$$E_c'(t) = \phi\beta S(t) I_h(t) + \phi\beta_m S(t) I_c(t) - \frac{1}{\tau} E_c(t), \qquad I_c'(t) = \frac{1}{\tau} E_c(t) - \frac{1}{T_m} I_c(t),$$

highlighting that tracing acts by *creating* a controlled infectious compartment with smaller transmission coefficient and faster removal [701]. In the (I, Q) template, this is equivalent to moving a traced fraction into a low-transmission quarantine/monitoring class (here $Q \equiv I_c$) with $\beta_Q = \beta_m \ll \beta$ and $\gamma_Q = 1/T_m$, while the tracing delay (time between reporting and tracing) is parameterized separately in their general model [701].

10.2 Optimal control theory for epidemics

Once interventions have been embedded into epidemic models as explicit control variables, the natural next step is to determine how these controls should be chosen over time. This transforms the epidemic system into a controlled dynamical system and shifts the modeling objective from prediction to optimization. To this end, let $\mathbf{x}(t) \in \mathbb{R}^n$ denote the epidemiological state variables and $\mathbf{u}(t) \in \mathcal{U}$ the admissible control functions representing intervention intensities. The epidemic evolves according to

$$\frac{d\mathbf{x}}{dt} = \mathbf{f}(\mathbf{x}(t), \mathbf{u}(t), \boldsymbol{\theta}), \qquad \mathbf{x}(0) = \mathbf{x}_0, \tag{10.14}$$

where $\boldsymbol{\theta}$ represents biological and behavioral parameters. The policymaker selects $\mathbf{u}(t)$ over a finite horizon $[0, T]$ to minimize an objective functional

$$J(\mathbf{u}) = \int_0^T L(\mathbf{x}(t), \mathbf{u}(t))\, dt + \Phi(\mathbf{x}(T)). \tag{10.15}$$

The running cost L typically encodes the health burden of infection together with economic or social costs of intervention, while the terminal cost Φ may represent the final epidemic size or long-term damage. A commonly used structure penalizes prevalence and control effort simultaneously,

$$L(\mathbf{x}, \mathbf{u}) = C_I I(t) + C_D D(t) + \frac{1}{2} \sum_i C_{u_i} u_i(t)^2, \tag{10.16}$$

where quadratic penalties on control variables regularize the solution and prevent unrealistically abrupt switching. Under standard regularity conditions on $\mathbf{f}$ and compactness of the admissible control set, existence of an optimal control can be established using classical results in deterministic control theory [702].

Necessary conditions for optimality are obtained through Pontryagin's Maximum Principle. Define the Hamiltonian

$$\mathcal{H}(\mathbf{x}, \mathbf{u}, \boldsymbol{\lambda}, t) = L(\mathbf{x}, \mathbf{u}) + \boldsymbol{\lambda}^\top \mathbf{f}(\mathbf{x}, \mathbf{u}), \tag{10.17}$$

where $\boldsymbol{\lambda}(t)$ is the adjoint vector. If $\mathbf{u}^*(t)$ is optimal with associated state trajectory $\mathbf{x}^*(t)$, then there exists a costate trajectory satisfying

$$\frac{d\mathbf{x}^*}{dt} = \frac{\partial \mathcal{H}}{\partial \boldsymbol{\lambda}}, \tag{10.18}$$

$$\frac{d\boldsymbol{\lambda}}{dt} = -\frac{\partial \mathcal{H}}{\partial \mathbf{x}}, \tag{10.19}$$

together with terminal condition $\boldsymbol{\lambda}(T) = \nabla_{\mathbf{x}} \Phi(\mathbf{x}(T))$. The optimal control must minimize the Hamiltonian pointwise in time,

$$\mathbf{u}^*(t) = \arg\min_{\mathbf{u} \in \mathcal{U}} \mathcal{H}(\mathbf{x}^*(t), \mathbf{u}, \boldsymbol{\lambda}(t), t). \tag{10.20}$$

The adjoint variables admit a natural interpretation as shadow prices: they quantify the marginal future cost generated by an incremental increase in each state variable. In epidemic systems, the adjoint associated with the infectious compartment measures the downstream burden caused by one additional infection at time t, linking present control decisions to future epidemic consequences [703].

To illustrate, consider an SIR model with transmission control $u(t)$ acting multiplicatively on the transmission rate,

$$\frac{dS}{dt} = -\beta(1-u(t))SI, \tag{10.21}$$

$$\frac{dI}{dt} = \beta(1-u(t))SI - \gamma I, \tag{10.22}$$

$$\frac{dR}{dt} = \gamma I, \tag{10.23}$$

and objective functional

$$J(u) = \int_0^T \left(C_I I(t) + \frac{1}{2} C_u u(t)^2 \right) dt. \tag{10.24}$$

The Hamiltonian becomes

$$\mathcal{H} = C_I I + \frac{1}{2} C_u u^2 + \lambda_S(-\beta(1-u)SI) + \lambda_I(\beta(1-u)SI - \gamma I) + \lambda_R(\gamma I). \tag{10.25}$$

Differentiating with respect to u yields the stationarity condition

$$\frac{\partial \mathcal{H}}{\partial u} = C_u u + \beta SI(\lambda_S - \lambda_I). \tag{10.26}$$

Projection onto the admissible interval $[0, u_{\max}]$ produces

$$u^*(t) = \min\left\{u_{\max}, \max\left\{0, -\frac{\beta SI(\lambda_S - \lambda_I)}{C_u}\right\}\right\}. \tag{10.27}$$

The structure of this solution reveals that optimal intervention intensity depends jointly on epidemiological prevalence and the adjoint gap $(\lambda_S - \lambda_I)$, which captures the marginal future cost of shifting individuals between compartments. When intervention costs are small relative to infection costs, the optimal policy may saturate at the upper bound, producing bang–bang dynamics. When intervention costs are large, interior solutions arise, yielding smooth and time-varying mitigation intensity. Such structural properties are well documented in optimal epidemic control analyses [681].

Analytical solutions are rare for nonlinear epidemic systems, and numerical approaches are generally required. Indirect methods solve the coupled state–adjoint system using forward–backward iteration schemes, while direct transcription methods discretize the control and convert the problem into nonlinear programming formulations [704]. Rolling-horizon strategies inspired by model predictive control further allow policies to adapt dynamically as new data become available.

Optimal control theory thus provides a principled mechanism for transforming epidemiological models into policy design tools. It formalizes trade-offs between present intervention effort and future epidemic burden, quantifies marginal values of state variables, and reveals structural features such as threshold-based or time-concentrated control. In doing so, it establishes the mathematical core upon which more elaborate frameworks—incorporating economics, spatial structure, evolutionary dynamics, and uncertainty—can be constructed.

10.3 Economic-epidemiological coupled models

Intervention policies are rarely evaluated solely by their epidemiological consequences. In practice, the most contested decisions during pandemics arise because the same actions that reduce transmission also disrupt production, income, schooling, and social welfare. Economic-epidemiological coupled models make these trade-offs explicit by endogenizing behavior and placing epidemic dynamics inside a welfare, utility, or resource-allocation framework. This line of work sits within the broader tradition of economic epidemiology, which emphasizes that individual choices respond to perceived risk and policy constraints, thereby feeding back into transmission itself [180].

A convenient starting point is to allow the force of infection to depend on economic activity. In a baseline SIR structure, new infections occur at rate βSI, but the effective transmission parameter can be decomposed into contacts generated through consumption activity, labor activity, and residual contacts unrelated to market behavior. One representative specification writes the infection flow as

$$\dot{S}(t) = -S(t)I(t)\Big(\beta_c\, c_S(t)c_I(t) + \beta_\ell\, \ell_S(t)\ell_I(t) + \beta_o\Big), \tag{10.28}$$

where $c_g(t)$ and $\ell_g(t)$ denote consumption and labor choices for group $g \in \{S, I\}$, and the coefficients $\beta_c, \beta_\ell, \beta_o$ map activity to infectious contact. This formulation captures the empirical idea that the scale and composition of interactions change with economic behavior and policy restrictions, and it provides a transparent channel by which interventions operate: distancing rules, closures, and work-from-home mandates constrain $c(t)$ and $\ell(t)$, thereby reducing effective contacts [187].

To evaluate policies, a welfare objective is introduced. A common approach is to maximize discounted utility from consumption and leisure while penalizing deaths, morbidity, or healthcare overload. Writing $D(t)$ for cumulative deaths, and $C(t)$, $L(t)$ for aggregate consumption and labor, a reduced-form planner objective can be represented as

$$\max_{\mathbf{u}(\cdot)} \int_0^T e^{-\rho t}\, U\big(C(t), L(t)\big)\, dt - \chi\, D(T), \tag{10.29}$$

subject to the epidemic dynamics and the economic resource constraints that link $C(t)$ and $L(t)$ to production. The control vector $\mathbf{u}(t)$ may encode interventions directly, such as mandated reductions in labor supply in specific sectors, caps on gathering sizes that constrain consumption opportunities, or testing and isolation capacity that reduces infectious participation in market activity. In this setting, epidemic mitigation is not an external add-on; it is an endogenous allocation problem in which infection risk is a byproduct of choices that also generate economic value.

Coupled models also make clear why decentralized behavior can diverge from socially optimal outcomes. When individuals reduce activity to protect themselves, they partially internalize risk, but they do not fully internalize the externality that their activity increases infection risk for others. This wedge motivates policy instruments that resemble Pigouvian regulation: restrictions, taxes, subsidies for risk-reducing behavior, or publicly provided testing and isolation. A formal planning formulation emphasizes that optimal mitigation often involves a time-varying intensity that balances the convex costs of suppression against future infection burdens, rather than a constant policy level. Within such frameworks, quarantine, testing, and tracing can appear as substitutes for broad lockdowns because they target infectious contacts more precisely, reducing the welfare loss per unit of transmission reduction [705].

Empirical evidence supports the central premise that economic activity and disease spread are tightly coupled. High-frequency data analyses have shown that changes in mobility and market participation correlate with changes in infection dynamics, and that policy interventions shift both behavior and transmission simultaneously [706]. This observation has two methodological implications. First, models that treat $\beta(t)$ as purely exogenous risk misattributing behavior-driven contact reductions to biological changes, or vice versa. Second, policy evaluation must account for behavioral adaptation: if perceived risk rises, voluntary distancing may amplify formal restrictions; if fatigue or misinformation spreads, compliance may erode, weakening the same policy over time.

Behavioral feedback can be incorporated in different ways. A reduced-form approach makes contact rates depend on perceived prevalence, for example, $\beta(t) = \beta_0\, g(I(t))$ with $g'(\cdot) < 0$, capturing voluntary risk avoidance. A more micro-founded approach models individuals choosing activity to maximize private utility under infection risk, generating an equilibrium contact pattern. Both approaches align with evidence that disease dynamics and human behavior co-evolve, and that ignoring this coupling can systematically bias intervention impact estimates [133]. In optimization terms, the presence of feedback creates time inconsistency risks for simple policy rules and increases the value of adaptive strategies that update interventions as behavioral responses change.

Finally, economic-epidemiological coupling naturally broadens what counts as an intervention. Policies are not limited to reducing β; they can also shift the feasibility set of economic choices, reallocate risk across sectors, and alter incentives for protective behavior. This perspective clarifies why distributional considerations matter: interventions that reduce aggregate contacts may impose uneven welfare losses across occupations and income groups, and those asymmetries can change both compliance and epidemic outcomes. In subsequent sections, these coupled formulations will be extended toward structured populations, spatial targeting, and uncertainty-aware policy design, preserving the core insight that epidemic control is inseparable from the incentives and constraints shaping human behavior.

10.4 Vaccination strategy optimization

Vaccination is among the most powerful intervention instruments because it alters the epidemic system by directly reducing susceptibility, infectiousness, and/or the probability of severe outcomes, thereby reshaping both transmission and burden. In compartmental terms, vaccination induces controlled flows out of the susceptible state or into partially protected classes, but its policy relevance comes from the fact that vaccine supply is typically limited, rollout is staggered over time, and protection may be imperfect. These features turn vaccination into an allocation problem: who should be vaccinated first, at what pace, and with what targeting principle, given explicit objectives and constraints. The classical theory of immunization programs begins with the observation that the population-level effect of vaccination exceeds the direct protection of vaccinated individuals, because immune individuals provide indirect protection to others by interrupting transmission chains [707]. In homogeneous mixing models, this indirect effect yields a familiar threshold in terms of effective coverage, but real-world vaccination decisions almost always operate in heterogeneous populations where such thresholds become vector- and matrix-valued and where prioritization can dominate aggregate coverage as a determinant of outcomes.

A useful baseline is the effective reproduction number under vaccination. In the simplest homogeneous SIR setting with a vaccine that reduces susceptibility by factor $(1-\varepsilon)$ and coverage p, the post-vaccination transmission potential is approximated by $R_v = R_0(1-\varepsilon p)$, suggesting a control target $\varepsilon p > 1 - 1/R_0$ when the objective is elimination rather than mitigation. The core limitation of this scalar picture is that it hides heterogeneity in contacts, risk, and clinical severity. Once the population is stratified into groups (e.g., by age, risk, occupation, or geography), vaccination can be represented as a group-specific coverage vector $\mathbf{p} = (p_1, \ldots, p_m)$, and transmission is governed by a next-generation matrix K whose spectral radius determines the invasion threshold. Vaccination modifies K by reducing the contribution of vaccinated individuals to generating new infections. For a susceptibility-reducing vaccine, one convenient representation is

$$K(\mathbf{p}) = \mathrm{diag}\big(1-\varepsilon p_1, \ldots, 1-\varepsilon p_m\big)\, K, \tag{10.30}$$

and control of transmission can be formalized through the condition $\rho\big(K(\mathbf{p})\big) < 1$, where $\rho(\cdot)$ denotes spectral radius [223]. This formulation makes clear that the impact of vaccinating group i depends not only on its size, but also on how strongly it couples to other groups through contact structure and infectiousness.

Optimizing vaccination policies requires specifying what outcome is being minimized. If the objective is transmission suppression or minimizing cumulative infections, the optimization naturally targets reductions in $\rho(K(\mathbf{p}))$ or in incidence integrals. If the objective is minimizing deaths, hospitalizations, or years of life lost, the vaccine allocation problem must incorporate group-specific severity and vaccine effects on clinical progression, not only infection risk. In practice, the objective is often a weighted combination, reflecting that the policy trade-off between indirect protection (vaccinating high-transmission groups) and direct protection (vaccinating high-risk groups) depends on vaccine properties, epidemic phase, and societal preferences. The broad structure of the constrained allocation problem can be written as

$$\min_{\mathbf{p}(t)} \int_0^T \mathcal{B}\big(\mathbf{x}(t), \mathbf{p}(t)\big)\, dt \qquad \text{subject to} \qquad \sum_{i=1}^{m} N_i\, \dot{p}_i(t) \le s(t), \quad 0 \le p_i(t) \le 1, \tag{10.31}$$

where N_i is the size of group i, $s(t)$ is the vaccine supply rate, and $\mathcal{B}$ encodes the burden being minimized. Even when the underlying epidemic model is deterministic, the presence of rollout constraints couple timing and targeting: the same allocation can have different impact depending on whether doses arrives before, during, or after the growth phase.

For structured populations, one analytically informative approach is to interpret vaccination as a control that perturbs the dominant eigenvalue of the next-generation operator. Small changes in p_i induce changes in $\rho(K(\mathbf{p}))$ that can be approximated by eigenvalue sensitivity, yielding marginal-impact scores that reflect each group's role in sustaining transmission. These scores rationalize why vaccinating groups with high contact rates can produce large indirect benefits, especially when the objective is to minimize infections rather than severe outcomes. Applied work for influenza illustrated how contact-driven heterogeneity can shift optimal allocations away from naive severity-only prioritization and toward vaccinating groups central to transmission for certain objectives [708]. Conversely, when vaccine efficacy is lower in older adults or when vaccines primarily prevent disease severity rather than infection, direct protection of high-risk groups can dominate, and optimal policies may prioritize older ages even if they contribute less to transmission.

The COVID-19 experience emphasized the dependence of prioritization on vaccine characteristics and preexisting immunity. When vaccines substantially block infection and onward transmission, prioritizing groups with many contacts can reduce incidence and indirectly protect the vulnerable, but when the dominant effect is reducing severe disease, prioritizing the elderly or clinically vulnerable tends to minimize mortality. Modeling studies comparing age-stratified strategies under varying assumptions about efficacy profiles and naturally acquired immunity highlighted that the mortality-minimizing and incidence-minimizing allocations can diverge, and that serostatus-aware prioritization can increase the effective impact of limited supply [709]. These results reinforce a general principle: the optimal policy is not a universal ranking of groups, but a solution to a constrained optimization problem whose shape depends on (i) the epidemic state, (ii) vaccine effects on susceptibility, infectiousness, and severity, (iii) contact structure, and (iv) the chosen social objective.

Finally, vaccination optimization must be interpreted in the broader context of epidemic control. Vaccine allocation rarely acts alone; it interacts with non-pharmaceutical measures, testing and isolation, and behavior change. Vaccination can relax the need for broad suppression, but the timing of that relaxation depends on how quickly effective protection accumulates in the groups that most affect the objective. These considerations connect vaccination strategy design to the general theory of infectious disease control and its emphasis on coupling between transmission dynamics, heterogeneity, and policy constraints [118]. In subsequent sections, the same optimization logic will be extended to explicitly account for spatial structure and network connectivity, and to incorporate uncertainty and robustness when vaccine effectiveness, uptake, and variant evolution are imperfectly known.

10.5 Spatial and network intervention strategies

Spatial structure and network connectivity fundamentally shape both epidemic trajectories and the effectiveness of interventions. When transmission is embedded in a geography of patches connected by mobility, or in a graph of individuals linked by contacts, the same average reduction in transmission can yield dramatically different outcomes depending on *where* and *on whom* the intervention acts. This motivates intervention strategies that are explicitly spatially targeted, connectivity-aware, and adaptive to the evolving distribution of risk. In such settings, interventions act not only by scaling a local transmission rate, but also by modifying the couplings that transport infection across space and through contact networks [710–712].

A standard metapopulation representation partitions the population into locations indexed by $i \in \{1, \ldots, m\}$, each with compartmental variables such as $S_i(t)$, $I_i(t)$, $R_i(t)$, coupled through mobility flows. One convenient form models movement by a matrix $M(t)$ whose entry $M_{ij}(t)$ represents per-capita flow from patch i to patch j. A minimal controlled formulation may be written as

$$\frac{dS_i}{dt} = -\lambda_i(t)S_i(t) + \sum_{j \neq i} M_{ji}(t)S_j(t) - \sum_{j \neq i} M_{ij}(t)S_i(t), \tag{10.32}$$

$$\frac{dI_i}{dt} = \lambda_i(t)S_i(t) - \gamma I_i(t) + \sum_{j \neq i} M_{ji}(t)I_j(t) - \sum_{j \neq i} M_{ij}(t)I_i(t), \tag{10.33}$$

with $\lambda_i(t)$ denoting the local force of infection, possibly itself dependent on local policy. Travel restrictions, cordon sanitaires, and mobility reduction orders enter naturally as controls that scale mobility intensities, for example,

$$M_{ij}(t) = M_{ij}^0 \big(1 - u_{ij}(t)\big), \qquad 0 \leq u_{ij}(t) \leq 1, \tag{10.34}$$

thereby reducing importation and exportation risk by weakening specific edges in the mobility network. A central result in this literature is that invasion risk in metapopulation systems exhibits threshold behavior controlled jointly by local transmission and network connectivity, so that targeted reduction of mobility along highly influential connections can shift a system from a regime of widespread spatial invasion to one of contained, local outbreaks [713].

The practical impact of such controls depends on the empirical structure of mobility networks, which are typically heterogeneous across scales and dominated by heavy-tailed flows between hubs. Multiscale mobility models show that coupling patterns can generate rapid long-range seeding even when most movement is local, implying that coarse interventions that treat space as homogeneous can severely underestimate invasion speed and overestimate the effectiveness of uniform travel reduction [145]. This motivates spatially selective policies, such as restricting mobility between highly connected hubs while maintaining mobility in less connected regions, or implementing localized suppression in patches that are both epidemiologically hot and structurally central for dissemination.

Network-based formulations make the same point at the individual level. Let A be an adjacency matrix describing contacts, and consider a simple susceptible–infectious–susceptible (SIS) process on a network. In many mean-field and spectral approximations, the epidemic threshold depends on the largest eigenvalue $\lambda_{\max}(A)$, yielding a condition of the form

$$\frac{\beta}{\gamma} > \frac{1}{\lambda_{\max}(A)} \tag{10.35}$$

for persistence. Interventions that reduce transmission along edges, remove edges, or temporarily deactivate nodes can be interpreted as operations that reduce $\lambda_{\max}(A)$, thereby raising the threshold and suppressing spread. This spectral viewpoint provides a mathematically explicit bridge between intervention design and network structure, and it clarifies why targeting highly connected individuals or contacts can dominate uniform reductions when connectivity is heterogeneous [144].

Connectivity-aware targeting can be implemented in several mathematically equivalent ways. One can prioritize interventions on nodes with high degree or high centrality, reduce weights on edges that lie on many paths between communities, or impose localized restrictions that sever bridges between densely connected subgraphs. In immunization and contact-reduction contexts, a key theoretical insight is that targeted strategies can achieve large reductions in epidemic potential using a smaller intervention "budget" than random or uniform strategies, particularly in networks with broad degree distributions [714]. When translated into policy language, this corresponds to prioritizing high-contact occupations, high-mobility individuals, and settings that act as mixing amplifiers, while leaving lower-impact edges less constrained.

Empirical evidence from modern pandemics suggests that interventions that alter movement and contact structure are not merely theoretical constructs but measurable levers that co-determine transmission dynamics. Analyses combining mobility data with epidemic indicators have shown that reductions in movement and changes in connectivity patterns can align closely with changes in growth rates, implying that policy actions that reshape connectivity are often among the dominant drivers of time-varying transmission in practice [256]. This empirical coupling strengthens the case for incorporating mobility and contact-network structure explicitly into the intervention model, rather than treating $\beta(t)$ as a purely exogenous, location-independent function.

Spatial and network intervention strategies introduce a characteristic computational challenge: once controls are defined over patches, edges, or communities, the dimension of the control variable grows rapidly, and feasibility constraints become structurally complex. In practice, optimization therefore often proceeds by parameterizing controls in low-dimensional families, such as regional tiers, community-level triggers, or edge classes (e.g., airport-to-airport vs. local commuting), while preserving the structural insight that interventions are most effective when aligned with connectivity. This leads naturally to adaptive, trigger-based policies where restrictions are activated when local indicators cross thresholds and deactivated as risk declines, with the trigger itself chosen to balance stability against responsiveness. In subsequent sections, these ideas will be integrated with uncertainty-aware and robust control, since the benefits of targeting depend critically on imperfect observability of where transmission is occurring and on time variation in mobility and behavior.

10.6 Multi-strain and evolution-aware policy

When pathogens evolve on the same timescale as epidemic spread, intervention policies cannot be evaluated purely by their short-term effect on incidence: they also reshape the selective environment in which variants arise and compete. This coupling is central for influenza-like antigenic evolution, for antimicrobial resistance, and for immune-escape dynamics in rapidly evolving viruses. A unifying perspective is that epidemiological dynamics determine the ecological conditions for selection, while immune landscapes and interventions determine the fitness differences among strains, jointly producing the phylodynamic patterns observed in surveillance data [129,715,716].

A convenient modeling starting point extends compartmental dynamics to multiple strains indexed by $i \in \{1, \dots, m\}$, with strain-specific transmission and clearance parameters and with immunity-mediated interaction between strains. In a minimal formulation with a single susceptible pool and multiple infectious classes, one may write

$$\frac{dS}{dt} = -S\sum_{i=1}^{m} \beta_i(t) I_i(t), \tag{10.36}$$

$$\frac{dI_i}{dt} = S\beta_i(t) I_i(t) - \gamma_i I_i(t) + \sum_{j \neq i} \mu_{ji} I_j(t) - \sum_{j \neq i} \mu_{ij} I_i(t), \tag{10.37}$$

where μ_{ij} represents mutation or effective transition from strain i to strain j. The control dependence enters through $\beta_i(t)$ and, potentially, γ_i if treatment affects clearance. For example, non-pharmaceutical interventions can be represented by $\beta_i(t) = \beta_i^0(1 - u(t))$, while strain-targeted pharmaceutical pressure can be represented by $\gamma_i(t) = \gamma_i^0 + a(t)\eta_i$, where η_i captures treatment efficacy against strain i. In many-strain systems with cross-immunity, richer formulations track host immune histories or susceptibility classes; a well-studied alternative uses status-based representations to manage combinatorial immune structure while preserving tractability for large m [717].

Even in the simple system above, the evolutionary effect of interventions becomes transparent by considering relative growth rates. Define $I(t)=\sum_i I_i(t)$ and frequencies $f_i(t)=I_i(t)/I(t)$. Ignoring mutation for intuition, strain i has instantaneous growth rate

$$r_i(t)=\beta_i(t)S(t)-\gamma_i, \tag{10.38}$$

and the strain composition evolves approximately by a replicator equation,

$$\frac{df_i}{dt}\approx f_i(t)\big(r_i(t)-\bar{r}(t)\big), \qquad \bar{r}(t)=\sum_k f_k(t)r_k(t), \tag{10.39}$$

so policies that change $S(t)$ or $\beta_i(t)$ change not only the epidemic trajectory but also the direction and strength of selection among strains. In influenza, for example, ecological and immunological conditions jointly restrict diversity and shape drift patterns, implying that interventions which alter immune selection or transmission opportunities can alter evolutionary outcomes even when their aggregate effect on incidence is similar [718].

Evolution-aware policy design therefore extends the objective of control beyond immediate burden to include future risk generated by selection and mutation. One modeling route augments a standard cost functional with a term penalizing evolutionary hazard. If $h(t)$ denotes the instantaneous hazard of generating or establishing a variant of concern, a planner objective can be written as

$$J(\mathbf{u})=\int_0^T\Big[B\big(\mathbf{x}(t),\mathbf{u}(t)\big)+\eta\,h(t)\Big]dt+\Phi\big(\mathbf{x}(T)\big), \tag{10.40}$$

where B measures contemporaneous health and policy costs and η weights the value placed on reducing evolutionary risk. A simple mechanistic proxy takes $h(t)$ proportional to prevalence and replication opportunities, for example, $h(t)=\sum_i \kappa_i I_i(t)$, where κ_i captures the per-infection propensity of strain i to generate problematic mutants or seed immune escape. This structure formalizes a key intuition: aggressive suppression can reduce mutation supply by reducing infections, but if suppression differentially disadvantages one strain more than another, it can increase the relative advantage of escape-capable strains, shifting composition even as totals decline.

Vaccination introduces particularly subtle evolutionary effects because vaccines vary in whether they block infection, reduce transmission, or primarily prevent disease severity. Theory shows that "imperfect" vaccines that reduce within-host growth or disease but do not fully block transmission can relax natural selection against high-virulence strains, potentially permitting more harmful strains to circulate in partially vaccinated populations [719]. Experimental evidence from Marek's disease in poultry supports this mechanism: vaccination that prevents severe disease but allows infection and onward transmission can enable very virulent strains to transmit that would otherwise be too lethal to persist [720]. These results do not imply that vaccination is undesirable; rather, they motivate explicit modeling of vaccine mode-of-action, vaccine coverage heterogeneity, and the evolutionary endpoints being penalized in the objective. In practical terms, evolution-aware optimization tends to favor strategies that jointly reduce incidence (limiting mutation supply), maintain high protection in the most vulnerable groups, and avoid long periods in which transmission continues at substantial scale under strong directional immune selection.

The central implication is that policies should be assessed not only by how they move the epidemic state today, but by how they shape the adaptive landscape for tomorrow. Multi-strain control problems can therefore be viewed as optimal control of an eco-evolutionary system, where epidemiological state variables and evolutionary composition variables co-determine risk. This framing naturally connects to subsequent robustness and adaptivity questions, since uncertainty about the genotype–phenotype map of escape, the distribution of fitness effects of mutations, and the response of behavior to policies can dominate quantitative recommendations even when qualitative evolutionary mechanisms are clear.

10.7 Use cases

The abstract machinery of intervention modeling and optimization becomes most convincing when it is grounded in concrete historical and operational episodes where policy choices are constrained by time, information, and feasibility. This section surveys several canonical use cases that illustrate how formal models translate into decisions, what kinds of objectives and constraints arise in practice, and which methodological choices tend to determine success or failure.

A first benchmark comes from the 1918 influenza pandemic, where many cities implemented layered non-pharmaceutical interventions such as school closures, bans on public gatherings, and isolation measures, but differed sharply in timing, duration, and intensity. Retrospective empirical analysis of U.S. cities indicates that earlier and more sustained packages of interventions were associated with lower peak mortality and, in many cases, lower cumulative mortality, highlighting the nonlinear payoff to acting before exponential growth saturates healthcare and public compliance [721]. Mechanistic modeling that reconstructs transmission under city-level policies further supports the view that interventions worked primarily by reducing effective contact rates and that premature relaxation could permit substantial rebound, an early example of why optimization should incorporate not only the magnitude of suppression but also the timing of switching [722]. In modern terms, these analyses show a policy problem with partially observed state, delayed signals, and a sharp penalty on peak burden, a structure that naturally leads to feedback-oriented control rather than one-shot planning.

A second use case concerns containment and mitigation planning for pandemic influenza in the era of antivirals and rapid response. When vaccines are unavailable or delayed, interventions such as antiviral treatment, prophylaxis of contacts, and targeted social distancing can be framed as constrained allocation problems in which a limited stockpile must be deployed to reduce the probability of sustained transmission. Modeling work on containing emerging influenza at its source demonstrated that the feasibility of containment depends jointly on transmissibility, detection delays, and the speed and coverage of targeted prophylaxis, making explicit the operational role of logistics constraints inside the epidemic dynamics rather than as an external afterthought [723]. Related simulation studies explored combinations of pharmaceutical and non-pharmaceutical measures across large synthetic populations and showed that strategies that look similar in average transmission reduction can differ drastically in peak suppression and in the timing of resurgence, reinforcing that the optimization target must align with the true policy objective (peak control, cumulative infections, or time-to-vaccine) rather than a generic reduction in β [724]. These influenza planning models also illustrate a recurring design pattern: when interventions are costly, targeted measures that concentrate resources on high-impact transmission channels can dominate uniform measures, but only if detection and delivery systems are fast enough to exploit that targeting.

A third use case is the 2003 SARS epidemic, in which classic levers of case finding, isolation, and contact tracing played a decisive role. SARS is an instructive example because interventions act primarily through identifying and removing infectious individuals from contact networks, rather than through population-wide suppression, and because delays between symptom onset, detection, and isolation can change the qualitative regime from controllable to explosive. Modeling analyses of the Hong Kong outbreak quantified transmission dynamics and evaluated the effective impact of isolation and behavioral change, illustrating how a control variable such as the isolation rate can be mapped onto a measurable operational quantity (time from onset to hospitalization) and how improvements in that quantity translate into reductions in the effective reproduction number [70]. From an optimization standpoint, SARS emphasizes that the best "control" may be investment in surveillance and response capacity that reduces delay and increases targeting precision, even if the formal control variable in the model is written as a rate $\delta(t)$.

A fourth use case comes from the 2014–2015 West African Ebola epidemic, where the dominant intervention levers included case isolation capacity, safe burial practices, and community behavior change, all operating under severe resource and trust constraints. Early outbreak analyses stressed that observed growth rates implied urgent scale-up of isolation and burial interventions to reduce transmission, and that delays in expanding treatment and response infrastructure were themselves a driver of epidemic size [86]. Subsequent modeling work illustrated how spatial heterogeneity, under-ascertainment, and behavioral adaptation complicated inference and policy evaluation, and how intervention effectiveness depended critically on matching the control mechanism to the dominant transmission routes in each setting [725]. Ebola also provides a particularly clear example of the difference between nominal and robust control: when data are sparse and reporting fractions uncertain, policies that merely track the estimated state can oscillate or overshoot, whereas policies designed with explicit safety margins against capacity exceedance can be operationally superior even if they are not point-optimal for a best-fit parameter set.

A final use case is COVID-19, where the scale, heterogeneity, and rapid policy iteration of the pandemic made it a stress test for the entire modeling-to-decision pipeline. Early in 2020, model-based analyses were used to compare mitigation and suppression strategies and to anticipate healthcare demand under alternative policy bundles, formalizing the trade-off between reducing transmission rapidly and sustaining interventions long enough to avoid rebound in a largely susceptible population. As vaccines became available, vaccination strategy became a real-time allocation problem under supply constraints, heterogeneous risk, and variant-driven changes in transmission, with policy decisions effectively implementing time-dependent controls on susceptibility and severity across groups. COVID-19 thus illustrates that intervention optimization is rarely a single problem solved once; it is a sequence of coupled problems whose objectives, constraints, and uncertainty sets evolve with pathogen biology, immunity landscapes, and social response. Across these use cases, the persistent lesson is that formal modeling is most valuable when it preserves the operational meaning of controls, when objectives reflect what decision-makers truly value (often peak and capacity outcomes, not only cumulative totals), and when uncertainty and delay are treated as core system features rather than as small perturbations.

CHAPTER

Uncertainty quantification, sensitivity analysis, and model comparison

11

Epidemiological models used to support scientific understanding and inform consequential decisions such as situational awareness, forecasting, and intervention planning. However, as already discussed in length and repeatedly in this book, epidemic data are noisy, delayed, censored, and shaped by changing surveillance practices; meanwhile, mechanistic assumptions and data-driven pipelines introduce additional layers of uncertainty from the modeling and implementation layers. As a result, point predictions can be misleadingly confident, and policy conclusions can be brittle to modeling choices. This motivates three tightly linked tasks that should accompany essentially any serious modeling effort: uncertainty quantification (UQ), sensitivity analysis (SA), and model comparison (MC). Together, they provide a disciplined way to (i) report what the model implies *and* how unsure we are, (ii) identify which assumptions and inputs matter most, and (iii) decide which model families are better supported by the data for a given purpose [686].

Formally, we use a common abstract representation that covers extended SIR-style state-space models, as well as many data-driven time series and machine-learning approaches. This formalization can be slightly too abstract in order to allow it to capture any modeling practice discussed in this book, but as such, it provides a common ground for others to use more specific definitions. Let D denote observed data (e.g., case counts, hospitalizations, wastewater signals, covariates), and let $\mathcal{M}$ be a model class. A model $\mathcal{M}$ induces a probability law for data and latent quantities through parameters $\theta \in \Theta$ and (optionally) latent states X:

$$p(D, X \mid \theta, \mathcal{M}) = p(D \mid X, \theta, \mathcal{M})\, p(X \mid \theta, \mathcal{M}), \tag{11.1}$$

where $p(X \mid \theta, \mathcal{M})$ may encode mechanistic dynamics (e.g., compartmental transitions) or a learned temporal process, and $p(D \mid X, \theta, \mathcal{M})$ represents measurement and reporting mechanisms. A *quantity of interest* (QoI) is any functional

$$Q = g(X, \theta; \mathcal{M}), \tag{11.2}$$

such as R_t, peak incidence timing, cumulative infections, or a forecast trajectory. Importantly, even when a model produces a single trajectory or estimate, the analysis question is typically about the *distribution* of plausible Q given limited and imperfect information.

First, UQ is the task of characterizing the uncertainty in QoIs implied by the model and the data. Formally, UQ aims to construct (exactly or approximately) a distribution for Q conditional on D under $\mathcal{M}$. In a Bayesian formulation, uncertainty about θ and X is represented through the posterior

$$p(\theta, X \mid D, \mathcal{M}) \propto p(D, X \mid \theta, \mathcal{M})\, p(\theta \mid \mathcal{M}), \tag{11.3}$$

Epidemic Modeling. https://doi.org/10.1016/B978-0-44-355525-1.00017-5

and the induced posterior distribution of the QoI is

$$p(Q \mid D, \mathcal{M}) = \int \delta\big(Q - g(X, \theta; \mathcal{M})\big)\, p(\theta, X \mid D, \mathcal{M})\, d\theta\, dX. \tag{11.4}$$

For forecasting and decision support, a central object is the *posterior predictive* distribution

$$p(\widetilde{D} \mid D, \mathcal{M}) = \int p(\widetilde{D} \mid \theta, \mathcal{M})\, p(\theta \mid D, \mathcal{M})\, d\theta, \tag{11.5}$$

which enables calibrated probabilistic predictions and principled uncertainty intervals [726,727]. Frequentist UQ can be expressed analogously via sampling distributions and confidence sets, but the practical message is the same: because uncertainty is intrinsic (measurement noise, process stochasticity, parameter non-identifiability, and misspecification), it must be made explicit to avoid overconfident conclusions [118,686].

Second, even after uncertainty is quantified, it is rarely uniform across inputs and assumptions. SA studies how variation in model inputs changes model outputs. Let $U = (U_1, \ldots, U_d)$ collect uncertain inputs (parameters, initial conditions, hyperparameters, reporting-rate assumptions, feature/lag choices, etc.), and consider the mapping $Q = f(U)$ induced by $\mathcal{M}$ (possibly after conditioning on D). SA seeks quantitative measures of influence, ranging from local derivatives to global variance decompositions. A canonical global approach decomposes the output variance and attributes it to inputs and their interactions:

$$\mathrm{Var}(Q) = \sum_{i=1}^{d} V_i + \sum_{i<j} V_{ij} + \cdots, \tag{11.6}$$

leading to Sobol-type sensitivity indices such as

$$S_i = \frac{V_i}{\mathrm{Var}(Q)}, \qquad S_i^{\mathrm{T}} = 1 - \frac{\mathrm{Var}\big(\mathbb{E}[Q \mid U_{-i}]\big)}{\mathrm{Var}(Q)}, \tag{11.7}$$

where S_i measures the first-order (main-effect) contribution of U_i and S_i^{T} measures the total contribution, including interactions [728]. In epidemiological terms, SA helps distinguish, for example, whether uncertainty in peak timing is driven primarily by transmission parameters, reporting dynamics, behavioral covariates, or structural assumptions; it also guides data collection by revealing which uncertainties most constrain actionable conclusions [729].

Third, epidemiological practice rarely offers a single uncontested model: one may compare mechanistic variants (e.g., age structure, waning immunity, time-varying transmission), classical time-series baselines, and ML predictors trained on different signals. MC provides formal criteria for selecting among candidate models $\{\mathcal{M}_k\}_{k=1}^{K}$ or combining them, based on their explanatory and predictive adequacy for a given task. In Bayesian terms, a key quantity is the *marginal likelihood* (model evidence)

$$p(D \mid \mathcal{M}_k) = \int p(D \mid \theta_k, \mathcal{M}_k)\, p(\theta_k \mid \mathcal{M}_k)\, d\theta_k, \tag{11.8}$$

from which posterior model probabilities and Bayes factors follow [730]. In information-theoretic and predictive paradigms, comparison is often performed via penalized fit criteria such as AIC or BIC,

$$\mathrm{AIC} = -2\log\hat{L} + 2p, \qquad \mathrm{BIC} = -2\log\hat{L} + p\log n, \tag{11.9}$$

or via out-of-sample predictive evaluation such as cross-validation and proper scoring rules [731,732]. These tools make explicit an unavoidable trade-off: more flexible models can fit historical data better but may generalize worse, especially under distribution shift (e.g., changing testing or intervention regimes). Model comparison, therefore, is not only about "picking a winner", but rather it is about matching model complexity to the inferential or forecasting objective and quantifying the robustness of conclusions to modeling choices [726,732].

Importantly, UQ, SA, and MC are best viewed as a single workflow rather than separate add-ons. UQ produces distributions over QoIs; SA explains which uncertainties dominate those distributions; MC evaluates competing model assumptions and provides principled ways to select or average models. In the remainder of this chapter, we develop each component in detail, emphasizing practical procedures that apply across mechanistic extended SIR models and modern data-driven time-series and ML approaches.

11.1 A taxonomy of uncertainty in epidemiological models

The scientifically relevant output is not a single trajectory but a predictive distribution over future quantities of interest (QoIs), together with a clear accounting of where uncertainty comes from [733,734]. Formally, let M denote a model class, let $y_{1:T}$ be the observed surveillance data up to time T, and let z_{T+h} be a QoI at horizon h (e.g., incidence next week, ICU occupancy, or the peak time). A convenient abstraction that covers both mechanistic and data-driven models is

$$p(z_{T+h} \mid y_{1:T}, M) = \int p(z_{T+h} \mid x_{1:T}, \theta, M)\, p(x_{1:T}, \theta \mid y_{1:T}, M)\, \mathrm{d}x_{1:T}\, \mathrm{d}\theta, \tag{11.10}$$

where $x_{1:T}$ denotes latent states (e.g., unobserved infections, susceptible fractions, time-varying transmission) and θ denotes unknown parameters and hyperparameters. For ML models, $x_{1:T}$ may be absent, but θ still represents fitted parameters and choices that affect predictive uncertainty [735,736]. Eq. (11.10) makes explicit that uncertainty can enter through multiple channels. In practice, failing to represent a major channel typically leads to (i) overconfident intervals, (ii) unstable inference under small data perturbations, or (iii) contradictory conclusions across plausible models [737–739].

Although taxonomies vary across disciplines, for infectious-disease modeling the following decomposition is particularly useful; it maps cleanly onto extended SIR models, partially observed Markov process (POMP) formulations, and forecasting pipelines [740].

1. **Intrinsic (process) stochasticity.** Transmission, recovery, and contact events are random at the individual level; even with perfectly known parameters, finite-population epidemics exhibit trajectory variability. This is why stochastic SIR/SEIR variants and state-space formulations are common for inference and forecasting [741,742]. A practical consequence is that single-trajectory fits can misrepresent tail risks (e.g., unusually early peaks or flare-ups), which matters for capacity planning.

2. **Measurement and reporting uncertainty (observation process).** Surveillance is an imperfect, delayed, and sometimes biased measurement of latent infections. Common issues include under-ascertainment (only a fraction of infections are tested/reported), time-varying case definitions, reporting delays, and misclassification (e.g., false negatives/positives). During COVID-19, substantial under-ascertainment (especially early) changed inferred epidemic size and growth [743]; and time-varying RT-PCR sensitivity created strong dependence on the delay from exposure/symptom onset [744]. Delays alone can shift apparent peaks by days to weeks, motivating nowcasting methods that explicitly model "occurred-but-not-yet-reported" events [618].
3. **Parameter and initial-condition uncertainty.** Even if the model form is accepted, key parameters (e.g., transmission rates, mixing modifiers, reporting fractions, dispersion/overdispersion) are learned from finite, noisy data and can be weakly identifiable. Early in outbreaks, limited data and changing surveillance can produce wide posterior uncertainty and strong sensitivity to priors/assumptions [745]. A canonical example is inference for the effective reproduction number: estimates depend on assumptions about generation/serial interval distributions and observation delays [746–748]. Even when point estimates look stable, uncertainty in these inputs can dominate credible intervals for R_t.
4. **Structural (model-form) uncertainty.** Different plausible mechanistic assumptions (e.g., SEIR vs. SIR, heterogeneous mixing, age structure, time-varying contacts, immunity waning) or different data-driven hypothesis classes (e.g., ARIMA vs. state-space vs. gradient-boosted trees) can fit historical data similarly while yielding different counterfactuals and forecasts. This is a central reason why R_0 is not a pathogen "constant" but a model- and context-dependent construct [749]. During COVID-19, estimated non-pharmasutical interventions effects were scrutinized precisely because conclusions can shift under alternative model structures and priors [128].
5. **Scenario/exogenous-input uncertainty.** Forecasts depend on future drivers that are not determined by the disease process alone: behavior change, policy responses, school schedules, testing regimes, and mobility. Mechanistic models often represent these drivers through time-varying parameters (e.g., $\beta(t)$) or explicit covariates; data-driven models inherit the same issue as distribution shift. This uncertainty is often *irreducible* at forecast time and should be communicated as scenario-conditional predictions rather than a single unconditional curve [745].
6. **Computational/approximation uncertainty.** Inference algorithms (MCMC, particle filters, variational approximations, bootstraps) introduce Monte Carlo error and may fail silently under misspecification or weak identifiability. POMP settings are especially illustrative: likelihood evaluation is itself approximate, and algorithmic choices (particles, resampling, tuning) affect estimates [741,742]. In ML forecasting, approximate Bayesian methods, ensembling, and calibration heuristics similarly introduce method-dependent uncertainty [735,736].

Items (1) and parts of (4) are often treated as aleatory (intrinsic variability), while (2)–(6) are largely *epistemic* (reducible in principle with better data/models) [733,734]. This distinction is operationally useful: it clarifies whether effort should go into collecting better data (e.g., serology to reduce under-ascertainment [745]), improving the observation model (e.g., nowcasting delays [618]), or expanding model classes/ensembles to hedge structural uncertainty [750].

The taxonomy above is not merely descriptive: it suggests what to vary (inputs, priors, model components), what to hold fixed, and which outputs to stress-test. Sensitivity analysis formalizes the propagation of uncertainty in inputs/assumptions to uncertainty in QoIs, while model comparison evaluates which model classes (or ensembles) deliver better-calibrated predictive distributions for the intended decision [750,751]. Subsequent sections develop these ideas systematically; for now, the key point is that *credible inference in epidemiology requires making the dominant uncertainty channels explicit rather than implicit* [738,739,749].

Fig. 11.1 operationalizes the taxonomy in this subsection by separating the latent epidemic process from the *measurement pipeline* that produces surveillance data. The top panel displays the daily number of new infections generated by a stochastic SIR model. Across repeated simulations, we vary (i) the random transmission/recovery events (process stochasticity) and (ii) a small set of epidemiologically meaningful parameters (here, the initial reproduction number and mean infectious period), producing a distribution over plausible latent trajectories. The solid curve is the Monte Carlo median of latent incidence, while the shaded band summarizes uncertainty across runs (a central credible/uncertainty interval). This panel corresponds to uncertainty sources (1) and (3) in the taxonomy: even before considering measurement, there is irreducible path-to-path variability and finite-information uncertainty in parameters and initial conditions [752].

The bottom panel shows what a routine surveillance stream, reported cases per day, might look like when those same latent infections are filtered through an explicit observation model. First, only a fraction ρ of infections is assumed to be detected and reported (under-ascertainment), reflecting incomplete testing, asymptomatic infections, and changing care-seeking behavior. Second, each infection is shifted forward in time by a reporting delay drawn from a delay distribution (here parameterized by a mean and standard deviation), representing the lag from infection to symptom onset, testing, and entry into the reporting system. Third, the resulting daily counts are subject to overdispersed measurement noise (here via a negative-binomial observation model) to capture extra-Poisson variability and day-to-day reporting irregularities. These mechanisms instantiate uncertainty source (2) in the taxonomy (measurement/reporting uncertainty), and they are exactly the features targeted by nowcasting and delay-adjustment methods used in practice [618]. The two colored example trajectories in the bottom panel hold the latent epidemic fixed (one representative latent run) but change only observation assumptions: Example A uses a lower reporting fraction and longer mean delay, while Example B uses a higher reporting fraction and shorter mean delay. The differences are not cosmetic: changing ρ rescales the observed peak height, and changing the delay distribution shifts and smooths the observed curve, altering apparent growth rates and peak timing. Consequently, downstream quantities that practitioners often compute from reported data—such as growth rates and effective reproduction numbers—can differ materially unless the observation process is modeled or corrected [149,748]. In other words, the figure illustrates the core message of this subsection: the uncertainty we care about for decision-making is the uncertainty of QoIs conditional on imperfect observations, and a meaningful uncertainty decomposition must distinguish intrinsic epidemic variability from the uncertainties introduced by surveillance and reporting.

Fig. 11.2 summarizes four recurring questions in practical UQ for epidemic models: *(i)* Are predictions sensible before seeing data (prior predictive)? *(ii)* How much do key epidemiological assumptions affect inferred quantities such as R_t? *(iii)* Are parameters actually identifiable from the available observations? and *(iv)* Are reported uncertainty intervals calibrated? Top-left (prior vs. posterior predictive). The blue band and line show the *prior predictive* distribution of reported cases, obtained by sampling parameters from the prior and propagating them through the mechanistic model and observation model. The orange band and line show the *posterior predictive* distribution after conditioning on the observed data (green curve). This panel makes two points central to this

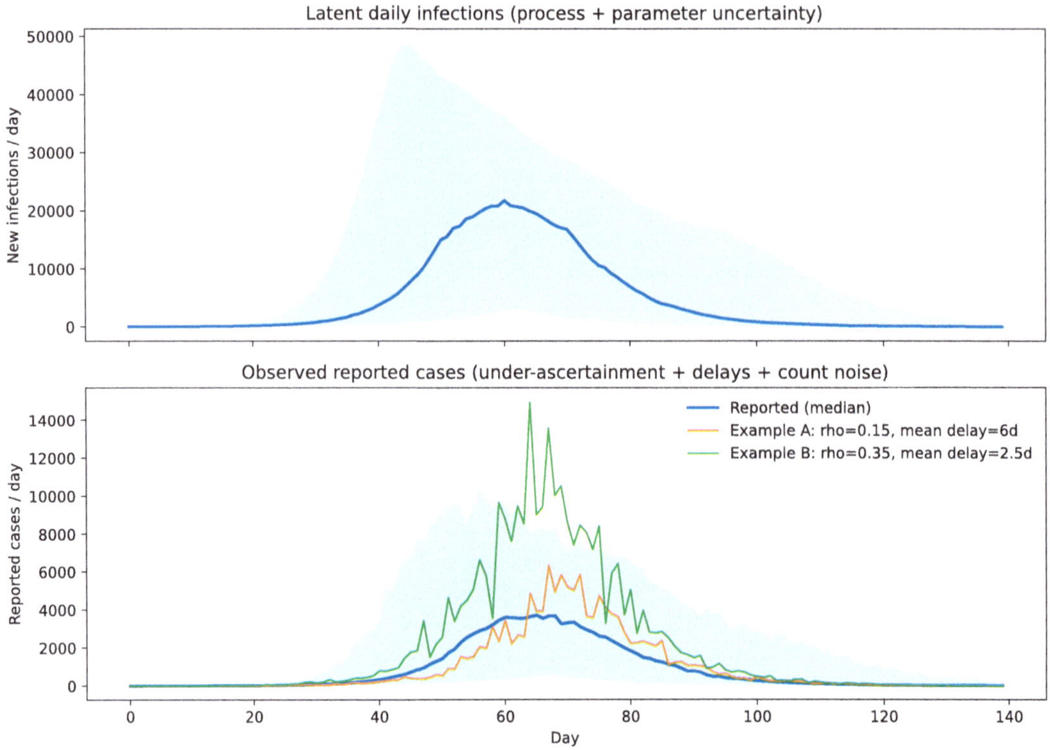

FIGURE 11.1

Latent versus observed epidemic curves under realistic uncertainty. Top: latent daily infections from a stochastic SIR simulation with mild parameter variability; the line is the Monte Carlo median, and the shaded region is a central uncertainty band. Bottom: reported cases generated from the same latent process but passed through an observation model with under-ascertainment, reporting delays, and overdispersed count noise. Two illustrative realizations (Examples A/B) show how different reporting fractions and delay distributions can produce markedly different observed trajectories, even when the underlying epidemic is similar.

subsection's taxonomy. First, uncertainty is not only about noisy observations: even before fitting, plausible parameter values can imply a wide range of epidemic shapes (parameter uncertainty). Second, conditioning on data should shrink uncertainty toward trajectories consistent with surveillance, but only to the extent permitted by the information content of the data and the adequacy of the observation model (measurement/reporting uncertainty) [739]. In practice, prior predictive checks help detect unrealistic priors or misspecified mechanisms early, before they contaminate inference and forecasts [738]. Top-right (R_t sensitivity to the generation interval). The effective reproduction number is not observed directly; it is inferred from incidence data through assumptions about the generation interval (or serial interval) distribution. This panel overlays R_t trajectories inferred under multiple plausible generation-interval specifications. Differences between curves quantify a form of epistemic uncertainty induced by assumptions external to the surveillance stream: even with the same observed cases, alternative

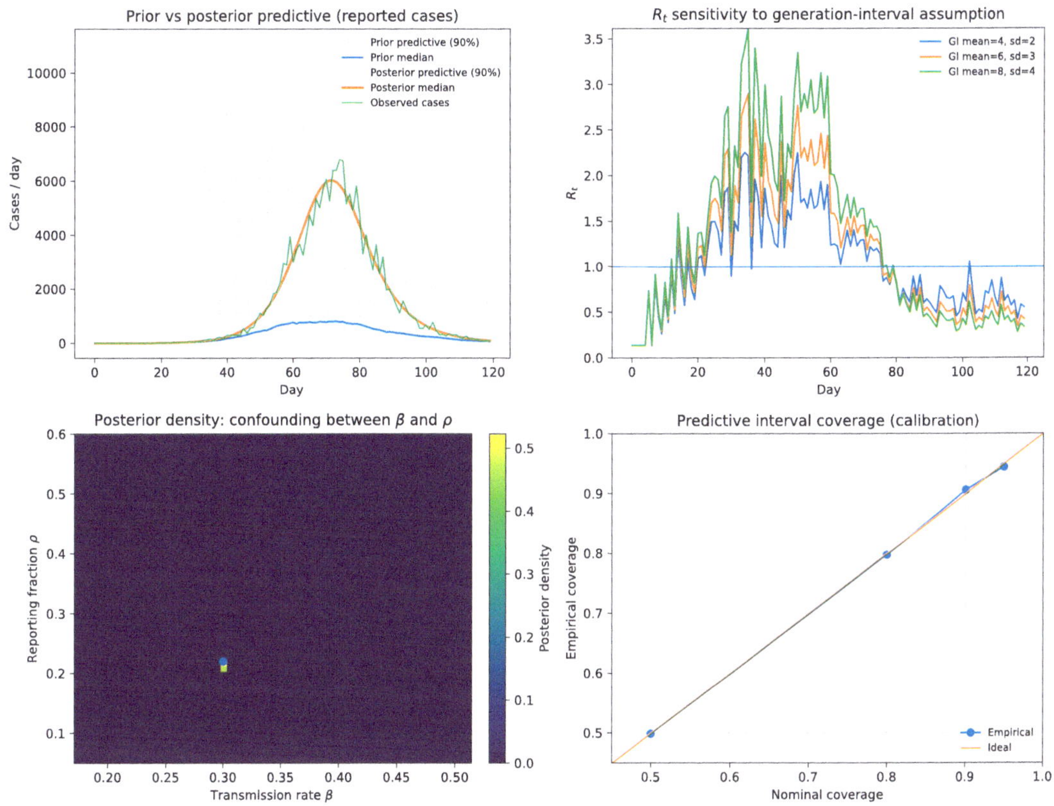

FIGURE 11.2

Four diagnostic views of uncertainty quantification in epidemic modeling. *Top-left:* prior predictive (blue) and posterior predictive (orange) distributions for reported cases, compared to observed data (green); this distinguishes uncertainty present before conditioning on data from uncertainty remaining after inference. *Top-right:* inferred R_t trajectories under different plausible generation-interval assumptions, illustrating assumption-induced (epistemic) uncertainty in derived quantities. *Bottom-left:* joint posterior density of transmission β and reporting fraction ρ, showing confounding/non-identifiability typical of partially observed epidemics. *Bottom-right:* calibration check via predictive-interval coverage; agreement with the diagonal indicates calibrated uncertainty.

but defensible generation-interval choices yield different R_t levels and uncertainty bands, especially during rapid growth [747,748]. This directly illustrates the taxonomy point that uncertainty arises not only from stochasticity and noise, but also from modeling choices that are often treated as fixed inputs. Bottom-left (posterior confounding between transmission and reporting). The heatmap depicts the joint posterior density of the transmission rate β and the reporting fraction ρ. A common identifiability issue in partially observed epidemics is that higher transmission coupled with lower reporting can generate similar observed case curves as lower transmission coupled with higher reporting. Such confounding produces ridges or strong posterior dependence, signaling that the data constrain only certain *combinations* of parameters rather than each parameter individually. This is a concrete example of *practical non-identifiability*: the model can fit the data well while remaining uncertain about mechanistically meaningful parameters [742,745]. Recognizing these dependencies is crucial because downstream QoIs (e.g., inferred attack rates, counterfactual impacts) may be sensitive to where along the ridge the posterior concentrates. Bottom-right (predictive interval coverage / calibration). Here, we assess calibration by comparing nominal predictive-interval levels (e.g., 50%, 80%, 90%, 95%) to the empirical frequency with which held-out observations fall inside those intervals. Points close to the diagonal indicate well-calibrated uncertainty quantification; systematic deviations indicate over- or under-confidence [753]. This panel links the taxonomy to practice: a model may look reasonable and yield narrow intervals, but if those intervals do not achieve their advertised coverage out of sample, the uncertainty representation is unreliable for decision-making. Taken together, these four panels connect directly to the uncertainty taxonomy: the prior/posterior predictive views reflect parameter and process uncertainty and the role of the observation model; the R_t overlay visualizes assumption-driven (structural/input) uncertainty; the β–ρ heatmap exposes identifiability limits; and the coverage plot evaluates whether the full pipeline produces calibrated predictive uncertainty [733,738].

11.2 Identifiability and confounding as a bridge between uncertainty quantification and sensitivity analysis

A recurring theme in epidemic inference is that good fit does not imply well-determined parameters. In partially observed dynamical systems, multiple distinct parameter configurations can generate nearly indistinguishable distributions over the observed data. In such cases, wide posteriors, strong posterior correlations, or flat likelihood surfaces are often information limits imposed by the data and the observation process, rather than a symptom of "bad inference" [754,755]. This makes identifiability a natural conceptual bridge between UQ and SA – UQ characterizes what remains uncertain after conditioning on data, while SA explains which mechanisms and assumptions create that uncertainty and how it propagates to quantities of interest. Identifiability comes in two main forms. Structural identifiability is a property of the model under idealized conditions: with noise-free observations of the relevant outputs, do the parameters map uniquely to the implied outputs? A convenient formalization is that, for a model class inducing an observation distribution $p(y_{1:T} \mid \theta)$, a parameterization is structurally identifiable if

$$p(y_{1:T} \mid \theta_1) = p(y_{1:T} \mid \theta_2) \;\; \forall y_{1:T} \quad \Longrightarrow \quad \theta_1 = \theta_2, \tag{11.11}$$

up to known symmetries or re-parameterizations [755]. Practical identifiability, by contrast, concerns what can be learned from finite, noisy, discretely sampled, and biased surveillance data. Even structurally identifiable models can be practically non-identifiable because the likelihood is nearly flat along certain directions in parameter space, producing wide credible intervals and strong posterior dependence [754]. These flat directions are not merely a nuisance: they delineate which combinations of parameters the data can constrain and which combinations remain effectively unconstrained. A closely related phenomenon is sloppiness, where the mapping from parameters to observables is highly anisotropic: a small number of "stiff" parameter combinations strongly affect the likelihood, while many other combinations have weak or compensating effects. Sloppiness manifests as elongated ridges in the likelihood or posterior and as a wide spectrum of eigenvalues in local sensitivity or Fisher-information analyses [756,757]. This connects directly to SA because those ridges correspond to near-invariances of the model output: moving along a sloppy direction changes parameters substantially while changing observables only minimally. Confounding is especially common in epidemiology because most surveillance streams are only indirect and imperfect measurements of the latent infection process. A representative example is the trade-off between transmission and reporting. If the observed cases satisfy approximately $y_t \approx \rho_t \, i_t$, where i_t is latent incidence and ρ_t is a reporting fraction, then increased transmission can be offset by reduced reporting to yield similar observed trajectories. Allowing ρ_t to vary over time increases this ambiguity further and can make it difficult to distinguish changes in $\beta(t)$ from changes in detection, testing, or reporting practices unless external information is available [743]. A second common confounder links transmission with initial conditions: early in an outbreak, limited time series can permit many pairs (β, I_0) that reproduce the same initial growth, especially when susceptible depletion is negligible over the fitted window [745]. A third confounder arises from generation-interval assumptions, which are required to infer quantities such as R_t from incidence. Different plausible generation-interval choices can yield different R_t levels and timing even when the observed incidence curve is fixed, introducing assumption-driven epistemic uncertainty [747,748]. In each case, the uncertainty is not only about noise; it is about the non-uniqueness of mechanistic explanations compatible with the same observations.

To emphasize these ideas, consider a deterministic SIR model with latent daily incidence $i_t(\beta, \gamma, I_0)$ and an observation model for reported cases

$$y_t \sim \mathrm{NegBin}\big(\mu_t, k\big), \qquad \mu_t = \rho \, i_t(\beta, \gamma, I_0), \tag{11.12}$$

where $\rho \in (0, 1)$ is the reporting fraction and k controls overdispersion. Suppose we fit parameters using data only up to time T, yielding a likelihood $p(y_{1:T} \mid \beta, \rho)$ (with γ, I_0, k fixed for simplicity). Because y_t depends on the product $\rho \, i_t(\beta, \gamma, I_0)$, there typically exists a continuum of pairs (β, ρ) that produce similar $\mu_{1:T}$ and therefore similar likelihood. Formally, for many datasets, one can find (β_A, ρ_A) and (β_B, ρ_B) such that

$$\rho_A \, i_t(\beta_A, \gamma, I_0) \approx \rho_B \, i_t(\beta_B, \gamma, I_0), \tag{11.13}$$

for $t = 1, \ldots, T$, which implies that the fitted window alone cannot reliably separate transmission intensity from detection intensity. Yet, forecasts depend on the latent dynamics beyond T, not only on the mean observation curve over the fitted interval. If $\beta_A > \beta_B$ and $\rho_A < \rho_B$ are chosen to satisfy Eq. (11.13), then the two parameter sets can agree closely on $y_{1:T}$ while implying different latent infection burdens, different susceptible depletion rates, and therefore different future trajectories for $t > T$. In particular, the higher-transmission/lower-reporting explanation typically implies more cumulative infections by time T, a smaller remaining susceptible pool, and a faster transition toward peak and decline under the same mechanistic assumptions. The lower-transmission/higher-reporting explanation, on the other hand, implies fewer latent infections by time T, a larger susceptible pool, and potentially a later peak or more prolonged tail. Both explanations may be compatible with the observed data up to T; the divergence appears in forecasts precisely because the fitted data do not uniquely identify the latent state trajectory.

This example clarifies how identifiability connects UQ and SA. UQ makes the ridge explicit by producing a posterior over (β, ρ) (or a profile likelihood) that is diffuse along the confounded direction, while SA explains why that direction is weakly informed: the observation model (presented in

Eq. (11.12)) is approximately invariant under compensating changes in transmission and reporting over the fitted window. The same logic extends to confounding between $\beta(t)$ and time-varying reporting, between β and initial conditions, and between inferred R_t and generation-interval assumptions. Practically, the resolution is rarely to force narrower intervals; it is to add information that breaks the symmetry, for example by incorporating additional data streams with different ascertainment and delay mechanisms (hospitalizations, deaths, serology, wastewater), by strengthening the observation model (explicit delay distributions and reporting dynamics), or by reporting scenario-conditional predictions that transparently reflect the plausible range of mechanisms consistent with the data [754].

11.3 A sensitivity analysis toolbox

SA complements uncertainty quantification by explaining how uncertainty in inputs, assumptions, and data-processing choices propagates to uncertainty in QoIs. In epidemic modeling, SA supports several practical goals: diagnosing which parameters or assumptions drive forecast spread, prioritizing data collection to reduce decision-relevant uncertainty, identifying fragile conclusions, and detecting regimes where nonlinearities or interactions dominate [758–760]. Because epidemic models range from low-dimensional deterministic SIR extensions to high-dimensional stochastic simulators and ML pipelines, no single SA method is universally appropriate. Instead, it is useful to organize methods by interpretability and computational cost, and to choose tools that match the model's dimensionality, stochasticity, and the available compute budget.

We write the model output as a mapping $Q = f(U)$, where Q is a QoI and $U = (U_1, \ldots, U_d)$ collects uncertain inputs. In mechanistic models, U typically includes transmission and progression parameters, initial conditions, and observation-model parameters. In data-driven models, U can also include preprocessing choices, feature sets, hyperparameters, and training window definitions; SA is still well-defined as long as the pipeline defines a mapping from inputs to outputs [758].

Local SA focuses on infinitesimal or small perturbations around a baseline input $u^\star$. For differentiable models, the basic object is the gradient $\nabla f(u^\star)$, from which one can form normalized sensitivities or elasticities to compare parameters on different scales. A common choice is the log-sensitivity

$$E_i(u^\star) = \frac{\partial \log Q}{\partial \log U_i}(u^\star) = \frac{U_i^\star}{Q^\star}\frac{\partial Q}{\partial U_i}(u^\star), \tag{11.14}$$

which measures the percent change in Q induced by a percent change in U_i [759,761]. Local SA is attractive because it is inexpensive and highly interpretable; it is often the right first step for extended SIR models, where gradients can be obtained by finite differences or, for ODE models, by forward sensitivity equations or adjoint methods [761]. However, local SA can be misleading when the model is strongly nonlinear, when uncertainties are large, when there are thresholds (e.g., near $R_t \approx 1$), or when interactions between parameters dominate. It also depends on the choice of baseline $u^\star$, which may itself be uncertain or poorly identified.

Global SA addresses these limitations by exploring the input space more broadly and quantifying influence under an explicit distribution for U. Variance-based methods decompose the output variance into contributions attributable to each input and their interactions. Under mild regularity conditions, one can write an ANOVA-like decomposition of $\mathrm{Var}(Q)$, leading to Sobol' indices [762]:

$$S_i = \frac{\mathrm{Var}(\mathbb{E}[Q \mid U_i])}{\mathrm{Var}(Q)}, \qquad S_i^T = 1 - \frac{\mathrm{Var}(\mathbb{E}[Q \mid U_{-i}])}{\mathrm{Var}(Q)}. \tag{11.15}$$

Here, S_i measures the main-effect contribution of U_i to output variance, while S_i^T captures the total contribution, including interactions. Sobol' indices are robust and policy-relevant because they directly answer questions like "what fraction of forecast uncertainty is driven by reporting delay uncertainty versus transmission uncertainty?" Yet they can be computationally expensive, especially when d is large or when each model evaluation is costly.

Morris screening provides a pragmatic middle ground for high-dimensional problems: it estimates distributions of "elementary effects" by sampling trajectories through the input space [763,764]. It is typically used to identify a small subset of influential inputs before applying more expensive methods. Morris is especially useful when d is large (many candidate parameters, covariates, or hyperparameters), when the model is moderately expensive, or when the goal is qualitative ranking rather than precise variance attribution [758].

Stochasticity changes both the interpretation and the design of SA. When the model output is random conditional on inputs, Q is better viewed as a random variable with conditional mean and variance, and SA must specify whether it targets sensitivities of $\mathbb{E}[Q \mid U]$, $\mathrm{Var}(Q \mid U)$, tail probabilities, or full predictive distributions [765]. In epidemic contexts, this distinction matters: an input might have little effect on the mean forecast but a large effect on tail risk (e.g., probability of exceeding ICU capacity). Practically, stochastic simulators require replicated runs per input setting or variance-reduction techniques; for expensive models, emulators (e.g., Gaussian-process surrogates) can make global SA feasible [765].

A useful way to summarize the toolbox is as a guidance chart keyed by dimensionality, compute budget, and stochasticity. Table 11.1 provides a concrete decision aid. The intent is not to prescribe a single "best" method, but to encourage a staged workflow: start with inexpensive local or screening methods to understand the model qualitatively, then apply variance-based global methods to the reduced set of influential inputs when quantitative attribution is needed [758].

Sensitivity analysis methods are easiest to adopt when they are expressed as concrete computational patterns. In what follows, we give a set of specific SA "recipes" that recur across extended SIR models and data-driven pipelines. Each recipe makes explicit the quantity of interest (QoI), the perturbation or sampling design, and the summary statistic reported. The examples are intentionally minimal; later sections can refine them (e.g., adding observation delays, multiple data streams, or age structure) without changing the core logic.

Local sensitivities are the most interpretable entry point. The next listing computes finite-difference sensitivities and elasticities for two QoIs—attack rate and peak time—with respect to parameters such as the transmission rate β and recovery rate γ. The key idea is to perturb one input slightly around a baseline and measure the corresponding change in the QoI. Elasticities normalize units so that parameters on different scales are comparable: an elasticity of 2 means that a 1% increase in the parameter produces an approximately 2% increase in the QoI near the baseline. This approach is fast and works well for deterministic mechanistic models, but it reflects only local behavior around the chosen baseline and can miss interactions and nonlinear threshold effects.

Table 11.1 Guidance chart for choosing sensitivity analysis methods in epidemiological modeling.

Setting	Recommended SA method(s)	Compute cost	Typical use
Low d ($\lesssim 10$), deterministic ODE (extended SIR)	Local gradients; elasticities; forward sensitivities / adjoints	Low	Fast "first look"; highly interpretable; baseline-dependent; may miss interactions [759,761].
Low–moderate d ($\lesssim 20$), nonlinear but cheap	Sobol' indices (first + total order)	Medium–High	Quantifies interactions; requires many model evaluations; best when a QoI variance decomposition is desired [762].
High d ($\gtrsim 20$), moderate cost per run	Morris screening (elementary effects) then refine	Low–Medium	Identifies influential inputs and nonlinear/interaction-prone factors; use as a filter before Sobol' [763,764].
Stochastic model (SIR with demographic noise; particle-filter likelihood; agent-based)	Replicated local SA for $\mathbb{E}[Q \mid U]$; Morris screening; surrogate-assisted Sobol'	Medium–Very High	Decide whether targeting mean, variance, or tail risk; consider emulators to reduce Monte Carlo burden [765].
Decision-oriented QoIs (thresholds / tail probabilities)	Distributional / moment-targeted SA; variance decomposition of risk metrics	Medium–High	Focus on exceedance probabilities and worst-case outcomes; mean-only SA may hide tail sensitivity [758].

```
import numpy as np

def sir_euler(beta, gamma, N, I0, T):
    S, I, R = float(N - I0), float(I0), 0.0
    new_inf = np.zeros(T, dtype=float)
    for t in range(T):
        inf = beta * S * I / N
        rec = gamma * I
        dS = min(inf, S)
        dR = min(rec, I)
        S -= dS
        I += dS - dR
        R += dR
        new_inf[t] = dS
    return new_inf

def qoi_peak_time(incidence):
    return int(np.argmax(incidence))

def qoi_attack_rate(incidence, N):
    return float(np.sum(incidence) / N)

def fd_sensitivity(f, u, i, rel_step=1e-3):
```

```
    """Central finite difference w.r.t u[i]."""
    u = np.array(u, dtype=float)
    h = rel_step * (abs(u[i]) + 1e-12)
    u1, u2 = u.copy(), u.copy()
    u1[i] += h
    u2[i] -= h
    return (f(u1) - f(u2)) / (2.0 * h)

def elasticity(f, u, i, rel_step=1e-3):
    """Log-sensitivity E_i = (u_i / f(u)) * df/du_i."""
    base = f(u)
    dfd = fd_sensitivity(f, u, i, rel_step=rel_step)
    return (u[i] / (base + 1e-12)) * dfd

# Example usage
N, I0, T = 1_000_000, 25, 140
beta, gamma = 0.30, 1/7  # baseline

def attack_rate_from_params(u):
    beta_u, gamma_u = u
    inc = sir_euler(beta_u, gamma_u, N=N, I0=I0, T=T)
    return qoi_attack_rate(inc, N)

u0 = np.array([beta, gamma], dtype=float)
E_beta_AR  = elasticity(attack_rate_from_params, u0, i=0)
E_gamma_AR = elasticity(attack_rate_from_params, u0, i=1)

print("Elasticities for attack rate:")
print({"E_beta": E_beta_AR, "E_gamma": E_gamma_AR})
```

Listing 11.1: Local (finite-difference) sensitivities and elasticities for a deterministic SIR QoI.

Often the most consequential sensitivities in practice are not only in mechanistic parameters but in the observation model: how many infections are detected, and how quickly they appear in the reported data. The next listing performs a simple one-at-a-time (OAT) analysis: it keeps the latent epidemic fixed and varies the reporting fraction ρ and the mean reporting delay, then records how the observed peak height and peak time change. This is a newcomer-friendly way to see why observation assumptions can shift peaks and trends, and it directly links to the uncertainty taxonomy by isolating reporting effects from disease dynamics. OAT analyses are easy to explain and cheap to run, but they do not capture interactions unless the grid is expanded or combined with more systematic designs.

```
import numpy as np
from scipy.stats import gamma as gamma_dist

def discretize_gamma_pmf(mean_days, sd_days, max_day=30):
    k = (mean_days / sd_days) ** 2
    theta = (sd_days ** 2) / mean_days
    d = np.arange(max_day + 1)
```

```
    cdf_hi = gamma_dist.cdf(d + 0.5, a=k, scale=theta)
    cdf_lo = gamma_dist.cdf(np.maximum(d - 0.5, 0.0), a=k, scale=theta)
    pmf = np.maximum(cdf_hi - cdf_lo, 0.0)
    return pmf / pmf.sum()

def apply_observation_model(latent_incidence, rho, delay_pmf, T):
    mu = rho * np.convolve(latent_incidence, delay_pmf, mode="full")[:T]
    return mu

latent = sir_euler(beta=0.30, gamma=1/7, N=1_000_000, I0=25, T=140)
T = len(latent)

def qoi_observed_peak(mu):
    return float(np.max(mu))

def qoi_observed_peak_time(mu):
    return int(np.argmax(mu))

baseline = {"rho": 0.20, "delay_mean": 5.5, "delay_sd": 2.0}
delay0 = discretize_gamma_pmf(baseline["delay_mean"], baseline["delay_sd"])
mu0 = apply_observation_model(latent, baseline["rho"], delay0, T)

grid_rho = [0.10, 0.15, 0.20, 0.30, 0.40]
grid_delay_mean = [2.5, 4.0, 5.5, 7.0]

results = []
for rho in grid_rho:
    for dm in grid_delay_mean:
        delay = discretize_gamma_pmf(dm, baseline["delay_sd"])
        mu = apply_observation_model(latent, rho, delay, T)
        results.append({
            "rho": rho, "delay_mean": dm,
            "peak": qoi_observed_peak(mu),
            "peak_time": qoi_observed_peak_time(mu)
        })

results = sorted(results, key=lambda r: (r["peak_time"], r["rho"]))
print("First 10 sensitivity rows:", results[:10])
```

Listing 11.2: One-at-a-time (OAT) sensitivity for observation assumptions (reporting fraction and delay).

When the number of uncertain inputs is large, running a full global method immediately is often wasteful. Morris screening is designed for this situation: it provides a cheap, global "importance scan" that ranks inputs and flags nonlinear or interaction-prone factors. The listing below treats the simulator as a black box: you only need a function that maps parameters to a QoI. Morris constructs randomized trajectories through the parameter space and computes elementary effects (finite changes along one coordinate at a time). The mean absolute effect μ is an importance score, while the standard deviation

σ indicates nonlinearity or interactions. In practice, Morris is commonly used to reduce from dozens of candidates to a manageable subset before applying more expensive variance-based methods.

```
import numpy as np

def scale_to_bounds(x01, bounds):
    return np.array([lo + xi*(hi-lo) for xi, (lo,hi) in zip(x01, bounds)], dtype=float)

def morris_trajectories(d, r, p, rng):
    delta = p / (2*(p-1))
    levels = np.linspace(0, 1, p)
    trajs = []
    for _ in range(r):
        x = rng.choice(levels[:-1], size=d)  # avoid 1 to allow +delta
        order = rng.permutation(d)
        X = [x.copy()]
        for j in order:
            x_new = x.copy()
            x_new[j] = min(1.0, x_new[j] + delta)
            X.append(x_new)
            x = x_new
        trajs.append(np.array(X))
    return trajs

def morris_elementary_effects(f, bounds, r=20, p=6, rng=None):
    rng = np.random.default_rng() if rng is None else rng
    d = len(bounds)
    trajs = morris_trajectories(d=d, r=r, p=p, rng=rng)
    delta = p / (2*(p-1))
    effects = [[] for _ in range(d)]

    for X01 in trajs:
        Y = []
        for x01 in X01:
            u = scale_to_bounds(x01, bounds)
            Y.append(f(u))
        Y = np.array(Y, dtype=float)

        for k in range(1, d+1):
            diff = X01[k] - X01[k-1]
            j = int(np.argmax(np.abs(diff)))
            ee = (Y[k] - Y[k-1]) / (delta * (bounds[j][1] - bounds[j][0]))
            effects[j].append(ee)

    mu_star = np.array([np.mean(np.abs(e)) for e in effects])
    sigma   = np.array([np.std(e, ddof=1) if len(e)>1 else 0.0 for e in effects])
    return mu_star, sigma
```

```
def black_box_qoi(u):
    beta, gamma, rho = u
    latent = sir_euler(beta=beta, gamma=gamma, N=1_000_000, I0=25, T=140)
    mu = rho * latent
    return float(np.argmax(mu))  # QoI: peak time

bounds = [
    (0.15, 0.45),     # beta
    (1/10, 1/4),      # gamma
    (0.05, 0.50)      # rho
]

mu_star, sigma = morris_elementary_effects(
    black_box_qoi, bounds, r=40, p=6, rng=np.random.default_rng(3)
)
print("Morris mu* (importance):", mu_star)
print("Morris sigma (nonlinearity/interactions):", sigma)
```

Listing 11.3: Morris screening (elementary effects) for a black-box epidemic simulator.

If the goal is not only ranking but quantitative attribution of forecast uncertainty, variance-based methods are the standard choice. Sobol' sensitivity indices decompose the output variance into contributions from each input alone and from interactions. The first-order index S_i measures how much of $\mathrm{Var}(Q)$ is explained by varying only U_i, while the total-order index S_i^T captures the full effect of U_i including all interactions with other inputs. The listing below implements a common Saltelli sampling scheme: it builds two random designs, A and B, and a set of hybrid designs $A_B^{(i)}$ in which only coordinate i is swapped. This requires many model evaluations and is therefore best used after screening or when d is modest and the model is not too expensive.

```
import numpy as np

def saltelli_matrices(bounds, N, rng):
    d = len(bounds)
    A01 = rng.random((N, d))
    B01 = rng.random((N, d))

    def scale(X01):
        X = np.empty_like(X01)
        for j, (lo, hi) in enumerate(bounds):
            X[:, j] = lo + X01[:, j] * (hi - lo)
        return X

    A = scale(A01)
    B = scale(B01)

    AB = []
    for i in range(d):
        X = A.copy()
```

```
        X[:, i] = B[:, i]
        AB.append(X)
    return A, B, AB

def sobol_indices(f, bounds, N=2000, rng=None):
    rng = np.random.default_rng() if rng is None else rng
    d = len(bounds)
    A, B, AB = saltelli_matrices(bounds, N, rng)

    YA = np.array([f(u) for u in A], dtype=float)
    YB = np.array([f(u) for u in B], dtype=float)
    VY = np.var(np.concatenate([YA, YB]), ddof=1) + 1e-12

    S1 = np.zeros(d)
    ST = np.zeros(d)

    for i in range(d):
        YABi = np.array([f(u) for u in AB[i]], dtype=float)
        S1[i] = np.mean(YB * (YABi - YA)) / VY
        ST[i] = 0.5 * np.mean((YA - YABi) ** 2) / VY

    return S1, ST

def qoi_attack_rate_from_u(u):
    beta, gamma, I0 = u
    inc = sir_euler(beta=beta, gamma=gamma, N=1_000_000, I0=int(I0), T=180)
    return float(np.sum(inc) / 1_000_000)

bounds = [
    (0.15, 0.45),     # beta
    (1/10, 1/4),      # gamma
    (5, 200)          # I0 (seed size)
]

S1, ST = sobol_indices(qoi_attack_rate_from_u, bounds, N=1500, rng=np.random.default_rng(11))
print("First-order Sobol S1:", S1)
print("Total order Sobol ST:", ST)
```

Listing 11.4: Sobol' variance-based sensitivity indices (first-order and total-order) via Saltelli sampling.

Finally, many epidemiological models are stochastic even when inputs are fixed, because they represent demographic randomness, contact variability, or stochastic observation processes. In that case, it is useful to separate uncertainty due to uncertain inputs from uncertainty due to intrinsic randomness. The nested design below does exactly that: it samples an input setting U (outer loop), runs multiple stochastic replicates conditional on that U (inner loop), and decomposes the total variance into two terms: $\mathrm{Var}_U(\mathbb{E}[Q \mid U])$ (input-driven variability) and $\mathbb{E}_U(\mathrm{Var}[Q \mid U])$ (process-driven variability). This decomposition is directly aligned with the taxonomy from earlier in the chapter, where process stochasticity and parameter uncertainty are distinct uncertainty sources and should not be conflated.

```
import numpy as np

def stochastic_sir_tau(beta, gamma, N, I0, T, rng):
    S, I, R = N - I0, I0, 0
    new_inf = np.zeros(T, dtype=float)
    for t in range(T):
        lam_inf = beta * S * I / N
        lam_rec = gamma * I
        dS = rng.poisson(lam_inf)
        dR = rng.poisson(lam_rec)
        dS = min(dS, S)
        dR = min(dR, I)
        S -= dS
        I += dS - dR
        R += dR
        new_inf[t] = dS
        if I <= 0:
            break
    return new_inf

def qoi_peak(new_inf):
    return float(np.max(new_inf))

def variance_decomposition(f_stochastic, sampler_U, R_outer=200, R_inner=10, rng=None):
    rng = np.random.default_rng() if rng is None else rng
    q_means = np.zeros(R_outer)
    q_vars  = np.zeros(R_outer)

    for m in range(R_outer):
        u = sampler_U(rng)
        qs = np.array([f_stochastic(u, rng) for _ in range(R_inner)], dtype=float)
        q_means[m] = np.mean(qs)
        q_vars[m]  = np.var(qs, ddof=1)

    var_input   = np.var(q_means, ddof=1)  # Var_U(E[Q|U])
    var_process = np.mean(q_vars)          # E_U(Var[Q|U])
    var_total   = var_input + var_process
    return {"Var_total": var_total, "Var_input": var_input, "Var_process": var_process}

def f_stochastic(u, rng):
    beta, gamma = u
    inc = stochastic_sir_tau(beta, gamma, N=1_000_000, I0=25, T=140, rng=rng)
    return qoi_peak(inc)

def sample_U(rng):
    beta = rng.uniform(0.15, 0.45)
    gamma = rng.uniform(1/10, 1/4)
```

```
    return (beta, gamma)

out = variance_decomposition(
    f_stochastic, sample_U, R_outer=150, R_inner=20, rng=np.random.default_rng(21)
)
print(out)
print("Fraction input-driven:", out["Var_input"] / out["Var_total"])
print("Fraction process-driven:", out["Var_process"] / out["Var_total"])
```

Listing 11.5: Variance decomposition with replicated stochastic simulations (separating input uncertainty from process noise).

Fig. 11.3 illustrates a variance-based (global) sensitivity analysis using Sobol' indices for an epidemiologically meaningful quantity of interest: the day on which the mean reported-case curve reaches its maximum. The modeling pipeline behind this QoI has two layers. First, a mechanistic SIR model generates a latent incidence trajectory governed by transmission and recovery parameters. Second, a simplified observation model converts latent infections into reported cases by applying a reporting fraction and a reporting-delay distribution. The input vector therefore spans both mechanistic parameters, (β, γ), and observation-related parameters, (ρ, μ_d), along with the initial number of infections I_0. This explicit inclusion of observation parameters reflects the central message of uncertainty-aware epidemic modeling: uncertainty in surveillance and reporting can influence not only levels but also timing-related summaries. The bars in the figure report two complementary Sobol measures. The first-order index $S1$ quantifies the fraction of the QoI variance explained by varying one input alone, averaged over uncertainty in all other inputs; it is a main-effect measure. The total-order index ST quantifies the total contribution of an input to output variance, including all interaction terms with other inputs. When ST and $S1$ are close, the influence of that input is largely additive and can be understood through its direct effect. When ST is substantially larger than $S1$, a large portion of the input's influence is expressed through interactions and nonlinear dependence on other parameters. In this example, β and γ exhibit a noticeable $ST - S1$ gap, indicating that peak timing is shaped not only by their individual effects but also by how they combine with other uncertain inputs (such as reporting and initialization) to determine the epidemic speed and the timing of the observed peak.

Fig. 11.4 illustrates the simplest form of sensitivity analysis: a local, first-order linearization around a baseline parameter set. Let $Q(U)$ denote a scalar quantity of interest (here, the peak height of the mean reported-case curve) as a function of inputs $U = (U_1, \ldots, U_d)$. Local sensitivity approximates the effect of small perturbations by the first-order Taylor expansion

$$Q(U_1, \ldots, U_i, \ldots, U_d) \approx Q(U^\star) + \left.\frac{\partial Q}{\partial U_i}\right|_{U^\star} (U_i - U_i^\star),$$

with all other inputs fixed at $U^\star$. Each panel varies one input while keeping the remaining inputs at baseline; the dashed line is the linear approximation implied by the local derivative, and the solid line is the actual model response. The panels highlight both the usefulness and the limitations of linear sensitivity. Varying the transmission rate β increases peak height, while increasing the recovery rate γ reduces it, reflecting how faster removal diminishes epidemic growth. The reporting fraction ρ produces an almost perfectly linear scaling of the reported peak, since it multiplies the latent incidence in the observation model. In contrast, the mean reporting delay μ_d has only a weak and slightly non-

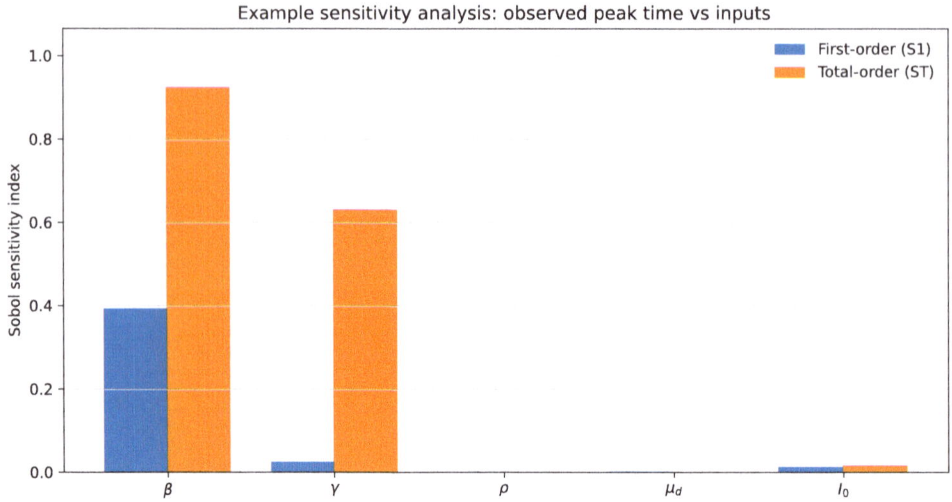

FIGURE 11.3

The quantity of interest is the day of the peak of the mean reported-case curve produced by a mechanistic SIR incidence model combined with a simple observation model. Inputs are the transmission rate β, recovery rate γ, reporting fraction ρ, mean reporting delay μ_d, and initial infections I_0. Bars show first-order indices (S1) and total-order indices (ST). S1 measures the fraction of output variance attributable to changing one input alone, while ST measures the total contribution of that input including all interactions with other inputs. When ST is much larger than S1 (as for β and γ in this example), interactions and nonlinearities contribute substantially to uncertainty in the peak timing. This is precisely the regime where local, derivative-based sensitivity can miss important effects and where global SA is preferred.

monotonic effect on peak height in this configuration; the linear approximation captures the local slope at the baseline but cannot represent the subtle curvature seen over a wider neighborhood. This is a typical outcome: local linear SA is fast and interpretable, but it is reliable only for sufficiently small perturbations and can miss nonlinearities and interactions that become important in global uncertainty propagation.

11.4 Model comparison

Model comparison in epidemiology is often framed as the search for a single "best" model. In practice, however, models are built and deployed for distinct tasks, and the criteria that define success differ across those tasks. A model that provides a compelling mechanistic explanation of observed patterns may yield poor forecasts if key drivers change, while a model that excels at short-horizon prediction may be unsuitable for causal interpretation or counterfactual reasoning [737,738]. The central principle of this section is therefore that model comparison should be conducted as an assessment of *fitness for purpose*: comparison criteria should be matched to the intended use, the available data streams, and the decision context.

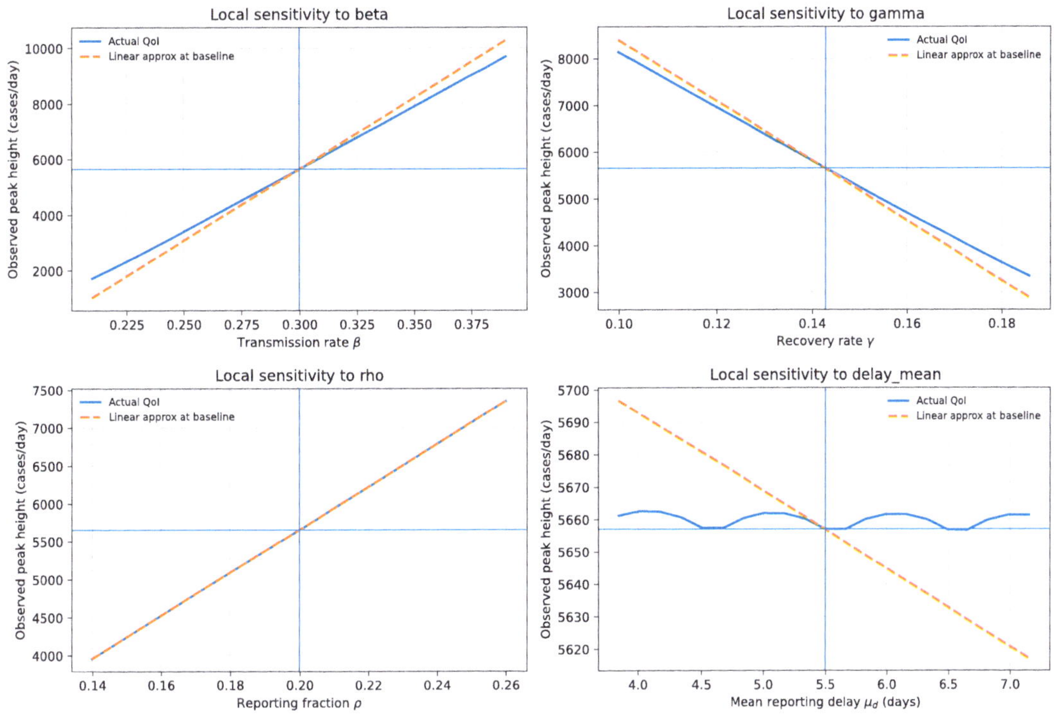

FIGURE 11.4

Local (linear) sensitivity analysis of an epidemic observation pipeline with respect to four inputs. The quantity of interest (QoI) is the peak height (maximum) of the mean reported-case curve produced by a deterministic SIR incidence model combined with a reporting model. In each panel, one parameter is varied while all others are fixed at a baseline value (vertical and horizontal reference lines). The solid curve shows the actual QoI evaluated on a small neighborhood around the baseline, while the dashed curve shows the first-order (linear) approximation obtained from the local derivative at the baseline.

A useful starting point is to distinguish three broad families of objectives. Nowcasting aims to estimate the current state of the epidemic (e.g., present incidence) when reporting is delayed or incomplete. Forecasting aims to predict future outcomes (incidence, hospital occupancy, mortality) over specified horizons. Attribution (or causal explanation) aims to quantify the effects of interventions, behavior changes, or other covariates on transmission and outcomes. These tasks overlap but are not equivalent, and they induce different trade-offs between bias, variance, and interpretability. In particular, excellent in-sample fit is not sufficient evidence of forecasting skill, and excellent short-term forecast accuracy is not sufficient evidence of causal validity.

This separation is mirrored by the distinction between in-sample fit and out-of-sample predictive performance. In-sample fit is typically assessed using likelihood-based criteria, residual diagnostics, or posterior predictive checks. Such tools are valuable for detecting gross misspecification and for assessing whether a model can reproduce key features of the observed data under the assumed observation process [739]. Yet, epidemic time series are nonstationary and policy-responsive, so a model can fit the past well and still fail under regime changes, new variants, evolving behavior, or testing-policy shifts. Out-of-sample evaluation is therefore essential for forecasting, and should be carried out with rolling-origin or forward-chaining protocols that mimic real-time deployment [766]. Moreover, because epidemic forecasts are often used to manage risk, point-error metrics alone are insufficient: probabilistic forecasts should be evaluated with proper scoring rules that reward calibrated uncertainty, such as the logarithmic score or CRPS [751].

Decision utility provides a third axis that can dominate purely statistical measures. In public health settings, the cost of underprediction (e.g., insufficient ICU capacity) can be far larger than the cost of overprediction (e.g., temporary surplus). Model comparison can therefore be framed in terms of task-specific loss functions and decision-theoretic criteria rather than generic accuracy measures [767]. This perspective makes it explicit that a model with slightly worse mean error may still be preferred if it better represents tail risk or yields more reliable trigger policies under uncertainty.

Parsimony and interpretability further complicate comparison. Mechanistic compartmental models often prioritize interpretability and causal coherence: parameters corresponding (at least approximately) to transmissibility, infectious periods, mixing patterns, and intervention effects. This structure supports attribution and counterfactual analysis but can introduce bias if key mechanisms are missing or if parameters are forced to absorb unmodeled drivers. Data-driven models may achieve strong short-horizon predictive accuracy by exploiting flexible function classes, but the learned relationships can be brittle under distribution shift and may not support causal claims [737,768]. Robustness under distribution shift is therefore a critical criterion in epidemics, where covariate distributions, reporting processes, and dominant variants can change rapidly. Evaluation protocols that include stress tests across regimes, locations, and policy phases help reveal whether a model is learning stable signals or merely interpolating within a narrow historical domain [753].

These considerations motivate a simple decision matrix that maps tasks to preferred comparison criteria. The matrix in Table 11.2 is not exhaustive, but it captures recurring best practices for epidemic modeling workflows and provides a template that can be adapted to the needs of a particular application.

The matrix also clarifies why no single leaderboard is appropriate. A nowcasting model should be judged primarily on its handling of reporting delays and its real-time calibration; a forecasting model should be judged on prospective probabilistic performance and robustness; and an attribution model should be judged on causal assumptions, interpretability, and sensitivity to confounding. When multiple objectives must be served simultaneously, multi-criteria evaluation or ensembles can be more appropriate than selecting a single winner [737,750]. In subsequent sections, we develop the tools that implement these principles—proper scoring rules and calibration for predictive evaluation, and sensitivity and robustness analyses for attribution and structural uncertainty.

A purpose-driven comparison workflow benefits from separating three questions: (i) does the model describe the observed data adequately under its assumed observation process, (ii) does it predict well in the way we will actually deploy it, and (iii) does it support the decisions we intend to make with its outputs. The methods below are organized around these questions and are meant to be applied as a menu rather than a mandatory checklist.

In-sample adequacy and parsimony

Mechanistic models are often first screened for internal consistency and adequacy using likelihood-based fit and residual diagnostics, followed by parsimony-aware criteria. If a model admits a likelihood

Table 11.2 Decision matrix for model comparison: preferred criteria depend on the modeling task.

Task	Primary evaluation target	Preferred comparison criteria	Typical pitfalls
Nowcasting	Current latent incidence and recent observations under reporting delay	Delay-model adequacy; calibration of nowcast intervals; coverage of revisions/backfill; proper scores on reconstructed truth when available	Ignoring reporting-delay dynamics; overconfident intervals near "now"; treating revised counts as ground truth without modeling backfill [618].
Forecasting	Future QoIs (incidence, hospital/ICU occupancy, mortality) over a horizon	Out-of-sample proper scoring rules (log score, CRPS); calibration (PIT/reliability); horizon-specific performance; robustness across phases/regions; ensemble benefit	Ranking by in-sample fit; relying only on point metrics (MAE/RMSE); failure under regime change; uncalibrated uncertainty [750,751].
Attribution / causal explanation	Intervention effects; counterfactual outcomes; mechanisms	Causal identifiability assumptions; sensitivity to confounding and priors; model transparency; posterior predictive checks for mechanistic plausibility; robustness to alternative specifications	Interpreting predictive associations causally; weak identifiability between $\beta(t)$ and reporting/behavior; policy endogeneity; omitted-variable bias [739,745,749].

$\ell(\hat{\theta}) = \log p(y_{1:T} \mid \hat{\theta})$ with k effective free parameters, then Akaike's Information Criterion (AIC) and Schwarz's Bayesian Information Criterion (BIC) are

$$\text{AIC} = -2\ell(\hat{\theta}) + 2k, \qquad \text{BIC} = -2\ell(\hat{\theta}) + k\log(T),$$

providing simple penalties for over-parameterization [769]. For Bayesian fits with posterior draws, WAIC and LOO approximate out-of-sample predictive performance using the pointwise log-likelihood and an effective-parameter penalty [770,771]. These criteria are helpful for quickly rejecting models that are too complex for the available information, but they should not be mistaken for forecasting guarantees in nonstationary epidemics.

```
import numpy as np

def aic_bic(loglik_hat, k, n_obs):
    aic = -2.0 * loglik_hat + 2.0 * k
    bic = -2.0 * loglik_hat + k * np.log(max(n_obs, 1))
    return {"AIC": aic, "BIC": bic}
```

Listing 11.6: AIC/BIC from a fitted likelihood (quick parsimony check).

Out-of-sample prediction with deployment-faithful backtesting

Forecast comparison should be performed with a protocol that mirrors how forecasts are made in real time. Rolling-origin (also called forward-chaining) evaluation fits each model on data up to a cutoff t and evaluates predictions for horizons $h = 1, \ldots, H$ on future observations. This prevents information leakage and reveals horizon-dependent trade-offs (a model can be excellent at $h=1$ and poor at $h=21$). For mechanistic models, the refit step may mean re-estimating parameters; for ML models, it typically means retraining on the expanding window.

```
import numpy as np

def rolling_origin_splits(y, min_train=60, step=7, horizon=14):
    T = len(y)
    cutoffs = range(min_train, T - horizon + 1, step)
    for t in cutoffs:
        train_idx = np.arange(0, t)
        test_idx  = np.arange(t, t + horizon)
        yield train_idx, test_idx

def evaluate_point_forecasts(y_true, y_pred):
    err = y_true - y_pred
    mae  = np.mean(np.abs(err))
    rmse = np.sqrt(np.mean(err**2))
    return {"MAE": mae, "RMSE": rmse}
```

Listing 11.7: Rolling-origin backtesting for multiple horizons.

Probabilistic scoring rules for forecast distributions

Epidemic forecasts are rarely used as single numbers; they are used as distributions (or quantiles) because capacity planning and policy triggers depend on uncertainty. Proper scoring rules compare predictive distributions to realized outcomes while discouraging hedging. Two widely used choices are the logarithmic score and the continuous ranked probability score (CRPS) [753]. For count outcomes, the log score is natural when the predictive distribution is explicit (e.g., Poisson or negative binomial), while CRPS can be computed from forecast samples, ensembles, or predictive quantiles.

```
import numpy as np
from scipy.special import gammaln

def nb_logpmf(y, mu, k):
    y  = np.asarray(y, dtype=float)
    mu = np.asarray(mu, dtype=float)
    k  = float(k)
    p  = k / (k + mu + 1e-12)        # success prob
    n  = k                           # number of successes
    return (gammaln(y+n) - gammaln(n) - gammaln(y+1.0)
            + n*np.log(p) + y*np.log(1.0-p+1e-12))

def mean_log_score_nb(y_true, mu_pred, k):
    return float(np.mean(nb_logpmf(y_true, mu_pred, k)))
```

Listing 11.8: Log score for a negative binomial predictive distribution (counts).

```
import numpy as np

def crps_samples(y, samples):
    s = np.asarray(samples, dtype=float)
    y = float(y)
    term1 = np.mean(np.abs(s - y))
    term2 = 0.5 * np.mean(np.abs(s[:, None] - s[None, :]))
    return float(term1 - term2)

def mean_crps(y_true, samples_mat):
    return float(np.mean([crps_samples(y_true[t], samples_mat[t]) for t in range(len(y_true))
        ]))
```

Listing 11.9: CRPS from predictive samples (ensemble-style forecasts).

Calibration diagnostics: are uncertainty bands honest?

A model can have low error but still be dangerously overconfident. Calibration checks whether nominal predictive statements match empirical frequencies. Two common diagnostics are interval coverage and probability integral transform (PIT) histograms [753]. Coverage asks whether a nominal 90% predictive interval contains the truth roughly 90% of the time under backtesting. PIT uses the predictive CDF evaluated at the observation; for calibrated continuous forecasts, PIT values should be approximately uniform (for discrete counts, randomized PIT is often used).

```
import numpy as np

def interval_coverage(y_true, lo, hi):
    y_true = np.asarray(y_true)
    lo = np.asarray(lo)
    hi = np.asarray(hi)
    return float(np.mean((y_true >= lo) & (y_true <= hi)))
```

Listing 11.10: Predictive interval coverage under backtesting.

```
import numpy as np

def pit_from_samples(y_true, samples_mat, rng=None):
    rng = np.random.default_rng() if rng is None else rng
    y_true = np.asarray(y_true, dtype=float)
    S = np.asarray(samples_mat, dtype=float)  # (T, M)
    T, M = S.shape
    pit = np.zeros(T, dtype=float)
    for t in range(T):
        # randomized tie breaking
        lt = np.mean(S[t] < y_true[t])
        eq = np.mean(S[t] == y_true[t])
        pit[t] = lt + rng.random() * eq
```

```
    return pit
```

Listing 11.11: PIT values from predictive samples (useful for calibration plots).

Decision utility: comparing models by the decisions they support

When forecasts inform actions, comparison should account for the decision loss. A model that slightly worsens average error may still be preferable if it reduces costly failures (e.g., underpredicting hospital load). A simple operational approach is to define an asymmetric loss that weights underprediction more heavily than overprediction, or to evaluate a trigger policy (e.g., "activate surge plan if predicted occupancy exceeds threshold") and measure realized regret [734].

```
import numpy as np

def asymmetric_loss(y_true, y_pred, under_weight=5.0, over_weight=1.0):
    y_true = np.asarray(y_true, dtype=float)
    y_pred = np.asarray(y_pred, dtype=float)
    err = y_pred - y_true
    loss = np.where(err < 0, under_weight * (-err), over_weight * err)
    return float(np.mean(loss))
```

Listing 11.12: Asymmetric loss for capacity planning (underprediction penalized more).

Robustness under distribution shift and regime changes

Epidemic data are shaped by changing testing policies, behavior, variants, and interventions; distribution shift is the norm rather than the exception. A practical stress test is to stratify evaluation by regime (e.g., waves, policy phases, or pre-/post-testing expansion) and compare scores within each stratum. Large performance swings suggest a model is relying on non-robust correlations or a stationarity assumption that does not hold [737]. For mechanistic models, robustness can also be probed by comparing fits and forecasts under alternative observation-model assumptions (delay distributions, time-varying reporting), which directly links model comparison to UQ and SA.

```
import numpy as np

def regime_scores(y_true, y_pred, regimes):
    regimes = np.asarray(regimes)
    out = {}
    for r in np.unique(regimes):
        mask = regimes == r
        err = y_true[mask] - y_pred[mask]
        out[str(r)] = {
            "MAE": float(np.mean(np.abs(err))),
            "RMSE": float(np.sqrt(np.mean(err**2))),
            "n": int(np.sum(mask))
        }
    return out
```

Listing 11.13: Regime-stratified evaluation to probe distribution shift.

The synthetic data used in Fig. 11.5 are generated from a deliberately "messy but realistic" reporting pipeline built on top of a mechanistic SIR incidence process. Specifically, we simulate $T = 230$ daily observations with population size $N = 10^6$ and recovery rate $\gamma = 1/7$. The latent daily infections are produced by an SIR model initialized at $I_0 = 35$ and subjected to a mid-course regime change at day $t_{\text{change}} = 125$: the transmission rate is set to $\beta_1 = 2.2\gamma$ before the change and $\beta_2 = 1.35\gamma$ after the change, mimicking an intervention or behavioral shift that reduces transmission. Observed reported cases are then generated by applying a time-varying ascertainment process with both a structural change and a weekly reporting artifact. The baseline reporting fraction increases from $\rho_1 = 0.18$ to $\rho_2 = 0.30$ after t_{change}, capturing expanded testing or surveillance, and a day-of-week multiplier introduces weekend dips (multipliers $\{1.05, 1.05, 1.00, 1.00, 0.95, 0.70, 0.65\}$). Finally, counts are drawn from a negative binomial observation model with dispersion $k = 25$, so that variability exceeds Poisson noise. This construction ensures that the data contain nonstationarity (regime change), systematic reporting variation (weekly pattern), and overdispersion, all of which commonly invalidate one-size-fits-all comparisons. All four panels are computed under a deployment-faithful rolling-origin backtest. Starting from an initial training window of 70 days, we refit each model every 14 days and evaluate $H = 14$-day-ahead forecasts; each forecast is represented by a predictive sample of size $M = 150$ per horizon day, enabling probabilistic evaluation. The compared baselines are intentionally heterogeneous. The refit mechanistic SIR baseline estimates constant (β, ρ, I_0) on each training window by a simple grid search that minimizes mean squared error between observed cases and the model mean $\mu_t = \rho \cdot \text{Inc}_t$; predictive samples are generated by drawing negative binomial counts around the forecast mean. The two data-driven forecasters use the reported-case series as input, with 14 autoregressive lags and day-of-week indicators: one is a ridge-regularized linear model ("AR-lags (ridge)"), and the other is a nonlinear gradient-boosted regression tree ("GBRT (lags)"). The seasonal-naive baseline repeats the observation from one week ago. For the three non-mechanistic methods, predictive samples are generated by residual bootstrap: we compute in-sample one-step residuals on the training window, then form forecast samples by recursively forecasting the mean and adding resampled residuals at each horizon.

Panel (1) reports the continuous ranked probability score (CRPS) as a function of forecast horizon, computed from predictive samples and averaged across all rolling cutoffs; lower values indicate better probabilistic forecasts. In this data case, the mechanistic SIR baseline attains the lowest CRPS across essentially all horizons, reflecting robustness to the regime change compared with purely autoregressive extrapolation. The lag-based models degrade as the horizon grows, with a pronounced loss of skill in the mid-to-long range, illustrating that models optimized for short-term pattern continuation can struggle when the data-generating process changes.

Panel (2) contrasts in-sample fit with out-of-sample forecast skill by plotting a fit proxy (training residual RMSE) against the 7-day-ahead CRPS. The key observation is that strong fit does not imply strong prediction. In the displayed results, the nonlinear GBRT achieves the best in-sample RMSE (rank 1 in Panel (4)), yet it does not dominate out-of-sample performance at 7 days. Conversely, the SIR model is not the best by in-sample RMSE (rank 3) but is the best by 7-day CRPS (rank 1), with substantially lower 7-day CRPS than the other baselines (approximately half to two-thirds of their values in the plotted instance). The ridge AR model provides a cautionary example: it is not the worst in-sample (rank 2) but is the worst at 7-day probabilistic forecasting (rank 4), consistent with the idea that in-sample agreement can mask instability under distribution shift.

Panel (3) evaluates uncertainty quality through the sharpness-calibration trade-off at nominal 90% prediction intervals. For each model, we compute empirical coverage $\hat{c}$ by checking whether the realized outcome falls within the sample-based 90% interval, and we summarize calibration by the absolute deviation $|\hat{c} - 0.9|$; sharpness is summarized by the average interval width. The displayed results highlight that sharper intervals are not automatically better: the SIR baseline produces the narrowest intervals (rank 1 in average width) but has the worst calibration error (rank 4), indicating overconfident uncertainty under misspecification (here, constant parameters and simplified observation assumptions confronted with time-varying reporting). The seasonal-naive baseline exhibits the opposite behavior: it has the widest intervals (rank 4 in width) but the best calibration error (rank 1), demonstrating how a crude model can sometimes be more honest about uncertainty than a sharper but misspecified alternative. The lag-based ML models fall between these extremes, with intermediate widths and intermediate calibration ranks.

Panel (4) synthesizes these outcomes into a ranked heatmap across four criteria: in-sample RMSE, 7-day CRPS, 90% coverage error, and average 90% interval width. The resulting rank pattern makes the section's thesis concrete. The GBRT is best at fitting (rank 1) but not best at 7-day probabilistic forecasting (rank 2). The SIR model is best for 7-day forecast skill (rank 1) and sharpness (rank 1) but worst for calibration (rank 4), cautioning against equating mechanistic plausibility with reliable uncertainty without explicit calibration checks. The seasonal-naive baseline is best calibrated (rank 1) but worst at fit (rank 4) and sharpness (rank 4). The ridge AR model is never best and is worst for 7-day CRPS (rank 4), reflecting brittleness in this particular shifted, artifact-laden data regime. Together, the panels show that model comparison is not the search for a single winner: the preferred model depends on whether the task prioritizes forecasting skill at a given horizon, uncertainty calibration, parsimonious explanation, or operational reliability under shift.

Deepen your knowledge

Exercise 1: Uncertainty sources and what they do to decisions

Task

For each question below, identify the dominant uncertainty sources and propose a minimal uncertainty-aware modeling move. For each, write: (i) the quantity of interest (QoI), (ii) at least two uncertainty sources (parameter, process, observation, structural), and (iii) one decision or operational risk that is sensitive to that uncertainty.

1. Q1: "You must decide whether to activate surge staffing if predicted hospital admissions in 10 days exceed 120/day. What uncertainty sources matter most?"
2. Q2: "You estimate R_t from reported cases, but testing volume changes sharply mid-wave. What uncertainty sources can mislead your estimate?"
3. Q3: "You want a 14-day forecast of cases for a city, but a new variant may be emerging. Which uncertainty sources should be explicitly represented?"
4. Q4: "You fit an extended SEIR model to deaths and cases jointly. The model fits well, but forecasts disagree across plausible delay assumptions. What is the uncertainty here?"
5. Q5: "You compare two models and both match past data. One yields narrow forecast intervals and one yields wide intervals. Which could still be the better model, and why?"

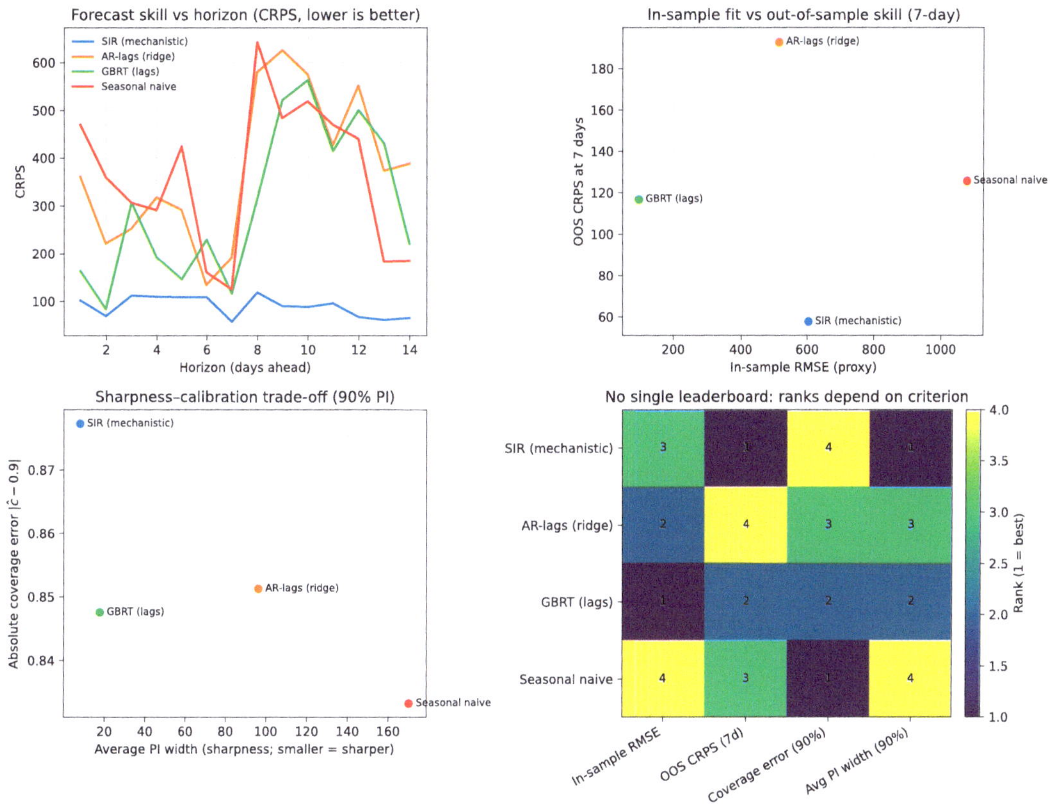

FIGURE 11.5

Model comparison as fitness for purpose is illustrated using a controlled epidemic example with a regime change and reporting artifacts. Panel (1) shows probabilistic forecast skill (CRPS) as a function of horizon, highlighting that a model can be competitive at short horizons and poor at longer ones. Panel (2) contrasts in-sample fit with out-of-sample skill (7-day CRPS), demonstrating that strong in-sample agreement does not guarantee predictive performance under nonstationarity. Panel (3) visualizes the sharpness–calibration trade-off: narrower prediction intervals (greater sharpness) are not beneficial if empirical coverage deviates from nominal coverage. Panel (4) summarizes the central message with a rank heatmap across multiple criteria (fit, forecast skill, calibration, and interval width), showing that there is no single universal leaderboard.

Solution (worked)

- **Q1:** QoI: admissions at $t + 10$ and $\mathbb{P}(\text{admissions} > 120)$. Key uncertainties: observation (reporting delays/backfill), process (stochasticity), structural (future contact changes). Sensitive decision: false negatives are costly (underprediction), so calibration of tails matters more than mean error.
- **Q2:** QoI: R_t trajectory. Key uncertainties: observation (time-varying ascertainment, test-seeking behavior), structural (generation-interval assumption). Risk: spurious changes in R_t that reflect testing rather than transmission.

- **Q3:** QoI: 14-day predictive distribution. Key uncertainties: structural (variant-driven transmissibility shift), parameter (time-varying $\beta(t)$), observation (weekend effects). Risk: distribution shift breaks purely autoregressive extrapolation; scenario ensembles may be necessary.
- **Q4:** QoI: near-term deaths/cases forecasts. Key uncertainties: structural/observation (delay distributions, reporting fraction), identifiability confounding (e.g., $\beta(t)$ vs. reporting). Risk: forecasts depend on non-identifiable assumptions rather than data.
- **Q5:** QoI: full predictive distribution. Key uncertainties: calibration vs sharpness. The model with wider intervals may be better if it is calibrated (honest uncertainty) and supports safer decisions; narrow but miscalibrated intervals can be operationally dangerous.

Exercise 2: Identifiability check by thought experiment (two parameter sets)

Task

Consider an SIR/SEIR-type model with a reporting fraction ρ and unknown initial infections I_0. You observe only reported cases y_t and assume $y_t \approx \rho\,\mathrm{Inc}_t$ (possibly with noise). Construct two parameter sets, θ_A and θ_B, that could plausibly produce similar likelihood for $y_{1:T}$ but different forecasts for $y_{T+1:T+14}$. For each set, specify: (i) which parameters differ, (ii) why they can be confounded in-sample, and (iii) which forecast feature diverges (peak time, peak height, tail behavior).

1. Q1: Build (β, ρ) confounding examples.
2. Q2: Build (I_0, β) confounding examples.
3. Q3: Build a generation-interval (or γ) assumption change that preserves in-sample fit but alters forecast horizon.
4. Q4: Explain which additional data stream (e.g., hospitalizations, serology) would break each confounding.
5. Q5: State a diagnostic you would plot to detect practical non-identifiability (profile likelihood, posterior ridge, or predictive equivalence class).

Solution (worked)

A typical construction is to keep the product $\rho \cdot \mathrm{Inc}_t$ approximately invariant in-sample while changing the latent epidemic dynamics. For example, θ_A with higher β and lower ρ can match observed levels but implies faster depletion and earlier turning points than θ_B with lower β and higher ρ. Similarly, a larger I_0 paired with a smaller β may match early growth in y_t but produce different near-future curvature. Changing the assumed generation interval (or γ) can keep an inferred R_t trajectory similar in-sample yet shift the implied mapping from recent growth to future incidence. Additional data streams that track severity (hospitalizations), cumulative attack rate (serology), or reporting intensity (testing volume) provide the extra information needed to separate these effects.

Exercise 3: Sensitivity analysis method selection by constraints

Task

For each scenario, choose a sensitivity analysis method and justify it in one or two sentences. Then specify: (i) the QoI, (ii) the input dimension d, (iii) whether the simulator is stochastic, and (iv) the computational budget (number of model runs).

1. Q1: “A deterministic SEIR model with $d = 6$ parameters; you need a quick screen for which parameters affect peak hospital occupancy.”
2. Q2: “A stochastic agent-based model with $d = 20$ uncertain inputs; you can afford ~ 500 runs; interactions may matter.”
3. Q3: “A mechanistic model with $d = 10$ and an expensive likelihood; you suspect only 2–3 parameters dominate forecast uncertainty.”
4. Q4: “A fitted ML forecaster with $d = 30$ input features; you want local explanations for why next-week predictions are high.”
5. Q5: “A policy simulator where the QoI is a threshold event (capacity exceedance). You care about tails rather than mean.”

Solution (worked)

A plausible mapping is: local derivatives/elasticities for Q1 (cheap, interpretable); Morris screening for Q2 (budget-friendly global screening); sequential screening then Sobol’ for Q3 (screen first, decompose variance second); local feature attribution (e.g., gradients or perturbation-based) for Q4; and reliability/rare-event-oriented sensitivity for Q5 (focus on exceedance probabilities, not only variance).

Exercise 4: Build and evaluate a calibrated predictive interval

Task

You produce probabilistic forecasts for daily cases using predictive samples $\{y_{t+h}^{(m)}\}_{m=1}^{M}$. Write code to compute: (i) CRPS, (ii) 90% predictive interval width, and (iii) empirical 90% coverage on a backtest. Then answer: Which is preferable for decision-making: the sharpest interval or the best-calibrated interval? Explain with one concrete risk example (e.g., underprediction of hospital load).

1. Q1: Compute CRPS from samples and interpret what lower means.
2. Q2: Compute empirical coverage and diagnose overconfidence vs underconfidence.
3. Q3: Compare two models: one sharp but miscalibrated, one wide but calibrated.
4. Q4: Suggest one fix for miscalibration (inflation, model averaging, improved observation model).
5. Q5: State which metric you would prioritize for surge planning and why.

Solution (code skeleton)

```
import numpy as np

def crps_samples(y, s):
    s = np.asarray(s, dtype=float)   # (M,)
    y = float(y)
    term1 = np.mean(np.abs(s - y))
    term2 = 0.5 * np.mean(np.abs(s[:, None] - s[None, :]))
    return float(term1 - term2)

def interval_width(s, level=0.90):
    s = np.asarray(s, dtype=float)
    a = (1.0 - level) / 2.0
    lo = np.quantile(s, a)
    hi = np.quantile(s, 1.0 - a)
```

```
    return float(hi - lo), float(lo), float(hi)

def coverage(y_true, samples_mat, level=0.90):
    covered = 0
    widths = []
    crps = []
    for t in range(len(y_true)):
        w, lo, hi = interval_width(samples_mat[t], level=level)
        widths.append(w)
        crps.append(crps_samples(y_true[t], samples_mat[t]))
        covered += int((y_true[t] >= lo) and (y_true[t] <= hi))
    return {
        "CRPS": float(np.mean(crps)),
        "AvgWidth": float(np.mean(widths)),
        "Coverage": float(covered / len(y_true))
    }
```

Listing 11.14: CRPS, interval width, and empirical coverage from predictive samples.

Exercise 5: Model comparison as fitness for purpose (decision matrix)

Task

For each question below, choose the most appropriate comparison criteria and justify why. Write: (i) the task type (nowcasting / forecasting / attribution), (ii) the primary metric(s) you would report, and (iii) one criterion you would *not* prioritize (and why).

1. Q1: "Estimate today's incidence despite reporting delays; decision-makers will use it to trigger contact tracing scale-up."
2. Q2: "Forecast hospital admissions 14 days ahead for staffing; missing a surge is much worse than overpreparing."
3. Q3: "Quantify the effect of school masking on transmission using observational data."
4. Q4: "Pick a model for weekly public communication where interpretability and stability matter."
5. Q5: "A model wins AIC/WAIC but has miscalibrated predictive intervals. Is it acceptable for operational forecasting?"

Solution (worked)

Nowcasting (Q1) should prioritize delay/backfill handling and near-now calibration; forecasting (Q2) should prioritize prospective proper scores, tail risk and decision-weighted loss; attribution (Q3) should prioritize causal assumptions, identifiability, and robustness to alternative specifications rather than pure predictive accuracy; communication (Q4) should emphasize interpretability and robustness; and (Q5) highlights that parsimony-based fit criteria are insufficient when uncertainty calibration is central to safe decisions.

CHAPTER

Computational considerations in pandemic models

12

This chapter is about the parts of epidemic modeling that rarely show up in equations, but often decide whether a model is useful in practice: runtime, memory, data plumbing, software design, and the workflow that turns an idea into results that can be trusted and communicated. To this end, "computational considerations" are the engineering constraints and choices that shape what you can simulate, how quickly you can iterate, and how reliably other people can reproduce what you did. In a pandemic, where decisions are time-sensitive and uncertainty is high, these considerations become part of the science rather than a separate technical afterthought [772,773]. The chapter treats computation as a set of design decisions under pressure. One may have a mathematically elegant model that is too slow to calibrate, too brittle to maintain, or too opaque to explain. In contrast, another may have a fast and clean implementation that omits key mechanisms and produces misleading confidence. Computational thinking helps navigate these tensions.

To be exact, in epidemic modeling, in particular, and scientific modeling, in general, computational considerations commonly refer to everything that affects the feasibility, reliability, and interpretability of running the model on real hardware with real data. This includes algorithmic complexity (how runtime grows with population size, number of compartments, or network edges), numerical stability (whether the solver behaves well across parameter ranges), and the structure of the codebase (how easy it is to test, modify, and review) [774–777]. It also includes practicalities such as parallelization, checkpointing long runs, managing random seeds, and handling data versions so that results can be traced back to inputs. These considerations differ by model class. Deterministic compartment models are often cheap per run but can still become computationally heavy when embedded in calibration loops or extended with age structure, spatial resolution, and time-varying parameters. Agent-based and network models can represent heterogeneity more directly, but they may require careful optimization to avoid prohibitively slow runs. Stochastic models introduce additional cost because uncertainty is not just a byproduct—it is estimated through repeated simulations, which turns one run into an ensemble, often multiplying runtime by orders of magnitude. Moreover, computational considerations also include the human side – what your team can build, validate, and explain. A model that only one person understands is a risk during a rapidly evolving outbreak. Decisions about tooling, documentation, modularity, and testing are computational decisions because they determine how quickly you can incorporate new evidence, fix errors, and answer urgent questions without breaking the model.

Most pandemic modeling projects follow a lifecycle with four main phases: prototype, calibration, scenario runs, and reporting. The prototype phase is about speed of learning: you build the simplest working model that captures the core mechanisms relevant to the policy question. Computationally, this phase favors clarity and rapid iteration. You might accept slower code or rough data handling because the goal is to explore structure, confirm that outputs behave sensibly, and identify what

Epidemic Modeling. https://doi.org/10.1016/B978-0-44-355525-1.00018-7

data the model actually needs. The prototype is also where early computational choices can become "locked in", so it is worth keeping the implementation modular, even when moving fast. Afterward, calibration is where computational reality hits hardest. Fitting epidemic models typically involves searching a high-dimensional parameter space, often with noisy likelihoods, delays in observation, and changing reporting practices. Calibration is not just running the model once; it is running it thousands to millions of times under different parameters, often with stochastic replications. This is where performance engineering, vectorization, parallel computation, and efficient likelihood evaluation pay off. Calibration is also where reproducibility begins to matter sharply: small changes in code, data preprocessing, or random seeds can shift posterior distributions and undermine trust. Next, scenario runs translate a calibrated model into decision support. Here, the model is run repeatedly under intervention assumptions: school closures, vaccination rollouts, behavior change, new variants, seasonal forcing, and combinations of these. Computationally, scenario work emphasizes throughput and organization. You want standardized run configurations, consistent outputs, and automation that prevents one-off manual changes from creeping into results. Because scenario sets can be large, you also need to manage compute budgets: which scenarios deserve full stochastic ensembles, which can be approximated, and which should be discarded early because they are implausible or irrelevant. Finally, reporting is where computational choices become visible to non-modelers. Tables, plots, uncertainty intervals, and narrative summaries need to be generated reliably, ideally from the same pipeline that ran the simulations. Reporting also includes provenance: documenting data sources, parameter assumptions, model version, and run identifiers so results can be reproduced later. A model that cannot be rerun to regenerate a key figure is not just inconvenient – it is scientifically fragile, especially when findings are scrutinized by external reviewers or used in public communication. In this regard, each simulation and model development starts with the right programming language and the ecosystem it brings to the (development) table.

12.1 Programming language and library choices

It all starts with how you explain the computer what you want. So, the programming language plays a central role in the development process. Programming language and library decisions in epidemic modeling are rarely about ideology and almost always about iteration speed, team skills, computational budget, and the kinds of uncertainty quantification you intend to do. A common pattern is to prototype in a high-level language (Python, R, Julia), where the scientific "shape" of the model can be changed quickly, then selectively optimize the hottest parts of the computation (a simulator kernel, likelihood evaluation, particle filter step, or contact-network update) in a high-performance core (C++/Rust/Fortran) once profiling shows clear bottlenecks. The goal is not to "rewrite everything", but to keep the outer workflow productive while making the inner loop fast enough for calibration and scenario ensembles.

Indeed, Python and R excel at rapid prototyping since they have rich ecosystems for data handling, plotting, and off-the-shelf inference routines. Julia occupies an interesting middle: it can feel like a high-level scripting language while generating fast code, which makes it attractive when the team wants both expressive model code and performance without leaving one language. By contrast, C++/Rust/Fortran are typically chosen when the simulator itself is the bottleneck, when you need fine control over memory and parallelism, or when the model structure (large agent populations, detailed contact networks, high-resolution spatial grids) makes interpreted or high-overhead execution too expensive. In practice, a "two-language" architecture is common: Python/R for the pipeline and experimentation, a compiled core for the simulator.

In Python, a frequent baseline is NumPy/SciPy for vectorized numerics and ODE solving, pandas for tabular data, and xarray for labeled multi-dimensional outputs (helpful for scenario ensembles indexed by time, region, and parameter draws). When the model needs acceleration, Numba can JIT-compile tight loops (e.g., agent updates, transition sampling, or likelihood kernels) with minimal code changes. JAX enables automatic differentiation and compilation (XLA), which is especially useful for gradient-based inference and fast batched simulation, while PyTorch is common when the model is embedded in a learning system (surrogate modeling, amortized inference, or policy learning). In R, deSolve is widely used for deterministic compartment models; pomp supports simulation-based inference workflows such as particle filters and iterated filtering; EpiModel provides network-based epidemic modeling; and tidyverse underpins data wrangling and visualization. In Julia, DifferentialEquations.jl provides a comprehensive solver suite for deterministic and stochastic differential equation models, and Turing.jl supports probabilistic programming for Bayesian inference. In C++, Stan is often used for inference on ODE-based and state-space epidemic models, and custom simulators are parallelized with OpenMP (shared memory) or MPI (distributed memory) for large ensembles or agent-based models.

However, there is a limit how much one can get from Python. The most defensible trigger is profiling: switch or optimize only when measurements show a specific component dominates runtime (for example, 80% of runtime is in the transition loop for agents, or likelihood evaluation dominates calibration). A second trigger is scale: when your model's runtime grows beyond what allows meaningful iteration (e.g., calibration needs 10^5 simulator evaluations and each run is seconds instead of milliseconds), you are no longer doing science efficiently. A third trigger is inference needs: if you require gradients (for HMC, variational inference, or efficient optimization), you may benefit from JAX, Stan, or Julia AD-enabled solvers; if you require many stochastic replications (particle MCMC, bootstrap filters, ensemble forecasting), you may need compiled kernels and parallel execution. A good rule is to keep the "experiment harness" in the language where your team moves fastest, and move only the computational hot spots into accelerated libraries or compiled modules.

For instance, even a simple SIR model illustrates stylistic and computational differences. The numerical method is the same (solving a system of ODEs), but ecosystems differ in defaults, ergonomics, and how easily you can extend the code toward inference. Below are comparable implementations in Python (SciPy), R (deSolve), and Julia (DifferentialEquations.jl). The examples use normalized state variables ($S + I + R = 1$) and do not include births/deaths, or observation models; these are typically added later during calibration.

```
import numpy as np
from scipy.integrate import solve_ivp

def sir_rhs(t, y, beta, gamma):
S, I, R = y
N = S + I + R
dSdt = -beta * S * I / N
dIdt = beta * S * I / N - gamma * I
```

```
dRdt = gamma * I
return [dSdt, dIdt, dRdt]

beta = 0.3
gamma = 0.1

y0 = [0.99, 0.01, 0.0]
t_span = (0.0, 160.0)
t_eval = np.linspace(t_span[0], t_span[1], 500)

sol = solve_ivp(
sir_rhs,
t_span,
y0,
args=(beta, gamma),
t_eval=t_eval,
rtol=1e-8,
atol=1e-10
)

S, I, R = sol.y
```

Listing 12.1: Python implementation of the SIR model using SciPy's solve_ivp.

```
library(deSolve)

sir_rhs <- function(t, y, parms) {
with(as.list(c(y, parms)), {
N <- S + I + R
dS <- -beta * S * I / N
dI <- beta * S * I / N - gamma * I
dR <- gamma * I
list(c(dS, dI, dR))
})
}

parms <- c(beta = 0.3, gamma = 0.1)
y0 <- c(S = 0.99, I = 0.01, R = 0.0)
times <- seq(0, 160, length.out = 500)

out <- ode(y = y0, times = times, func = sir_rhs, parms = parms,
rtol = 1e-8, atol = 1e-10)
```

Listing 12.2: R implementation of the SIR model using `deSolve::ode`.

```
using DifferentialEquations

function sir_rhs!(du, u, p, t)
```

```
S, I, R = u
beta, gamma = p
N = S + I + R
du[1] = -beta * S * I / N
du[2] = beta * S * I / N - gamma * I
du[3] = gamma * I
end

u0 = [0.99, 0.01, 0.0]
tspan = (0.0, 160.0)
p = (0.3, 0.1)

prob = ODEProblem(sir_rhs!, u0, tspan, p)
sol = solve(prob; reltol=1e-8, abstol=1e-10, saveat=range(0,160,length=500))

S = sol[1, :]
I = sol[2, :]
R = sol[3, :]
```

Listing 12.3: Julia implementation of the SIR model using `DifferentialEquations.jl`.

The Python/R/Julia ODE formulations are concise and readable, and they integrate smoothly with plotting and data analysis. They are often "fast enough" for deterministic models and small calibration loops, especially if the likelihood is cheap. JAX (or PyTorch) becomes attractive when you want compilation, batching, and automatic differentiation, particularly for gradient-based inference or to amortize computation over many runs. C++ becomes attractive when the model is dominated by large-scale discrete updates (agent-based/network models), when memory layout matters, or when you need predictable performance with explicit control over threads (OpenMP) or distributed execution (MPI).

```
import jax
import jax.numpy as jnp

def sir_step(y, params):
S, I, R = y
beta, gamma, dt = params
N = S + I + R
dS = -beta * S * I / N
dI = beta * S * I / N - gamma * I
dR = gamma * I
return jnp.array([S + dtdS, I + dtdI, R + dt*dR])

@jax.jit
def simulate(y0, beta, gamma, dt, n_steps):
params = jnp.array([beta, gamma, dt])
def body(y, _):
y_next = sir_step(y, params)
return y_next, y_next
_, ys = jax.lax.scan(body, y0, xs=None, length=n_steps)
```

```
return ys # shape: (n_steps, 3)

y0 = jnp.array([0.99, 0.01, 0.0])
ys = simulate(y0, beta=0.3, gamma=0.1, dt=0.5, n_steps=320)
S, I, R = ys[:,0], ys[:,1], ys[:,2]
```

Listing 12.4: JAX implementation of SIR with JIT-compiled Euler steps and batched parameters.

For large agent-based simulations or network models, the bottleneck is often a tight loop that updates many individuals and accumulates events. Below is a minimal C++ illustration of an Euler-stepped SIR (again, for demonstration). In real epidemic engines, the C++ advantage is less about this toy ODE and more about efficient event processing, memory layout, and multi-threading (OpenMP) across agents or replicate runs.

```
#include <array>
#include <vector>

struct Params { double beta, gamma, dt; };

std::vector<std::array<double,3>> simulate_sir(
std::array<double,3> y0, Params p, int n_steps
) {
std::vector<std::array<double,3>> ys;
ys.reserve(n_steps);

auto y = y0;
for (int k = 0; k < n_steps; ++k) {
    double S = y[0], I = y[1], R = y[2];
    double N = S + I + R;

    double dS = -p.beta * S * I / N;
    double dI =  p.beta * S * I / N - p.gamma * I;
    double dR =  p.gamma * I;

    y = {S + p.dt*dS, I + p.dt*dI, R + p.dt*dR};
    ys.push_back(y);
}
return ys;
}
```

Listing 12.5: C++ implementation of SIR with an explicit Euler integrator (illustrative core).

12.2 Software architecture for epidemic models

Software architecture shapes what an epidemic model can become over its lifetime. Early prototypes more often than not begin as a single script, but pandemic modeling quickly evolves into a workflow that mixes mechanistic assumptions, noisy and delayed observations, multiple data sources, and repeated inference runs. Without a deliberate architecture, small changes such as a new age group, a revised reporting delay, and a policy switch can ripple unpredictably through the codebase. These produce results that are difficult to verify or reproduce, deeming the project hard to maintain, use, or further develop. Thus, a good architecture does not merely make code "clean"; it makes claims easier to audit by forcing assumptions into explicit modules with stable interfaces.

A core architectural principle is to separate the *process model* (i.e., the latent epidemic dynamics) from the *observation model* (i.e., how data arise from those dynamics) and from the *inference engine* (i.e., how parameters and latent states are estimated). In practice, many errors and misunderstandings come from conflating these layers. For instance, a rise in reported cases could reflect increased transmission (process), improved testing (observation), or a change in the calibration window (inference). Hence, a clear separation forces the model to express each possibility in the right place. Formally, the process model defines state variables and transition rules: compartment flows (SIR/SEIR variants), agent interactions, mobility coupling, seasonality, waning immunity, vaccination, and variant replacement. It should expose a forward simulator that supports logic in the form of given parameters and initial state, which produces latent trajectories. This simulator should not "know" about reporting artifacts such as weekend effects, changing test sensitivity, or backfill revisions; those belong in the observation layer. Following this layer, the observation model maps latent states to the data streams available: reported cases, hospital admissions, ICU occupancy, deaths, seroprevalence, wastewater, and syndromic surveillance. It accounts for delays, under-ascertainment, reporting noise, censoring, and data definitions that change over time. Architecturally, the observation model should produce a likelihood function (or an equivalent scoring rule) and should be able to generate synthetic observations for simulation-based validation. A useful design is to treat each data stream as its own observation component with consistent inputs (latent trajectory, parameters, metadata) and outputs (log-likelihood contribution, simulated measurements, diagnostics). Lastly, the inference engine wraps the simulator and likelihood into a fitting procedure where maximum likelihood with numerical optimization, Bayesian inference, particle filtering, iterated filtering, approximate Bayesian computation, or variational methods are utilized. The inference layer should be the only component responsible for sampling, gradient evaluation, parameter transformations, priors, convergence diagnostics, and posterior predictive checks. Keeping inference separate makes it easier to swap methods without rewriting the model, and it helps prevent a common failure mode where calibration logic leaks into the model itself (e.g., hard-coded parameter bounds, ad hoc smoothing, or implicit priors hidden in preprocessing).

A practical pattern is to define three interfaces:

- `simulate(params, state0, schedule, tgrid) -> latent_trajectory`
- `loglik(latent_trajectory, data, obs_params) -> scalar`
- `fit(model, data, prior, algorithm_settings) -> fitted_object`

Even when implementation details differ across languages, these conceptual boundaries provide a stable mental model for both developers and reviewers.

Getting more into the technical details, the choice of state representation is both a computational and a scientific design decision because it affects performance, extensibility, and what questions the model can answer. The simplest representation is a dense vector of compartment counts or proportions per group, typically used in deterministic ODE models or stochastic compartment models with a small number of strata. This representation is simple conceptually and also programmatically, but can be computationally inefficient for a large number of possible states. Dense vectors are convenient for linear algebra, automatic differentiation, and solver interfaces; they also make serialization and checkpointing easy. They are less convenient when the state space becomes large and mostly empty, as in detailed network models with many potential event types but few active at a time. In a complementary manner, sparse or structured representations become important when the model includes large, irregular objects: contact networks, individual-level attributes, household or workplace membership, or spatial graphs with sparse connectivity. In those settings, the state may be a collection of arrays (structure-of-arrays) or sparse matrices for adjacency and mixing. The architecture should avoid scattering state across ad hoc objects that are expensive to copy, difficult to checkpoint, and hard to validate. A good compromise is to define a `State` container that stores the canonical state in a small set of well-defined arrays, plus optional caches that can be recomputed.

A related architectural decision is whether to store dynamics as event logs or aggregated trajectories. Aggregated trajectories store counts over time (e.g., I_t and daily incidence), which is compact and aligns with many observation models. Event logs record each transition event (infection, symptom onset, hospitalization) with timestamps and metadata. Event logs enable richer analyses (generation intervals, contact tracing simulations, individual-level attribution, intervention compliance auditing) but can be large and expensive to store. A common approach is dual output: always compute aggregates for inference and reporting, and optionally enable event logging via a flag for diagnostic or research runs. Event logging should be implemented as a pluggable sink so the core simulator does not become entangled with I/O. To this end, when designing the simulation, it is also important to plan for determinism and randomness. Stochastic models require explicit random number generation that is reproducible across platforms and parallel execution. Rather than calling global random number generators (RNGs) inside many functions, a robust approach is to pass an RNG handle (or seed stream) into the simulator and ensure each parallel run has a distinct and traceable seed. This belongs at the architecture level because it affects reproducibility, debugging, and scientific credibility.

Due to the dynamic nature of pandemics, models are often also changing rapidly over time, which requires from such systems to be highly extensible. Extensibility is not only about convenience; it prevents subtle errors when the model is modified under time pressure. Epidemic models evolve: new variants appear, vaccination changes susceptibility and severity, policies target subpopulations, and data streams shift. The architecture should make extensions explicit and local, so that adding a mechanism changes a small number of files and triggers tests. A practical extensibility technique is to enforce that all modifications are either (i) new state variables, (ii) new transitions, or (iii) new parameter transforms. If a change does not fit into one of these categories, it may be a sign that responsibilities are mixed or that the architecture needs a clearer abstraction. First, for compartment models, a common extensibility pattern is to represent transitions as a set of named flows rather than hard-coded equations. For example, define compartments as an ordered list and define transitions as objects: `from`, `to`, `rate_function`, `metadata`. The simulator builds the derivative (e.g., ODE) or event rates (e.g., CTMC) from this transition list. This reduces the risk of missing terms when adding a new compartment (e.g., introducing E or splitting I into symptomatic/asymptomatic). It also encourages consistent bookkeeping of incidence measures (which transitions count as cases, hospital admissions, deaths) needed by the observation model. Second, group structure (age, region, risk class) is often where model size explodes. Architecturally, it is useful to treat grouping as an axis in the state rather than as separate code paths. Dense tensor state (e.g., `state[compartment, group]`) makes group expansions systematic: adding groups becomes a configuration change plus updated mixing matrices, rather than duplicating equations. Mixing should be a first-class component, separate from the disease progression logic. For instance, define a `ContactModel` that provides effective contact rates between groups, potentially time-varying due to policies or behavior. Third, interventions typically enter as time-varying modifications of parameters or contact structure: $\beta(t)$ changes under mandates, mixing matrices change under school closures, hospitalization probabilities change with treatment improvements, vaccination changes susceptibility and severity. A robust pattern is to model interventions as a schedule of actions that transform parameters over time. The simulator consumes a schedule that can be evaluated at any time step, rather than sprinkling conditional statements throughout the code. This makes scenario generation systematic: scenarios differ by schedules, not by forked code.

12.3 Numerical methods and stability

Numerical methods are the hidden machinery of epidemic models: they determine whether a model run is trustworthy, whether calibration is stable, and whether scenario comparisons reflect epidemiology or numerical artifacts. Two implementations of the same conceptual model can produce meaningfully different outputs if they use different solvers, different tolerances, or different discretizations. This is particularly important in pandemic settings, where models are repeatedly re-run under changing assumptions and where small differences can be amplified by exponential growth. A numerically robust model is not only faster and less fragile; it is also easier to validate because failures appear as clear violations of invariants rather than as subtle, hard-to-reproduce discrepancies.

12.3.1 ODE solvers

Deterministic compartment models are often expressed as systems of ODEs. The numerical task is to approximate the continuous-time trajectory $y(t)$ governed by $\dot{y} = f(t, y, \theta)$, where y contains compartment counts (or proportions) and θ are parameters. In practice, the solver choice matters because epidemic models can exhibit rapid transients (early exponential growth, abrupt intervention changes) and multiple time scales (fast infection dynamics coupled with slower immunity waning, demography, or behavior), a combination that frequently produces stiff dynamics. When the system is not stiff, and parameters change smoothly, explicit methods such as adaptive Runge–Kutta schemes are typically efficient as they advance the solution using information from the current state and adjust step sizes to keep local error under control. The difficulty is that the same explicit methods can become unstable in stiff regimes unless step sizes are made very small, which can turn calibration into an infeasible computational burden. Implicit methods, including Backward Differentiation Formula (BDF) based solvers and other stiff integrators, trade more expensive steps (they solve an equation involving the future state) for substantially better stability properties in stiff settings. For pandemic modeling, this trade often pays off once latency (or progression) rates are fast, when there is strong coupling across groups, or when intervention schedules introduce sharp corners in the dynamics. Technically, error control through absolute and relative tolerances is where numerical analysis meets inference. Loose tolerances can inject numerical noise that the inference algorithm may interpret as stochasticity or as evidence for time-varying parameters. Excessively strict tolerances can slow down calibration without improving decision-relevant outputs. A practical workflow is to pick baseline tolerances, then

Table 12.1 Practical guide to numerical choices in epidemic models.

Situation / feature	Likely symptom	Recommended approach
Non-stiff ODE, smooth parameters	Fast and stable with RK methods	Adaptive explicit RK (e.g., Dormand–Prince), moderate tolerances
Stiff ODE (multiple time scales)	Tiny steps / slow explicit solver	Implicit solver (BDF / Rosenbrock), check Jacobian options
Discontinuous interventions	Smearing or bias around change points	Break integration at change times; enforce step at discontinuities
Large calibration loop	Runtime dominated by simulator calls	Profile; tune tolerances; consider compiled kernels or batching
Need gradients for inference	Slow finite differences / unstable gradients	AD-friendly solver or discretization; JAX/Stan/Julia AD stack
CTMC with very high event rate	SSA too slow	Tau-leaping / hybrid methods; validate against SSA on smaller cases
Near-boundary dynamics (small counts)	Negative counts or extinction artifacts	Exact SSA or bounded tau-leaping (binomial), smaller τ

confirm that key outputs (peak timing, peak magnitude, and cumulative outcomes, for example) are stable when tolerances are tightened. This sensitivity check is particularly important when the model is used inside a likelihood evaluation loop, because small numerical irregularities can degrade optimizer and sampler behavior. Moreover, time-varying parameters introduce another stability consideration for step-size control near discontinuities. Epidemic models commonly encode interventions via a time-varying transmission parameter $\beta(t)$ or a changing contact matrix. If parameters change abruptly (for instance, in step functions modeling), an adaptive solver may take a step that straddles the discontinuity and effectively smear the intervention over time, producing biased timing and magnitude. The most robust remedy is operational rather than mathematical: split the integration interval at known change points so the solver is forced to step exactly at the discontinuity. Smooth ramps are typically easier for solvers to handle, but they still merit inspection to ensure the solver is not taking steps so large that it misses important curvature.

Due to the fact that these solver decisions are often made under time pressure, especially during calibration and scenario production, it is helpful to maintain a short "default playbook" that links common modeling situations to robust numerical strategies. Table 12.1 summarizes typical decision points spanning ODE and CTMC settings, emphasizing symptoms that practitioners actually observe (slow runs, unstable trajectories, negative states) and the corresponding remedies.

12.3.2 Discrete-time approximations and their pitfalls

Discrete-time epidemic models replace continuous-time dynamics with updates over a time step Δt. They are attractive because surveillance data are typically reported daily (sometimes hourly) and because discrete updates can be simpler to implement and faster to run, particularly for large ensembles. The risk is that seemingly small discretization choices can introduce systematic bias or even qualitatively incorrect behavior, especially during phases of rapid growth.

The most common starting point is a forward Euler update [778], $y_{t+\Delta t} = y_t + \Delta t\, f(t, y_t, \theta)$, where y is the signal over time, Δt is step in time and $F(t, y_t, \theta)$ is the dynamic. This is easy to code and can work well when Δt is small enough, and rates are moderate. However, epidemic hazards can become large during surges, and Euler updates can subtract more individuals than remain in a compartment, producing negative states or oscillations. Such failures are not merely cosmetic: they distort peak timing and size, and during calibration, they can create spurious likelihood improvements that arise from numerical artifacts rather than epidemiology. As such, a subtler but equally important pitfall arises when continuous-time rates are interpreted as discrete-time probabilities. If the infection hazard over a step is λ, then the probability of infection during that step is not $\lambda\Delta t$ unless $\lambda\Delta t$ is small. Under the assumption of a constant hazard within the interval, the correct conversion is $p = 1 - \exp(-\lambda\Delta t)$. Using $p \approx \lambda\Delta t$ can materially overestimate transition probabilities when hazards are large, leading to biased dynamics precisely when accuracy matters most. Robust discrete-time simulators therefore compute stepwise transition probabilities from hazards using exponential transforms, or they use sub-stepping when hazards vary substantially within a day.

12.3.3 Continuous-time Markov chains simulation

Stochastic compartment models and many mechanistic inference approaches rely on continuous-time Markov chains (CTMCs) [779]. In a CTMC, transitions occur as random events with state-dependent intensities (rates). CTMC simulation is attractive because it preserves the discrete nature of individuals and naturally represents demographic stochasticity, which is especially important in early outbreak phases, in small subpopulations, and for extinction probabilities. The computational cost, however, can be substantial when event rates are high or when the model includes many groups and transitions. As a promising candidate, Gillespie's stochastic simulation algorithm (SSA) simulates one event at a time. At state x, each transition channel k has rate $a_k(x)$. The time to the next event is exponential with rate $a_0(x) = \sum_k a_k(x)$, and the event type is chosen with probability $a_k(x)/a_0(x)$. This procedure is exact for the CTMC, which makes it a useful gold standard for validation, but it becomes slow when $a_0(x)$ is large (large populations, intense transmission, or highly stratified models). In such settings, algorithmic refinements can reduce overhead: the next-reaction method can exploit sparse propensity updates, and efficient sampling schemes (e.g., alias-based selection) can accelerate choosing among many channels. However, when exact SSA is too slow for the required ensemble size, tau-leaping offers an approximation: instead of simulating each event, it approximates the number of events of each type within a time interval τ by drawing from Poisson (or related) distributions, effectively leaping over many events at once. This can dramatically improve throughput when propensities change slowly over short intervals. The cost is approximation error and an increased risk of violating constraints, particularly near boundaries where compartments are small. Practical tau-leaping implementations therefore include safeguards such as adaptive selection of τ, bounded (e.g., binomial) variants for transitions out of finite compartments, and rejection or correction steps when proposed updates would leave the feasible region.

Many real pandemic models land between these extremes and benefit from hybrid strategies. High-count components may be treated deterministically (or via diffusion approximations), while low-count or decision-critical events are simulated exactly. Hybridization preserves key stochastic features where they matter while keeping runtime manageable for large ensembles. From an implementation perspective, hybrid methods are easiest to support when the process model is expressed in terms of transition channels, allowing multiple numerical backends to operate on the same model specification. The numerical approximation level also interacts with inference. If the simulator introduces approximation error comparable to observation noise, parameter estimates can become biased and uncertainty intervals can become misleadingly narrow or wide. A defensible approach is to validate approximate simulators against SSA on smaller instances or shorter windows and to ensure that approximation error is small relative to the assumed observation model.

12.3.4 Handling constraints and validation checks

Epidemic model states are constrained: compartment counts should be nonnegative, and totals should satisfy conservation laws (e.g., $S + E + I + R = N$ in closed-population models, or consistent accounting with births, deaths, and migration in open models). Numerical methods can violate these constraints due to discretization error, loose tolerances, or approximation schemes. Treating such violations as mere numerical nuisances is risky because they can silently contaminate downstream inference and scenario comparisons. First, nonnegativity violations are the clearest failure mode. For ODE solvers, nonnegativity is not automatically guaranteed unless the solver and formulation preserve it; in practice, stability-oriented solver choices and reasonable tolerances often suffice, while ad hoc clamping can hide deeper problems and break conservation. For discrete-time simulation and tau-leaping, maintaining feasibility typically requires explicit safeguards: sampling transitions out of compartments with bounded distributions (e.g., binomial draws), limiting leap sizes, or rejecting updates that would overdraw a compartment. Second, conservation laws and accounting checks should be automated and treated as unit tests for the numerical layer. In closed models, total population drift beyond a small tolerance usually indicates a missing flow term or a bug in indexing. In multi-group models, conservation may be expected within each region or only globally, depending on the mobility formulation; the validation code should reflect that design choice explicitly. In addition, other structural invariants provide additional early warnings. Cumulative variables (infections, deaths, vaccinations) should be monotone; computing them as integrals or cumulative sums of nonnegative increments avoids spurious decreases due to numerical noise. Incidence derived from $-dS/dt$ should not become negative in standard formulations; when it does, the cause is often a discretization issue or an intervention schedule applied with an inconsistent ordering. These checks are most useful when they are routine rather than occasional. Numerical stability can change when new mechanisms are added, and pandemic models often evolve rapidly. Table 12.2 provides a compact checklist of invariants and common failure signatures that can be integrated into continuous testing and run-time diagnostics.

12.4 Performance profiling and optimization

The computational budget directly constrains what kinds of uncertainty quantification are feasible, how often models can be recalibrated as data updates, and how many scenarios can be explored before reporting deadlines. In practice, the most expensive part of the workflow is rarely a single simulation. Cost accumulates because calibration, sensitivity analysis, and ensemble forecasting multiply simulator calls by orders of magnitude. A productive approach towards computational efficiency begins with a simple rule: optimize only what you have measured. Many models are slowed down by a small number of "hotspots", and intuition is often wrong about where they are. Profiling provides a map of where time and memory go, which makes optimization targeted and auditable. It also protects against premature micro-optimizations that reduce readability without meaningful gains.

Table 12.2 Validation checks for numerical stability and common failure modes.

Check	How to test	Typical causes if it fails
Nonnegativity	Assert $y_i(t) \geq 0$ for all states/times	Step too large; tau too large; incorrect hazard-to-prob conversion; bug in transitions
Conservation law	Track $S + E + I + R$ (or total by group)	Missing flow term; indexing bug; solver tolerances too loose; rounding/clamping
Monotone cumulatives	Ensure cumulative variables never decrease	Computed from noisy differences; numerical oscillations; incorrect reset logic
Step-size sensitivity	Compare outputs under halved Δt	Discretization error dominates; instability in explicit updates
Tolerance sensitivity	Compare outputs under tighter rtol/atol	Solver error dominates; stiffness; discontinuities not handled
SSA vs approximate	Compare tau-leaping mean to SSA mean	τ too aggressive; boundary effects; propensities vary within τ

12.4.1 How to profile and interpret bottlenecks

CPU profiling answers the question: which functions consume wall-clock time? In epidemic models, typical CPU hotspots include the state transition loop (agent updates, compartment flows across many groups), likelihood evaluation (especially if repeated for multiple data streams), and repeated solver calls during calibration. Thus, a common and useful first step is coarse profiling, which is to measure total runtime per simulation and per calibration iteration, then break down the pipeline into major stages such as preprocessing, simulation, observation mapping, likelihood computation, and output serialization. This top-level timing often reveals bottlenecks that are invisible when looking only at micro-level function timings. Once a dominant stage is identified, fine-grained profiling can attribute time to specific functions and lines of code. The most informative interpretation is not simply which line is slow, but why it is slow. A slow line of code can indicate excessive programming language-related overhead, repeated allocations, cache-unfriendly memory access patterns, or algorithmic complexity that scales poorly with the number of groups, edges, or ensemble members. Profiling should therefore be repeated at multiple scales: a small toy run to debug functionality and a representative large run to reveal scaling behavior. If a hotspot grows superlinearly (i.e., quicker than linear) with population size or number of groups, algorithmic changes may matter more than low-level tuning.

Memory profiling answers a different question: where is memory allocated, how large are the peak allocations, and how often are large arrays created and discarded? Epidemic models can become memory-limited when they store full trajectories for many compartments, groups, and ensemble members, or when they keep multiple copies of large tensors (e.g., contact matrices, mobility flows) across scenarios. Memory issues also show up indirectly as CPU slowdowns due to garbage collection and paging. A model that "fits" in memory but triggers frequent allocations can still be slow because allocation and cache misses dominate.

I/O profiling is often neglected, but it matters in scenario pipelines. Writing large outputs (full time series for each run), reading large mobility or contact datasets repeatedly, or serializing intermediate results at high frequency can dominate runtime, especially on networked storage. A common sign is that CPU utilization is low while runtime is high. The remedy is rarely "faster disks"; it is usually a data strategy: write less, write later, compress selectively, and avoid reloading static datasets inside loops.

Fig. 12.1 summarizes how runtime grows with model size (e.g., number of groups, agents, edges, or particles). The main purpose of this plot is not to report absolute timing, which varies by hardware, but to reveal scaling behavior. A roughly linear slope indicates that the implementation is dominated by operations proportional to state size, while curvature suggests superlinear components such as dense mixing ($O(G^2)$) or other pairwise interactions. This figure is therefore a quick diagnostic for deciding whether further optimization should focus on low-level kernel speedups or on algorithmic changes that reduce complexity.

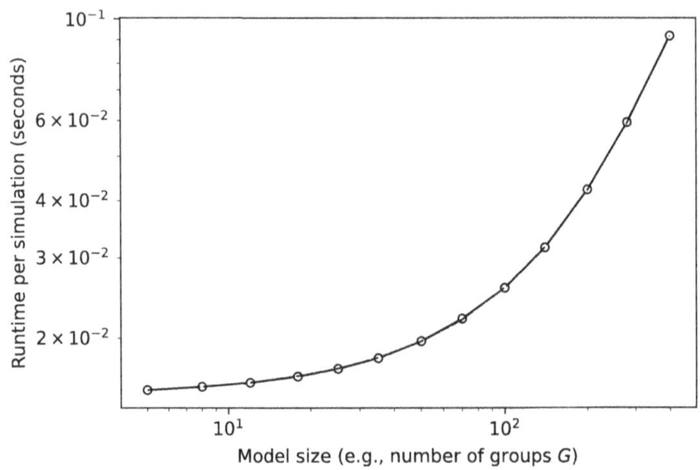

FIGURE 12.1

Scaling curve: runtime versus model size (log–log axes).

12.4.2 Vectorization, batching, and avoiding Python loops

In Python-based epidemic models, a common bottleneck is the Python loop in the inner simulation step. Loops over individuals, groups, or edges incur interpreter overhead and often prevent efficient use of vectorized numerical kernels. Vectorization replaces explicit loops with array operations implemented in optimized native code, which can yield large speedups when computations are regular and can be expressed as linear algebra or element-wise operations. Avoiding Python loops does not mean eliminating all loops; it means ensuring the hot loop is not at the Python level. If the dynamics are irregular (agent-based, network-driven), pure vectorization can become unnatural. In those cases, moving the hot loop into a JIT-compiled function (Numba) or a compiled extension (C++/Rust) can provide similar benefits while preserving the natural control flow of the model.

For compartment models with many groups, vectorization often means representing the state as arrays (e.g., `state[compartment, group]`) and expressing flows using broadcasting and matrix multiplication. For example, infection pressure may be computed via a mixing matrix times the infectious vector, rather than looping over source and target groups. Similarly, batching means running multiple parameter sets or multiple stochastic replicates simultaneously by adding an extra batch dimension to the state tensor. This is particularly effective when the same operations are applied across replicates, enabling SIMD-friendly computations and making accelerator backends (JAX, PyTorch) more effective. A useful heuristic is to treat per-step operations as kernels. If a kernel can be expressed as a small set of array operations, vectorize it. If it requires branching logic per agent or per edge, consider a JIT or compiled core. In both cases, the outer orchestration (scenario loops, configuration management, logging) can remain in Python, while the computational core is optimized.

12.4.3 Preallocation, caching, sparse operations, and memory locality

Many slowdowns come not from arithmetic but from memory behavior: repeated allocations, poor cache use, and moving data between formats. Preallocation is a simple but powerful practice: allocate arrays for trajectories, temporary buffers, and intermediate computations once, then reuse them across time steps and runs. This reduces pressure on the memory allocator and garbage collector and improves cache predictability.

Caching is equally important when the model repeatedly uses quantities that do not change within a run or across runs. Examples include fixed contact matrices, static mobility graphs, age-specific severity multipliers, and parameter transforms that are reused at every time step. Caching should be done carefully to avoid hidden state that breaks reproducibility; a good pattern is to compute caches from a configuration hash or from explicit initialization steps, so that caches are invalidated when inputs change.

Sparse operations can be decisive when mixing structures are large but mostly zeros, as in spatial models with local mobility or network adjacency. Using sparse matrices can reduce both memory and computation, but only if the operations match the sparse strengths. Sparse matrix–vector multiplication is efficient; repeated conversions between dense and sparse formats are not. It is therefore useful to choose a canonical representation and stick with it throughout the hot path. In a similar manner, for block-structured mixing (e.g., age-by-location), exploiting Kronecker or block-diagonal structure can outperform generic sparse representations.

Memory locality matters because modern CPUs are frequently bandwidth-limited. Contiguous arrays, "structure-of-arrays" layouts for agent attributes, and predictable iteration orders can reduce cache misses. Conversely, object-heavy representations (lists of Python objects, nested dictionaries, pointer-chasing graphs) can make the simulation memory-latency dominated. Even without changing languages, choosing arrays and stable layouts can yield substantial gains.

These memory-centric optimizations also improve parallel performance. Threads contend less when data are contiguous and when each thread works on a separate chunk of memory. Poor locality can erase the benefits of multi-threading, leading to disappointing scaling.

Fig. 12.2 shows the memory footprint of a representative workload over time (typically a calibration iteration, a large scenario batch, or a particle-filter update). This plot helps distinguish steady-state memory use from transient spikes caused by allocations such as storing full trajectories, buffering I/O, or creating temporary arrays inside the simulation loop. In optimization work, the actionable signal is often not the baseline level but the spikes and drift, which can indicate avoidable copies, insufficient preallocation, or output strategies that are too verbose for ensemble workloads.

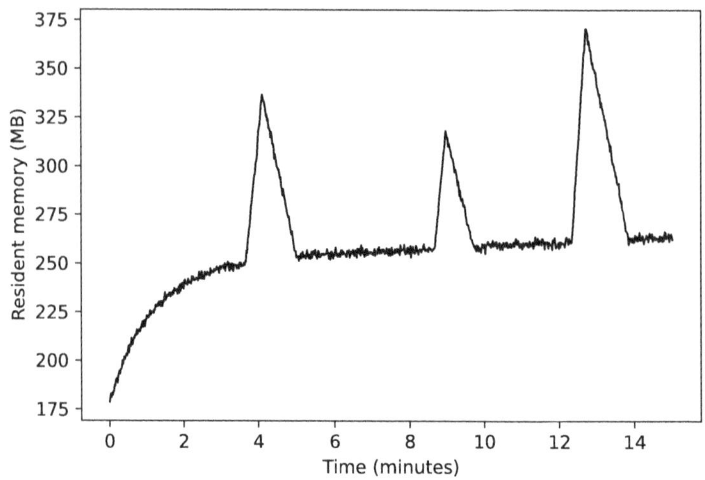

FIGURE 12.2

Memory footprint over time (resident set size trace).

12.4.4 Algorithmic improvements: reducing state dimension, exploiting structure in mixing matrices

The highest-impact optimizations are often algorithmic rather than micro-level. Reducing the effective state dimension can change runtime by orders of magnitude. In stratified compartment models, this can mean aggregating groups that are not decision-relevant, reducing age bands, merging similar regions, or removing compartments that do not influence observed outputs. In agent-based models, it can mean switching to a metapopulation approximation for parts of the system where individual detail is unnecessary.

Exploiting structure in mixing matrices is a particularly common opportunity. Mixing matrices often have patterns: they can be low-rank (dominated by a few contact modes), block structured (within- and between-group mixing), or separable across dimensions (age $\times$ location). Instead of multiplying a full dense matrix at every time step, one can use block operations, low-rank factorizations, or Kronecker products. These transformations preserve model meaning while reducing computational cost. Importantly, they also improve interpretability by making the assumed structure explicit.

Algorithmic changes should be guided by sensitivity analysis. If a dimension reduction changes key outputs within acceptable bounds, it may be a justified simplification. If it changes results materially, the added fidelity may be worth the cost. The key is to treat performance work as part of model design: choose the simplest computational structure that answers the question with the required fidelity and uncertainty characterization.

A convenient way to turn profiling results into action is to keep a small "optimization playbook" that connects the dominant bottleneck you observe to the most appropriate intervention. Table 12.3 summarizes when a technique tends to deliver real gains in epidemic pipelines, what it costs in code complexity or assumptions, and what to watch for so that performance improvements do not quietly harm reproducibility or scientific meaning. In practice, teams revisit this table repeatedly as a model matures: early iterations often prioritize readability and correctness, while later iterations apply one or two high-leverage techniques to unlock larger ensembles and more frequent recalibration.

12.5 Calibration and inference at scale

A pandemic model becomes decision-relevant only after it is connected to data, and that connection is typically the most computationally demanding part of the entire workflow. Unlike chapter 8, where we discussed these methods, in this section, we are only concerned about their computational aspect. Forward simulation might be fast, but fitting is rarely a one-time thing: it is thousands to millions of simulator evaluations, each embedded inside an optimizer, a sampler, or a sequential update. The central scaling challenge is therefore not just making the model run quickly, but making the inference loop stable, reproducible, and fast enough to iterate as new data arrive, assumptions change, or stakeholders request additional scenarios.

Inference at scale also forces clarity about what is being modeled. Many epidemic models are naturally formulated as state-space models: a latent disease process evolving over time, coupled with an observation process that maps latent states to reported cases, admissions, or deaths. Once the model is written this way, the computational questions become concrete: how to evaluate or approximate the likelihood; how to propagate uncertainty through latent trajectories; and how to fit parameters while preserving realistic observation noise, reporting delays, and time-varying interventions.

12.5.1 Likelihood-based fitting

A common formulation is a partially observed Markov process (POMP): latent states $x_{0:T}$ evolve according to a Markov transition $p_\theta(x_t \mid x_{t-1})$, while observations $y_{1:T}$ are generated from $p_\theta(y_t \mid x_t)$. The likelihood

$$p_\theta(y_{1:T}) = \int p_\theta(x_{0:T}, y_{1:T})\, dx_{0:T}$$

is generally intractable for nonlinear, non-Gaussian epidemic models, which is why simulation-based likelihood approximations are so central.

Particle filters (sequential Monte Carlo for state estimation) approximate the filtering distributions $p_\theta(x_t \mid y_{1:t})$ with weighted particles and provide an unbiased estimator of the likelihood (under standard conditions). In epidemic settings, particle filtering is attractive because it handles nonlinear dynamics, non-Gaussian observation models, and time-varying parameters. Its scaling challenges are equally familiar: particle degeneracy when the observation model is sharp; weight collapse when the process model is misspecified; and computational cost that grows linearly with the number of particles and time points. Practical implementations at scale typically rely on (i) careful choice of observation models (e.g., negative binomial rather than overly narrow Poisson when reporting is variable), (ii) tempering or bridging strategies when observations are highly informative, and (iii) efficient resampling and propagation kernels that avoid Python overhead and minimize allocations.

Table 12.3 Optimization playbook for epidemic simulation and inference pipelines.

Technique	Best when	What it changes	Trade-offs / risks	Typical gain
Vectorization (NumPy/xarray)	Per-step computations are regular (grouped compartments, matrix operations)	Replaces Python loops with array kernels (broadcasting, matvec)	Harder debugging; broadcasting mistakes; may increase temporary arrays	2–20×
Batching (extra run dimension)	Many runs share the same structure (scenario sweeps, posterior draws, particles)	Runs multiple simulations/likelihoods simultaneously using one set of kernels	Higher memory footprint; requires careful shape discipline; may constrain control flow	2–50×
Numba JIT (kernel compilation)	Hot loop has branching/irregular logic (ABM kernels, hazard sampling)	Compiles Python-like loops to native code; keeps algorithmic structure	Type restrictions; compilation overhead; need stable data structures for speed	2–50×
JAX / accelerator compilation	Need fast batched execution and/or gradients through simulation or likelihood	Traces computation into a compiled graph; enables AD and accelerator backends	Functional style; static shapes; debugging differs; careful numerical checks needed	2–100× (workload dependent)
Preallocation and buffer reuse	Long runs with repeated temporary arrays (trajectories, intermediates)	Allocates once and reuses memory across time steps / iterations	More disciplined code; risk of accidental overwrites if not carefully managed	1.2–5×
Caching invariant quantities	Matrices/transforms reused across steps or runs (mixing, severity multipliers)	Moves repeated computation to initialization; reuses computed results	Cache invalidation must be explicit; hidden state can harm reproducibility	1.2–10×
Sparse operations (CSR/CSC matvec)	Mixing/mobility graph is large but sparse; local coupling	Stores and multiplies using sparse structures; reduces arithmetic and memory	Sparse-dense conversion kills gains; limited benefit if matrix not truly sparse	2–100×
Exploit block / Kronecker structure	Mixing separable (age×location) or block-structured (schools/workplaces)	Replaces generic matvec with structured multiplies; reduces complexity	Adds modeling assumptions; structure must be validated against data and intent	2–50×
Low-rank mixing approximation	Contact patterns dominated by few modes (e.g., principal components)	Approximates dense mixing by $UV^\top$ to reduce cost from $O(G^2)$ to $O(rG)$	Approximation error; may distort subgroup dynamics if rank too small	2–50×
Reduce state dimension (groups/-compartments)	Some strata are weakly identifiable or not decision-relevant	Aggregates compartments or groups; reduces per-step state and parameters	Potential fidelity loss; must justify aggregation with sensitivity analysis	2–100×
Adaptive output thinning / streaming	I/O dominates; storing full trajectories for large ensembles	Stores summaries or thinned time series; streams metrics instead of full state	Harder post-hoc debugging; must decide outputs up front; risk losing detail	1.5–50×

continued on next page

Table 12.3 (*continued*)

Technique	Best when	What it changes	Trade-offs / risks	Typical gain
Parallelization across runs (multi-processing/cluster)	Many independent scenarios/replicates; per-run cost moderate to high	Distributes runs across cores/nodes; improves throughput	Overhead in orchestration; I/O contention; reproducibility via seeds needed	Up to $\approx p\times$ (limited by overhead)
Parallelization within a run (OpenMP/MPI)	Single run is heavy (ABM/network, very large N or E)	Splits inner loop across threads/nodes; reduces wall time per run	Race conditions; reductions and logging require care; bandwidth-limited scaling	2–20× (architecture dependent)

Once particle filters are available, several inference families become accessible. Iterated filtering and related methods use repeated particle filtering passes to optimize parameters in POMP models, and are often effective when gradients are unavailable or unreliable. Particle MCMC (such as particle marginal Metropolis–Hastings) uses particle filters inside MCMC to sample from the posterior while maintaining correctness, but it is computationally expensive because each MCMC step requires a full particle filter, and mixing can be poor if the likelihood estimator variance is large. This leads to a key scaling principle: when using particle MCMC, the number of particles should be tuned to control estimator variance, because too few particles yields noisy acceptance decisions and too many waste computation. In practice, one often begins with optimization (to locate plausible regions) and then performs posterior sampling with tuned particle counts and parallel chains.

Standard MCMC remains central for models where the likelihood is tractable (e.g., deterministic ODE models with explicit observation noise) or where gradients can be computed reliably. Hamiltonian Monte Carlo (HMC) can be highly efficient in such settings, but it is sensitive to numerical issues: discontinuities in intervention schedules, non-smooth transforms, and solver adaptivity can all produce problematic geometry. Sequential Monte Carlo samplers (SMC in parameter space) provide an alternative that is naturally parallelizable across particles and can adapt to multimodality, but requires careful temperature schedules and resampling strategies to avoid degeneracy. Variational inference (VI) can scale to large models and datasets by trading exactness for speed; its usefulness depends on whether the chosen variational family can represent posterior correlations common in epidemic models (e.g., β-ascertainment trade-offs, reporting delay-growth trade-offs).

Across these methods, the same engineering patterns recur: vectorization and batching across particles or parameter draws, parallelization across chains or scenarios, and careful handling of randomness and seeds. Just as importantly, inference code benefits from a strict separation between process simulation, observation likelihood, and inference driver, so that alternative inference algorithms can be swapped without rewriting the model.

A compact way to make practical non-identifiability visible is to inspect a two-parameter slice of the likelihood (or posterior). Fig. 12.3 shows such a slice for transmission intensity β and ascertainment ρ. The elongated high-likelihood region illustrates a common ridge: higher transmission coupled with lower ascertainment can match observed case counts similarly well to lower transmission coupled with higher ascertainment. This kind of structure is not merely a statistical curiosity; it determines which quantities are learnable from the available data streams and motivates either additional measurements (e.g., serology, wastewater) or stronger, explicit prior information.

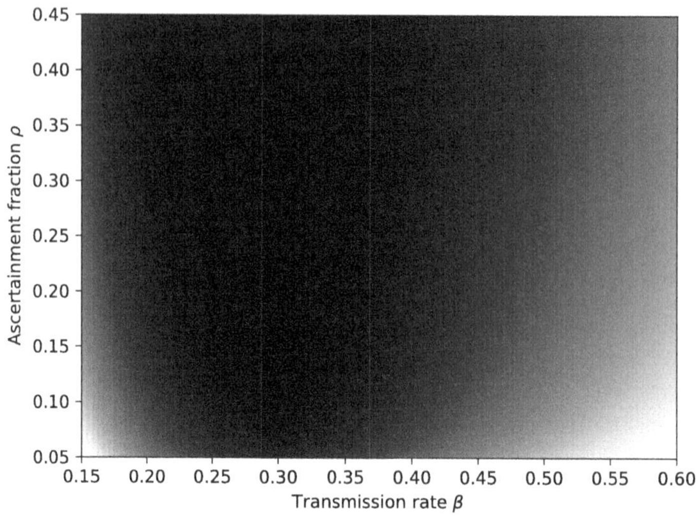

FIGURE 12.3

Likelihood surface over (β, ρ) exhibiting a practical identifiability ridge.

12.5.2 Gradient-based calibration with autodiff

When gradients are available and stable, they can dramatically reduce the number of simulator evaluations needed for calibration. In deterministic ODE-based models, gradients can be obtained through sensitivity equations, automatic differentiation (AD) of the numerical solver, or adjoint methods. Adjoint methods are particularly attractive when the parameter dimension is large: rather than propagating sensitivities for each parameter forward in time, the adjoint integrates a backward system whose cost is closer to the cost of the forward solve (up to constant factors). This can make gradient-based optimization and HMC feasible for models that would otherwise require expensive finite differences.

The practical challenge is that *differentiable* is not the same as *stable for inference*. Intervention schedules with step changes, if modeled as discontinuous functions, can produce gradients that are undefined or uninformative. Piecewise dynamics and event-triggered policies can create non-smooth likelihood surfaces. Even the numerical solver can introduce non-smoothness: adaptive step decisions and tolerance thresholds can cause small discontinuities in outputs as parameters vary. In these situations, smoothing interventions (e.g., logistic ramps), enforcing solver steps at known change points, or using fixed-step integrators during inference can yield gradients that lead to better optimizer and sampler behavior.

Differentiable simulators extend beyond ODEs. Discrete-time state-space models can often be differentiated directly, and some stochastic simulators can be reparameterized (or relaxed) to support gradient estimators. However, fully differentiable stochastic epidemic simulation remains delicate: naive differentiation through random sampling is not meaningful without reparameterization tricks or likelihood-ratio estimators, and these can have high variance. A common pragmatic approach is hybrid: use gradients where the model is smooth and deterministic (e.g., transmission dynamics), and use simulation-based methods where discreteness is essential (e.g., observation delays, discrete event components), combining them in a way that preserves validity.

12.5.3 Surrogate models and emulators for expensive simulators

When forward simulation is expensive (agent-based models, detailed network models, high-resolution spatial coupling), direct calibration may be computationally prohibitive even with parallelization. Surrogate models provide a way to approximate the mapping from parameters and interventions to outputs of interest, enabling fast inference and scenario exploration. The most common surrogates in epidemic modeling include Gaussian process emulators for low-dimensional parameter spaces, neural surrogates for higher-dimensional settings, polynomial chaos or low-rank expansions when the response is smooth, and reduced-order mechanistic approximations (e.g., metapopulation approximations of an ABM).

A surrogate strategy is only as good as its training design and its uncertainty handling. The surrogate must be trained on a space-filling design over parameters and intervention schedules, and it must be validated on held-out simulations to ensure it does not hallucinate behavior in poorly covered regions. Because surrogates introduce approximation error, they should not simply replace the simulator without accounting for that error; otherwise, inference can become overconfident. A practical pattern is to include a discrepancy term or to inflate observation noise to reflect surrogate uncertainty, and to update the surrogate iteratively: start with a coarse surrogate, fit a preliminary posterior, then add simulator evaluations in regions of high posterior density or high surrogate uncertainty (active learning). This "surrogate-in-the-loop" approach often yields large savings while keeping approximation risk visible and testable.

Surrogates are also useful even when full calibration is not the goal. They can accelerate scenario sweeps, enable real-time interactive dashboards, and support sensitivity analyses that would be infeasible with the full simulator. In pandemic operations, these uses can be as valuable as formal inference.

Because inference methods trade computational cost against statistical efficiency, it is often more informative to report effective sample size per unit time than raw runtime or raw sample count. Fig. 12.4 compares effective sample size per hour across representative inference strategies under a common model and dataset. The key operational message is that algorithmic choices which improve mixing (e.g., gradient-informed proposals or well-tempered SMC) can dominate wall-clock performance even when each iteration is more expensive, especially in epidemic posteriors with strong correlations and ridges.

Deepen your knowledge

Computational choices in epidemic modeling are rarely neutral: they shape what can be calibrated, how uncertainty is represented, and how quickly a team can respond to new data or stakeholder questions. The most robust projects are not the ones with the most sophisticated mathematics, but the ones that combine clear modeling assumptions with disciplined engineering practices. In particular, successful teams learn to distinguish between changes that improve scientific fidelity and changes that merely add complexity, and they invest early in reproducibility so that results remain trustworthy as the model evolves.

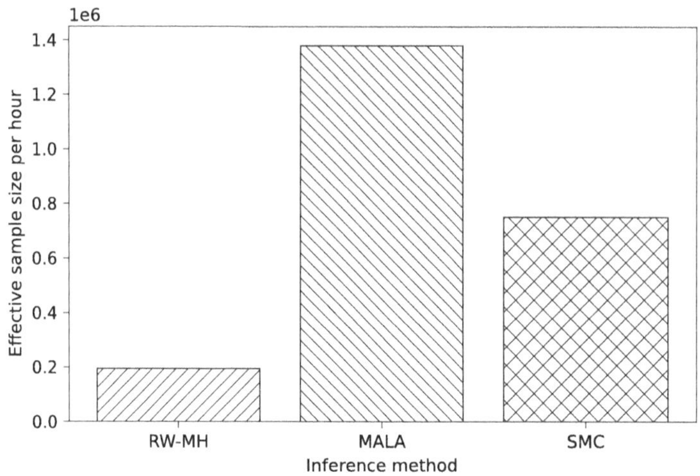

FIGURE 12.4

Effective sample size per hour for several inference strategies on the same calibrated model.

A practical way to translate the chapter into action is to keep a one-page checklist for model development and operations. The goal is not to impose bureaucracy, but to make good habits automatic: profile before optimizing, pin dependencies before publishing results, automate scenario runs so they are repeatable, and test invariants so numerical bugs cannot silently propagate. The checklist below is intended to be copied into a repository and treated as a living document: it should be updated as the model grows, as new data streams are added, and as inference methods become more demanding.

Architecture and Assumptions

- Do separate the process model, observation model, and inference engine with explicit interfaces.
- Do keep intervention logic in a schedule/config layer rather than scattered conditional statements.
- Do make priors and constraints explicit; document which parameters are learned from data versus fixed.
- Don't hard-code data quirks (weekday effects, backfill rules) into the process model.
- Don't mix calibration heuristics into simulation code (implicit priors, silent clipping, ad hoc smoothing).

Numerical Methods and Stability

- Do choose solvers based on stiffness and discontinuities; align steps with known regime changes.
- Do run step-size and tolerance sensitivity checks on representative scenarios.
- Do enforce constraints with principled methods (hazard-to-prob conversions, bounded draws), not silent clamps.
- Don't trust a discrete-time approximation without confirming outputs are stable under smaller Δt.
- Don't treat solver settings as "implementation details" when they affect inference geometry and fit quality.

Profiling and Optimization

- Do profile first (CPU, memory, I/O) on a realistic workload before changing code.
- Do optimize the hot path: remove Python from inner loops via vectorization, batching, JIT, or compiled cores.
- Do preallocate buffers and reuse arrays inside time-stepping and particle loops.
- Do exploit structure (sparsity, block structure, low-rank mixing) when scientifically justified.
- Don't micro-optimize non-bottlenecks; it reduces readability without improving throughput.
- Don't write full trajectories for every run by default; stream summaries and thin outputs for ensembles.

Calibration and Inference at Scale

- Do match the inference method to the model class: POMP/particle filters for nonlinear state-space models; HMC for smooth, differentiable likelihoods; SMC samplers for multimodality; VI for fast approximation/initialization.
- Do tune particle methods using diagnostics (ESS, log-likelihood variance) rather than guesswork.
- Do report mixing and convergence (ESS/time, $\hat{R}$, trace plots; PF degeneracy statistics).
- Do run posterior predictive checks and simulation-based calibration when feasible.
- Don't over-interpret point estimates when the posterior shows ridges or strong correlations (practical non-identifiability).
- Don't assume gradients help unless the simulator and intervention schedule are numerically smooth enough.

Reproducibility and Experiment Management

- Do pin dependencies (lockfiles) and record environment details (OS, BLAS/MKL, GPU/driver versions if used).
- Do make experiments config-driven; store the resolved config with every run output.
- Do record provenance: code version (git commit), random seeds, data snapshot/version, and run metadata.
- Do automate runs via a CLI (and/or a workflow tool) so results can be regenerated non-interactively.
- Don't rely on notebooks as the sole source of truth; treat them as clients of a tested package.
- Don't overwrite outputs; use run IDs and immutable result folders for auditability.

Testing and Validation (Invariants)

- Do test invariants automatically: nonnegativity, conservation laws, monotone cumulative counts, bounded probabilities.
- Do include regression tests for fixed-seed runs to detect unintended changes.
- Do cross-check simplified limits (ODE vs CTMC mean-field; SSA vs tau-leaping on small instances).
- Don't suppress warnings about negative states or drifting totals; treat them as failing tests.
- Don't accept a "good fit" without verifying that fitted simulations reproduce key data structure.

Operational Habits (the "Last Mile")

- Do define a small set of primary outputs tied to the decision question (not everything the simulator can produce).
- Do maintain a standard reporting script that regenerates figures/tables from stored outputs.
- Do communicate uncertainty clearly: intervals, scenario assumptions, and sensitivity to key modeling choices.
- Don't change model structure and calibration settings simultaneously without an ablation to isolate effects.
- Don't ship results without a documented data cut date and a reproducible run configuration.

If the checklist is used as intended, it becomes a practical guardrail as it prevents computational shortcuts from turning into scientific errors, and it keeps model outputs auditable even as the project scales from a prototype into a production forecasting and scenario engine.

CHAPTER

13 Closing remarks and future directions

The arc of this book begins with memory and ends with humility. We began with pandemics as historical events, but also as mirrors held up to human societies: they reveal not only pathogens and hosts, but fear and cooperation, inequality and governance, fragility and adaptation. From there, we moved through the many languages by which I invited you to think about contagion. We saw how compartmental models teach us to think in flows and states; how spatial and multi-strain models remind us that epidemics move, branch, and evolve; how agent-based simulations show that collective outcomes can emerge from countless local acts; and how data-driven methods reveal that prediction can sometimes arise not from explicit assumptions about mechanism, but from patterns hidden in observation. As our journey continued into calibration, multimodal data, intervention design, uncertainty, and computation, I hope it became clear that a model is never just an equation or an algorithm. It is a disciplined way of deciding what to see, what to ignore, and what kind of truth we are willing to call useful.

If there is one philosophical thread I hope we have followed together, it is that epidemiological modeling is not a march toward a final, supreme formalism. It is, instead, an education in pluralism. Along the way, we encountered models that ask very different questions: mechanistic models ask what must be true for transmission to unfold; statistical models ask what the data can support; machine learning asks what can be predicted; simulation asks what can emerge; and uncertainty analysis asks whether our confidence deserves to survive contact with reality. Each modeling family offers us something precious, but none offers us everything. There is, therefore, no one model to rule them all-and perhaps that is exactly the lesson we were meant to discover [780]. The temptation to seek a single ruling framework is intellectually seductive, even mythic, but I believe the real wisdom of epidemiology lies elsewhere. It lies in comparison, in synthesis, in doubt, and in knowing when one model must be corrected by another. If this book has accomplished anything, I hope it has shown that the deepest value of modeling is not that it conquers uncertainty, but that it teaches us how to live responsibly within it.

In these closing remarks, I want to circle back to the main points in the book, extracting the main points both new and seasoned modelers should keep in mind. Simply put, every modeling paradigm illuminates one part of the epidemiological landscape while leaving another part in shadow. Mechanistic models taught us to think in terms of processes: who infects whom, how transmission unfolds, where feedback enters, and why interventions matter. Their great strength is not merely that they produce curves, but that they give those curves meaning. They offer interpretability, causal language, and a way to reason about counterfactuals. When we ask what might happen under vaccination, isolation, behavior change, or the emergence of a new strain, mechanistic models remain uniquely valuable because they are built around the logic of the epidemic itself, rather than only around its observed traces. At the same time, we also learned that interpretability alone is not enough. A model may be conceptually elegant and still fail if its assumptions are too rigid, its parameters too uncertain, or its observation layer too distant

Epidemic Modeling. https://doi.org/10.1016/B978-0-44-355525-1.00019-9

from the data we actually collect. This is where statistical and time-series models taught us a different form of discipline. They reminded us that surveillance data are not transparent windows into reality, but noisy, delayed, biased, and operationally shaped measurements. In that setting, simpler statistical baselines are not intellectually modest because they are weak, but because they are often honest. They force us to confront seasonality, overdispersion, reporting artifacts, and forecast calibration directly. Again and again, we saw that strong forecasting practice often begins not with the most sophisticated model, but with the model that best respects the measurement process. As our journey moved into machine learning and deep learning, another lesson emerged. Flexibility is powerful, but it is never free. These methods can discover nonlinearities, interactions, latent representations, and multimodal relationships that would be cumbersome to specify by hand. When data are abundant, when signals are rich, and when evaluation is careful, ML and DL can deliver substantial predictive gains. Yet they also taught us caution. A model that learns complex structure may also learn accidental structure. It may absorb leakage, chase nonstationary artifacts, or perform impressively in one regime only to fail in the next. From them, I hope we learned not only that predictive power matters, but that it must always be paired with validation under realistic temporal conditions, with uncertainty assessment, and with humility about what has truly been learned. The arrival of large language models pushed this lesson even further by expanding what we are willing to count as data. Textual reports, policy bulletins, clinical narratives, expert summaries, and even fragmented public discourse can now be translated into machine-readable representations. This is a remarkable development, because much of public-health reality has always existed in language before it existed in tables. LLMs, therefore, do not simply add another predictive tool; they enlarge the epistemic boundary of epidemiological modeling itself. They allow us to incorporate narrative context into analytic pipelines. But here too, the promise comes with a warning: when a model speaks fluently, it can tempt us to mistake coherence for truth. Their value is real, but it depends on constraints, verification, and thoughtful integration with more traditional forms of modeling. Seen together, these paradigms do not form a ladder in which newer methods replace older ones. They form a conversation. Mechanistic models explain. Statistical models stabilize. Machine learning models exploit predictive structure. Deep learning models learn representations. LLMs translate human language into computable form. Each answers a different weakness in the others. Each one also inherits a different kind of risk.

Despite the progress described throughout this book, several limitations recur across modeling paradigms. These limitations are not confined to any single framework. Rather, they reflect structural difficulties in epidemic data, epidemic processes, and the relationship between models and decision-making. Recognizing these challenges is essential because they shape both the reliability of current methods and the priorities for future research.

A first persistent challenge is non-stationarity. Epidemic systems change over time as a result of interventions, behavioral adaptation, immunity, pathogen evolution, seasonality, and changes in reporting practice. As a consequence, relationships learned from one time period often do not remain stable in the next. This affects mechanistic, statistical, and machine learning models alike. Parameters drift, covariate effects change, and patterns that were previously predictive may lose their value. Distribution shift is therefore not a peripheral issue, but a central feature of epidemiological modeling.

A second challenge is that surveillance data are incomplete representations of the underlying epidemic process. Reported cases, hospitalizations, deaths, wastewater signals, syndromic indicators, mobility streams, and genomic samples are all shaped by measurement processes. These processes introduce missingness, reporting delays, under-ascertainment, backfill, administrative artifacts, and systematic biases. In practice, model performance often depends as much on how well the observation process is handled as on the predictive or mechanistic core of the model itself. Any framework that ignores the distinction between latent epidemic dynamics and observed data risks producing misleading inferences.

A third limitation is weak transportability. Models calibrated in one region, one time period, or one outbreak context often do not transfer reliably to another. Differences in contact structure, healthcare systems, surveillance quality, public behavior, intervention regimes, demography, and pathogen characteristics can all degrade performance when a model is moved across settings. This problem is especially important in public health, where there is strong practical pressure to reuse models quickly in new contexts. Good performance in one jurisdiction should therefore not be treated as evidence of general validity without explicit external evaluation.

A fourth challenge concerns interpretability. As models become more flexible and higher-dimensional, they may achieve stronger predictive accuracy while becoming more difficult to explain. This is especially visible in complex machine learning, deep learning, and hybrid systems, but it can also arise in large simulation models with many interacting assumptions. Limited interpretability makes it harder to understand failure modes, communicate results to decision-makers, and distinguish meaningful mechanisms from statistical fit. In epidemiological settings, this is not merely a technical inconvenience. It affects trust, transparency, and the ability to use models responsibly in policy contexts.

A fifth recurring issue is uncertainty quantification and calibration. Epidemiological forecasts and scenario analyses are often used under conditions where uncertainty matters as much as point estimates. However, many models either underestimate uncertainty or report it in forms that are poorly calibrated. Sources of uncertainty arise from parameter estimation, stochastic variability, structural misspecification, observation error, and future interventions or behavioral changes. A model that produces accurate point predictions but poorly calibrated intervals may still be unfit for decision support. For this reason, uncertainty must be treated as a primary modeling object rather than as a secondary output.

A final challenge is evaluation under realistic real-time conditions. Many models appear stronger in retrospective studies than they would in deployment because the evaluation protocol does not faithfully reproduce the information available at prediction time. Leakage from revised data, finalized labels, future covariates, or non-causal preprocessing can create overly optimistic results. Realistic evaluation requires rolling-origin backtesting, historical data snapshots when relevant, horizon-specific assessment, and metrics that reflect the intended operational use. Without such protocols, model comparison becomes unreliable, and progress is difficult to interpret.

In addition to modeling challenges, progress in epidemiological modeling does not depend only on better algorithms. It also depends on the quality and structure of the scientific infrastructure surrounding those algorithms. Surveillance systems determine what can be observed, how quickly it becomes available, and how much measurement error is introduced before modeling even begins. Datasets require clear documentation, stable schemas, version control, and transparent revision histories if they are to support valid calibration and evaluation. Without such foundations, even technically sophisticated models can produce results that are difficult to reproduce, compare, or interpret. In this sense, data pipelines and reporting systems are not merely supporting tools. They are part of the modeling problem itself. This point has several practical consequences. First, epidemiological modeling benefits from standardized benchmarks that allow methods to be compared under shared tasks, shared data definitions, and shared scoring rules. Second, open and well-curated datasets are essential for cumulative scientific progress, because they allow results to be independently tested, extended, and challenged. Third, reporting pipelines should be versioned whenever possible, since backfill, deduplication, and retrospective corrections can materially change both inference and forecast evaluation. Fourth, reproducible code and transparent workflows should be treated as a baseline scientific requirement rather than an optional convenience. This includes fixed preprocessing steps, documented hyperparameter choices, archived model configurations, and evaluation scripts that can be rerun by others. Finally, transparent evaluation frameworks are necessary so that model performance is assessed under realistic operational constraints rather than under idealized retrospective conditions. Importantly, these considerations shift attention from models in isolation to the broader scientific ecosystem in which they operate. A strong epidemiological forecasting system is not defined only by the quality of its equations or learning architecture, but also by the reliability of its data streams, the clarity of its benchmark tasks, the reproducibility of its implementation, and the transparency of its evaluation. Future progress will therefore require investment not only in methodological innovation, but also in shared infrastructure that makes modeling more reliable, comparable, and scientifically cumulative.

These limitations and challenges are fruitful ground for future works, developments, and discoveries. Several research directions appear especially promising for the next phase of epidemiological modeling.

Real-time and adaptive forecasting under distribution shift

A major priority is the development of forecasting systems that can adapt as epidemic conditions change. Many current pipelines still rely on periodic retraining or fixed model structures, even though outbreaks are shaped by interventions, behavioral responses, seasonal forcing, pathogen evolution, and reporting changes. Future work should therefore focus on online learning, change-point-aware forecasting, dynamic ensembling, and regime-sensitive model updating. Another important direction is selective adaptation: instead of refitting an entire model at every step, future methods should identify which components need to change, such as observation models, transmission parameters, mobility effects, or textual context layers. Research in this area should also examine how to detect when a model is no longer reliable, and how to trigger fallback strategies before large forecast failures occur.

Multimodal surveillance and joint data integration

A second promising direction is the construction of models that treat surveillance as a genuinely multimodal problem. Clinical reports, hospital burden, wastewater measurements, genomics, mobility, weather, digital behavior, and policy text each provide only a partial view of epidemic activity. Future models should aim to integrate these streams through common latent representations or joint state-space formulations, while explicitly accounting for different sampling frequencies, reporting delays, missingness patterns, and levels of noise. This requires more than simply concatenating features. It requires methods that can reason about which modality is informative at which stage of an outbreak, and how contradictions between sources should be resolved. A particularly important research challenge is the design of fusion methods that remain robust when one or more modalities become sparse, delayed, or unreliable.

Foundation models for public-health data

Foundation models are likely to become an important research area, but their value in epidemiology will depend on how they are adapted to domain-specific constraints. Future work should investigate pretraining on large collections of public-health time series, outbreak reports, genomic summaries, mobility panels, and other health-relevant corpora, followed by task-specific adaptation for forecasting, anomaly detection, nowcasting, and scenario analysis. At the same time, it will be important to move beyond generic prompting. Domain-adapted foundation models should incorporate epidemiological priors, explicit temporal structure, uncertainty estimation, and mechanisms for grounding outputs in trusted sources. For LLM-based systems in particular, promising work directions include retrieval-augmented forecasting, schema-constrained extraction, multimodal reasoning over text and numeric signals, and methods that separate useful contextual synthesis from unsupported generation.

Human–AI collaboration for epidemiological analysis

Another important direction is the design of systems in which AI assists epidemiologists rather than attempting to replace them. This includes tools for structured literature synthesis, outbreak report extraction, hypothesis generation, model specification support, calibration assistance, and scenario drafting. The key research question is not only whether an AI system can produce an answer, but whether it can improve the speed, quality, and transparency of expert analysis. This suggests a shift toward human-in-the-loop workflows with explicit checkpoints, editable intermediate representations, audit trails, and interfaces that allow domain experts to inspect and correct the reasoning process. In practice, useful systems may be those that help experts compare assumptions, surface missing data problems, or generate structured model variants, rather than those that operate as fully automated black boxes.

Benchmark design and prospective evaluation

Methodological progress will remain difficult to interpret unless evaluation becomes more realistic and more standardized. Future work should therefore invest in benchmark suites that include archived data snapshots, revision histories, multiple pathogens, multiple spatial scales, and clear operational targets. Evaluation should distinguish retrospective fit from prospective utility and should include scenarios with backfill, missingness, regime shifts, and limited covariate availability. There is also a need for broader metric design. In addition to standard forecast error measures, future benchmarks should include calibration, robustness to shift, timeliness, stability across horizons, and subgroup performance across regions or population strata. Prospective collaborative forecasting platforms remain especially valuable because they test methods under live conditions rather than under reconstructed hindsight.

Decision-focused modeling rather than prediction alone

A final major direction is to move from prediction-centered modeling toward decision-centered modeling. Public-health agencies often do not need a forecast in isolation; they need support for action under uncertainty. Future models should therefore be designed to connect naturally with downstream decisions such as hospital staffing, vaccine allocation, surge planning, genomic sampling, school policy, or travel screening. This implies a greater role for utility-aware loss functions, robust optimization, scenario-conditioned forecasting, and methods that explicitly represent trade-offs between epidemiological outcomes, economic costs, and operational constraints. It also suggests closer integration between forecasting models, causal inference tools, and policy simulation frameworks. In many settings, the most useful model will not be the one that minimizes average prediction error, but the one that most reliably improves decisions.

As we close this book, I would like to leave you with a hopeful conclusion. Pandemics are among the most difficult phenomena we try to understand: they unfold across biology, behavior, space, time, institutions, and uncertainty. Yet this should not be a cause for pessimism. It is, instead, a reminder of how much progress becomes possible when we approach hard problems with intellectual honesty, methodological diversity, and a willingness to learn across disciplines. Epidemiological modeling is strongest not when it claims certainty, but when it brings structure to uncertainty and helps us reason more clearly, act more carefully, and prepare more wisely. If there is an enduring message in these pages, it is that the future of epidemiological modeling is rich with possibilities. I hope this book has shown that modeling is not merely a technical exercise, but a constructive human endeavor – one that connects mathematics, computation, data, and judgment in the service of better decisions. Much remains to be done, but there is every reason to move forward with confidence: not because the challenges are small, but because our ability to study them, together and rigorously, has never been greater.

References

[1] Alin Gabriel Ionescu, Mihaela Ionescu, Adina Dorina Glodeanu, Teodor Nicusor Sas, Cristin Constantin Vere, A short history of pandemics, Istorie/History (2022) 79.

[2] Dara Grennan, What is a pandemic?, JAMA 321 (9) (2019) 910.

[3] David M. Morens, Gregory K. Folkers, Anthony S. Fauci, What is a pandemic?, The Journal of Infectious Diseases 200 (2009) 1018–1021.

[4] Cyril Caminade, Kelly M. McIntyre, Anne E. Jones, Impact of recent and future climate change on vector-borne diseases, Annals of the New York Academy of Sciences 1436 (2019) 157–173.

[5] A. Marm Kilpatrick, Sarah E. Randolph, Drivers, dynamics, and control of emerging vector-borne zoonotic diseases, The Lancet 380 (2012) 1946–1955.

[6] Nathan D. Wolfe, Claire P. Dunavan, Jared Diamond, Origins of major human infectious diseases, Nature 447 (2007) 279–283.

[7] Henry O. Ukuhor, The interrelationships between antimicrobial resistance, covid-19, past, and future pandemics, Journal of Infection and Public Health 14 (2021) 53–60.

[8] M. Oliveira, G. Mason-Buck, D. Ballard, W. Branicki, A. Amorim, Biowarfare, bioterrorism and biocrime: a historical overview on microbial harmful applications, Forensic Science International 314 (2020) 110366.

[9] Sue Turale, Chutima Meechamnan, Wipada Kunaviktikul, Challenging times: ethics, nursing and the covid-19 pandemic, International Nursing Review 67 (2020) 164–167.

[10] C. Widodo Pitoyo, The collateral impacts of pandemic, Acta Medica Indonesiana 52 (2020) 315–317.

[11] Mary Grace Keating Duley, The next pandemic: anticipating an overwhelmed health care system, The Yale Journal of Biology & Medicine 78 (5) (2005) 355.

[12] Georgina Pujolar, Aida Oliver-Anglès, Ingrid Vargas, María-Luisa Vázquez, Changes in access to health services during the covid-19 pandemic: a scoping review, International Journal of Environmental Research and Public Health 19 (3) (2022) 1749.

[13] William H. McNeill, Plagues and Peoples, Anchor Press/Doubleday, Garden City, NY, 1976.

[14] Lester K. Little (Ed.), Plague and the End of Antiquity: The Pandemic of 541–750, Cambridge University Press, Cambridge, 2008.

[15] Michael Goodich, The Black Death, 1346–1353: The Complete History, Boydell Press, 2006.

[16] Eugenia Tognotti, Lessons from the history of quarantine, from plague to influenza A, Emerging Infectious Diseases 19 (2) (2013) 254–259.

[17] Frank M. Snowden, Epidemics and Society: From the Black Death to the Present, Yale University Press, New Haven, CT, 2024.

[18] John Reynolds, Fortress Europe, global migration & the global pandemic, 2020.

[19] Alfred W. Crosby, Virgin soil epidemics as a factor in the aboriginal depopulation in America, The William and Mary Quarterly: A Magazine of Early American Historyy (1976) 289–299.

[20] Alfred W. Crosby, Virgin soil epidemics as a factor in the aboriginal depopulation in America, The William and Mary Quarterly 33 (2) (1976) 289–299, Third Series.

[21] Louis P. Cain, Elyce J. Rotella, Urbanization, sanitation, and mortality in the progressive era, 1899–1929, in: Standard of Living: Essays on Economics, History, and Religion in Honor of John E. Murray, Springer, 2022, pp. 1–17.

[22] Christopher Hamlin, Cholera: The Biography, Oxford University Press, Oxford, 2011.

[23] Louis Pasteur, Germ theory and its applications to medicine & on the antiseptic principle of the practice of surgery, Great Minds Series (1996).

[24] John M. Barry, The Great Influenza: The Epic Story of the Deadliest Plague in History, Viking, New York, 2006.

[25] Anderson Warwick, Natural histories of infectious disease: ecological vision in twentieth-century biomedical science, Osiris 19 (2004) 39–61.

[26] Anish P. Mahajan, Jennifer N. Sayles, Vishal A. Patel, Robert H. Remien, Sharif R. Sawires, Daniel J. Ortiz, Greg Szekeres, Thomas J. Coates, Stigma in the hiv/aids epidemic: a review of the literature and recommendations for the way forward, AIDS 22 (2008) S67–S79.

[27] Garrett Laurie, The Coming Plague: Newly Emerging Diseases in a World Out of Balance, Farrar, Straus and Giroux, New York, 1995.

[28] Paul Farmer, AIDS and Accusation: Haiti and the Geography of Blame, University of California Press, Berkeley, CA, 1995.

[29] David Quammen, Spillover: Animal Infections and the Next Human Pandemic, W.W. Norton & Company, New York, 2013.

[30] Michael A. Peters, The coming pandemic era, 2022.

[31] Gideon Towett, R. Sterling Snead, Knarik Grigoryan, Julia Marczika, Geographical and practical challenges in the implementation of digital health passports for cross-border covid-19 pandemic management: a narrative review and framework for solutions, Globalization and Health 19 (1) (2023) 98.

[32] World Health Organization, International Health Regulations (2005), third edition, World Health Organization, Geneva, 2016.

[33] World Health Organization, Who director-general's opening remarks at the media briefing on covid-19 – 11 March 2020, March 2020 Accessed 2025-12-15.

[34] Richard Horton, Richard Horton. The COVID-19 Catastrophe. What's Gone Wrong and How to Stop It Happening Again?, Polity Press, Cambridge, 2021.

[35] John Francis Lazenby, The Peloponnesian War, Routledge, 2004.

[36] Thucydides, History of the Peloponnesian War, Volume I: Books 1–2, Number 108 in Loeb Classical Library, Harvard University Press, Cambridge, MA, 1919.

[37] J.C.F. Poole, A.J. Holladay, Thucydides and the plague of Athens, The Classical Quarterly 29 (2) (1979) 282–300.

[38] Denys Lionel Page, Thucydides' description of the great plague at Athens, The Classical Quarterly 3 (3–4) (1953) 97–119.

[39] Burke A. Cunha, The cause of the plague of Athens: plague, typhoid, typhus, smallpox, or measles?, Infectious Disease Clinics of North America 18 (1) (2004) 29–43.

[40] David M. Morens, Robert J. Littman, Epidemiology of the plague of Athens, Transactions of the American Philological Association (1974-) 122 (1992) 271–304.

[41] Francisco Javier Martínez, Political consequences of the plague of Athens, Graeco-Latina Brunensia 22 (1) (2017) 135–146.

[42] Costas Tsiamis, Historical epidemiology of the medieval eastern Mediterranean, in: A Companion to the Environmental History of Byzantium (Brill Companions to the Byzantine World 13), Leiden/Boston, 2024, pp. 109–136.

[43] Paul Slack, The Impact of Plague in Tudor and Stuart England, Clarendon Press, Oxford, 1987.

[44] Monica H. Green, Taking "pandemic" seriously: making the black death global, The Medieval Globe 1 (2014) 27–61.

[45] Samuel K. Cohn Jr., The Black Death Transformed: Disease and Culture in Early Renaissance Europe, Arnold, London, 2003.

[46] Kirsten I. Bos, Verena J. Schuenemann, G. Brian Golding, et al., A draft genome of Yersinia pestis from victims of the black death, Nature 478 (7370) (2011) 506–510.

[47] Bruce M.S. Campbell, Bruce M.S. Campbell, "The Great Transition. Climate, Disease and Society in the Late-Medieval World", Cambridge University Press, Cambridge, 2016.

[48] David Herlihy, The Black Death and the Transformation of the West, Harvard University Press, Cambridge, MA, 1997.
[49] James B. Kaper, Jr. Morris, J. Glenn, Myron M. Levine, Cholera, Clinical Microbiology Reviews 8 (1) (1995) 48–86.
[50] John J. Mekalanos, Eric J. Rubin, Matthew K. Waldor, Cholera: molecular basis for emergence and pathogenesis, FEMS Immunology and Medical Microbiology 18 (4) (1997) 241–248.
[51] Mark Harrison, Contagion: How Commerce Has Spread Disease, Yale University Press, New Haven, CT, 2013.
[52] David Arnold, Colonizing the Body: State Medicine and Epidemic Disease in Nineteenth-Century India, University of California Press, Berkeley, CA, 1994.
[53] Ramamurthy Thandavarayan, Amit Ghosh, A re-look at cholera pandemics from early times to now in the current era of epidemiology, Journal of Disaster Research 16 (1) (2021) 110–117.
[54] Charles E. Rosenberg, The Cholera Years: The United States in 1832, 1849, and 1866, 2 edition, University of Chicago Press, Chicago, 1988.
[55] Antoni Trilla, Guillem Trilla, Carolyn Daer, The 1918 "Spanish flu" in Spain, Clinical Infectious Diseases 8 (11) (2008) 668–673.
[56] Jeffrey K. Taubenberger, David M. Morens, 1918 influenza: the mother of all pandemics, Emerging Infectious Diseases 12 (1) (2006) 15–22.
[57] David M. Morens, Jeffrey K. Taubenberger, Anthony S. Fauci, Predominant role of bacterial pneumonia as a cause of death in pandemic influenza: implications for pandemic influenza preparedness, The Journal of Infectious Diseases 198 (7) (2008) 962–970.
[58] Niall P.A.S. Johnson, Juergen Mueller, Updating the accounts: global mortality of the 1918–1920 "Spanish" influenza pandemic, Bulletin of the History of Medicine 76 (1) (2002) 105–115.
[59] Richard J. Hatchett, Carter E. Mecher, Marc Lipsitch, Public health interventions and epidemic intensity during the 1918 influenza pandemic, Proceedings of the National Academy of Sciences 104 (18) (2007) 7582–7587.
[60] H. Irene Hall, Ruiguang Song, Philip Rhodes, Joseph Prejean, Qian An, Lisa M. Lee, John Karon, Ron Brookmeyer, Edward H. Kaplan, Matthew T. McKenna, et al., Estimation of hiv incidence in the United States, JAMA 300 (5) (2008) 520–529.
[61] Kevin M. De Cock, Harold W. Jaffe, James W. Curran, The evolving epidemiology of hiv/aids, AIDS 26 (10) (2012) 1205–1213.
[62] Paul M. Sharp, Beatrice H. Hahn, Origins of hiv and the aids pandemic, Cold Spring Harbor Perspectives in Medicine 1 (1) (2011) a006841.
[63] Michael S. Gottlieb, Robert Schroff, H.M. Schanker, Joel D. Weisman, Peng T. Fan, Robert A. Wolf, Andrew Saxon, Pneumocystis carinii pneumonia and mucosal candidiasis in previously healthy homosexual men: evidence of a new acquired cellular immunodeficiency, The New England Journal of Medicine 305 (24) (1981) 1425–1431.
[64] Richard Parker, Peter Aggleton, Hiv and aids-related stigma and discrimination: a conceptual framework and implications for action, Social Science & Medicine 57 (1) (2003) 13–24.
[65] Frank J. Palella Jr, Kathleen M. Delaney, Anne C. Moorman, Mark O. Loveless, Jack Fuhrer, Glen A. Satten, Diane J. Aschman, Scott D. Holmberg, HIV Outpatient Study Investigators, Declining morbidity and mortality among patients with advanced human immunodeficiency virus infection, The New England Journal of Medicine 338 (13) (1998) 853–860.
[66] Roy M. Anderson, Christophe Fraser, Azra C. Ghani, et al., Epidemiology, transmission dynamics and control of sars: the 2002–2003 epidemic, Philosophical Transactions - Royal Society. Biological Sciences 359 (1447) (2004) 1091–1105.
[67] Marc Lipsitch, Ted Cohen, Ben Cooper, et al., Transmission dynamics and control of severe acute respiratory syndrome, Science 300 (5627) (2003) 1966–1970.

[68] David L. Heymann, Guénaël R. Rodier, Sars: a global response to an international public health emergency, Philosophical Transactions of the Royal Society of London. Series B, Biological Sciences 359 (1447) (2004) 1127–1129.
[69] Christl A. Donnelly, Azra C. Ghani, Gabriel M. Leung, et al., Epidemiological determinants of spread of causal agent of severe acute respiratory syndrome in Hong Kong, The Lancet 361 (9371) (2003) 1761–1766.
[70] S. Riley, C. Fraser, C.A. Donnelly, A.C. Ghani, L.J. AbuRaddad, A.J. Hedley, G.M. Leung, L.M. Ho, T.H. Lam, T.Q. Thach, P. Chau, K.P. Chan, S.V. Lo, P.Y. Leung, T. Tsang, W. Ho, K.H. Lee, E.M.C. Lau, N.M. Ferguson, R.M. Anderson, Transmission dynamics of the etiological agent of SARS in Hong Kong: impact of public health interventions, Science 300 (5627) (2003) 1961–1966.
[71] Nelson Lee, David Hui, Alan Wu, et al., A major outbreak of severe acute respiratory syndrome in Hong Kong, The New England Journal of Medicine 348 (20) (2003) 1986–1994.
[72] J.S.M. Peiris, C.M. Chu, Vincent C.C. Cheng, et al., Clinical progression and viral load in a community outbreak of coronavirus-associated sars pneumonia: a prospective study, The Lancet 361 (9371) (2003) 1767–1772.
[73] Ignatius T.S. Yu, Yuguo Li, Thomas W. Wong, et al., Evidence of airborne transmission of the severe acute respiratory syndrome virus, The New England Journal of Medicine 350 (17) (2004) 1731–1739.
[74] Harvey V. Fineberg, Pandemic preparedness and response — lessons from the h1n1 influenza of 2009, The New England Journal of Medicine 370 (14) (2014) 1335–1342.
[75] Novel Swine-Origin Influenza A (H1N1) Virus Investigation Team, Emergence of a novel swine-origin influenza A (h1n1) virus in humans, The New England Journal of Medicine 360 (25) (2009) 2605–2615.
[76] Gavin J.D. Smith, Dhanasekaran Vijaykrishna, Justin Bahl, Samantha J. Lycett, Michael Worobey, Oliver G. Pybus, Siu Kit Ma, Chung Lam Cheung, Jayna Raghwani, Samir Bhatt, J.S. Malik Peiris, Yi Guan, Andrew Rambaut, Origins and evolutionary genomics of the 2009 swine-origin h1n1 influenza A epidemic, Nature 459 (2009) 1122–1125.
[77] Rebecca J. Garten, et al., Antigenic and genetic characteristics of swine-origin 2009 a(h1n1) influenza viruses circulating in humans, Science 325 (5937) (2009) 197–201.
[78] Seema Jain, Laurie Kamimoto, Anna M. Bramley, et al., Hospitalized patients with 2009 h1n1 influenza in the United States, April–June 2009, The New England Journal of Medicine 361 (20) (2009) 1935–1944.
[79] Kathy Hancock, Vic Veguilla, Xiuhua Lu, et al., Cross-reactive antibody responses to the 2009 pandemic h1n1 influenza virus, The New England Journal of Medicine 361 (2009) 1945–1952.
[80] Christophe Fraser, Christl A. Donnelly, Simon Cauchemez, William P. Hanage, Maria D. Van Kerkhove, et al., Pandemic potential of a strain of influenza A (h1n1): early findings, Science 324 (5934) (2009) 1557–1561.
[81] Writing Committee of the WHO Consultation on Clinical Aspects of Pandemic (H1N1) 2009 Influenza, Clinical aspects of pandemic 2009 influenza A (h1n1) virus infection, The New England Journal of Medicine 362 (18) (2010) 1708–1719.
[82] Fatimah S. Dawood, A. Danielle Iuliano, Carrie Reed, Martin I. Meltzer, David K. Shay, et al., Estimated global mortality associated with the first 12 months of 2009 pandemic influenza A h1n1 virus circulation: a modelling study, Lancet. Infectious Diseases 12 (9) (2012) 687–695.
[83] The Lancet Global Health. The West Africa Ebola virus disease outbreak: 10 years on. The Lancet Global Health, 2024.
[84] Sylvain Baize, Delphine Pannetier, Lisa Oestereich, Thomas Rieger, Lamine Koivogui, N'Faly Magassouba, et al., Emergence of Zaire Ebola virus disease in Guinea, The New England Journal of Medicine 371 (15) (2014) 1418–1425.
[85] Almudena Marí Saéz, Sabrina Weiss, Kathrin Nowak, Vincent Lapeyre, Fee Zimmermann, Ariane Düx, et al., Investigating the zoonotic origin of the West African Ebola epidemic, EMBO Molecular Medicine 7 (2015) 17–23.
[86] WHO Ebola Response Team, Ebola virus disease in West Africa—the first 9 months of the epidemic and forward projections, The New England Journal of Medicine 371 (16) (2014) 1481–1495.

[87] WHO Ebola Response Team, After Ebola in West Africa—unpredictable risks, preventable epidemics, The New England Journal of Medicine 375 (6) (2016) 587–596.
[88] Monica Gandhi, Deborah S. Yokoe, Diane V. Havlir, Asymptomatic transmission, the Achilles' heel of current strategies to control covid-19, The COVID-19 Reader 382 (22) (2020) 2158–2160.
[89] Ben Hu, Hua Guo, Peng Zhou, Zheng-Li Shi, Characteristics of sars-cov-2 and covid-19, Nature Reviews. Microbiology 19 (3) (2021) 141–154.
[90] Aakriti Gupta, Mahesh V. Madhavan, Kartik Sehrawat, Neha Nair, Sahil Mahajan, Tuje S. Sehrawat, Behnood Bikdeli, Neha Ahluwalia, John C. Ausiello, Elaine Y. Wan, et al., Extrapulmonary manifestations of covid-19, Nature Medicine 26 (7) (2020) 1017–1032.
[91] Trisha Greenhalgh, Jose L. Jimenez, Kimberly A. Prather, Zeynep Tufekci, David Fisman, Robert Schooley, Ten scientific reasons in support of airborne transmission of sars-cov-2, The Lancet 397 (10285) (2021) 1603–1605.
[92] Shirley Sze, Daniel Pan, C.R. Nevill, L.J. Gray, C.A. Martin, J. Nazareth, J.S. Minhas, P. Divall, K. Khunti, K.R. Abrams, et al., Ethnicity and clinical outcomes in covid-19: a systematic review and meta-analysis, EClinicalMedicine 29 (2020) 100630.
[93] Haidong Wang, Katherine R. Paulson, Spencer A. Pease, Stefanie Watson, Haley Comfort, Peng Zheng, Aleksandr Y. Aravkin, Catherine Bisignano, Ryan M. Barber, Tahiya Alam, et al., Estimating excess mortality due to the covid-19 pandemic: a systematic analysis of covid-19-related mortality, 2020–21, The Lancet 399 (10334) (2022) 1513–1536.
[94] William Msemburi, Ariel Karlinsky, Victoria Knutson, Serge Aleshin-Guendel, Somnath Chatterji, Jon Wakefield, The who estimates of excess mortality associated with the covid-19 pandemic, Nature 613 (7942) (2023) 130–137.
[95] Fernando P. Polack, Stephen J. Thomas, Nicholas Kitchin, Judith Absalon, Alejandra Gurtman, Stephen Lockhart, John L. Perez, Gonzalo Pérez Marc, Edson D. Moreira, Cristiano Zerbini, et al., Safety and efficacy of the bnt162b2 mrna covid-19 vaccine, The New England Journal of Medicine 383 (27) (2021) 2603–2615.
[96] Victoria Pilkington, Sarai Mirjam Keestra, Andrew Hill, Global covid-19 vaccine inequity: failures in the first year of distribution and potential solutions for the future, Frontiers in Public Health 10 (2022) 821117.
[97] Héctor J. Pérez, Rainer Reisenzein, On Jon snow's death: plot twist and global fandom in game of thrones, Culture & Psychology 26 (3) (2020) 384–400.
[98] Peter Vinten-Johansen, Howard Brody, Nigel Paneth, Stephen Rachman, Michael Rip, David Zuck, Cholera, Chloroform, and the Science of Medicine: a Life of John Snow, Oxford University Press, 2003.
[99] John Snow, On the Mode of Communication of Cholera, 2 edition, John Churchill, London, 2023.
[100] David R. Bellhouse, A new look at Halley's life table, Journal of the Royal Statistical Society. Series A. Statistics in Society 174 (3) (2011) 823–832.
[101] Halley Edmond, An estimate of the degrees of the mortality of mankind, drawn from curious tables of the births and funerals at the city of breslau, with an attempt to ascertain the price of annuities upon lives, Philosophical Transactions of the Royal Society of London 17 (1693) 596–610.
[102] Nicolas Bacaër, Daniel Bernoulli, d'alembert and the inoculation of smallpox (1760), in: A Short History of Mathematical Population Dynamics, Springer, 2011, pp. 21–30.
[103] Daniel Bernoulli, Essai d'une nouvelle analyse de la mortalité causée par la petite vérole, et des avantages de l'inoculation pour la prévenir, Histoire de l'Académie Royale des Sciences (Paris) (1760) 1–45, Published in the 1760 volume, printed in 1766.
[104] Jacqueline Isero, Quarantine in 18th and 19th century England: epidemics and empires, The UC Santa Barbara Undergraduate Journal of History 1 (2) (2021).
[105] Alexander D. Langmuir, William farr: founder of modern concepts of surveillance, International Journal of Epidemiology 5 (1) (1976) 13–18.
[106] William Farr, Vital Statistics: A Memorial Volume of Selections from the Reports and Writings of William Farr, The Sanitary Institute, London, 1977.

[107] Adewunmi James Falode, Joshua Olusegun Bolarinwa, Moses Yakubu, History of pandemics in the twentieth and twenty-first century, Synesis: Journal of Humanities and Social Sciences 2 (1) (2021) 9–26.
[108] William Heaton Hamer, The Milroy Lectures on Epidemic Diseases in England: The Evidence of Variability and of Persistency of Type, Bedford Press, London, 1906.
[109] Ronald Ross, The Prevention of Malaria, 2 edition, John Murray, London, 1911.
[110] William Ogilvy Kermack, Anderson G. McKendrick, A contribution to the mathematical theory of epidemics, Proceedings of the Royal Society of London. Series A, Containing Papers of a Mathematical and Physical Character 115 (772) (1927) 700–721.
[111] William Ogilvy Kermack, Anderson G. McKendrick, Contributions to the mathematical theory of epidemics. II.—the problem of endemicity, Proceedings of the Royal Society of London. Series A, Containing Papers of a Mathematical and Physical Character 138 (834) (1932) 55–83.
[112] William Ogilvy Kermack, Anderson G. McKendrick, Contributions to the mathematical theory of epidemics. III.—further studies of the problem of endemicity, Proceedings of the Royal Society of London. Series A, Containing Papers of a Mathematical and Physical Character 141 (843) (1933) 94–122.
[113] Lukas Engelmann, A box, a trough and marbles: how the Reed-frost epidemic theory shaped epidemiological reasoning in the 20th century, History and Philosophy of the Life Sciences 43 (3) (2021) 105.
[114] Lukas Engelmann, A box, a trough and marbles: how the Reed-frost epidemic theory shaped epidemiological reasoning in the 20th century, History and Philosophy of the Life Sciences 43 (2021) 105.
[115] Alfredo Morabia, Snippets from the past: the evolution of wade hampton frost's epidemiology as viewed from the American journal of hygiene/epidemiology, American Journal of Epidemiology 178 (7) (2013) 1013–1019.
[116] Alexander D. Langmuir, The surveillance of communicable diseases of national importance, The New England Journal of Medicine 268 (1963) 182–192.
[117] Stephen B. Thacker, Ruth L. Berkelman, Public health surveillance in the United States, Epidemiologic Reviews 10 (1988) 164–190.
[118] Roy M. Anderson, Robert M. May, Infectious Diseases of Humans: Dynamics and Control, Oxford University Press, Oxford, 1992.
[119] Abba B. Gumel, Shigui Ruan, Troy Day, James Watmough, Fred Brauer, Pauline van den Driessche, Dave Gabrielson, Chris Bowman, Murray E. Alexander, Sten Ardal, et al., Modelling strategies for controlling sars outbreaks, Proceedings - Royal Society. Biological Sciences 271 (1554) (2004) 2223–2232.
[120] Chris Dye, Nigel Gay, Modeling the sars epidemic, Science 300 (5627) (2003) 1884–1885.
[121] David L. Heymann, The international response to the outbreak of sars in 2003, Philosophical Transactions - Royal Society. Biological Sciences 359 (1447) (2004) 1127–1129.
[122] Erdem Acar, Gamze Bayrak, Yongsu Jung, Ikjin Lee, Palaniappan Ramu, Suja Shree Ravichandran, Modeling, analysis, and optimization under uncertainties: a review, Structural and Multidisciplinary Optimization 64 (5) (2021) 2909–2945.
[123] Teddy Lazebnik, The economical-ecological benefits of matching non-matching socks, arXiv preprint, arXiv:2602.18221, 2026.
[124] Neil M. Ferguson, Derek A.T. Cummings, Christophe Fraser, James C. Cajka, Philip C. Cooley, Donald S. Burke, Strategies for mitigating an influenza pandemic, Nature 442 (2006) 448–452.
[125] Marcel Salathé, Linus Bengtsson, Todd J. Bodnar, Devon D. Brewer, John S. Brownstein, Caroline Buckee, Ellsworth M. Campbell, Ciro Cattuto, Shashank Khandelwal, Patricia L. Mabry, Alessandro Vespignani, Digital epidemiology, PLoS Computational Biology 8 (7) (2012) e1002616.
[126] Pedro Soto-Acosta, Covid-19 pandemic: shifting digital transformation to a high-speed gear, Information Systems Management 37 (4) (2020) 260–266.
[127] Olcay Akman, Sudipa Chauhan, Aditi Ghosh, Sara Liesman, Edwin Michael, Anuj Mubayi, Rebecca Perlin, Padmanabhan Seshaiyer, Jai Prakash Tripathi, The hard lessons and shifting modeling trends of covid-19 dynamics: multiresolution modeling approach, Bulletin of Mathematical Biology 84 (1) (2022) 3.

[128] Seth Flaxman, Swapnil Mishra, Axel Gandy, H. Juliette T. Unwin, Thomas A. Mellan, Helen Coupland, et al., Estimating the effects of non-pharmaceutical interventions on COVID-19 in Europe, Nature 584 (2020) 257–261.

[129] Bryan T. Grenfell, Oliver G. Pybus, Julia R. Gog, James L.N. Wood, Janet M. Daly, John A. Mumford, Edward C. Holmes, Unifying the epidemiological and evolutionary dynamics of pathogens, Science 303 (5656) (2004) 327–332.

[130] Oliver G. Pybus, Christophe Fraser, Andrew Rambaut, Evolutionary epidemiology: preparing for an age of genomic plenty, Philosophical Transactions - Royal Society. Biological Sciences 368 (1614) (2013) 20120193.

[131] Fred Brauer, Carlos Castillo-Chavez, Zhilan Feng, et al., Mathematical Models in Epidemiology, vol. 32, Springer, 2019.

[132] Sander Greenland, Rhian Daniel, Neil Pearce, Outcome modelling strategies in epidemiology: traditional methods and basic alternatives, International Journal of Epidemiology 45 (2) (2016) 565–575.

[133] Sebastian Funk, Marcel Salathé, Vincent A.A. Jansen, Modelling the influence of human behaviour on the spread of infectious diseases: a review, Journal of the Royal Society Interface 7 (50) (2010) 1247–1256.

[134] Eli P. Fenichel, Carlos Castillo-Chavez, Maria G. Ceddia, Gerardo Chowell, Paula A. Gonzalez Parra, Garth Holloway, Richard Horan, Benjamin Morin, Adaptive human behavior in epidemiological models, Proceedings of the National Academy of Sciences 108 (15) (2011) 6306–6311.

[135] Jay J. Van Bavel, Katherine Baicker, Paulo S. Boggio, Valerio Capraro, Aleksandra Cichocka, Mina Cikara, Molly J. Crockett, Alia J. Crum, Karen M. Douglas, James N. Druckman, John Drury, Oeindrila Dube, Naomi Ellemers, Eli J. Finkel, James H. Fowler, Michele Gelfand, Shihui Han, S. Alexander Haslam, Jolanda Jetten, Shinobu Kitayama, Dean Mobbs, Lucy E. Napper, Dominic J. Packer, Gordon Pennycook, Ellen Peters, Robert E. Petty, David G. Rand, Stephen D. Reicher, Simone Schnall, Azim Shariff, Linda J. Skitka, Sandra S. Smith, Cass R. Sunstein, Nassim Tabri, Joshua A. Tucker, Sander van der Linden, Paul van Lange, Kim A. Weeden, Michael J.A. Wohl, Jamil Zaki, Sean R. Zion, Robb Willer, Using social and behavioural science to support COVID-19 pandemic response, Nature Human Behaviour 4 (2020) 460–471.

[136] Mathilde Frérot, Annick Lefebvre, Simon Aho, Patrick Callier, Karine Astruc, Ludwig Serge Aho Glélé, What is epidemiology? Changing definitions of epidemiology 1978-2017, PLoS ONE 13 (12) (2018) e0208442.

[137] Odo Diekmann, Johan A.P. Heesterbeek, Johan A.J. Metz, On the definition and the computation of the basic reproduction ratio R0 in models for infectious diseases in heterogeneous populations, Journal of Mathematical Biology 28 (4) (1990) 365–382.

[138] J.A.P. Heesterbeek, A brief history of r_0 and a recipe for its calculation, Acta Biotheoretica 50 (3) (2002) 189–204.

[139] Michael Gilpin, Metapopulation Dynamics: Empirical and Theoretical Investigations, Academic Press, 2012.

[140] Joël Mossong, Niel Hens, Mark Jit, Philippe Beutels, Kari Auranen, Rafael Mikolajczyk, Marco Massari, Stefania Salmaso, Gianpaolo Scalia Tomba, Jacco Wallinga, Janneke Heijne, Malgorzata Sadkowska-Todys, Magdalena Rosińska, W. John Edmunds, Social contacts and mixing patterns relevant to the spread of infectious diseases, PLoS Medicine 5 (3) (2008) e74.

[141] Kiesha Prem, Alex R. Cook, Mark Jit, Projecting social contact matrices in 152 countries using contact surveys and demographic data, PLoS Computational Biology 13 (9) (2017) e1005697.

[142] Richard A. Stein, Super-spreaders in infectious diseases, International Journal of Infectious Diseases 15 (8) (2011) e510–e513.

[143] Amir Teicher, Super-spreaders: a historical review, Lancet. Infectious Diseases 23 (10) (2023) e409–e417.

[144] Romualdo Pastor-Satorras, Claudio Castellano, Piet Van Mieghem, Alessandro Vespignani, Epidemic processes in complex networks, Reviews of Modern Physics 87 (3) (2015) 925–979.

[145] Duygu Balcan, Vittoria Colizza, Bruno Gonçalves, Hao Hu, Jose J. Ramasco, Alessandro Vespignani, Multiscale mobility networks and the spatial spreading of infectious diseases, Proceedings of the National Academy of Sciences 106 (51) (2009) 21484–21489.
[146] Shixiang Zhu, Alexander Bukharin, Liyan Xie, Mauricio Santillana, Shihao Yang, Yao Xie, High-resolution spatio-temporal model for county-level covid-19 activity in the us, ACM Transactions on Management Information Systems (TMIS) 12 (4) (2021) 1–20.
[147] D. Temitope, Timothy Oyedotun, Stephan Moonsammy, Spatiotemporal variation of covid-19 and its spread in South America: a rapid assessment, Annals of the Association of American Geographers 111 (6) (2021) 1868–1879.
[148] Jacco Wallinga, Peter Teunis, Different epidemic curves for severe acute respiratory syndrome reveal similar impacts of control measures, American Journal of Epidemiology 160 (6) (2004) 509–516.
[149] Anne Cori, Neil M. Ferguson, Christophe Fraser, Simon Cauchemez, A new framework and software to estimate time-varying reproduction numbers during epidemics, American Journal of Epidemiology 178 (9) (2013) 1505–1512.
[150] Matt J. Keeling, Pejman Rohani, Modeling Infectious Diseases in Humans and Animals, Princeton University Press, Princeton, NJ, 2008.
[151] Arturo Casadevall, Liise-anne Pirofski, Host-pathogen interactions: basic concepts of microbial commensalism, colonization, infection, and disease, Infection and Immunity 68 (12) (2000) 6511–6518.
[152] Merrill Singer, Pathogen-pathogen interaction: a syndemic model of complex biosocial processes in disease, Virulence 1 (1) (2010) 10–18.
[153] David J. Weber, William A.Rutala, Biological Basis of Infectious Disease Epidemiology, Oxford University Press, New York, 2001.
[154] Justin Lessler, Nicholas G. Reich, Ron Brookmeyer, Trish M. Perl, Kenrad E. Nelson, Derek A.T. Cummings, Incubation periods of acute respiratory viral infections: a systematic review, Lancet. Infectious Diseases 9 (5) (2009) 291–300.
[155] Christophe Fraser, Steven Riley, Roy M. Anderson, Neil M. Ferguson, Factors that make an infectious disease outbreak controllable, Proceedings of the National Academy of Sciences 101 (16) (2004) 6146–6151.
[156] Thomas P. Weber, Nikolaos I. Stilianakis, Inactivation of influenza A viruses in the environment and modes of transmission: a critical review, The Journal of Infection 57 (5) (2008) 361–373.
[157] World Health Organization, Compendium of WHO and other UN guidance in health and environment, 2024 update, World Health Organization, 2024.
[158] Alan S. Perelson, Ruy M. Ribeiro, Modeling the within-host dynamics of hiv infection, BMC Biology 11 (1) (2013) 96.
[159] Zhilan Feng, Jorge Velasco-Hernandez, Brenda Tapia-Santos, Maria Conceição, A. Leite, A model for coupling within-host and between-host dynamics in an infectious disease, Nonlinear Dynamics 68 (3) (2012) 401–411.
[160] James O. Lloyd-Smith, S.J. Schreiber, P.E. Kopp, W.M. Getz, Superspreading and the effect of individual variation on disease emergence, Nature 438 (2005) 355–359.
[161] Troy Day, Stephen R. Proulx, A general theory for the evolutionary dynamics of virulence, The American Naturalist 163 (4) (2004) E40–E63.
[162] Samuel Alizon, Amy Hurford, Nicole Mideo, Minus van Baalen, Virulence evolution and the trade-off hypothesis: history, current state of affairs and the future, Journal of Evolutionary Biology 22 (2) (2009) 245–259.
[163] Kenneth M. Murphy, Casey Weaver, Janeway's Immunobiology, 9 edition, Garland Science, 2016.
[164] Joel Henrique Ellwanger, José Artur Bogo Chies, Zoonotic spillover: understanding basic aspects for better prevention, Genetics and Molecular Biology 44 (2021) e20200355.
[165] Bharti Bhatia, Sudipta Sonar, Seema Khan, Jayanta Bhattacharya, Pandemic-proofing: intercepting zoonotic spillover events, Pathogens 13 (12) (2024) 1067.

[166] Raina K. Plowright, Colin R. Parrish, Hamish McCallum, Peter J. Hudson, Albert I. Ko, Andrea L. Graham, James O. Lloyd-Smith, Pathways to zoonotic spillover, Nature Reviews. Microbiology 15 (2017) 502–510.

[167] Hans Heesterbeek, Roy M. Anderson, Viggo Andreasen, Shweta Bansal, Daniela De Angelis, Chris Dye, Ken T.D. Eames, W. John Edmunds, Simon D.W. Frost, Sebastian Funk, T. Deirdre Hollingsworth, Thomas House, Valerie Isham, Petra Klepac, Justin Lessler, James O. Lloyd-Smith, C. Jessica, E. Metcalf, Denis Mollison, Lorenzo Pellis, Juliet R.C. Pulliam, Mick Roberts, Cécile Viboud, Isaac Newton Institute IDD Collaboration, Modeling infectious disease dynamics in the complex landscape of global health, Science 347 (6227) (2015) aaa4339.

[168] Noa Vardi, Teddy Lazebnik, The causal role of lockdowns in covid-19: conclusions from daily epidemiological, psychological, and sociological data, Psychiatric Quarterly 94 (2) (2023) 321–341.

[169] Gregory Knell, Michael C. Robertson, Erin E. Dooley, Katie Burford, Karla S. Mendez, Health behavior changes during covid-19 pandemic and subsequent "stay-at-home" orders, International Journal of Environmental Research and Public Health 17 (17) (2020) 6268.

[170] Deborah Salon, Matthew Wigginton Conway, Denise Capasso da Silva, Rishabh Singh Chauhan, Sybil Derrible, Abolfazl Mohammadian, Sara Khoeini, Nathan Parker, Laura Mirtich, Ali Shamshiripour, et al., The potential stickiness of pandemic-induced behavior changes in the United States, Proceedings of the National Academy of Sciences 118 (27) (2021) e2106499118.

[171] Lucia Savadori, Marco Lauriola, Risk perception and protective behaviors during the rise of the covid-19 outbreak in Italy, Frontiers in Psychology 11 (2021) 577331.

[172] Amos Tversky, Daniel Kahneman, Judgment under uncertainty: heuristics and biases, Science 185 (4157) (1974) 1124–1131.

[173] Icek Ajzen, The theory of planned behavior, Organizational Behavior and Human Decision Processes 50 (2) (1991) 179–211.

[174] Thomas J. Bollyky, Michael Bang Petersen, A practical agenda for incorporating trust into pandemic preparedness and response, Bulletin of the World Health Organization 102 (6) (2024) 440.

[175] Michael Siegrist, Alexandra Zingg, The role of public trust during pandemics, European psychologist (2014).

[176] World Health Organization Regional Office for Europe, Vaccination and trust: How concerns arise and the role of communication in mitigating crises, Technical report, World Health Organization Regional Office for Europe, Copenhagen, 2017.

[177] Soroush Vosoughi, Deb Roy, Sinan Aral, The spread of true and false news online, Science 359 (6380) (2018) 1146–1151.

[178] World Health Organization Regional Office for Europe, Pandemic fatigue: reinvigorating the public to prevent COVID-19: policy framework for supporting pandemic prevention and management, Technical report, World Health Organization Regional Office for Europe, Copenhagen, 2020.

[179] Pierre-Yves Geoffard, Tomas J. Philipson, Rational epidemics and their public control, International Economic Review 37 (3) (1996) 603–624.

[180] Tomas J. Philipson, Economic epidemiology and infectious diseases, Technical Report 7037, National Bureau of Economic Research, 1999.

[181] Teddy Lazebnik, Labib Shami, Svetlana Bunimovich-Mendrazitsky, Spatio-temporal influence of non-pharmaceutical interventions policies on pandemic dynamics and the economy: the case of covid-19, Economic Research-Ekonomska Istraživanja 35 (1) (2022) 1833–1861.

[182] Astha Sanjeev Gupta, Jaydeep Mukherjee, Long-term changes in consumers' shopping behavior post-pandemic: an exploratory study, International Journal of Retail & Distribution Management 50 (12) (2022) 1518–1534.

[183] Elliane Irani, Atsadaporn Niyomyart, Ronald L. Hickman Jr., Family caregivers' experiences and changes in caregiving tasks during the covid-19 pandemic, Clinical Nursing Research 30 (7) (2021) 1088–1097.

[184] Sung-Eun Kang, YooHee Hwang, Choong-Ki Lee, Yae-Na Park, Roles of travel and leisure in quality of life during the covid-19 pandemic, Leisure Studies 41 (3) (2022) 326–340.

[185] Sebastian Galiani, Pandemic economics, Journal of Economic Behavior & Organization 193 (2021) 269.

[186] Mark Gersovitz, Jeffrey S. Hammer, The economical control of infectious diseases, The Economic Journal 114 (492) (2004) 1–27.

[187] Martin S. Eichenbaum, Sergio Rebelo, Mathias Trabandt, The macroeconomics of epidemics, The Review of Financial Studies 34 (11) (2021) 5149–5187.

[188] Labib Shami, Teddy Lazebnik, Financing and managing epidemiological-economic crises: are we ready for another outbreak?, Journal of Policy Modeling 45 (1) (2023) 74–89.

[189] Stefan Pichler, Katherine Wen, Nicolas R. Ziebarth, Covid-19 emergency sick leave has helped flatten the curve in the United States, Health Affairs 39 (12) (2020) 2197–2204.

[190] Hua Zhang, Chunyan Luo, Cheng Yang, Yanyan Ouyang, Xiuxian Deng, Risk-driven or authority-based? Unraveling public policy compliance during the pandemic in China, Frontiers in Public Health 13 (2025) 1718039.

[191] Austan Goolsbee, Chad Syverson, Fear, lockdown, and diversion: comparing drivers of pandemic economic decline 2020, Journal of Public Economics 193 (2021) 104311.

[192] Raj Chetty, John N. Friedman, Michael Stepner, Opportunity Insights Team, The economic impacts of covid-19: evidence from a new public database built using private sector data, The Quarterly Journal of Economics 139 (2) (2024) 829–889.

[193] Jill E. Yavorsky, Yue Qian, Amanda C. Sargent, The gendered pandemic: the implications of covid-19 for work and family, in: Working in America, Routledge, 2022, pp. 305–317.

[194] International Monetary Fund, World economic outlook, April 2020, Technical report, International Monetary Fund, April 2020.

[195] Richard Baldwin, Beatrice Weder di Mauro (Eds.), Survivalist Economics in the Time of COVID-19, CEPR Press, 2021.

[196] Teddy Lazebnik, Labib Shami, Svetlana Bunimovich-Mendrazitsky, Intervention policy influence on the effect of epidemiological crisis on industry-level production through input–output networks, Socio-Economic Planning Sciences 87 (2023) 101553.

[197] Teddy Lazebnik, Evaluating supply chain resilience during pandemic using agent-based simulation, Physica A: Statistical Mechanics and its Applications 674 (2025) 130780.

[198] David Baqaee, Emmanuel Farhi, Supply and demand in disaggregated Keynesian economies with an application to the covid-19 crisis, American Economic Review 112 (5) (2022) 1397–1436.

[199] Stefanie Stantcheva, Inequalities in the times of a pandemic, Technical Report 29657, National Bureau of Economic Research, 2022.

[200] International Labour Organization, Ilo monitor: Covid-19 and the world of work. Fourth edition. Updated estimates and analysis, Technical report, International Labour Organization, May 2020.

[201] D. Acemoglu, V. Chernozhukov, I. Werning, M.D. Whinston, Optimal targeted lockdowns in a multigroup sir model, American Economic Review: Insights 3 (4) (2021) 487–502.

[202] W. Kip Viscusi, Joseph E. Aldy, The value of a statistical life: a critical review of market estimates throughout the world, Journal of Risk and Uncertainty 27 (1) (2003) 5–76.

[203] Michael Greenstone, Vishan Nigam, Does social distancing matter?, Technical Report 2020-26, Becker Friedman Institute for Research in Economics, 2020.

[204] Inga Holmdahl, Caroline O. Buckee, Wrong but useful—what covid-19 epidemiologic models can and cannot tell us, The New England Journal of Medicine 383 (4) (2020) 303–305.

[205] Andrea Saltelli, Gabriele Bammer, Isabelle Bruno, Erica Charters, Monica Di Fiore, Emmanuel Didier, Wendy Nelson Espeland, John Kay, Samuele Lo Piano, Deborah Mayo, Roger Pielke Jr, Tommaso Portaluri, Theodore M. Porter, Arnald Puy, Ismael Rafols, Jerome R. Ravetz, Erik Reinert, Daniel Sarewitz, Philip B. Stark, Andrew Stirling, Jeroen van der Sluijs, Paolo Vineis, Five ways to ensure that models serve society: a manifesto, Nature 582 (7813) (2020) 482–484.

[206] World Health Organization, Communicating risk in public health emergencies: a WHO guideline for emergency risk communication (erc) policy and practice, Technical report, World Health Organization, Geneva, 2018.
[207] Nicholas P. Jewell, Joseph A. Lewnard, On the use of the reproduction number for sars-cov-2: estimation, misinterpretations and relationships with other ecological measures, Journal of the Royal Statistical Society. Series A. Statistics in Society 185 (Supplement_1) (2022) S16–S27.
[208] Teddy Lazebnik, Uri Itai, Bounding pandemic spread by heat spread, Journal of Engineering Mathematics 138 (1) (2023) 6.
[209] Alexandra Nikolaeva, Jenny Versnel, Analytical observational study evaluating global pandemic preparedness and the effectiveness of early covid-19 responses in Ethiopia, Nigeria, Singapore, South Korea, Sweden, Taiwan, UK and USA, BMJ Open 12 (2) (2022) e053374.
[210] Caitlin M. Rivers, Sarah Polishchuk, Julie P. Potter, Deron Mande, Imtiaz Biswas, et al., Using outbreak science to strengthen the use of models during epidemics, Nature Communications 10 (2019) 3102.
[211] Ruth McCabe, Mara D. Kont, Nora Schmit, Charles Whittaker, Alessandra Løchen, Patrick G.T. Walker, Azra C. Ghani, Neil M. Ferguson, Peter J. White, Christl A. Donnelly, Oliver J. Watson, Communicating uncertainty in epidemic models, Epidemics 37 (2021) 100520.
[212] Marina Banuet-Martinez, Yang Yang, Behnaz Jafari, Avneet Kaur, Zahid A. Butt, Helen H. Chen, Svetlana Yanushkevich, Iain R. Moyles, Jane M. Heffernan, Chapin S. Korosec, Monkeypox: a review of epidemiological modelling studies and how modelling has led to mechanistic insight, Epidemiology and Infection 151 (2023) e121.
[213] Moritz U.G. Kraemer, Joseph L.-H. Tsui, Serina Y. Chang, Spyros Lytras, Mark P. Khurana, Samantha Vanderslott, Sumali Bajaj, Neil Scheidwasser, Jacob Liam Curran-Sebastian, Elizaveta Semenova, et al., Artificial intelligence for modelling infectious disease epidemics, Nature 638 (8051) (2025) 623–635.
[214] Herbert W. Hethcote, The mathematics of infectious diseases, SIAM Review 42 (4) (2000) 599–653.
[215] Linda J.S. Allen, An introduction to stochastic epidemic models, in: Fred Brauer, Pauline van den Driessche, Jianhong Wu (Eds.), Mathematical Epidemiology, in: Lecture Notes in Mathematics, vol. 1945, Springer, Berlin, 2008.
[216] Teddy Lazebnik, Computational applications of extended sir models: a review focused on airborne pandemics, Ecological Modelling 483 (2023) 110422.
[217] Teddy Lazebnik, Svetlana Bunimovich-Mendrazitsky, Leonid Shaikhet, Novel method to analytically obtain the asymptotic stable equilibria states of extended sir-type epidemiological models, Symmetry 13 (7) (2021) 1120.
[218] Ariel Alexi, Ariel Rosenfeld, Teddy Lazebnik, A security games inspired approach for distributed control of pandemic spread, Advanced Theory and Simulations 6 (2) (2023) 2200631.
[219] Teddy Lazebnik, Ariel Alexi, High resolution spatio-temporal model for room-level airborne pandemic spread, Mathematics 11 (2) (2023) 426.
[220] Odo Diekmann, Johan A.P. Heesterbeek, Michael G. Roberts, The construction of next-generation matrices for compartmental epidemic models, Journal of the Royal Society Interface 7 (47) (2010) 873–885.
[221] Carles Bretó, Daihai He, Edward L. Ionides, Aaron A. King, Time series analysis via mechanistic models, Annals of Applied Statistics 3 (1) (2009) 319–348.
[222] Aaron A. King, Dao Nguyen, Edward L. Ionides, Statistical inference for partially observed Markov processes via the R package pomp, Journal of Statistical Software 69 (12) (2016) 1–43.
[223] Pauline van den Driessche, James Watmough, Reproduction numbers and sub-threshold endemic equilibria for compartmental models of disease transmission, Mathematical Biosciences 180 (2002) 29–48.
[224] Andrew McMahon, Nicole C. Robb, et al., Reinfection with sars-cov-2: discrete sir (susceptible, infected, recovered) modeling using empirical infection data, JMIR Public Health and Surveillance 6 (4) (2020) e21168.
[225] Alessandro S. Barros, Suani T.R. Pinho, Stochastic dynamics for reinfection by transmitted diseases, Physical Review E 95 (6) (2017) 062135.

[226] Alex Pelliccione, Sharareh Modaressi, Bruce Fireman, Evan Layefsky, Nicola P. Klein, Mitchell N. Luu, Ousseny Zerbo, Trends in gonorrhea and chlamydia testing and infections across the covid-19 pandemic in adolescents and young adults in an integrated health system, Journal of Adolescent Health 75 (6) (2024) 952–957.

[227] Justin Lessler, Nicholas G. Reich, Ron Brookmeyer, Trish M. Perl, Kenrad E. Nelson, Derek AT Cummings, Incubation periods of acute respiratory viral infections: a systematic review, Lancet. Infectious Diseases 9 (5) (2009) 291–300.

[228] Victor Virlogeux, Juan Yang, Vicky J. Fang, Luzhao Feng, Tim K. Tsang, Hui Jiang, Peng Wu, Jiandong Zheng, Eric H.Y. Lau, Ying Qin, et al., Association between the severity of influenza A (h7n9) virus infections and length of the incubation period, PLoS ONE 11 (2) (2016) e0148506.

[229] Tove L. Fitzgerald, David N. Durrheim, Tony D. Merritt, Christopher Birch, Thomas Tran, Measles with a possible 23 day incubation period, Communicable Diseases Intelligence Quarterly Report 36 (3) (2012) E277–E280.

[230] Teddy Lazebnik, Svetlana Bunimovich-Mendrazitsky, The signature features of covid-19 pandemic in a hybrid mathematical model—implications for optimal work–school lockdown policy, Advanced Theory and Simulations 4 (5) (2021) 2000298.

[231] Ellen Rafferty, Wade McDonald, Weicheng Qian, Nathaniel D. Osgood, Alexander Doroshenko, Evaluation of the effect of chickenpox vaccination on shingles epidemiology using agent-based modeling, PeerJ 6 (2018) e5012.

[232] Jacco Wallinga, Marc Lipsitch, How generation intervals shape the relationship between growth rates and reproductive numbers, Proceedings of the Royal Society of London. Series B, Biological Sciences 274 (1609) (2007) 599–604.

[233] Alun L. Lloyd, Destabilization of epidemic models with the inclusion of realistic distributions of infectious periods, Proceedings of the Royal Society of London. Series B, Biological Sciences 60 (1) (2001) 59–71.

[234] John C. Butcher, A.D. Heard, Stability of numerical methods for ordinary differential equations, Numerical Algorithms 31 (1) (2002) 59–73.

[235] John Charles Butcher, Numerical Methods for Ordinary Differential Equations, John Wiley & Sons, 2016.

[236] George Macdonald, Epidemiology and Control of Malaria, Oxford University Press, London, 1957.

[237] Shigui Ruan, Dongmei Xiao, John C. Beier, On the delayed Ross–Macdonald model for malaria transmission, Bulletin of Mathematical Biology 70 (4) (2008) 1098–1114.

[238] David J.D. Earn, Pejman Rohani, Benjamin M. Bolker, Bryan T. Grenfell, A simple model for complex dynamical transitions in epidemics, Science 287 (5453) (2000) 667–670.

[239] Maurice S. Bartlett, Deterministic and stochastic models for recurrent epidemics, in: Proceedings of the Third Berkeley Symposium on Mathematical Statistics and Probability, Berkeley, vol. 4, 1956, p. 109.

[240] Mick Roberts, Viggo Andreasen, Alun Lloyd, Lorenzo Pellis, Nine challenges for deterministic epidemic models, Epidemics 10 (2015) 49–53.

[241] Wilfried Gabriel, Reinhard Bürger, Survival of small populations under demographic stochasticity, Theoretical Population Biology 41 (1) (1992) 44–71.

[242] Håkan Andersson, Tom Britton, Stochastic Epidemic Models and Their Statistical Analysis, Lecture Notes in Statistics, Springer, New York, 2002.

[243] Mario Pineda-Krch, Gillespiessa: implementing the gillespie stochastic simulation algorithm in R, Journal of Statistical Software 25 (2008) 1–18.

[244] Mick G. Roberts, J.A.P. Heesterbeek, Characterizing the next-generation matrix and basic reproduction number in ecological epidemiology, Journal of Mathematical Biology 66 (4) (2013) 1045–1064.

[245] Odo Diekmann, Johan Andre Peter Heesterbeek, Michael G. Roberts, The construction of next-generation matrices for compartmental epidemic models, Journal of the Royal Society Interface 7 (47) (2010) 873–885.

[246] Johannes Alt, László Erdős, Torben Krüger, Spectral radius of random matrices with independent entries, Probability and Mathematical Physics 2 (2) (2021) 221–280.

[247] Matt J. Keeling, Ken T.D. Eames, Networks and epidemic models, Journal of the Royal Society Interface 2 (4) (2005) 295–307.

[248] Denis Mollison, Spatial contact models for ecological and epidemic spread, Journal of the Royal Statistical Society, Series B, Statistical Methodology 39 (3) (1977) 283–326.

[249] Matt J. Keeling, The effects of local spatial structure on epidemiological invasions, Proceedings of the Royal Society of London. Series B, Biological Sciences 266 (1421) (1999) 859–867.

[250] Vittoria Colizza, Alain Barrat, Marc Barthélemy, Alessandro Vespignani, The role of the airline transportation network in the prediction and predictability of global epidemics, Proceedings of the National Academy of Sciences 103 (7) (2006) 2015–2020.

[251] Lars Hufnagel, Dirk Brockmann, Theo Geisel, Forecast and control of epidemics in a globalized world, Proceedings of the National Academy of Sciences 101 (42) (2004) 15124–15129.

[252] Duygu Balcan, Vittoria Colizza, Bruno Gonçalves, Hao Hu, José J. Ramasco, Alessandro Vespignani, Multiscale mobility networks and the spatial spreading of infectious diseases, Proceedings of the National Academy of Sciences 106 (51) (2009) 21484–21489.

[253] Dirk Brockmann, Dirk Helbing, The hidden geometry of complex, network-driven contagion phenomena, Science 342 (6164) (2013) 1337–1342.

[254] Filippo Simini, Marta C. González, Amos Maritan, Albert-László Barabási, A universal model for mobility and migration patterns, Nature 484 (7392) (2012) 96–100.

[255] Kyra H. Grantz, Hannah R. Meredith, Derek A.T. Cummings, C. Jessica E. Metcalf, Bryan T. Grenfell, John R. Giles, Sarah Mehta, Shanon Solomon, Alain Labrique, Nishant Kishore, Caroline O. Buckee, Amy Wesolowski, The use of mobile phone data to inform analysis of COVID-19 pandemic epidemiology, Nature Communications 11 (2020) 4961.

[256] Moritz U.G. Kraemer, Chia-Hung Yang, Bernardo Gutierrez, Chieh-Hsi Wu, Brennan Klein, David M. Pigott, et al., The effect of human mobility and control measures on the covid-19 epidemic in China, Science 368 (6490) (2020) 493–497.

[257] Serina Chang, Emma Pierson, Pang Wei Koh, Jaline Gerardin, Beth Redbird, David Grusky, Jure Leskovec, Mobility network models of COVID-19 explain inequities and inform reopening, Nature 589 (7840) (2021) 82–87.

[258] A. Berke, R. Doorley, L. Alonso, V. Arroyo, M. Pons, K. Larson, Using mobile phone data to estimate dynamic population changes and improve the understanding of a pandemic: a case study in andorra, PLoS ONE 17 (4) (2022) e0264860.

[259] Jean-Christophe Foltête, Patrick Giraudoux, A graph-based approach to investigating the influence of the landscape on population spread processes, Ecological Indicators 18 (2012) 684–692.

[260] Emily S. Minor, Dean L. Urban, Graph theory as a proxy for spatially explicit population models in conservation planning, Ecological Applications 17 (6) (2007) 1771–1782.

[261] Bryan T. Grenfell, John Harwood, (meta)population dynamics of infectious diseases, Trends in Ecology & Evolution 12 (10) (1997) 395–399.

[262] Bryan T. Grenfell, Ottar N. Bjørnstad, Jennifer Kappey, Travelling waves and spatial hierarchies in measles epidemics, Nature 414 (2001) 716–723.

[263] Vittoria Colizza, Alain Barrat, Marc Barthélemy, Alessandro Vespignani, The role of the airline transportation network in the prediction and predictability of global epidemics, Proceedings of the National Academy of Sciences 103 (7) (2006) 2015–2020.

[264] Lars Hufnagel, Dirk Brockmann, Theo Geisel, Forecast and control of epidemics in a globalized world, Proceedings of the National Academy of Sciences 101 (42) (2004) 15124–15129.

[265] Marta C. González, Cesar A. Hidalgo, Albert-László Barabási, Understanding individual human mobility patterns, Nature 453 (7196) (2008) 779–782.

[266] M. Bognanni, H. Doug, D. Kolliner, K. Mitman, Economics and epidemics: evidence from an estimated spatial econ-sir model, SSRN Electronic Journal (2020).

[267] R. Goel, R. Sharma, Mobility based sir model for pandemics – with case study of covid-19, in: 2020 IEEE/ACM International Conference on Advances in Social Networks Analysis and Mining (ASONAM), 2020.
[268] Ruud van de Bovenkamp, Piet Van Mieghem, Survival time of the susceptible-infected-susceptible infection process on a graph, Physical Review E 92 (3) (2015) 032806.
[269] Eric Cator, Piet Van Mieghem, Susceptible-infected-susceptible epidemics on the complete graph and the star graph: exact analysis, Physical Review E, Statistical, Nonlinear, and Soft Matter Physics 87 (1) (2013) 012811.
[270] István Z. Kiss, Joel C. Miller, Péter L. Simon, Mathematics of Epidemics on Networks, Springer, Cham, 2017.
[271] M.E.J. Newman, Spread of epidemic disease on networks, Physical Review E 66 (1) (2002) 016128.
[272] Romualdo Pastor-Satorras, Alessandro Vespignani, Epidemic spreading in scale-free networks, Physical Review Letters 86 (14) (2001) 3200–3203.
[273] Erik Volz, SIR dynamics in random networks with heterogeneous connectivity, Journal of Mathematical Biology 56 (3) (2008) 293–310.
[274] Joel C. Miller, A note on a paper by Erik Volz: SIR dynamics in random networks, Journal of Mathematical Biology 62 (3) (2011) 349–358.
[275] Petter Holme, Jari Saramäki, Temporal networks, Physics Reports 519 (3) (2012) 97–125.
[276] J.D. Murray, Mathematical Biology I: An Introduction, 3 edition, Springer, New York, 2002.
[277] Akira Okubo, Simon A. Levin, Diffusion and Ecological Problems: Modern Perspectives, 2 edition, Springer, New York, 2001.
[278] Roman M. Cherniha, Vasyl V. Davydovych, A reaction–diffusion system with cross-diffusion: Lie symmetry, exact solutions and their applications in the pandemic modelling, European Journal of Applied Mathematics 33 (5) (2022) 785–802.
[279] Wenzhang Huang, Traveling waves for a biological reaction-diffusion model, Journal of Dynamics and Differential Equations 16 (3) (2004) 745–765.
[280] Jianhong Wu, Xingfu Zou, Traveling wave fronts of reaction-diffusion systems with delay, Journal of Dynamics and Differential Equations 13 (3) (2001) 651–687.
[281] Maud El-Hachem, Scott W. McCue, Wang Jin, Yihong Du, Matthew J. Simpson, Revisiting the Fisher–Kolmogorov–Petrovsky–Piskunov equation to interpret the spreading–extinction dichotomy, Proceedings of the Royal Society A. Mathematical, Physical and Engineering Sciences 475 (2229) (2019) 20190378.
[282] R.A. Fisher, The wave of advance of advantageous genes, Annals of Eugenics 7 (4) (1937) 355–369.
[283] A. Kolmogorov, I. Petrovsky, N. Piskunov, Étude de l'équation de la diffusion avec croissance de la quantité de matière et son application à un problème biologique, Bulletin of Moscow University, Mathematics and Mechanics 1 (1937) 1–25.
[284] Nobuyuki Shigesada, Kohkichi Kawasaki, Biological Invasions: Theory and Practice, Oxford University Press, Oxford, 1998.
[285] Teddy Lazebnik, Transforming norm-based to graph-based spatial representation for spatio-temporal epidemiological models, Engineering Applications of Artificial Intelligence 166 (2026) 113619.
[286] Masashi Kamo, Akira Sasaki, The effect of cross-immunity and seasonal forcing in a multi-strain epidemic model, Physica D: Nonlinear Phenomena 165 (3–4) (2002) 228–241.
[287] Zaynab A. Mohammed, Mathematical models for multi-strain infectious disease spread: a case study approach, Eurasian Journal of Humanities and Education Research (EJHER) (2025) 64–75.
[288] Miguel Fudolig, Reka Howard, The local stability of a modified multi-strain sir model for emerging viral strains, PLoS ONE 15 (12) (2020) e0243408.
[289] Teddy Lazebnik, Svetlana Bunimovich-Mendrazitsky, Generic approach for mathematical model of multi-strain pandemics, PLoS ONE 17 (4) (2022) e0260683.
[290] Ariel Alexi, Ariel Rosenfeld, Teddy Lazebnik, Multi-species prey–predator dynamics during a multi-strain pandemic, Chaos: An Interdisciplinary Journal of Nonlinear Science 33 (7) (2023).

[291] Asiyeh Ebrahimzadeh, Raheleh Khanduzi, Amin Jajarmi, Collocation method with flood-based metaheuristic optimizer for optimal control on a multi-strain covid-19 model, An International Journal of Optimization and Control: Theories & Applications 15 (2) (2025).

[292] Weiyi Tang, Jenna Kim, Raphael T.C. Lee, Sebastian Maurer-Stroh, Laurent Renia, Matthew Z. Tay, Sars-cov-2: lessons in virus mutation prediction and pandemic preparedness, Current Opinion in Immunology 95 (2025) 102560.

[293] Md Mamun-Ur-Rashid Khan, Jun Tanimoto, Influence of waning immunity on vaccination decision-making: a multi-strain epidemic model with an evolutionary approach analyzing cost and efficacy, Infectious Disease Modelling 9 (3) (2024) 657–672.

[294] Sergey Kryazhimskiy, Ulf Dieckmann, Simon A. Levin, Jonathan Dushoff, On state-space reduction in multi-strain pathogen models, with an application to antigenic drift in influenza A, PLoS Computational Biology 3 (8) (2007) e159.

[295] Viggo Andreasen, Juan Lin, Simon A. Levin, The dynamics of cocirculating influenza strains conferring partial cross-immunity, Journal of Mathematical Biology 35 (7) (1997) 825–842.

[296] Julia R. Gog, Bryan T. Grenfell, Dynamics and selection of many-strain pathogens, Proceedings of the National Academy of Sciences 99 (26) (2002) 17209–17214.

[297] Laith J. Abu-Raddad, Neil M. Ferguson, The impact of cross-immunity, mutation and stochastic extinction on pathogen diversity, Proceedings of the Royal Society of London. Series B, Biological Sciences 271 (1556) (2004) 2431–2438.

[298] Julia R. Gog, Jonathan Swinton, A status-based approach to multiple strain dynamics, Journal of Mathematical Biology 44 (2) (2002) 169–184.

[299] Paul S. Wikramaratna, Adam Kucharski, Sunetra Gupta, Viggo Andreasen, Angela R. McLean, Julia R. Gog, Five challenges in modelling interacting strain dynamics, Epidemics 10 (2015) 31–34.

[300] Neil M. Ferguson, Alison P. Galvani, Robin M. Bush, Ecological and immunological determinants of influenza evolution, Nature 422 (2003) 428–433.

[301] Katia Koelle, Sarah Cobey, Bryan Grenfell, Mercedes Pascual, Epochal evolution shapes the phylodynamics of interpandemic influenza A (h3n2) in humans, Science 314 (5807) (2006) 1898–1903.

[302] Wenhan Shao, Xinxin Li, Mohsan Ullah Goraya, Song Wang, Ji-Long Chen, Evolution of influenza A virus by mutation and re-assortment, International Journal of Molecular Sciences 18 (8) (2017) 1650.

[303] Julia M. Smyk, Natalia Szydłowska, Weronika Szulc, Anna Majewska, Evolution of influenza viruses—drug resistance, treatment options, and prospects, International Journal of Molecular Sciences 23 (20) (2022) 12244.

[304] William T. Harvey, Alessandro M. Carabelli, Ben Jackson, Ravindra K. Gupta, Emma C. Thomson, Ewan M. Harrison, Catherine Ludden, Richard Reeve, Andrew Rambaut, Sharon J. Peacock, David L. Robertson, Sars-cov-2 variants, spike mutations and immune escape, Nature Reviews. Microbiology 19 (7) (2021) 409–424.

[305] Nicholas G. Davies, Sam Abbott, Rosanna C. Barnard, Christopher I. Jarvis, Adam J. Kucharski, James D. Munday, Carl A.B. Pearson, Timothy W. Russell, Damien C. Tully, Alex D. Washburne, Tom Wenseleers, Amy Gimma, William Waites, Kerry L.M. Wong, Kevin van Zandvoort, Justin D. Silverman, Karla Diaz-Ordaz, Ruth Keogh, Rosalind M. Eggo, Sebastian Funk, Mark Jit, Katherine E. Atkins, W. John Edmunds, CMMID COVID-19 Working Group, COVID-19 Genomics UK (COG-UK) Consortium, Estimated transmissibility and impact of sars-cov-2 lineage b.1.1.7 in England, Science 372 (6538) (2021) eabg3055.

[306] Malay Sarkar, Irappa Madabhavi, Covid-19 mutations: an overview, World Journal of Methodology 14 (3) (2024) 89761.

[307] Yong-Sik Bong, David Brown, Ezra Chung, Neeti Ananthaswamy, Renxiang Chen, Evan Lewoczko, William Sabbers, Athena C. Patterson-Orazem, Zachary Dorsey, Yiqing Zou, et al., S6p mutation in delta and omicron variant spike protein significantly enhances the efficacy of mrna covid-19 vaccines, Frontiers in Immunology 15 (2025) 1495561.

[308] Oliver G. Pybus, Andrew Rambaut, Evolutionary analysis of the dynamics of viral infectious disease, Nature Reviews. Genetics 10 (8) (2009) 540–550.

[309] Erik M. Volz, Sergei L. Kosakovsky Pond, Melissa J. Ward, Andrew J. Leigh Brown, Simon D.W. Frost, Phylodynamics of infectious disease epidemics, Genetics 183 (4) (2009) 1421–1430.

[310] Trevor Bedford, Marc A. Suchard, Philippe Lemey, Gytis Dudas, Victoria Gregory, Alan J. Hay, John W. McCauley, Colin A. Russell, Derek J. Smith, Andrew Rambaut, Integrating influenza antigenic dynamics with molecular evolution, eLife 3 (2014) e01914.

[311] Derek J. Smith, Andrew S. Lapedes, Joost C. de Jong, Theo M. Bestebroer, Guus F. Rimmelzwaan, Albert D.M.E. Osterhaus, Ron A.M. Fouchier, Mapping the antigenic and genetic evolution of influenza virus, Science 305 (5682) (2004) 371–376.

[312] Igor M. Rouzine, Evolutionary mechanisms of the emergence of the variants of concern of sars-cov-2, Viruses 17 (2) (2025) 197.

[313] Carlos Maluquer de Motes, David O. Ulaeto, Mpox poses an ever-increasing epidemic and pandemic risk, Nature Medicine 31 (6) (2025) 1743–1746.

[314] Charles M. Macal, Michael J. North, Tutorial on agent-based modelling and simulation, Journal of Simulation 4 (3) (2010) 151–162.

[315] Steven F. Railsback, Volker Grimm Steven, F. Railsback, Volker Grimm, Agent-Based and Individual-Based Modeling. a Practical Introduction, 2 edition, Princeton University Press, 2019.

[316] Joshua M. Epstein, Robert Axtell, Growing Artificial Societies: Social Science from the Bottom up, Brookings Institution Press, 1997.

[317] Ewout Ter Hoeven, Jan Kwakkel, Vincent Hess, Thomas Pike, Boyu Wang, Jackie Kazil, et al., Mesa 3: agent-based modeling with python in 2025, Journal of Open Source Software 10 (107) (2025) 7668.

[318] Ariel Alexi, Teddy Lazebnik, Labib Shami, Microfounded tax revenue forecast model with heterogeneous population and genetic algorithm approach, Computational Economics 63 (5) (2024) 1705–1734.

[319] Robert L. Axtell, J. Doyne Farmer, Agent-based modeling in economics and finance: past, present, and future, Journal of Economic Literature 63 (1) (2025) 197–287.

[320] Adelinde M. Uhrmacher, Danny Weyns, Multi-Agent Systems Applications, CRC Press, 1999.

[321] Carol Lynn Curchoe, All models are wrong; some are useful, American Journal of Public Health 37 (10) (2017) 2389–2391.

[322] Jennifer L. Hicks, Thomas K. Uchida, Ajay Seth, Apoorva Rajagopal, Scott L. Delp, Is my model good enough? Best practices for verification and validation of musculoskeletal models and simulations of movement, Journal of Biomechanical Engineering 137 (2) (2015) 020905.

[323] Chung-Yuan Huang, Yu-Shiuan Tsai, Tzai-Hung Wen, Simulations for epidemiology and public health education, Journal of Simulation 4 (1) (2010) 68–80.

[324] Sean Hoban, Giorgio Bertorelle, Oscar E. Gaggiotti, Computer simulations: tools for population and evolutionary genetics, Nature Reviews. Genetics 13 (2) (2012) 110–122.

[325] Brandon D.L. Marshall, Sandro Galea, Formalizing the role of agent-based modeling in causal inference and epidemiology, American Journal of Epidemiology 181 (2) (2015) 92–99.

[326] Eduard Babulak, Ming Wang, Discrete event simulation, The International Series in Engineering and Computer Science 1 (2010).

[327] Bilash Kanti Bala, Fatimah Mohamed Arshad, Kusairi Mohd Noh, et al., System Dynamics, Springer, 2021.

[328] Sameera Abar, Georgios K. Theodoropoulos, Pierre Lemarinier, Gregory M.P. O'Hare, Agent based modelling and simulation tools: a review of the state-of-art software, Computer Science Review 24 (2017) 13–33.

[329] S.P. Shepherd, A review of system dynamics models applied in transportation, Transportmetrica B: Transport Dynamics 2 (2) (2014) 83–105.

[330] Federico Cosenz, Guido Noto, Applying system dynamics modelling to strategic management: a literature review, Systems Research and Behavioral Science 33 (6) (2016) 703–741.

[331] Mehdi Zomorodian, Sai Hin Lai, Mehran Homayounfar, Shaliza Ibrahim, Seyed Ehsan Fatemi, Ahmed El-Shafie, The state-of-the-art system dynamics application in integrated water resources modeling, Journal of Environmental Management 227 (2018) 294–304.
[332] Rosaria Conte, Mario Paolucci, On agent-based modeling and computational social science, Frontiers in Psychology 5 (2014) 668.
[333] Adam J. McLane, Christina Semeniuk, Gregory J. McDermid, Danielle J. Marceau, The role of agent-based models in wildlife ecology and management, Ecological Modelling 222 (8) (2011) 1544–1556.
[334] Nissim Cohen, Teddy Lazebnik, Do institutions make street-level bureaucrats prosocial? Agent-based evidence shows that new public management does not, European Policy Analysis (2025).
[335] John F. Kain, John R. Meyer, Computer simulations, physio-economic systems, and intraregional models, The American Economic Review 58 (2) (1968) 171–181.
[336] William D. Coplin, Approaches to the social sciences through man-computer simulations, Simulation & Games 1 (4) (1970) 391–410.
[337] Donald L. DeAngelis, Stephanie G. Diaz, Decision-making in agent-based modeling: a current review and future prospectus, Frontiers in Ecology and Evolution 6 (2019) 237.
[338] Palash Sarkar, A brief history of cellular automata, ACM Computing Surveys 32 (1) (2000) 80–107.
[339] Marco A. Janssen, Elinor Ostrom, Chapter 30 governing social-ecological systems, Handbook of Computational Economics 2 (2006) 1465–1509.
[340] Peer-Olaf Siebers, Uwe Aickelin, Introduction to multi-agent simulation, in: Encyclopedia of Decision Making and Decision Support Technologies, IGI Global, 2008, pp. 554–564.
[341] Rita Parada Ramos, Sancho Moura Oliveira, Susana Margarida Vieira, Anders Lyhne Christensen, Evolving flocking in embodied agents based on local and global application of Reynolds' rules, PLoS ONE 14 (10) (2019) e0224376.
[342] Teddy Lazebnik, Orr Spiegel, Individual variation affects outbreak magnitude and predictability in multi-pathogen model of pigeons visiting dairy farms, Ecological Modelling 499 (2025) 110925.
[343] Feres A. Salem, Ubirajara F. Moreno, A multi-agent-based simulation model for the spreading of diseases through social interactions during pandemics, Journal of Control, Automation and Electrical Systems 33 (4) (2022) 1161–1176.
[344] Kerri-Ann Norton, Daniel Bergman, Harsh Vardhan Jain, Trachette Jackson, Advances in surrogate modeling for biological agent-based simulations: trends, challenges, and future prospects, Journal of Mathematical Biology 92 (1) (2026) 6.
[345] Teddy Lazebnik, Avner Friedman, Comparing partial differential equations and agent-based simulations in spatio-temporal modeling of cancer growth and shape, Journal of Computational and Applied Mathematics (2026) 117183.
[346] Hazel R. Parry, Mike Bithell, Large scale agent-based modelling: a review and guidelines for model scaling, Agent-Based Models of Geographical Systems (2012) 271–308.
[347] Alan M. Turing, On computable numbers, with an application to the Entscheidungsproblem, Proceedings of the London Mathematical Society s2–42 (1) (1936) 230–265.
[348] Alex Reprintsev, Turing completeness, in: Oracle SQL Revealed: Executing Business Logic in the Database Engine, Springer, 2018, pp. 235–242.
[349] Howard Gutowitz, Cellular Automata: Theory and Experiment, MIT Press, 1991.
[350] Stephen Wolfram, A New Kind of Science, Wolfram Media, 2006.
[351] Matthew Cook, Universality in elementary cellular automata, Complex Systems 15 (1) (2004) 1–40.
[352] Stephen Eubank, Hasan Guclu, V.S. Anil Kumar, Madhav Marathe, Aravind Srinivasan, Zoltán Toroczkai, Nan Wang, Modelling disease outbreaks in realistic urban social networks, Nature 429 (6988) (2004) 180–184.
[353] Onur Özgün, Yaman Barlas, Discrete vs. Continuous simulation: when does it matter, in: Proceedings of the 27th International Conference of the System Dynamics Society, vol. 6, 2009, pp. 1–22.

[354] Mark Brouard, David H. Parker, Sebastiaan Y.T. van de Meerakker, Taming molecular collisions using electric and magnetic fields, Chemical Society Reviews 43 (21) (2014) 7279–7294.

[355] Ting Lu, Dmitri Volfson, Lev Tsimring, Jeff Hasty, Cellular growth and division in the gillespie algorithm, Systems Biology 1 (1) (2004) 121–128.

[356] S. Mauch, M. Stalzer, An efficient method for computing steady state solutions with Gillespie's direct method, Journal of Chemical Physics 133 (14) (October 2010) 144108.

[357] Natalie S. Glance, Bernardo A. Huberman, The outbreak of cooperation, The Journal of Mathematical Sociology 17 (4) (1993) 281–302.

[358] Geoffrey Caron-Lormier, Roger W. Humphry, David A. Bohan, Cathy Hawes, Pernille Thorbek, Asynchronous and synchronous updating in individual-based models, Ecological Modelling 212 (3) (2008) 522–527.

[359] Yan W. Chen, Lu F. Zhang, Jian P. Huang, The Watts–Strogatz network model developed by including degree distribution: theory and computer simulation, Journal of Physics A: Mathematical and Theoretical 40 (29) (2007) 8237–8246.

[360] Albert-László Barabási, Réka Albert, Emergence of scaling in random networks, Science 286 (5439) (1999) 509–512.

[361] L. Di Lucchio, G. Modanese, The bass diffusion model: agent-based implementation on arbitrary networks, Mathematical and Computer Modelling of Dynamical Systems 30 (1) (2024) 364–384.

[362] Xiaosong Li, Wentong Cai, Stephen John Turner, Efficient neighbor searching for agent-based simulation on gpu, in: 2014 IEEE/ACM 18th International Symposium on Distributed Simulation and Real Time Applications, IEEE, 2014, pp. 87–96.

[363] David L. Krongauz, Amir Ayali, Gal A. Kaminka, Vision-based collective motion: a locust-inspired reductionist model, PLoS Computational Biology 20 (1) (2024) e1011796.

[364] Clifford A. Shaffer, Hanan Samet, Optimal quadtree construction algorithms, Computer Vision, Graphics, and Image Processing 37 (3) (1987) 402–419.

[365] Prudence M. Mavhemwa, Ignatius Nyangani, Uniform spatial subdivision to improve boids algorithm in a gaming environment, International Journal for Advance Research and Development 3 (10) (2018) 49–57.

[366] Lih-Yuan Deng, Dale Bowman, Developments in pseudo-random number generators, WIREs Computational Statistics 9 (5) (2017) e1404.

[367] Michael Georgeff, Barney Pell, Martha Pollack, Milind Tambe, Michael Wooldridge, The belief-desire-intention model of agency, in: International Workshop on Agent Theories, Architectures, and Languages, Springer, 1999, pp. 1–10.

[368] Leslie Pack Kaelbling, Michael L. Littman, Andrew W. Moore, Reinforcement learning: a survey, Journal of Artificial Intelligence Research 4 (1996) 237–285.

[369] Ashish Kumar Shakya, Gopinatha Pillai, Sohom Chakrabarty, Reinforcement learning algorithms: a brief survey, Expert Systems with Applications 231 (2023) 120495.

[370] Jianqing Fan, Zhaoran Wang, Yuchen Xie, Zhuoran Yang, A theoretical analysis of deep q-learning, in: Learning for Dynamics and Control, PMLR, 2020, pp. 486–489.

[371] Adi Shuchami, Teddy Lazebnik, Spatio-temporal SIR model of pandemic spread during warfare with optimal dual-use health care system administration using deep reinforcement learning, Disaster Medicine and Public Health Preparedness 19 (2025) e197.

[372] Teddy Lazebnik, Ariel Rosenfeld, Labib Shami, Economical–epidemiological analysis of the coffee trees rust pandemic, Chaos: An Interdisciplinary Journal of Nonlinear Science 34 (11) (2024).

[373] Yonatan Herskowitz, Svetlana Bunimovich-Mendrazitsky, Teddy Lazebnik, Mathematical model of coffee tree's rust control using snails as biological agents, Biosystems 229 (2023) 104916.

[374] Teddy Lazebnik, Data-driven hospitals staff and resources allocation using agent-based simulation and deep reinforcement learning, Engineering Applications of Artificial Intelligence 126 (November 2023) 106783.

[375] Pranjal Kumar, Large language models (llms): survey, technical frameworks, and future challenges, Artificial Intelligence Review 57 (10) (2024) 260.

[376] Teddy Lazebnik, Labib Shami, Investigating tax evasion emergence using dual large language model and deep reinforcement learning powered agent-based simulation, arXiv preprint, arXiv:2501.18177, 2025.

[377] Joon Sung Park, Joseph C. O'Brien, Carrie Jun Cai, Meredith Ringel Morris, Percy Liang, Michael S. Bernstein, Generative agents: interactive simulacra of human behavior, in: Proceedings of the 36th Annual ACM Symposium on User Interface Software and Technology (UIST '23), ACM, 2023, pp. 2:1–2:22.

[378] Jiakai Tang, Heyang Gao, Xuchen Pan, Lei Wang, Haoran Tan, Dawei Gao, Yushuo Chen, Xu Chen, Yankai Lin, Yaliang Li, Bolin Ding, Jingren Zhou, Jun Wang, Ji-Rong Wen, GenSim: a general social simulation platform with large language model based agents, in: Proceedings of the 2025 Conference of the Nations of the Americas Chapter of the Association for Computational Linguistics: Human Language Technologies (System Demonstrations), Albuquerque, New Mexico, Association for Computational Linguistics, April 2025, pp. 143–150.

[379] Jinghua Piao, Yuwei Yan, Jun Zhang, Nian Li, Junbo Yan, Xiaochong Lan, Zhihong Lu, Zhiheng Zheng, Jing Yi Wang, Di Zhou, Chen Gao, Fengli Xu, Fang Zhang Ke Rong, Jun Su, Yong Li, Agentsociety: Large-scale simulation of llm-driven generative agents advances understanding of human behaviors and society, 2025.

[380] Chen Gao, Xiaochong Lan, Nian Li, Yuan Yuan, Jingtao Ding, Zhilun Zhou, Fengli Xu, Yong Li, Large language models empowered agent-based modeling and simulation: a survey and perspectives, Humanities & Social Sciences Communications 11 (2024) 1259.

[381] Ariel Flint Ashery, Luca Maria Aiello, Andrea Baronchelli, Emergent social conventions and collective bias in LLM populations, Science Advances 11 (20) (2025) eadu9368.

[382] John J. Grefenstette, Shawn T. Brown, Roni Rosenfeld, Jay DePasse, Nathan T.B. Stone, Phillip C. Cooley, William D. Wheaton, Alona Fyshe, David D. Galloway, Anuroop Sriram, et al., Fred (a framework for reconstructing epidemic dynamics): an open-source software system for modeling infectious diseases and control strategies using census-based populations, BMC Public Health 13 (1) (2013) 940.

[383] Cliff C. Kerr, Robyn M. Stuart, Dina Mistry, Romesh G. Abeysuriya, Katherine Rosenfeld, Gregory R. Hart, Rafael C. Núñez, Jamie A. Cohen, Prashanth Selvaraj, Brittany Hagedorn, et al., Covasim: an agent-based model of covid-19 dynamics and interventions, PLoS Computational Biology 17 (7) (2021) e1009149.

[384] Andrea Saltelli, Marco Ratto, Terry Andres, Francesca Campolongo, Jessica Cariboni, Debora Gatelli, Michaela Saisana, Stefano Tarantola, Global Sensitivity Analysis: The Primer, John Wiley & Sons, Chichester, 2008.

[385] Federica Gugole, Luc E. Coffeng, Wouter Edeling, Benjamin Sanderse, Sake J. de Vlas, Daan Crommelin, Uncertainty quantification and sensitivity analysis of covid-19 exit strategies in an individual-based transmission model, PLoS Computational Biology 17 (9) (2021) e1009355.

[386] Andrea Saltelli, Paola Annoni, How to avoid a perfunctory sensitivity analysis, Environmental Modelling & Software 25 (12) (2010) 1508–1517.

[387] Emanuele Borgonovo, Marco Pangallo, Jan Rivkin, Leonardo Rizzo, Nicolaj Siggelkow, Sensitivity analysis of agent-based models: a new protocol, Computational and Mathematical Organization Theory 28 (2022) 52–94.

[388] Xuefei Lu, Emanuele Borgonovo, Global sensitivity analysis in epidemiological modeling, European Journal of Operational Research 304 (1) (2023) 9–24.

[389] Louis Melville Milne-Thomson, The Calculus of Finite Differences, American Mathematical Soc., 2000.

[390] Guus ten Broeke, George A.K. van Voorn, Arend Ligtenberg, Which sensitivity analysis method should I use for my agent-based model?, Journal of Artificial Societies and Social Simulation 19 (1) (2016) 5.

[391] Paul Glasserman, David D. Yao, Some guidelines and guarantees for common random numbers, Management Science 38 (6) (1992) 884–908.

[392] Emanuele Borgonovo, Sensitivity analysis with finite changes: an application to modified EOQ models, European Journal of Operational Research 200 (1) (2010) 127–138.

[393] Ilya M. Sobol', Global sensitivity indices for nonlinear mathematical models and their Monte Carlo estimates, Mathematics and Computers in Simulation 55 (1–3) (2001) 271–280.

[394] Toshimitsu Homma, Andrea Saltelli, Importance measures in global sensitivity analysis of nonlinear models, Reliability Engineering & System Safety 52 (1) (1996) 1–17.
[395] Max D. Morris, Factorial sampling plans for preliminary computational experiments, Technometrics 33 (2) (1991) 161–174.
[396] Francesca Campolongo, Jessica Cariboni, Andrea Saltelli, An effective screening design for sensitivity analysis of large models, Environmental Modelling & Software 22 (10) (2007) 1509–1518.
[397] Andrea Saltelli, Making best use of model evaluations to compute sensitivity indices, Computer Physics Communications 145 (2) (2002) 280–297.
[398] Emanuele Borgonovo, Marco Pangallo, Jan W. Rivkin, Leonardo Rizzo, Nicolaj Siggelkow, Sensitivity analysis for agent-based models: the state of the art and new challenges, European Journal of Operational Research 310 (1) (2023) 25–48.
[399] M.D. McKay, R.J. Beckman, W.J. Conover, A comparison of three methods for selecting values of input variables in the analysis of output from a computer code, Technometrics 21 (2) (1979) 239–245.
[400] Jon Herman, Will Usher, Salib: an open-source python library for sensitivity analysis, Journal of Open Source Software 2 (9) (2017) 97.
[401] Francesca Pianosi, Keith Beven, Jim Freer, Jim W. Hall, Jonathan Rougier, David B. Stephenson, Thorsten Wagener, Sensitivity analysis of environmental models: a systematic review with practical workflow, Environmental Modelling & Software 79 (2016) 214–232.
[402] John Norton, An introduction to sensitivity assessment of simulation models, Environmental Modelling & Software 69 (2015) 166–174.
[403] Arnaud Dell'Oca, Sensitivity analysis: an operational picture, Water Resources Research 59 (6) (2023) e2022WR033780.
[404] Saman Razavi, Anthony Jakeman, Andrea Saltelli, Clémentine Prieur, Bertrand Iooss, Emanuele Borgonovo, Elmar Plischke, Samuele Lo Piano, Takuya Iwanaga, William Becker, Stefano Tarantola, Joseph H.A. Guillaume, John Jakeman, Hoshin Gupta, Nicola Melillo, Giovanni Rabitti, Vincent Chabridon, Qingyun Duan, Xifu Sun, Stefán Smith, Holger R. Maier, The future of sensitivity analysis: an essential discipline for systems modeling and policy support, Environmental Modelling & Software 137 (2021) 104954.
[405] Andrea Saltelli, Paola Annoni, Introduction to sensitivity analysis, in: Sensitivity Analysis in Earth Observation Modelling, Springer, Cham, 2014, pp. 1–13.
[406] Cameron Zachreson, Julian Savulescu, Freya M. Shearer, Michael J. Plank, Simon Coghlan, Joel C. Miller, Kylie E.C. Ainslie, Nicholas Geard, Ethical frameworks should be applied to computational modelling of infectious disease interventions, PLoS Computational Biology 20 (3) (2024) e1011933.
[407] Victoria Loblay, Louise Freebairn, Jo-An Occhipinti, Conceptualising the value of simulation modelling for public engagement with policy: a critical literature review, Health Research Policy and Systems 21 (1) (2023) 123.
[408] Yangdi Jiang, Lucy Mosquera, Bei Jiang, Linglong Kong, Khaled El Emam, Measuring re-identification risk using a synthetic estimator to enable data sharing, PLoS ONE 17 (6) (2022) e0269097.
[409] Simeone Marino, Ruth Cassidy, Joseph Nanni, Yuxuan Wang, Yipeng Liu, Mingyi Tang, Yuan Yuan, Toby Chen, Anik Sinha, Balaji Pandian, Ivo D. Dinov, Michael L. Burns, Medical data sharing and synthetic clinical data generation – maximizing biomedical resource utilization and minimizing participant re-identification risks, npj Digital Medicine 8 (1) (2025) 526.
[410] Serina Chang, Emma Pierson, Pang Wei Koh, Jaline Gerardin, Beth Redbird, David Grusky, Jure Leskovec, Mobility network models of covid-19 explain inequities and inform reopening, Nature 589 (7840) (2021) 82–87.
[411] Michele Tizzoni, Elaine O. Nsoesie, Laetitia Gauvin, Márton Karsai, Nicola Perra, Shweta Bansal, Addressing the socioeconomic divide in computational modeling for infectious diseases, Nature Communications 13 (1) (2022) 2897.

[412] Inga Holmdahl, Caroline Buckee, Wrong but useful—what covid-19 epidemiologic models can and cannot tell us, The New England Journal of Medicine 383 (4) (2020) 303–305.
[413] David D. Kim, The role of simulation modelling in public health policy evaluation, The Lancet. Public Health 9 (3) (2024) e150–e151.
[414] Taiwo M. Adegoke, Akindeh M. Nji, Sheetal P. Silal, Innocent M. Ali, Andrillene L.D. Wondeu, Pascal F. Madeleine, Masso T. Sylvanie, Calvino F. Tah, Christophe Antonio-Nkondjio, Jude D. Bigoga, et al., A systematic review of epidemiological models for malaria transmission in sub-Saharan Africa, Malaria Journal (2026).
[415] Kelly McCain, Anna Vicco, Christian Morgenstern, Thomas Rawson, Tristan M. Naidoo, Sangeeta Bhatia, Dominic P. Dee, Patrick Doohan, Keith Fraser, Anna-Maria Hartner, et al., A systematic review and meta-analysis of Zika virus epidemiology, Nature Health (2026) 1–13.
[416] Lemuel R. Non, Alexandre R. Marra, Dilek Ince, Rise of the machines-artificial intelligence in healthcare epidemiology, Current Infectious Disease Reports 27 (1) (2025) 4.
[417] Elizaveta Savchenko, Teddy Lazebnik, Computer aided functional style identification and correction in modern Russian texts, Journal of Data, Information and Management 4 (1) (2022) 25–32.
[418] Dor Atias, Saar Ashri, Uri Goldbourt, Yael Benyamini, Ran Gilad-Bachrach, Tal Hasin, Yariv Gerber, Uri Obolski, Machine learning in epidemiology: an introduction, comparison with traditional methods, and a case study of predicting extreme longevity, Annals of Epidemiology (2025).
[419] Lulah Alnaji, Machine learning in epidemiology: neural networks forecasting of monkeypox cases, PLoS ONE 19 (5) (2024) e0300216.
[420] S.K. Towfek, Mohamed Elkanzi, A review on the role of machine learning in predicting the spread of infectious diseases, Metaheuristic Optimization Review 2 (1) (2024) 14–27.
[421] Benjamin Lucas, Behzad Vahedi, Morteza Karimzadeh, A spatiotemporal machine learning approach to forecasting covid-19 incidence at the county level in the USA, International Journal of Data Science and Analytics 15 (3) (2023) 247–266.
[422] Joohon Sung, John L. Hopper, Co-evolution of epidemiology and artificial intelligence: challenges and opportunities, International Journal of Epidemiology 52 (4) (2023) 969–973.
[423] Aikaterini Sakagianni, Christina Koufopoulou, Petros Koufopoulos, Georgios Feretzakis, Dimitris Kalles, Evgenia Paxinou, Pavlos Myrianthefs, Vassilios S. Verykios, The synergy of machine learning and epidemiology in addressing carbapenem resistance: a comprehensive review, Antibiotics 13 (10) (2024) 996.
[424] Vaia I. Kontopoulou, Athanasios D. Panagopoulos, Ioannis Kakkos, George K. Matsopoulos, A review of arima vs. Machine learning approaches for time series forecasting in data driven networks, Future Internet 15 (8) (2023) 255.
[425] Yi Song, Yu-Nong Gong, Kuan-Fu Chen, David K. Smith, Hassan Zaraket, Seweryn Bialasiewicz, Sarah Tozer, Paul K.S. Chan, Evelyn S.C. Koay, Hong Kai Lee, et al., Global epidemiology, seasonality and climatic drivers of the four human parainfluenza virus types, The Journal of Infection 90 (4) (2025) 106451.
[426] Dae-Hyup Koh, Monica Duong, Nodar Kipshidze, Virginia E. Pitzer, Jong-Hoon Kim, Time series data on typhoid fever incidence during outbreaks from 2000 to 2022, Scientific Data 12 (1) (2025) 94.
[427] Omnia M. Osama, Khder Alakkari, Mostafa Abotaleb, El-Sayed M. El-Kenawy, Forecasting global monkeypox infections using lstm: a non-stationary time series analysis, in: 2023 3rd International Conference on Electronic Engineering (ICEEM), IEEE, 2023, pp. 1–7.
[428] Daniel C. Medina, Sally E. Findley, Boubacar Guindo, Seydou Doumbia, Forecasting non-stationary diarrhea, acute respiratory infection, and malaria time-series in niono, Mali, PLoS ONE 2 (11) (2007) e1181.
[429] Ethan O. Romero-Severson, Ruy M. Ribeiro, Mario Castro, Noise is not error: detecting parametric heterogeneity between epidemiologic time series, Frontiers in Microbiology 9 (2018) 1529.
[430] Sandra Gudziunaite, Zana Shabani, Lisbeth Weitensfelder, Hanns Moshammer, Time series analysis in environmental epidemiology: challenges and considerations, International Journal of Occupational Medicine and Environmental Health 36 (6) (2023) 704.

[431] Ensheng Dong, Hongru Du, Lauren Gardner, An interactive web-based dashboard to track COVID-19 in real time, Lancet. Infectious Diseases 20 (5) (2020) 533–534.
[432] Peter McCullagh, John A. Nelder, Generalized Linear Models, 2 edition, Chapman and Hall/CRC, London, 1989.
[433] Ludwig Fahrmeir, Thomas Kneib, Stefan Lang, Brian Marx, Regression: Models, Methods and Applications, Springer, Berlin, Heidelberg, 2014.
[434] Rob J. Hyndman, George Athanasopoulos, Business Forecasting: Principles and Practice. OTexts, 3 edition, Melbourne, 1953, Online textbook.
[435] Leonhard Held, Niel Hens, Philip D. O'Neill, Jacco Wallinga, Handbook of Infectious Disease Data Analysis, Chapman and Hall/CRC, Boca Raton, 2017.
[436] Leonhard Held, Michael Höhle, Mathias Hofmann, A statistical framework for the analysis of multivariate infectious disease surveillance counts, Statistical Modelling 5 (3) (2005) 187–199.
[437] Johannes Bracher, Evan L. Ray, Tilmann Gneiting, Nicholas G. Reich, Evaluating epidemic forecasts in an interval format, PLoS Computational Biology 17 (2) (2021) e1008618.
[438] A. Colin Cameron, Pravin K. Trivedi, Regression Analysis of Count Data, 2 edition, Cambridge University Press, 2013.
[439] Joseph M. Hilbe, Negative Binomial Regression, 2 edition, Cambridge University Press, 2011.
[440] Rob J. Hyndman, George Athanasopoulos, Business Forecasting: Principles and Practice, OTexts, 3 edition, 1953.
[441] Kimberly F. Sellers, Galit Shmueli, A flexible regression model for count data, Annals of Applied Statistics 4 (2) (2010) 943–961.
[442] Diane Lambert, Zero-inflated Poisson regression, with an application to defects in manufacturing, Technometrics 34 (1) (1992) 1–14.
[443] John Mullahy, Specification and testing of some modified count data models, Journal of Econometrics 33 (3) (1986) 341–365.
[444] J. Aitchison, C.H. Ho, The multivariate Poisson-log normal distribution, Biometrika 76 (4) (1989) 643–653.
[445] R.W.M. Wedderburn, Quasi-likelihood functions, generalized linear models, and the Gauss—Newton method, Biometrika 61 (3) (1974) 439–447.
[446] Angel N. Desai, Moritz U.G. Kraemer, Sangeeta Bhatia, Anne Cori, Pierre Nouvellet, Mark Herringer, Emily L. Cohn, Malwina Carrion, John S. Brownstein, Lawrence C. Madoff, et al., Real-time epidemic forecasting: challenges and opportunities, Health Security 17 (4) (2019) 268–275.
[447] Sylvain Gandon, Troy Day, C. Jessica E. Metcalf, Bryan T. Grenfell, Forecasting epidemiological and evolutionary dynamics of infectious diseases, Trends in Ecology & Evolution 31 (10) (2016) 776–788.
[448] Vladislav Soukhovolsky, Anton Kovalev, Anne Pitt, Katerina Shulman, Olga Tarasova, Boris Kessel, The cyclicity of coronavirus cases: "waves" and the "weekend effect", Chaos, Solitons and Fractals 144 (March 2021) 110718.
[449] Elizabeth Buckingham-Jeffery, Roger Morbey, Thomas House, et al., Correcting for day of the week and public holiday effects: improving a national daily syndromic surveillance service for detecting public health threats, BMC Public Health 17 (2017) 477.
[450] Peter C. Young, D. Pedregal, Wlodek Tych, Dynamic harmonic regression, Journal of Forecasting 18 (6) (1999) 369–394.
[451] Alysha M. De Livera, Rob J. Hyndman, Ralph D. Snyder, Forecasting time series with complex seasonal patterns using exponential smoothing, Journal of the American Statistical Association 106 (496) (2011) 1513–1527.
[452] Robert B. Cleveland, William S. Cleveland, Jean E. McRae, Irma Terpenning, Uncertainty-aware seasonal-trend decomposition based on loess, IEEE Transactions on Visualization and Computer Graphics 6 (2025) 3–73.
[453] Howard S. Burkom, Sean Patrick Murphy, Galit Shmueli, Automated time series forecasting for biosurveillance, Statistics in Medicine 26 (22) (2007) 4202–4218.

[454] Felix Günther, Andreas Bender, Katharina Katz, Helmut Küchenhoff, Michael Höhle, Nowcasting the covid-19 pandemic in bavaria, Biometrical Journal 63 (3) (2021) 490–502.

[455] Long Ma, Zhihao Qiu, Piet Van Mieghem, Maksim Kitsak, Reporting delays: a widely neglected impact factor in covid-19 forecasts, PNAS Nexus 3 (6) (2024) 204.

[456] Zhirui He, Hongbing Tao, Epidemiology and arima model of positive-rate of influenza viruses among children in Wuhan, China: a nine-year retrospective study, International Journal of Infectious Diseases 74 (2018) 61–70.

[457] Xingyu Zhang, Tao Zhang, Alistair A. Young, Xiaosong Li, Applications and comparisons of four time series models in epidemiological surveillance data, PLoS ONE 9 (2) (2014) e88075.

[458] Meng Wang, Jinhua Pan, Xinghui Li, Mengying Li, Zhixi Liu, Qi Zhao, Linyun Luo, Haiping Chen, Sirui Chen, Feng Jiang, et al., Arima and arima-ernn models for prediction of pertussis incidence in mainland China from 2004 to 2021, BMC Public Health 22 (1) (2022) 1447.

[459] George E.P. Box, Gwilym M. Jenkins, Gregory C. Reinsel, Greta M. Ljung, Time Series Analysis: Forecasting and Control, 5 edition, John Wiley & Sons, Hoboken, NJ, 1972.

[460] R. Allard, Use of time-series analysis in infectious disease surveillance, Bulletin of the World Health Organization 76 (4) (1998) 327–333.

[461] Greta M. Ljung, George E.P. Box, On a measure of lack of fit in time series models, Biometrika 65 (2) (1978) 297–303.

[462] A.L. Schaffer, T.A. Dobbins, S.-A. Pearson, Interrupted time series analysis using autoregressive integrated moving average (ARIMA) models: a guide for evaluating large-scale health interventions, BMC Medical Research Methodology 21 (2021) 58.

[463] Clifford M. Hurvich, Chih-Ling Tsai, Regression and time series model selection in small samples, Biometrika 76 (2) (1989) 297–307.

[464] Rob J. Hyndman, Yeasmin Khandakar, Automatic time series forecasting: the forecast package for R, Journal of Statistical Software 27 (3) (2008) 1–22.

[465] Zhicheng Du, Lin Xu, Wangjian Zhang, Dingmei Zhang, Shicheng Yu, Yuantao Hao, Predicting the hand, foot, and mouth disease incidence using search engine query data and climate variables: an ecological study in Guangdong, China, BMJ Open 7 (10) (2017) e016263.

[466] George E.P. Box, George C. Tiao, Intervention analysis with applications to economic and environmental problems, Journal of the American Statistical Association 70 (349) (1975) 70–79.

[467] L. Liu, R.S. Luan, F. Yin, X.P. Zhu, Q. Lü, Predicting the incidence of hand, foot and mouth disease in Sichuan province, China using the ARIMA model, Epidemiology and Infection 144 (2016) 144–151.

[468] Xingyu Zhang, Tao Zhang, Alistair A. Young, Xiaosong Li, Applications and comparisons of four time series models in epidemiological surveillance data, PLoS ONE 9 (2) (2014) e88075.

[469] Sasikiran Kandula, Jeffrey Shaman, Near-term forecasts of influenza-like illness: an evaluation of autoregressive time series approaches, Epidemics 28 (2019) 100325.

[470] James Durbin, Siem Jan Koopman, Time Series Analysis by State Space Methods, 2 edition, Oxford University Press, 2012.

[471] Greg Welch, Gary Bishop, et al., An introduction to the Kalman filter, in: Advanced Data Assimilation for Geosciences, 2014.

[472] Masoud Khodarahmi, Vafa Maihami, A review on Kalman filter models, Archives of Computational Methods in Engineering 30 (1) (2023) 727–747.

[473] Andrew C. Harvey, Garry D.A. Phillips, Testing for structural change in simultaneous equation models, in: Statistical Analysis and Forecasting of Economic Structural Change, Springer, 1989, pp. 25–36.

[474] Mike West, Jeff Harrison, Bayesian Forecasting and Dynamic Models, Springer, 1997.

[475] Rob Hyndman, Anne Koehler, Keith Ord, Ralph Snyder, Forecasting with Exponential Smoothing: the State Space Approach, Springer, 2008.

[476] Rob J. Hyndman, Anne B. Koehler, Ralph D. Snyder, Simone Grose, A state space framework for automatic forecasting using exponential smoothing methods, International Journal of Forecasting 18 (3) (2002) 439–454.
[477] Michael Höhle, Matthias an der Heiden, Bayesian nowcasting during the STEC O104:H4 outbreak in Germany, 2011, Biometrics 70 (4) (2014) 993–1002.
[478] Sarah F. McGough, Michael A. Johansson, Marc Lipsitch, Nicolas A. Menzies, Nowcasting by Bayesian smoothing: a flexible, generalizable model for real-time epidemic tracking, PLoS Computational Biology 16 (4) (2020) e1007735.
[479] Felix Günther, Andreas Bender, Katharina Katz, Helmut Küchenhoff, Michael Höhle, Nowcasting the COVID-19 pandemic in bavaria, Biometrical Journal 63 (3) (2021) 490–502.
[480] Sharon K. Greene, Sarah F. McGough, Gretchen M. Culp, Laura E. Graf, Marc Lipsitch, Nicolas A. Menzies, Rebecca Kahn, Nowcasting for real-time COVID-19 tracking in New York city: an evaluation using reportable disease data from early in the pandemic, JMIR Public Health and Surveillance 7 (1) (2021) e25538.
[481] J.F. Lawless, Adjustments for reporting delays and the prediction of occurred but not reported events, Canadian Journal of Statistics 22 (1) (1994) 15–31.
[482] Long Ma, Z. Qiu, Piet Van Mieghem, Maksim Kitsak, Reporting delays: a widely neglected impact factor in COVID-19 forecasts, PNAS Nexus 3 (6) (2024) 204.
[483] Gopinath Rebala, Ajay Ravi, Sanjay Churiwala, An Introduction to Machine Learning, Springer, 2019.
[484] Yves Kodratoff, Introduction to Machine Learning, Elsevier, 2014.
[485] Labib Shami, Teddy Lazebnik, Implementing machine learning methods in estimating the size of the non-observed economy, Computational Economics 63 (4) (2024) 1459–1476.
[486] Liron Simon Keren, Alex Liberzon, Teddy Lazebnik, A computational framework for physics-informed symbolic regression with straightforward integration of domain knowledge, Scientific Reports 13 (1) (2023) 1249.
[487] Rebecca Roelofs, Vaishaal Shankar, Benjamin Recht, Sara Fridovich-Keil, Moritz Hardt, John Miller, Ludwig Schmidt, A meta-analysis of overfitting in machine learning, Advances in Neural Information Processing Systems 32 (2019).
[488] Shai Shalev-Shwartz, Shai Ben-David, Understanding Machine Learning: From Theory to Algorithms, Cambridge University Press, 2014.
[489] Gianluca Bontempi, Souhaib Ben Taieb, Yann-Aël Le Borgne, Machine learning strategies for time series forecasting, in: Marie-Aude Aufaure, Esteban Zimányi (Eds.), Business Intelligence, in: Lecture Notes in Business Information Processing, vol. 138, Springer, Berlin, Heidelberg, 2013, pp. 62–77.
[490] Peter McCullagh, John A. Nelder, Generalized Linear Models, 2 edition, Chapman and Hall/CRC, 1989.
[491] Isabelle Guyon, André Elisseeff, An introduction to variable and feature selection, Journal of Machine Learning Research 3 (2003) 1157–1182.
[492] A.E. Hoerl, R.W. Kennard, Ridge regression: biased estimation for nonorthogonal problems, Technometrics 12 (1) (1970) 55–67.
[493] Jerome Friedman, Trevor Hastie, Robert Tibshirani, Regularization paths for generalized linear models via coordinate descent, Journal of Statistical Software 33 (1) (2010) 1–22.
[494] Max Kuhn, Kjell Johnson, Applied Predictive Modeling, Springer, New York, 2013.
[495] Robert Tibshirani, Regression shrinkage and selection via the lasso, Journal of the Royal Statistical Society, Series B, Statistical Methodology 58 (1) (1996) 267–288.
[496] Shihao Yang, Mauricio Santillana, S.C. Kou, Accurate estimation of influenza epidemics using Google search data via Argo, Proceedings of the National Academy of Sciences 112 (47) (2015) 14473–14478.
[497] Leo Breiman, Random forests, Machine Learning 45 (1) (2001) 5–32.
[498] Teddy Lazebnik, Svetlana Bunimovich-Mendrazitsky, Decision tree post-pruning without loss of accuracy using the sat-pp algorithm with an empirical evaluation on clinical data, Data & Knowledge Engineering 145 (2023) 102173.

[499] Ashwini Venkatasubramaniam, Julian Wolfson, Nathan Mitchell, Timothy Barnes, Meghan JaKa, Simone French, Decision trees in epidemiological research, Emerging Themes in Epidemiology 14 (1) (2017) 11.
[500] Teddy Lazebnik, Zaher Bahouth, Svetlana Bunimovich-Mendrazitsky, Sarel Halachmi, Predicting acute kidney injury following open partial nephrectomy treatment using sat-pruned explainable machine learning model, BMC Medical Informatics and Decision Making 22 (1) (2022) 133.
[501] Julian Wolfson, Ashwini Venkatasubramaniam, Branching out: use of decision trees in epidemiology, Current Epidemiology Reports 5 (3) (2018) 221–229.
[502] Naizhuo Zhao, Katia Charland, Mabel Carabali, Elaine O. Nsoesie, Mathieu Maheu-Giroux, Erin Rees, Mengru Yuan, Cesar Garcia Balaguera, Gloria Jaramillo Ramirez, Kate Zinszer, Machine learning and Dengue forecasting: comparing random forests and artificial neural networks for predicting Dengue burden at national and sub-national scales in Colombia, PLoS Neglected Tropical Diseases 14 (2) (2020) e0008056.
[503] Tianqi Chen, Carlos Guestrin, XGBoost: a scalable tree boosting system, in: Proceedings of the 22nd ACM SIGKDD International Conference on Knowledge Discovery and Data Mining, 2016, pp. 785–794.
[504] Qiang Chang, et al., Internet search data-based epidemic forecasting: short-term prediction of the number of COVID-19 positive tests based on XGBoost and random forest models, BMC Public Health (2025).
[505] Alex J. Smola, Bernhard Schölkopf, A tutorial on support vector regression, Statistics and Computing 14 (3) (2004) 199–222.
[506] I.N. Tanawi, V. Vito, D. Sarwinda, H. Tasman, G.F. Hertono, Support vector regression for predicting the number of Dengue incidents in DKI Jakarta, Procedia Computer Science 179 (2021) 747–753.
[507] Carl Edward Rasmussen, Christopher K.I. Williams, Gaussian Processes in Machine Learning, MIT Press, 2004.
[508] Leah R. Johnson, Robert B. Gramacy, Jeremy Cohen, Erin Mordecai, Courtney Murdock, Jason Rohr, Sadie J. Ryan, Anna M. Stewart-Ibarra, Daniel Weikel, Phenomenological forecasting of disease incidence using heteroskedastic Gaussian processes: a Dengue case study, Annals of Applied Statistics (2018).
[509] Jakhar Deepack, Ishmeet Kaur, Artificial intelligence, machine learning and deep learning: definitions and differences, Clinical and Experimental Dermatology 45 (1) (2020) 131–132.
[510] Koosha Sharifani, Mahyar Amini, Machine learning and deep learning: a review of methods and applications, World Information Technology and Engineering Journal 10 (07) (2023) 3897–3904.
[511] Yann LeCun, Yoshua Bengio, Geoffrey Hinton, Deep learning, Nature 521 (7553) (2015) 436–444.
[512] Ian Goodfellow, Yoshua Bengio, Aaron Courville, Deep Learning, MIT Press, Cambridge, MA, 2016.
[513] Diederik P. Kingma, Jimmy Ba, Adam: a method for stochastic optimization, arXiv preprint, arXiv: 1412.6980, 2014.
[514] Diederik P. Kingma, Jimmy Ba, Adam: a method for stochastic optimization, CoRR, arXiv:1412.6980, 2015.
[515] Nitish Srivastava, Geoffrey E. Hinton, Alex Krizhevsky, Ilya Sutskever, Ruslan R. Salakhutdinov, Dropout: an effective approach to prevent neural networks from overfitting, Asian Journal of Research in Computer Science 15 (56) (2014) 1929–1958.
[516] Sergey Ioffe, Christian Szegedy, Batch normalization: accelerating deep network training by reducing internal covariate shift, in: Proceedings of the 32nd International Conference on Machine Learning (ICML), in: Proceedings of Machine Learning Research, vol. 37, PMLR, 2015, pp. 448–456.
[517] Sepp Hochreiter, Jürgen Schmidhuber, Long short-term memory, Neural Computation 9 (8) (1997) 1735–1780.
[518] Shaojie Bai, J. Zico Kolter, Vladlen Koltun, An empirical evaluation of generic convolutional and recurrent networks for sequence modeling, arXiv preprint, arXiv:1803.01271, 2018.
[519] Ashish Vaswani, Noam Shazeer, Niki Parmar, Jakob Uszkoreit, Llion Jones, Aidan N. Gomez, Łukasz Kaiser, Illia Polosukhin, Attention is all you need, in: Advances in Neural Information Processing Systems, vol. 30, 2017.
[520] David Salinas, Valentin Flunkert, Jan Gasthaus, Tim Januschowski, Deepar: probabilistic forecasting with autoregressive recurrent networks, International Journal of Forecasting 36 (3) (2020) 1181–1191.

[521] Bryan Lim, Sercan O. Arik, Nicolas Loeff, Tomas Pfister, Temporal fusion transformers for interpretable multi-horizon time series forecasting, International Journal of Forecasting 37 (4) (2021) 1748–1764.

[522] Boris N. Oreshkin, Dmitri Carpov, Nicolas Chapados, Yoshua Bengio, N-beats: neural basis expansion analysis for interpretable time series forecasting, in: International Conference on Learning Representations (ICLR), 2020.

[523] Balaji Lakshminarayanan, Alexander Pritzel, Charles Blundell, Simple and scalable predictive uncertainty estimation using deep ensembles, Advances in Neural Information Processing Systems 30 (2017).

[524] Ravikiran Keshavamurthy, Samuel Dixon, Karl T. Pazdernik, Lauren E. Charles, Predicting infectious disease for biopreparedness and response: a systematic review of machine learning and deep learning approaches, One Health 15 (2022) 100439.

[525] Bijaya Adhikari, Xiang Xu, Naren Ramakrishnan, B. Aditya Prakash, Epideep: exploiting embeddings for epidemic forecasting, in: Proceedings of the 25th ACM SIGKDD International Conference on Knowledge Discovery & Data Mining, ACM, 2019, pp. 577–586.

[526] Alexander Rodríguez, Anika Tabassum, Jiaming Cui, Jiajia Xie, Javen Ho, Pulak Agarwal, Bijaya Adhikari, B. Aditya Prakash, Deepcovid: an operational deep learning-driven framework for explainable real-time covid-19 forecasting, in: Proceedings of the Thirty-Fifth AAAI Conference on Artificial Intelligence (AAAI-21), AAAI Press, 2021, pp. 15393–15400.

[527] Amol Kapoor, Xue Ben, Luyang Liu, Bryan Perozzi, Matt Barnes, Martin Blais, Shawn O'Banion, Examining covid-19 forecasting using spatio-temporal graph neural networks, arXiv preprint, arXiv:2007.03113, 2020.

[528] Jeffrey L. Elman, Finding structure in time, Cognitive Science 14 (2) (1990) 179–211.

[529] Paul J. Werbos, Backpropagation through time: what it does and how to do it, Proceedings of the IEEE 78 (10) (1990) 1550–1560.

[530] Yoshua Bengio, Patrice Simard, Paolo Frasconi, Learning long-term dependencies with gradient descent is difficult, IEEE Transactions on Neural Networks 5 (2) (1994) 157–166.

[531] Razvan Pascanu, Tomas Mikolov, Yoshua Bengio, On the difficulty of training recurrent neural networks, in: Proceedings of the 30th International Conference on Machine Learning (ICML), in: Proceedings of Machine Learning Research, vol. 28, PMLR, 2013, pp. 1310–1318.

[532] Kyunghyun Cho, Bart van Merriënboer, Caglar Gulcehre, Dzmitry Bahdanau, Fethi Bougares, Holger Schwenk, Yoshua Bengio, Learning phrase representations using RNN encoder–decoder for statistical machine translation, in: Proceedings of the 2014 Conference on Empirical Methods in Natural Language Processing (EMNLP), Association for Computational Linguistics, 2014, pp. 1724–1734.

[533] Fisher Yu, Vladlen Koltun, Multi-scale context aggregation by dilated convolutions, in: International Conference on Learning Representations (ICLR), 2016.

[534] Vinay Kumar Reddy Chimmula, Lei Zhang, Time series forecasting of COVID-19 transmission in Canada using LSTM networks, Chaos, Solitons and Fractals 135 (2020) 109864.

[535] Emily L. Aiken, Andre T. Nguyen, Mauricio Santillana, Towards the use of neural networks for influenza prediction at multiple spatial resolutions, Science Advances (2021).

[536] Kyunghyun Lee, Jaideep Ray, Cosmin Safta, The predictive skill of convolutional neural networks models for disease forecasting, PLoS ONE 16 (7) (2021) e0254319.

[537] Dzmitry Bahdanau, Kyunghyun Cho, Yoshua Bengio, Alignart: non-autoregressive neural machine translation by jointly learning to estimate alignment and translate, in: Proceedings of the 2021 Conference on Empirical Methods in Natural Language Processing, 2021, arXiv:1409.0473.

[538] Peter Shaw, Jakob Uszkoreit, Ashish Vaswani, Self-attention with relative position representations, in: Proceedings of the 2018 Conference of the North American Chapter of the Association for Computational Linguistics: Human Language Technologies, Volume 2 (Short Papers), Association for Computational Linguistics, 2018, pp. 464–468.

[539] Haoyi Zhou, Shanghang Zhang, Jieqi Peng, Shuai Zhang, Jianxin Li, Hui Xiong, Wancai Zhang, Informer: beyond efficient transformer for long sequence time-series forecasting, in: Proceedings of the AAAI Conference on Artificial Intelligence, 2021.

[540] Haixu Wu, Jiehui Xu, Jianmin Wang, Mingsheng Long, Ad-autoformer: decomposition transformers with attention distilling for long sequence time-series forecasting, in: Advances in Neural Information Processing Systems 34: NeurIPS 2021, 2024, pp. 22419–22430.

[541] Wenjie Du, David Cote, Yan Liu, Saits: self-attention-based imputation for time series, Expert Systems with Applications 207 (2023) 117719.

[542] Seyed Mehran Kazemi, Rishab Goel, Siamak Eghbali, Jai Shankar Ramanan, Jaspreet Sahota, Sanjay Thakur, Stella Wu, Padhraic Smyth, Time2vec: learning a vector representation of time, CoRR, arXiv: 1907.05321, 2019.

[543] Liang Li, Yuewen Jiang, Biqing Huang, Long-term prediction for temporal propagation of seasonal influenza using transformer-based model, Journal of Biomedical Informatics 120 (2021) 103894.

[544] Jiande Wu, Shakhawat Tanim, MinJae Woo, Tanvir Ahammed, Amanda Marie Bleichrodt, Lior Rennert, A deep learning approach for enhancing pandemic prediction: a retrospective evaluation of transformer neural networks and multi-source data fusion for infectious disease forecasting, Epidemics 53 (2025) 100865.

[545] Neo Wu, Bradley Green, Xue Ben, Shawn O'Banion, Deep transformer models for time series forecasting: the influenza prevalence case, CoRR, arXiv:2001.08317, 2020.

[546] Thomas N. Kipf, Max Welling, Semi-supervised classification with graph convolutional networks, in: International Conference on Learning Representations (ICLR), 2017.

[547] Petar Veličković, Guillem Cucurull, Arantxa Casanova, Adriana Romero, Pietro Liò, Yoshua Bengio, Graph attention networks, in: International Conference on Learning Representations (ICLR), 2020.

[548] Yaguang Li, Rose Yu, Cyrus Shahabi, Yan Liu, Diffusion convolutional recurrent neural network with rank influence learning for traffic forecasting, in: International Conference on Learning Representations (ICLR), 2019.

[549] Bing Yu, Haoteng Yin, Zhanxing Zhu, Spatio-temporal graph convolutional networks: a deep learning framework for traffic forecasting, in: Proceedings of the Twenty-Seventh International Joint Conference on Artificial Intelligence (IJCAI), 2018, pp. 3634–3640.

[550] Aditya Grover, Jure Leskovec, node2vec: scalable feature learning for networks, in: Proceedings of the 22nd ACM SIGKDD International Conference on Knowledge Discovery and Data Mining (KDD), 2016, pp. 855–864.

[551] Songgaojun Deng, Shusen Wang, Huzefa Rangwala, Lijing Wang, Yue Ning, Cola-gnn: cross-location attention based graph neural networks for long-term ili prediction, in: Proceedings of the 29th ACM International Conference on Information and Knowledge Management (CIKM), 2020, pp. 245–254.

[552] Feng Xie, Zhong Zhang, Liang Li, Bin Zhou, Yusong Tan, Epignn: exploring spatial transmission with graph neural network for regional epidemic forecasting, in: CoRR, 2023.

[553] Yihong Ma, Patrick Gerard, Yijun Tian, Zhichun Guo, Nitesh V. Chawla, Hierarchical spatio-temporal graph neural networks for pandemic forecasting, in: Proceedings of the 31st ACM International Conference on Information and Knowledge Management (CIKM), 2022, p. 10.

[554] Satoki Fujita, Tatsuya Akutsu, Enhancing epidemic forecasting with a physics-informed spatial identity neural network, PLoS ONE 20 (9) (2025) e0331611.

[555] Simon Witzke, Noel Danz, Katharina Baum, Bernhard Y. Renard, Mobility data improve forecasting of covid-19 incidence trends using graph neural networks (extended abstract), in: epiDAMIK 2023: 6th epiDAMIK ACM SIGKDD International Workshop on Epidemiology Meets Data Mining and Knowledge Discovery, 2023.

[556] Truong Son Hy, Viet Bach Nguyen, Long Tran-Thanh, Risi Kondor, Temporal multiresolution graph neural networks for epidemic prediction, in: Proceedings of the 1st Workshop on Healthcare AI and COVID-19 (ICML 2022 Workshop), in: Proceedings of Machine Learning Research, vol. 184, PMLR, 2022, pp. 21–32.

[557] Rob J. Hyndman, George Athanasopoulos, Business Forecasting: Principles and Practice, OTexts, 2 edition, 1953.

[558] Christoph Bergmeir, Rob J. Hyndman, Bonsoo Koo, A note on the validity of cross-validation for evaluating autoregressive time series prediction, Computational Statistics & Data Analysis 120 (2018) 70–83.

[559] Rob J. Hyndman, George Athanasopoulos, Business Forecasting: Principles and Practice, 3 edition, OTexts, 1953.

[560] Kaiming He, Xiangyu Zhang, Shaoqing Ren, Jian Sun, Deep residual learning for image recognition, in: Proceedings of the IEEE Conference on Computer Vision and Pattern Recognition (CVPR), 2016, pp. 770–778.

[561] Roger Koenker, Quantile Regression, Cambridge University Press, 2005.

[562] Tilmann Gneiting, Adrian E. Raftery, Strictly proper scoring rules, prediction, and estimation, Journal of the American Statistical Association 102 (477) (2007) 359–378.

[563] Johannes Bracher, Evan L. Ray, Tilmann Gneiting, Nicholas G. Reich, Evaluating epidemic forecasts in an interval format, PLoS Computational Biology 17 (2) (2021) e1008618.

[564] Yupeng Chang, Xu Wang, Jindong Wang, Yuan Wu, Linyi Yang, Kaijie Zhu, Hao Chen, Xiaoyuan Yi, Cunxiang Wang, Yidong Wang, et al., A survey on evaluation of large language models, ACM Transactions on Intelligent Systems and Technology 15 (3) (2024) 1–45.

[565] Adir Solomon, Maxim Glebov, Teddy Lazebnik, Explainable surgical procedures recommender system leveraging large language models, ACM Transactions on Recommender Systems (2025).

[566] Michael S. Deiner, Russell Y. Deiner, Cherie Fathy, Natalie A. Deiner, Vagelis Hristidis, Stephen D. McLeod, Thomas J. Bukowski, Thuy Doan, Gerami D. Seitzman, Thomas M. Lietman, et al., Use of large language models to classify epidemiological characteristics in synthetic and real-world social media posts about conjunctivitis outbreaks: infodemiology study, Journal of Medical Internet Research 27 (2025) e65226.

[567] Louis Anthony Cox Jr., Causal reasoning about epidemiological associations in conversational ai, Global Epidemiology 5 (2023) 100102.

[568] Sergio Consoli, Peter Markov, Nikolaos I. Stilianakis, Lorenzo Bertolini, Antonio Puertas Gallardo, Mario Ceresa, Epidemic information extraction for event-based surveillance using large language models, in: Proceedings of Ninth International Congress on Information and Communication Technology, in: Lecture Notes in Networks and Systems, vol. 1011, Springer, Singapore, 2024, pp. 241–252.

[569] Joshua Harris, Timothy Laurence, Leo Loman, Fan Grayson, Toby Nonnenmacher, Harry Long, Loes WalsGriffith, Amy Douglas, Holly Fountain, Stelios Georgiou, Jo Hardstaff, Kathryn Hopkins, Y-Ling Chi, Galena Kuyumdzhieva, Lesley Larkin, Samuel Collins, Hamish Mohammed, Thomas Finnie, Luke Hounsome, Michael Borowitz, Steven Riley, Evaluating large language models for public health classification and extraction tasks, 2025.

[570] Jiacheng Xie, Ziyang Zhang, Shuai Zeng, Joel Hilliard, Guanghui An, Xiaoting Tang, Lei Jiang, Yang Yu, Xiufeng Wan, Dong Xu, Leveraging large language models for infectious disease surveillance—using a web service for monitoring COVID-19 patterns from self-reporting tweets: content analysis, Journal of Medical Internet Research 27 (2025) e63190.

[571] Kin On Kwok, Tom Huynh, Wan In Wei, Samuel Y.S. Wong, Steven Riley, Arthur Tang, Utilizing large language models in infectious disease transmission modelling for public health preparedness, Computational and Structural Biotechnology Journal 23 (2024) 3254–3257.

[572] Hongru Du, Yang Zhao, Jianan Zhao, Shaochong Xu, Xihong Lin, Yiran Chen, Lauren M. Gardner, Hao Frank Yang, Advancing real-time infectious disease forecasting using large language models, Nature Computational Science 5 (2025) 467–480.

[573] Chenghua Gong, Rui Sun, Yuhao Zheng, Juyuan Zhang, Tianjun Gu, Liming Pan, Linyuan Lv. EpiLLM: Unlocking the potential of large language models in epidemic forecasting, 2025.

[574] Matt J. Keeling, Pejman Rohani, Modeling Infectious Diseases in Humans and Animals, Princeton University Press, Princeton, NJ, 2011.

[575] George E.P. Box, Gwilym M. Jenkins, Gregory C. Reinsel, Greta M. Ljung, Time Series Analysis: Forecasting and Control, 5 edition, Wiley, Hoboken, NJ, 1972.
[576] Rob J. Hyndman, George Athanasopoulos, Business Forecasting: Principles and Practice, 3 edition, 1953, Online textbook.
[577] Trevor Hastie, Robert Tibshirani, Jerome Friedman, The Elements of Statistical Learning: Data Mining, Inference, and Prediction, 2 edition, Springer, New York, 2004.
[578] Andrew Gelman, John B. Carlin, Hal S. Stern, David B. Dunson, Aki Vehtari, Donald B. Rubin, Bayesian Data Analysis, 3 edition, CRC Press, Boca Raton, FL, 2013.
[579] Geir Evensen, Data Assimilation: The Ensemble Kalman Filter, Springer, Berlin, 2009.
[580] Edward L. Ionides, Carles Bretó, Aaron A. King, Inference for nonlinear dynamical systems, Proceedings of the National Academy of Sciences 103 (49) (2006) 18438–18443.
[581] Marc C. Kennedy, Anthony O'Hagan, Bayesian calibration of computer models, Journal of the Royal Statistical Society, Series B, Statistical Methodology 63 (3) (2001) 425–464.
[582] Albert Tarantola, Inverse Problem Theory and Methods for Model Parameter Estimation, Society for Industrial and Applied Mathematics (SIAM), Philadelphia, PA, 2005.
[583] Jorge Nocedal, Stephen J. Wright, Numerical Optimization, 2 edition, Springer, New York, 2008.
[584] Christophe Andrieu, Arnaud Doucet, Roman Holenstein, Particle Markov chain Monte Carlo methods, Journal of the Royal Statistical Society, Series B, Statistical Methodology 72 (3) (2010) 269–342.
[585] Jack Williams, Zacharoula Kalogiratou, Least squares and Chebyshev fitting for parameter estimation in odes, Advances in Computational Mathematics 1 (3) (1993) 357–366.
[586] Uri Ascher, Discrete least squares approximations for ordinary differential equations, SIAM Journal on Numerical Analysis 15 (3) (1978) 478–496.
[587] George A.F. Seber, C.J. Wild, Nonlinear Regression, Wiley, Hoboken, NJ, 2003.
[588] Donald W. Marquardt, An algorithm for least-squares estimation of nonlinear parameters, Journal of the Society for Industrial and Applied Mathematics 11 (2) (1963) 431–441.
[589] Peter J. Huber, Elvezio M. Ronchetti, Robust Statistics, 2 edition, Wiley, Hoboken, NJ, 2018.
[590] Mario Köppen, The curse of dimensionality, in: 5th Online World Conference on Soft Computing in Industrial Applications (WSC5), vol. 1, 2000, pp. 4–8.
[591] James Bergstra, Yoshua Bengio, Random search for hyper-parameter optimization, Journal of Machine Learning Research 13 (2012) 281–305.
[592] Jasper Snoek, Hugo Larochelle, Ryan P. Adams, Practical Bayesian optimization of machine learning algorithms, in: Advances in Neural Information Processing Systems (NeurIPS), 2012.
[593] Bobak Shahriari, Kevin Swersky, Ziyu Wang, Ryan P. Adams, Nando de Freitas, Taking the human out of the loop: a review of Bayesian optimization, Proceedings of the IEEE 104 (1) (2016) 148–175.
[594] Carl Edward Rasmussen, Christopher K.I. Williams, Gaussian Processes for Machine Learning, MIT Press, Cambridge, MA, 2008.
[595] John H. Holland, Adaptation in Natural and Artificial Systems, University of Michigan Press, Ann Arbor, MI, 1992.
[596] David E. Goldberg, Genetic Algorithms in Search, Optimization, and Machine Learning, Addison-Wesley, Reading, MA, 1989.
[597] Teddy Lazebnik, Pulling the carpet below the learner's feet: genetic algorithm to learn ensemble machine learning model during concept drift, Engineering Applications of Artificial Intelligence 152 (2025) 110772.
[598] Jinhan Kim, Shin Yoo, Software review: deap (distributed evolutionary algorithm in python) library, Genetic Programming and Evolvable Machines 20 (1) (2019) 139–142.
[599] A.E. Eiben, J.E. Smith, Introduction to Evolutionary Computing, Springer, Berlin, 2015.
[600] Sebastian Ruder, An overview of gradient descent optimization algorithms, arXiv preprint, arXiv:1609.04747, 2016.
[601] John A. Nelder, Roger Mead, A simplex method for function minimization, The Computer Journal 7 (4) (1965) 308–313.

[602] Andrew R. Conn, Katya Scheinberg, Luis N. Vicente, Introduction to Derivative-Free Optimization, Society for Industrial and Applied Mathematics (SIAM), Philadelphia, PA, 2009.

[603] James Kennedy, Russell Eberhart, Particle swarm optimization, in: Proceedings of ICNN'95 – International Conference on Neural Networks, 1995, pp. 1942–1948.

[604] Mark A. Beaumont, Wenyang Zhang, David J. Balding, Approximate Bayesian computation in population genetics, Genetics 162 (4) (2002) 2025–2035.

[605] Katalin Csilléry, Michael G.B. Blum, Oscar E. Gaggiotti, Olivier François, Approximate Bayesian computation (abc) in practice, Trends in Ecology & Evolution 25 (7) (2010) 410–418.

[606] Scott A. Sisson, Yanan Fan, Mark Beaumont, Handbook of Approximate Bayesian Computation, CRC Press, Boca Raton, FL, 2018.

[607] Arnaud Doucet, Adam M. Johansen, A tutorial on particle filtering and smoothing: fifteen years later, in: Dan Crisan, Boris Rozovskii (Eds.), Handbook of Nonlinear Filtering, Oxford University Press, 2009.

[608] Tim Rhodes, Kari Lancaster, Shelley Lees, Melissa Parker, Modelling the pandemic: attuning models to their contexts, BMJ Global Health 5 (6) (2020).

[609] Robert Northcott, Pandemic modeling, good and bad, Philosophy of Medicine 3 (1) (2022).

[610] Jan van de Kassteele, Paul H.C. Eilers, Jacco Wallinga, Nowcasting the number of new symptomatic cases during infectious disease outbreaks using constrained p-spline smoothing, Epidemiology 30 (5) (2019) 737–745.

[611] Nuria Oliver, Bruno Lepri, Harald Sterly, Renaud Lambiotte, Sébastien Deletaille, Marco De Nadai, et al., Mobile phone data for informing public health actions across the covid-19 pandemic life cycle, Science Advances 6 (23) (2020) eabc0764.

[612] James Hadfield, Colin Megill, Sidney M. Bell, John Huddleston, Barney Potter, Charlton Callender, Pavel Sagulenko, Trevor Bedford, Richard A. Neher, Nextstrain: real-time tracking of pathogen evolution, Bioinformatics 34 (23) (2018) 4121–4123.

[613] Erik M. Volz, Igor Siveroni, Bayesian phylodynamic inference with complex models, PLoS Computational Biology 14 (11) (2018) e1006546.

[614] Jana S. Huisman, Jérémie Scire, Lea Caduff, Xavier Fernandez-Cassi, Pravin Ganesanandamoorthy, Anina Kull, et al., Wastewater-based estimation of the effective reproductive number of SARS-CoV-2, Environmental Health Perspectives 130 (5) (2022) 057011.

[615] Emma Southall, Michael J. Tildesley, Louise Dyson, Prospects for detecting early warning signals in discrete event sequence data: application to epidemiological incidence data, PLoS Computational Biology 16 (9) (2020) e1007836.

[616] Leonhard Held, Niel Hens, Philip O'Neill, Jacco Wallinga (Eds.), Handbook of Infectious Disease Data Analysis, CRC Press, 2019.

[617] Sarah F. McGough, Michael A. Johansson, Marc Lipsitch, Nicolas A. Menzies, Nowcasting by Bayesian smoothing: a flexible, generalizable model for real-time epidemic tracking, PLoS Computational Biology 16 (4) (2020) e1007735.

[618] Michael Höhle, Bayesian nowcasting during the STEC O104:H4 outbreak in Germany, 2011, Biometrics 70 (4) (2014) 993–1002.

[619] Fanny Bergström, Felix Günther, Michael Höhle, Tom Britton, Bayesian nowcasting with leading indicators applied to covid-19 fatalities in Sweden, PLoS Computational Biology 18 (12) (2022) e1010767.

[620] Isabel Bergeri, Mairead G. Whelan, Harriet Ware, Lorenzo Subissi, Anthony Nardone, Hannah C. Lewis, et al., Global SARS-CoV-2 seroprevalence from January 2020 to April 2022: a systematic review and meta-analysis of standardized population-based studies, PLoS Medicine 19 (11) (2022) e1004107.

[621] Mayara Lisboa Bastos, Gamuchirai Tavaziva, Syed Kunal Abidi, Jonathon R. Campbell, Louis-Patrick Haraoui, James C. Johnston, Zhiyi Lan, Stephanie Law, Emily MacLean, Anete Trajman, Dick Menzies, Andrea Benedetti, Faiz Ahmad Khan, Diagnostic accuracy of serological tests for covid-19: systematic review and meta-analysis, BMJ 370 (2020) m2516.

[622] Jun Wei, Philippa C. Matthews, Nicole Stoesser, Thomas Maddox, Lorenzo Lorenzi, Ruth Studley, et al., Anti-spike antibody response to natural SARS-CoV-2 infection in the general population, Nature Communications 12 (1) (2021) 6250.

[623] Felix Günther, Andreas Bender, Katharina Katz, Helmut Küchenhoff, Michael Höhle, Nowcasting the COVID-19 pandemic in bavaria, Biometrical Journal 63 (3) (2021) 490–502.

[624] Sam Abbott, Joel Hellewell, Robin N. Thompson, Katharine Sherratt, Hamish P. Gibbs, Nikos I. Bosse, James D. Munday, et al., Estimating the time-varying reproduction number of SARS-CoV-2 using national and subnational case counts, Wellcome Open Research 5 (2020) 112.

[625] Alex Reinhart, Logan Brooks, Maria Jahja, Aaron Rumack, Jingjing Tang, Sumit Agrawal, et al., An open repository of real-time COVID-19 indicators, Proceedings of the National Academy of Sciences 118 (51) (2021) e2111452118.

[626] Jingjing Tang, Aaron Rumack, Bryan Wilder, Roni Rosenfeld, Real-time forecasting of data revisions in epidemic surveillance streams, PLoS Computational Biology 21 (11) (2025) e1013709.

[627] Amy Wesolowski, Caroline O. Buckee, Kari Engø-Monsen, C. Jessica E. Metcalf, Connecting mobility to infectious diseases: the promise and limits of mobile phone data, Journal of Infectious Diseases 214 (suppl_4) (2016) S414–S420.

[628] Uvencio José Giménez-Mujica, Oziel Gómez-Martínez, Jorge Velázquez-Castro, Ignacio Barradas, Andrés Fraguela-Collar, Local epidemic control through mobility restrictions, Mathematical Biosciences and Engineering 23 (2) (2026) 291–311.

[629] Xin Lu, Jiawei Feng, Shengjie Lai, Petter Holme, Shuo Liu, Zhanwei Du, Xiaoqian Yuan, Siqing Wang, Yunxuan Li, Xiaoyu Zhang, et al., Human mobility in epidemic modeling, Physics Reports 1157 (2026) 1–45.

[630] Jianhua Lin, Divergence measures based on the Shannon entropy, IEEE Transactions on Information Theory 37 (1) (2002) 145–151.

[631] Chaoming Song, Zehui Qu, Nicholas Blumm, Albert-László Barabási, Limits of predictability in human mobility, Science 327 (5968) (2010) 1018–1021.

[632] Dirk Brockmann, Lars Hufnagel, Theo Geisel, The scaling laws of human travel, Nature 439 (7075) (2006) 462–465.

[633] Luca Pappalardo, Filippo Simini, Salvatore Rinzivillo, Dino Pedreschi, Fosca Giannotti, Albert-László Barabási, Returners and explorers dichotomy in human mobility, Nature Communications 6 (2015) 8166.

[634] Christian M. Schneider, Vitaly Belik, Thomas Couronné, Zbigniew Smoreda, Marta C. González, Unravelling daily human mobility motifs, Journal of the Royal Society Interface 10 (84) (2013) 20130246.

[635] Hugo Barbosa, Marc Barthélemy, Gourab Ghoshal, Charlie R. James, Maxime Lenormand, Thomas Louail, Ronaldo Menezes, José J. Ramasco, Filippo Simini, Marc Tomasini, Human mobility: models and applications, Physics Reports 734 (2018) 1–74.

[636] Stephen Eubank, Hasan Guclu, V.S. Anil Kumar, Madhav V. Marathe, Anil Srinivasan, Zoltan Toroczkai, Nan Wang, Modelling disease outbreaks in realistic urban social networks, Nature 429 (2004) 180–184.

[637] Aidan Findlater, Isaac I. Bogoch, Human mobility and the global spread of infectious diseases: a focus on air travel, Trends in Parasitology 34 (9) (2018) 772–783.

[638] Andrew J. Tatem, David J. Rogers, Simon I. Hay, Global Transport Networks and Infectious Disease Spread, Advances in Parasitology, vol. 62, Elsevier, 2006, pp. 293–343.

[639] Frank Schlosser, Vedran Sekara, Dirk Brockmann, Manuel Garcia-Herranz, Human mobility in epidemic modeling, 2026, Preprint, December 24, 2021.

[640] Hamish Gibbs, Adam Musah, Aliyu Seidu, William K. Ampofo, Frank Asiedu-Bekoe, Darren J. Gray, Victor Adewole, James Cheshire, Michael Marks, Rosalind M. Eggo, et al., Call detail record aggregation methodology impacts infectious disease models informed by human mobility, PLoS Computational Biology 19 (8) (2023) e1011368.

[641] Mia Vu, Gregory Wellenius, Brian Williams, Royce J. Wilson, Google covid-19 community mobility reports: Anonymization process description, November 2020. Version dated November 4, 2020.

[642] Pierre Nouvellet, Sangeeta Bhatia, Anne Cori, Kylie E.C. Ainslie, Marc Baguelin, Samir Bhatt, Adhiratha Boonyasiri, Olivia Boyd, Lorenzo Cattarino, Louise V. Cooper, et al., Reduction in mobility and COVID-19 transmission, Nature Communications 12 (2021) 1090.
[643] Em Prestige, Pietro Coletti, Jantien Backer, Nicholas G. Davies, W. John Edmunds, Christopher I. Jarvis, Estimating social contact rates for the COVID-19 pandemic using Google mobility and pre-pandemic contact surveys, Epidemics 51 (2025) 100830.
[644] Nathan D. Grubaugh, Jason T. Ladner, Philippe Lemey, Oliver G. Pybus, Andrew Rambaut, Kristian G. Andersen, Edward C. Holmes, Tracking virus outbreaks in the twenty-first century, Nature Microbiology 4 (1) (2019) 10–19.
[645] Joshua Quick, Nathan D. Grubaugh, Stephen T. Pullan, Isabel M. Claro, Aaron D. Smith, Karthik Gangavarapu, Glauco Oliveira, Raul Robles-Sikisaka, Thomas F. Rogers, Nicholas A. Beutler, David R. Burton, Lia L. Lewis-Ximenez, Jaqueline G. de Jesus, Marta Giovanetti, Sarah C. Hill, Adam Black, Trevor Bedford, Miles W. Carroll, Marcio R.T. Nunes, Luiz C.J. Alcantara, Ester C. Sabino, Matthew Baylis, Nuno R. Faria, Matthew Loose, John T. Simpson, Oliver G. Pybus, Kristian G. Andersen, Nicholas J. Loman, Multiplex PCR method for MinION and illumina sequencing of Zika and other virus genomes directly from clinical samples, Nature Protocols 12 (6) (2017) 1261–1276.
[646] Nicola De Maio, Craig R. Walker, Rodrigo Borges, Lukas Weilguny, Greg Slodkowicz, Nick Goldman, Issues with SARS-CoV-2 sequencing data, Virus Evolution 6 (2) (2020) veaa064.
[647] Heng Li, Minimap2: pairwise alignment for nucleotide sequences, Bioinformatics 34 (18) (2018) 3094–3100.
[648] Kazutaka Katoh, Daron M. Standley, MAFFT multiple sequence alignment software version 7: improvements in performance and usability, Molecular Biology and Evolution 30 (4) (2013) 772–780.
[649] Yuelong Shu, John McCauley, GISAID: global initiative on sharing all influenza data—from vision to reality, Euro Surveillance 22 (13) (2017) 30494.
[650] Andrew Rambaut, Edward C. Holmes, Áine O'Toole, Verity Hill, John T. McCrone, Christopher Ruis, Louis du Plessis, Oliver G. Pybus, A dynamic nomenclature proposal for SARS-CoV-2 lineages to assist genomic epidemiology, Nature Microbiology 5 (11) (2020) 1403–1407.
[651] Kris V. Parag, Oliver G. Pybus, Céline H. Wu, Are outbreak trees growing faster? Detecting preferential sampling in phylodynamic inference, PLoS Computational Biology 16 (3) (2020) e1007733.
[652] Erik M. Volz, Simon D.W. Frost, Sampling through time and phylodynamic inference with coalescent and birth–death models, Journal of the Royal Society Interface 11 (101) (2014) 20140945.
[653] Michael D. Karcher, Julia A. Palacios, Trevor Bedford, Marc A. Suchard, Vladimir N. Minin, Quantifying and mitigating the effect of preferential sampling on phylodynamic inference, PLoS Computational Biology 12 (3) (2016) e1004789.
[654] Tanja Stadler, Denise Kühnert, Sebastian Bonhoeffer, Alexei J. Drummond, Birth–death skyline plot reveals temporal changes of epidemic spread in hiv and hepatitis C virus (hcv), Proceedings of the National Academy of Sciences 110 (1) (2013) 228–233.
[655] Áine O'Toole, Emily Scher, Anthony Underwood, Ben Jackson, Verity Hill, John T. McCrone, Rachel Colquhoun, et al., Assignment of epidemiological lineages in an emerging pandemic using the Pangolin tool, Virus Evolution 7 (2) (2021) veab064.
[656] Erik Volz, Sangeeta Mishra, Meera Chand, Jeffrey C. Barrett, Robin Johnson, Lily Geidelberg, Wes Hinsley, Daniel J. Laydon, Gavin Dabrera, Áine O'Toole, et al., Assessing transmissibility of SARS-CoV-2 lineage b. 1.1.7 in England, Nature 593 (2021) 266–269.
[657] Nicholas G. Davies, Sam Abbott, Rosanna C. Barnard, Christopher I. Jarvis, Adam J. Kucharski, James D. Munday, Carl A.B. Pearson, et al., Estimated transmissibility and impact of SARS-CoV-2 lineage b. 1.1.7 in England, Science 372 (6538) (2021) eabg3055.
[658] Sang Woo Park, Benjamin M. Bolker, David Champredon, David J.D. Earn, Michael Li, Joshua S. Weitz, Bryan T. Grenfell, Jonathan Dushoff, Reconciling early-outbreak estimates of the basic reproductive number and its uncertainty, Epidemics 36 (2021) 100468.

[659] Phil M. Choi, Ben J. Tscharke, Erica Donner, Jake W. O'Brien, Sharon C. Grant, Sarit L. Kaserzon, Rachel Mackie, Elissa O'Malley, Nicholas D. Crosbie, Kevin V. Thomas, et al., Wastewater-based epidemiology biomarkers: past, present and future, TrAC, Trends in Analytical Chemistry 105 (2018) 453–469.

[660] Maria Lorenzo, Yolanda Picó, Wastewater-based epidemiology: current status and future prospects, Current Opinion in Environmental Science & Health 9 (2019) 77–84.

[661] Gert Medema, Luc Heijnen, Goffe Elsinga, Ruud Italiaander, Arjan Brouwer, Presence of SARS-CoV-2 rna in sewage and correlation with reported COVID-19 prevalence in the early stage of the epidemic in the Netherlands, Environmental Science & Technology Letters 7 (7) (2020) 511–516.

[662] Jordan Peccia, Alessio Zulli, Douglas E. Brackney, Nathan D. Grubaugh, Eric H. Kaplan, Arnau Casanovas-Massana, et al., Measurement of SARS-CoV-2 rna in wastewater tracks community infection dynamics, Nature Biotechnology 38 (2020) 1164–1167.

[663] Alexander Crits-Christoph, Rachel S. Kantor, Matthew R. Olm, Olga N. Whitney, Basem Al-Shayeb, Yu Lou, Avi Flamholz, Lauren C. Kennedy, Hannah Greenwald, Anthony Hinkle, et al., Genome sequencing of sewage detects regionally prevalent SARS-CoV-2 variants, mBio 12 (1) (2021) e02703-20.

[664] Jeremy Ginsberg, Matthew H. Mohebbi, Rajan S. Patel, Lynnette Brammer, Mark S. Smolinski, Larry Brilliant, Detecting influenza epidemics using search engine query data, Nature 457 (2009) 1012–1014.

[665] Marcel Salathé, Clark C. Freifeld, Sumiko R. Mekaru, Annette F. Tomasulo, John S. Brownstein, influenza A (h7n9) and the importance of digital epidemiology, The New England Journal of Medicine 369 (5) (2013) 401–404.

[666] Ben Oppenheim, Meghan Gallivan, Nita K. Madhav, Nathan Brown, Valentyn Serhiyenko, Nathan D. Wolfe, Patrick Ayscue, Assessing global preparedness for the next pandemic: development and application of an epidemic preparedness index, BMJ Global Health 4 (1) (2019) e001157.

[667] David Lazer, Ryan Kennedy, Gary King, Alessandro Vespignani, The parable of Google flu: traps in big data analysis, Science 343 (6176) (2014) 1203–1205.

[668] Debashis Das, Kathleen Metzger, Ruth Heffernan, Stephen Balter, Deborah Weiss, Farzad Mostashari, Monitoring over-the-counter medication sales for early detection of disease outbreaks—New York city, Morbidity and Mortality Weekly Report 54 (Suppl) (2005) 41–46.

[669] Melanie Besculides, Ruth Heffernan, Farzad Mostashari, Deborah Weiss, Evaluation of school absenteeism data for early outbreak detection, New York city, BMC Public Health 5 (2005) 105.

[670] Arnaud Doucet, Nando de Freitas, Neil Gordon, Sequential Monte Carlo Methods in Practice, Springer, 2001.

[671] Jeffrey Shaman, Alicia Karspeck, Forecasting seasonal outbreaks of influenza, Proceedings of the National Academy of Sciences 109 (50) (2012) 20425–20430.

[672] Evan L. Ray, Nicha Wattanachit, Jarad Niemi, Azza H. Kanji, Katie House, Estee Y. Cramer, et al., Ensemble forecasts of coronavirus disease 2019 (COVID-19) in the U.S., medRxiv, 2020.

[673] Johannes Haushofer, C. Jessica E. Metcalf, Which interventions work best in a pandemic?, Science 368 (6495) (2020) 1063–1065.

[674] Vincenzo Alfano, Salvatore Ercolano, Mauro Pinto, Fighting the covid pandemic: national policy choices in non-pharmaceutical interventions, Journal of Policy Modeling 44 (1) (2022) 22–40.

[675] Amanda Kass, Philip Rocco, Local governments, pandemic aid, and community violence intervention, Urban Affairs Review 61 (1) (2025) 94–124.

[676] Luca L. Ghezzi, Carlo Piccardi, Pid control of a chaotic system: an application to an epidemiological model, Automatica 33 (2) (1997) 181–191.

[677] Waheed Ahmad, Muhammad Rafiq, Azhar Iqbal Kashif Butt, Momina Zainab, Naeed Ahmad, Dynamics of bi-susceptibility patterns in covid-19 outbreaks and associated abstain strategies, Modeling Earth Systems and Environment 11 (3) (2025) 198.

[678] Lara B. Aknin, Bernardo Andretti, Rafael Goldszmidt, John F. Helliwell, Anna Petherick, Jan-Emmanuel De Neve, Elizabeth W. Dunn, Daisy Fancourt, Elkhonon Goldberg, Sarah P. Jones, et al., Policy stringency

and mental health during the covid-19 pandemic: a longitudinal analysis of data from 15 countries, The Lancet. Public Health 7 (5) (2022) e417–e426.
[679] Jerzy Klamka, Controllability of dynamical systems. A survey, Bulletin of the Polish Academy of Sciences. Technical Sciences 61 (2) (2013).
[680] Rodrick Wallace, Robert G. Wallace, Introducing pandemic control theory, in: Neoliberal Ebola: Modeling Disease Emergence from Finance to Forest and Farm, Springer, 2016, pp. 69–80.
[681] Hildegard Behncke, Optimal control of deterministic epidemics, Optimal Control Applications & Methods 21 (6) (2000) 269–285.
[682] Cameron Nowzari, Victor M. Preciado, George J. Pappas, Analysis and control of epidemics: a survey of spreading processes on complex networks, IEEE Control Systems 36 (1) (2016) 26–46.
[683] Jacco Wallinga, Michiel van Boven, Marc Lipsitch, Optimizing infectious disease interventions during an emerging epidemic, Proceedings of the National Academy of Sciences 107 (2) (2010) 923–928.
[684] Fernando E. Alvarez, David Argente, Francesco Lippi, A simple planning problem for covid-19 lockdown, NBER Working Paper 26981, National Bureau of Economic Research, 2020.
[685] Rocío Balderrama, Javier Peressutti, Juan Pablo Pinasco, Federico Vázquez, Constanza Sánchez de la Vega, Optimal control for a SIR epidemic model with limited quarantine, Scientific Reports 12 (2022) 12583.
[686] Odo Diekmann, Hans Heesterbeek, Tom Britton (Eds.), Mathematical Tools for Understanding Infectious Disease Dynamics, Princeton University Press, 2012.
[687] Kiesha Prem, Yang Liu, Timothy W. Russell, Adam J. Kucharski, Rosalind M. Eggo, Nicholas Davies, Mark Jit, Petra Klepac, Centre for the Mathematical Modelling of Infectious Diseases COVID-19 Working Group, The effect of control strategies to reduce social mixing on outcomes of the covid-19 epidemic in Wuhan, China: a modelling study, The Lancet. Public Health 5 (5) (2020) e261–e270.
[688] Edwin van Leeuwen, Frank Sandmann, PHE Joint Modelling Group, Augmenting contact matrices with time-use data for fine-grained intervention modelling of disease dynamics: a modelling analysis, Statistical Methods in Medical Research 31 (9) (2021) 1704–1715.
[689] Edwin van Leeuwen, Frank Sandmann, Estimates of impact of interventions based on time use, Technical report, Public Health England (PHE), SAGE, 2020, On behalf of the PHE modelling group.
[690] Alberto Olivares, Ernesto Staffetti, Optimal control-based vaccination and testing strategies for COVID-19, Computer Methods and Programs in Biomedicine 211 (2021) 106411.
[691] Muhammad Dure Ahmad, Muhammad Usman, Adnan Khan, Mudassar Imran, Optimal control analysis of Ebola disease with control strategies of quarantine and vaccination, Infectious Diseases of Poverty 5 (1) (2016) 72.
[692] Liuyong Pang, Shigui Ruan, Sanhong Liu, Zhong Zhao, Xinan Zhang, Transmission dynamics and optimal control of measles epidemics, Applied Mathematics and Computation 256 (2015) 131–147.
[693] Imane Abouelkheir, Fadwa El Kihal, Mostafa Rachik, Ilias Elmouki, Optimal impulse vaccination approach for an SIR control model with short-term immunity, Mathematics 7 (5) (2019) 420.
[694] Salvatore Federico, Giorgio Ferrari, Maria-Laura Torrente, Optimal vaccination in a SIRS epidemic model, Economic Theory 77 (2024) 49–74.
[695] Ali Gharouni, Fred M. Abdelmalek, David J.D. Earn, Jonathan Dushoff, Benjamin M. Bolker, Testing and isolation efficacy: insights from a simple epidemic model, Bulletin of Mathematical Biology 84 (6) (2022) 66.
[696] Simiao Chen, Qiushi Chen, Juntao Yang, Lin Lin, Linye Li, Lirui Jiao, Pascal Geldsetzer, Chen Wang, Annelies Wilder-Smith, Till Baernighausen, Curbing the covid-19 pandemic with facility-based isolation of mild cases: a mathematical modeling study, Journal of Travel Medicine 28 (2) (2021) taaa226.
[697] Maria Soledad Aronna, Roberto Guglielmi, Lucas M. Moschen, A model for COVID-19 with isolation, quarantine and testing as control measures, Epidemics 34 (2021) 100437.
[698] Haitao Song, Ruifeng Wang, Shengqiang Liu, Zhen Jin, Daihai He, Global stability and optimal control for a COVID-19 model with vaccination and isolation delays, Results in Physics 42 (2022) 106011.

[699] Simone Sturniolo, William Waites, Tim Colbourn, David Manheim, Jasmina Panovska-Griffiths, Testing, tracing and isolation in compartmental models, PLoS Computational Biology 17 (3) (2021) e1008633.
[700] Julian Heidecke, Jan Fuhrmann, Maria Vittoria Barbarossa, A mathematical model to assess the effectiveness of test-trace-isolate-and-quarantine under limited capacities, PLoS ONE 19 (3) (2024) e0299880.
[701] Cameron Browne, Hayriye Gulbudak, Glenn Webb, Modeling contact tracing in outbreaks with application to Ebola, Journal of Theoretical Biology 384 (2015) 33–49.
[702] Lamberto Cesari, Optimization—Theory and Applications, Springer, 1983.
[703] Suzanne Lenhart, John T. Workman, Optimal Control Applied to Biological Models, Chapman & Hall/CRC, 2007.
[704] John T. Betts, Practical Methods for Optimal Control and Estimation Using Nonlinear Programming, SIAM, 2010.
[705] Fernando E. Alvarez, David Argente, Francesco Lippi, A simple planning problem for covid-19 lockdown, American Economic Review: Insights 3 (3) (2021) 367–382.
[706] Adda Jérôme, Economic activity and the spread of viral diseases: evidence from high frequency data, SSRN Electronic Journal 14 (4) (2016) 891–941.
[707] Paul Fine, Ken Eames, David L. Heymann, "herd immunity": a rough guide, Clinical Infectious Diseases 52 (7) (2011) 911–916.
[708] Jan Medlock, Alison P. Galvani, Optimizing influenza vaccine distribution, Science 325 (5948) (2009) 1705–1708.
[709] Kate M. Bubar, Kyle Reinholt, Stephen M. Kissler, Marc Lipsitch, Sarah Cobey, Yonatan H. Grad, Daniel B. Larremore, Model-informed covid-19 vaccine prioritization strategies by age and serostatus, Science 371 (6532) (2021) 916–921.
[710] Teddy Lazebnik, Ariel Alexi, Comparison of pandemic intervention policies in several building types using heterogeneous population model, Communications in Nonlinear Science and Numerical Simulation 107 (2022) 106176.
[711] Labib Shami, Teddy Lazebnik, Economic aspects of the detection of new strains in a multi-strain epidemiological–mathematical model, Chaos, Solitons and Fractals 165 (2022) 112823.
[712] Teddy Lazebnik, Gaddi Blumrosen, Advanced multi-mutation with intervention policies pandemic model, IEEE Access 10 (2022) 22769–22781.
[713] Vittoria Colizza, Alessandro Vespignani, Invasion threshold in heterogeneous metapopulation networks, Physical Review Letters 99 (14) (2007) 148701.
[714] Reuven Cohen, Shlomo Havlin, Daniel ben Avraham, Efficient immunization strategies for computer networks and populations, Physical Review Letters 91 (24) (2003) 247901.
[715] Teddy Lazebnik, Yehuda Samuel, Jonathan Tichon, Roi Lapid, Roni King, Tomer Nissimyan, Orr Spiegel, An empirically-parameterized spatio-temporal extended-sir model for combined dilution and vaccination mitigation for rabies outbreaks in wild jackals, Ecological Modelling 514 (2026) 111487.
[716] Bilge Taskın, Teddy Lazebnik, An epidemiological modeling take on religion dynamics, arXiv preprint, arXiv:2601.11436, 2026.
[717] Julia R. Gog, Bryan T. Grenfell, Dynamics and selection of many-strain pathogens, Proceedings of the National Academy of Sciences 99 (26) (2002) 17209–17214.
[718] Neil M. Ferguson, Alison P. Galvani, Robin M. Bush, Ecological and immunological determinants of influenza evolution, Nature 422 (2003) 428–433.
[719] Sylvain Gandon, Margaret J. Mackinnon, Sean Nee, Andrew F. Read, Imperfect vaccines and the evolution of pathogen virulence, Nature 414 (2001) 751–756.
[720] Andrew F. Read, Susan J. Baigent, Claire Powers, Lebo B. Kgosana, Lachlan Blackwell, Lesley P. Smith, David A. Kennedy, Stephen W. Walkden-Brown, Venugopal Nair, Imperfect vaccination can enhance the transmission of highly virulent pathogens, PLoS Biology 13 (7) (2015) e1002198.

[721] Howard Markel, Harvey B. Lipman, J. Alexander Navarro, Alexandra Sloan, Joseph R. Michalsen, Alexandra Minna Stern, Martin S. Cetron, Nonpharmaceutical interventions implemented by U.S. cities during the 1918–1919 influenza pandemic, JAMA 298 (6) (2007) 644–654.

[722] Martin C.J. Bootsma, Neil M. Ferguson, The effect of public health measures on the 1918 influenza pandemic in U.S. cities, Proceedings of the National Academy of Sciences 104 (18) (2007) 7588–7593.

[723] Ira M. Longini, Azhar Nizam, Shufu Xu, Kumnuan Ungchusak, Wanna Hanshaoworakul, Derek A.T. Cummings, M. Elizabeth Halloran, Containing pandemic influenza at the source, Science 309 (5737) (2005) 1083–1087.

[724] Timothy C. Germann, Kai Kadau, Ira M. Longini, Catherine A. Macken, Mitigation strategies for pandemic influenza in the United States, Proceedings of the National Academy of Sciences 103 (15) (2006) 5935–5940.

[725] Gerardo Chowell, Amna Tariq, James M. Hyman, A novel sub-epidemic modeling framework for short-term forecasting epidemic waves, BMC Medicine 17 (1) (2019) 164.

[726] Andrew Gelman, John B. Carlin, Hal S. Stern, David B. Dunson, Aki Vehtari, Donald B. Rubin, Bayesian Data Analysis, 3 edition, Chapman and Hall/CRC, 2013.

[727] Tilmann Gneiting, Adrian E. Raftery, Strictly proper scoring rules, prediction, and estimation, Journal of the American Statistical Association 102 (477) (2007) 359–378.

[728] Ilya M. Sobol', Global sensitivity indices for nonlinear mathematical models and their Monte Carlo estimates, Mathematics and Computers in Simulation 55 (1–3) (2001) 271–280.

[729] Andrea Saltelli, Marco Ratto, Terry Andres, Francesca Campolongo, Jessica Cariboni, Donato Gatelli, Michaela Saisana, Stefano Tarantola, Global Sensitivity Analysis: The Primer, John Wiley & Sons, 2007.

[730] Robert E. Kass, Adrian E. Raftery, Bayes factors, Journal of the American Statistical Association 90 (430) (1995) 773–795.

[731] M. Stone, An asymptotic equivalence of choice of model by cross-validation and Akaike's criterion, Journal of the Royal Statistical Society, Series B, Statistical Methodology 39 (1) (1977) 44–47.

[732] Kenneth P. Burnham, David R. Anderson, Model Selection and Multimodel Inference: A Practical Information-Theoretic Approach, 2 edition, Springer, 2002.

[733] Armen Der Kiureghian, Ove Ditlevsen, Aleatory or epistemic? does it matter?, Structural Safety 31 (2) (2009) 105–112.

[734] Ralph C. Smith, Uncertainty Quantification: Theory, Implementation, and Applications, SIAM, 2013.

[735] Yarin Gal, Zoubin Ghahramani, Dropout as a Bayesian approximation: representing model uncertainty in deep learning, in: Proceedings of the 33rd International Conference on Machine Learning, in: Proceedings of Machine Learning Research, vol. 48, 2016, pp. 1050–1059.

[736] Balaji Lakshminarayanan, Alexander Pritzel, Charles Blundell, Simple and scalable predictive uncertainty estimation using deep ensembles, Advances in Neural Information Processing Systems 30 (2017).

[737] Chris Chatfield, Model uncertainty, data mining and statistical inference, Journal of the Royal Statistical Society. Series A. Statistics in Society 158 (3) (1995) 419–444.

[738] George E.P. Box, Science and statistics, Journal of the American Statistical Association 71 (356) (1976) 791–799.

[739] Andrew Gelman, Aki Vehtari, Daniel Simpson, Charles C. Margossian, Bob Carpenter, Yuling Yao, Lauren Kennedy, Jonah Gabry, Paul-Christian Bürkner, Martin Modrák, Bayesian workflow, 2020.

[740] Emilia Vynnycky, Richard White, An Introduction to Infectious Disease Modelling, Oxford University Press, 2011.

[741] Edward L. Ionides, Anindya Bhadra, Yves Atchadé, Aaron A. King, Iterated filtering, The Annals of Statistics 39 (3) (2011) 1776–1802.

[742] Edward L. Ionides, Dao Nguyen, Yves Atchadé, Stilian Stoev, Aaron A. King, Inference for dynamic and latent variable models via iterated, perturbed Bayes maps, Proceedings of the National Academy of Sciences 112 (3) (2015) 719–724.

[743] Ruiyun Li, Sen Pei, Bin Chen, Yangyuan Song, Tao Zhang, Wan Yang, Jeffrey Shaman, Substantial undocumented infection facilitates the rapid dissemination of novel coronavirus (SARS-CoV-2), Science 368 (6490) (2020) 489–493.
[744] Lauren M. Kucirka, Stephen A. Lauer, Oliver Laeyendecker, Deniz Boon, Justin Lessler, Variation in false-negative rate of reverse transcriptase polymerase chain reaction–based SARS-CoV-2 tests by time since exposure, Annals of Internal Medicine 173 (4) (2020) 262–267.
[745] Marc Lipsitch, Steven Riley, Simon Cauchemez, Azra C. Ghani, Neil M. Ferguson, Managing and reducing uncertainty in an emerging influenza pandemic, The New England Journal of Medicine 361 (2) (2009) 112–115.
[746] Christophe Fraser, Estimating individual and household reproduction numbers in an emerging epidemic, PLoS ONE 2 (8) (2007) e758.
[747] Robin N. Thompson, Jake E. Stockwin, Rogier D. van Gaalen, Jonathan A. Polonsky, Zhian N. Kamvar, Patrick A. Demarsh, Eric Dahlqwist, Sheng Li, Emmanuelle Miguel, Thibaut Jombart, Justin Lessler, Simon Cauchemez, Anne Cori, Improved inference of time-varying reproduction numbers during infectious disease outbreaks, Epidemics 29 (2019) 100356.
[748] Katelyn M. Gostic, Lauren McGough, Edward B. Baskerville, Sam Abbott, Keya Joshi, Christine Tedijanto, Rebecca Kahn, Rene Niehus, James A. Hay, Pablo M. De Salazar, Joel Hellewell, Sophie Meakin, Nicolas A. Menzies, Neil M. Ferguson, Laura F. White, Seth Flaxman, Moritz U.G. Kraemer, Marc Lipsitch, Sarah Cobey, Practical considerations for measuring the effective reproductive number, r_t, PLoS Computational Biology 16 (12) (2020) e1008409.
[749] Paul L. Delamater, Erin J. Street, Timothy F. Leslie, Y. Tony Yang, Kathryn H. Jacobsen, Complexity of the basic reproduction number (r₀), Emerging Infectious Diseases 25 (1) (2019) 1–4.
[750] Estee Y. Cramer, Evan L. Ray, Velma K. Lopez, Johannes Bracher, Andrea Brennen, Alvaro J. Castro Rivadeneira, Aaron Gerding, Tilmann Gneiting, et al., Evaluation of individual and ensemble probabilistic forecasts of COVID-19 mortality in the United States, Proceedings of the National Academy of Sciences 119 (15) (2022) e2113561119.
[751] Claudia Czado, Tilmann Gneiting, Leonhard Held, Predictive model assessment for count data, Biometrics 65 (4) (2009) 1254–1261.
[752] Aaron A. King, Dao Nguyen, Edward L. Ionides, Statistical inference for partially observed Markov processes via the R package pomp, Journal of Statistical Software 69 (12) (2016) 1–43.
[753] Tilmann Gneiting, Adrian E. Raftery, Strictly proper scoring rules, prediction, and estimation, Journal of the American Statistical Association 102 (477) (2007) 359–378.
[754] Andreas Raue, Clemens Kreutz, Timo Maiwald, Jens Bachmann, Marcel Schilling, Ursula Klingmüller, Jens Timmer, Structural and practical identifiability analysis of partially observed dynamical models by exploiting the profile likelihood, Bioinformatics 25 (15) (2009) 1923–1929.
[755] Oana-Tana Chis, Julio R. Banga, Eva Balsa-Canto, Structural identifiability of systems biology models: a critical comparison of methods, PLoS ONE 6 (11) (2011) e27755.
[756] Ryan N. Gutenkunst, Joshua J. Waterfall, Fergal P. Casey, Kevin S. Brown, Calvin R. Myers, James P. Sethna, Universally sloppy parameter sensitivities in systems biology models, PLoS Computational Biology 3 (10) (2007) e189.
[757] Mark K. Transtrum, Benjamin B. Machta, Kevin S. Brown, Bryan C. Daniels, C.R. Myers, James P. Sethna, Perspective: sloppiness and emergent theories in physics, biology, and beyond, Journal of Chemical Physics 143 (1) (2015) 010901.
[758] Francesca Pianosi, Keith Beven, Jim Freer, Jim W. Hall, Jonathan Rougier, David B. Stephenson, Thorsten Wagener, Sensitivity analysis of environmental models: a systematic review with practical workflow, Environmental Modelling & Software 79 (2016) 214–232.
[759] Simeone Marino, Ian B. Hogue, Christian J. Ray, Denise E. Kirschner, A methodology for performing global uncertainty and sensitivity analysis in systems biology, Journal of Theoretical Biology 254 (1) (2008) 178–196.

[760] Sally M. Blower, Hadi Dowlatabadi, Sensitivity and uncertainty analysis of complex models of disease transmission: an hiv model, as an example, International Statistical Review 62 (2) (1994) 229–243.

[761] R.P. Dickinson, R.J. Gelinas, Sensitivity analysis of ordinary differential equation systems—a direct method, Journal of Computational Physics 21 (2) (1976) 123–143.

[762] Toshimitsu Homma, Andrea Saltelli, Importance measures in global sensitivity analysis of nonlinear models, Reliability Engineering & System Safety 52 (1) (1996) 1–17.

[763] Max D. Morris, Factorial sampling plans for preliminary computational experiments, Technometrics 33 (2) (1991) 161–174.

[764] Francesca Campolongo, Jessica Cariboni, Andrea Saltelli, An effective screening design for sensitivity analysis of large models, Environmental Modelling & Software 22 (10) (2007) 1509–1518.

[765] Jeremy E. Oakley, Anthony O'Hagan, Probabilistic sensitivity analysis of complex models: a Bayesian approach, Journal of the Royal Statistical Society, Series B, Statistical Methodology 66 (3) (2004) 751–769.

[766] Aaditya Karamchandani, Fatemeh Seyeddabbaghi, Marios Fokaefs, Streamlining epidemiological model generation using large language models, in: 2025 IEEE International Conference on Collaborative Advances in Software and COmputiNg (CASCON), IEEE, 2025, pp. 585–586.

[767] Peter D. Grünwald, Minimum Description Length Principle, MIT Press, 2017.

[768] Martin Arjovsky, Léon Bottou, Ishaan Gulrajani, David Lopez-Paz, Bayesian invariant risk minimization, in: arXiv preprint, 2022.

[769] Gideon Schwarz, Estimating the dimension of a model, The Annals of Statistics (1978) 461–464.

[770] Sumio Watanabe, Asymptotic equivalence of Bayes cross validation and widely applicable information criterion in singular learning theory, Journal of Machine Learning Research 11 (2010) 3571–3594.

[771] Aki Vehtari, Andrew Gelman, Jonah Gabry, Practical Bayesian model evaluation using leave-one-out cross-validation and WAIC, Statistics and Computing 27 (5) (2017) 1413–1432.

[772] Felicity Callard, Replication and reproduction: crises in psychology and academic labour, Review of General Psychology 26 (2) (2022) 199–211.

[773] Nasibeh Roozbeh, V. Ghasemi, E. Zerehi, M. Banaei, Design and development of a reproductive health software with emphasis on pre-, intra-and post-pregnancy care, Journal of Modern Medical Information Science 10 (4) (2025) 359–366.

[774] Diane Larsen-Freeman, Complexity theory, in: The Routledge Handbook of Second Language Acquisition, Routledge, 2013, pp. 73–87.

[775] Teddy Lazebnik, Svetlana Bunimovich-Mendrazitsky, More numerically accurate algorithm for stiff matrix exponential, Mathematics 12 (8) (2024) 1151.

[776] Nicholas J. Higham, Accuracy and Stability of Numerical Algorithms, SIAM, 2002.

[777] Teddy Lazebnik, Avi Rosenfeld, A new definition for feature selection stability analysis, Annals of Mathematics and Artificial Intelligence 92 (3) (2024) 753–770.

[778] B.N. Biswas, S. Chatterjee, S.P. Mukherjee, S. Pal, A discussion on Euler method: a review, Electronic Journal of Mathematics and its Applications 1 (2) (2013) 2090–2792.

[779] William J. Anderson, Continuous-Time Markov Chains: An Applications-Oriented Approach, Springer Science & Business Media, 1992.

[780] Craig Clark, Problems of good and evil in tolkien's the lord of the rings, Mallorn: The Journal of the Tolkien Society 35 (1997) 15–19.

Index

D

L

M

N

T

Printed and bound by CPI Group (UK) Ltd, Croydon, CR0 4YY
17/09/2026
02218533-0015